D0850735

THE **Evolution** OF **Knowledge**

THE
Evolution
OF
Knowledge

RETHINKING
SCIENCE
FOR THE
ANTHROPOCENE

Jürgen Renn

PRINCETON UNIVERSITY PRESS
Princeton & Oxford

Published by Princeton University Press
41 William Street, Princeton, New Jersey 08540
6 Oxford Street, Woodstock, Oxfordshire OX20 1TR

press.princeton.edu

Library of Congress Cataloging-in-Publication Data

Names: Renn, Jürgen, 1956– author.
Title: The evolution of knowledge : rethinking science for the Anthropocene / Jürgen Renn.
Description: Princeton : Princeton University Press, [2020] | Includes bibliographical
    references and index.
Identifiers: LCCN 2019016246 | ISBN 9780691171982 (hardcover)
Subjects: LCSH: Science—Philosophy. | Science—History.
Classification: LCC Q175 .R39275 2020 | DDC 500—dc23 LC record available at
    https://lccn.loc.gov/2019016246
ISBN (ebook): 9780691185675

British Library Cataloging-in-Publication Data is available

Editorial: Eric Crahan, Thalia Leaf
Production Editorial: Terri O'Prey
Text Design: Pamela L. Schnitter
Jacket/Cover Design: Pamela L. Schnitter
Production: Jacqueline Poirier
Publicity: Sara Henning-Stout, Katie Lewis
Copyeditors: Zachary Gresham, Beth Gianfagna

This book has been composed in Adobe Text Pro

Printed on acid-free paper. ∞

Printed in the United States of America

10 9 8 7 6 5 4 3 2 1

For Kathrin and Erika, Leonardo, Eleonora, and Louis

Contents

Explanatory Boxes

The Story of This Book

I can see no other escape from this dilemma (lest our true aim be lost for ever) than that some of us should venture to embark on a synthesis of facts and theories, albeit with second-hand and incomplete knowledge of some of them—and at the risk of making fools of ourselves.
—ERWIN SCHRÖDINGER, *What Is Life?*

A Long-Term Research Project and Its Roots

This book covers a time span from the origins of human thinking to the modern challenges of the Anthropocene. The Anthropocene is regarded here as the new geological epoch of humankind, defined by the profound and lasting impact of human activities on the Earth system. The Anthropocene is thus the ultimate context for a history of knowledge and the natural vanishing point for an investigation of cultural evolution from a global perspective. From this perspective, I have tried to bind multiple historical and geographical horizons together. This book deals with both the *longue durée* aspects of the evolution of knowledge and the accelerated changes in the development of knowledge that have brought us into the Anthropocene.

The foundation of the book is research pursued at the Max Planck Institute for the History of Science since 1994.[1] My studies and those of my colleagues have been dedicated from the beginning to an investigation of the history of science as part of a larger history of human knowledge. We have consistently emphasized the role of practical knowledge and historical continuity, even when focusing on the turning points of modern science. Our investigations include cross-cultural comparisons, in particular between Western, Chinese, and Islamic science, and a research program on the globalization of knowledge in history.

The research on which this book is based has been (and continues to be) a joint endeavor. It was born from a conceptual framework for a historical epistemology— understood as a historical theory of knowledge—developed with Peter Damerow, Peter McLaughlin, and Gideon Freudenthal on the basis of earlier work by Peter Damerow and Wolfgang Lefèvre on science and its relation to human labor and its societal organization. Wolfgang Lefèvre, Klaus Heinrich, and Yehuda Elkana taught me to see science within the broadest contexts of human history and to critically rethink the promises of its Enlightenment ideals and its potential to contribute to humanity's self-awareness.

The present work owes much to the thinking of Peter Damerow, to his leading role in our research team, and to our friendship and collaboration over more than thirty years. It also builds on the fundamental theoretical insights (drawing from philosophy, educational research, psychology, and cognitive science) collected in his 1996 book, *Abstraction and Representation*.[2] I have incorporated here some of the materials we prepared for a joint book on the history of mechanics—a book that could not be completed owing to his premature death in 2011.

Two Axes of Research

Our research at the Max Planck Institute for the History of Science is pursued along two main axes: the long-term transmission and transformation of knowledge, and the processes of knowledge transfer and globalization. Both of these aspects are essential, I believe, for understanding how we came to enter the Anthropocene, and both reveal patterns in the evolution of knowledge that have long been underestimated—despite their relevance in coping with challenges of the Anthropocene. As such, both are reflected in the structure of this book.

The history of science and technology has traditionally privileged innovation over the transmission, transformation, and transfer of knowledge. But it was often the less spectacular knowledge that led to the most celebrated discoveries and inventions. Some of this knowledge exhibits a striking stability and durability, often persisting over long stretches of time through phases of fundamental change. Similarly, the intercultural transfer and transformation of knowledge have shaped technological and scientific achievements since the dawn of human culture—a circumstance that is easily overlooked when focusing exclusively on the apparent points of convergence.

On the basis of our historical investigations, we have attempted to build up a theoretical language that helps us to describe all of these developmental and transmission processes, regardless of type or medium. For this purpose, we have drawn on insights from historical disciplines such as archaeology, political and economic history, the history of science and technology, and the history of art and religion, as well as from philosophical epistemology, cognitive science, the social and behavioral sciences, and in particular from sociology, economics, psychology, and social anthropology.

How can one possibly tackle such an ambitious and comprehensive research program—and how to present its outcome? Back in 1994, we decided to try an approach that may be likened to the attempt of biologists to understand general biological patterns by focusing on a single model organism such as *Drosophila melanogaster* or to the strategy of a film producer adapting a complex novel, with many interwoven narrative strands, into a simplified script for a movie: reduce the number of characters and narrative levels and concentrate on a few, carefully selected key figures and themes.[3] In our context, of course, it was not a question of selecting dramatis personae but rather of concentrating on a certain strand and domain of knowledge development that seemed particularly suited to investigations of both the long-term developments and the global transformations of knowledge.

The History of Mechanical Thinking

One narrative strand on which we have concentrated our efforts is the history of mechanics in its widest sense. What I mean is not so much the history of mechanics as a specific scientific discipline, but the history of mechanical knowledge, ranging from elementary, intuitive knowledge about a world in which gravity and pressure reign, to practical knowledge emerging from experience with instruments and tools, to theoretical forms of knowledge as documented in written texts. Mechanical knowledge has a history extending from its prehuman origins—via a long tradition of practical experience, natural philosophy, and classical mechanics—to the latest developments of science, including the new mechanics of relativity theory and quantum mechanics. Another remarkable and relevant feature of mechanical knowledge is the fact that it is not the exclusive boast of the Western tradition but has flourished in other cultures across time.

For all of these reasons, we decided (some twenty-five years ago) that mechanical knowledge would be the focus of my department's research program at the Max Planck Institute for the History of Science. Our larger aim was to probe the possibility of a historical theory of knowledge and human thinking. As it turned out, choosing mechanical knowledge was a fortunate decision, also because of the long-term collaboration it encouraged with the Museo Galileo in Florence, its charismatic director Paolo Galluzzi, and his team.

Specific investigations have been dedicated to specific time periods, from antiquity to modern science, and to varying levels of mechanical knowledge, from the use of simple machines to the formulation of highly abstract theories. By design, this research covers not just the European tradition but also developments in the Chinese and Islamic worlds.

Special attention has been paid not so much to individual discoveries and achievements as to the broader social processes enabling the transmission, accumulation, and innovation of mechanical knowledge, including its losses and the dramatic changes that the cognitive and social structures of this knowledge underwent through the millennia. The results of many of these investigations have been published in specialized studies that apply our new methods to the historical sources. Here I use these results as a background against which I outline a framework that may be useful for future studies in the history of knowledge.

Together with Peter Damerow, Gideon Freudenthal, and Peter McLaughlin, we first analyzed the emergence of classical mechanics with the aim of developing an understanding of conceptual transformations in the natural sciences. This work played a guiding role in our further research. The results were first published in 1991 in a joint book, *Exploring the Limits of Preclassical Mechanics*.[4] We coined the term "preclassical mechanics" to describe the extended, intermediate stage during the early modern period (between ca. 1500 and 1800) in which Aristotelian natural philosophy (which had shaped people's thinking about the physical world for centuries) was transformed into classical mechanics—not by a "scientific revolution," but by a process of conceptual reorganization that integrated practical knowledge within novel social circumstances.

We initially focused on protagonists like Galileo and Descartes and a few key themes, such as the law of fall and projectile motion. Later, this work was significantly extended by Jochen Büttner, Matthias Schemmel, and Matteo Valleriani, not only through further case studies but also with fundamental epistemological contributions such as the concepts of challenging objects, shared knowledge, and the structure of practical knowledge.[5] These concepts are cornerstones of this book. The broader context of early modern science also became the subject of collaborative work with other colleagues—in particular Rivka Feldhay and Pietro Omodeo—supported by the German-Israeli Fund and the Collaborative Research Centers 644 and 980 of Humboldt University and the Freie Universität of Berlin. The results of this cooperative work are published in a series of books dedicated to the historical epistemology of mechanics.[6]

Next to early modern science, the emergence of modern physics became a central subject of our investigation into transformation processes in the history of knowledge. We took the work of Albert Einstein on relativity theory as a focal point.[7] Later, a parallel endeavor was dedicated to the history of quantum physics, the other pillar of modern physics.[8] A deeper understanding of this strand in the emergence of modern physics had to go beyond Einstein's individual achievements; we had to take into account the broader system of knowledge involved in the transition from classical to modern physics, including the disciplinary organization of science, its relation to contemporary technical knowledge, the industrial and societal contexts in which it occurred, the work of other scientists, and the fact that this emergence can itself only be understood as part of a long-term development—one that was not completed with the publication of a few pathbreaking theoretical papers.

The research on Einstein's work involved (apart from Peter Damerow as an adviser) Michel Janssen, John Norton, Tilman Sauer, and John Stachel, all of them concerned not only with the historical aspects of this project but also with its implications for the general understanding of transformation processes of knowledge. The work of Einstein's contemporaries was explored together with Leo Corry and Matthias Schemmel, among others. The cultural contexts in which relativity theory emerged were studied within our research team by Giuseppe Castagnetti and Milena Wazeck. Later studies of the history of relativity theory were pursued in collaboration with Alexander Blum, Olaf Engler, Jean Eisenstaedt, Hanoch Gutfreund, Roberto Lalli, Robert Rynasiewicz, Matthias Schemmel, and others. I make extensive use of this work in this present book.

Longitudinal Studies of the History of Knowledge

The research on which this book relies is, however, not limited to mechanics. Crucial ingredients were provided by investigations into the history of Charles Darwin's evolutionary theory by Wolfgang Lefèvre and the history of chemistry by Ursula Klein, in particular Klein's work on the origins of modern chemistry in practical knowledge and on the use of chemical formulas as "paper tools" employed in the transformation of scientific knowledge. Her insights into the relation between the Scientific and the Industrial Revolutions (especially the role of "technoscience" and hybrid experts) have also, together with contributions by Wolfgang Lefèvre and Matteo Valleriani, greatly informed our understanding of the social preconditions and implications of scientific and technical knowledge.[9]

Two major longitudinal studies have also been pursued within the scope of our larger collective work: one on the historical epistemology of space and the other on the epistemic history of architecture, that is, the history of knowledge underlying architectural achievements. A research group within the framework of the TOPOI Excellence Cluster, headed by Matthias Schemmel, investigated the interaction of experience and reflection in the historical development of spatial knowledge from primate cognition to modern science.[10] The second systematic longitudinal study was dedicated to an epistemic history of architecture from the Neolithic age to the Renaissance.[11] It was based on a collaboration with another Max Planck Institute, the Bibliotheca Hertziana in Rome, and was led by Wilhelm Osthues and Hermann Schlimme.

Studies of Knowledge Circulation

Processes of cross-cultural knowledge transfer and circulation have been studied in a research project on the globalization of knowledge and its consequences. While jointly studying globalization processes of knowledge with a network of scholars from widely different disciplines and historical periods, a taxonomy was developed for the systematic analysis of historical processes of knowledge transfer and transformation. This project was pursued together with Peter Damerow, Kostas Gavroglu, Gerd Graßhoff, Malcolm Hyman, Daniel Potts, Mark Schiefsky, and Helge Wendt, among others. Several of the texts originally written jointly with Malcolm Hyman (four surveys first published in a 2012 book documenting the results of this research project) are integrated into the overall narrative here.[12] Tragically, Malcolm died before the completion of this work.

Our investigation of the globalization of knowledge has been complemented by cross-cultural comparisons and further detailed studies of knowledge transmission, in particular between Western, Chinese, and Islamic science. The comparisons with non-European science mainly rely, in the case of China, on joint work with Matthias Schemmel, Zhang Baichun, Tian Miao, and William Boltz, supported by a collaborative framework of the Max Planck Society and the Chinese Academy of Natural Sciences.[13] For the case of the Islamic world, after earlier work with Mohamed Abattouy and Paul Weinig, I have mainly relied on collaborative studies with Sonja Brentjes.[14] These studies were pursued within the wider context of the "Convivencia" project, a shared initiative of the Kunsthistorisches Institut in Florenz, the Max Planck Institute for the History of Science, the Max Planck Institute for Legal History, and the Max Planck Institute for Social Anthropology.

My participation in the Collaborative Research Center 980 ("Episteme in Motion") at the Freie Universität of Berlin has been an important stimulus for fleshing out concepts such as that of a knowledge economy, which is of central importance here. The globalization project also encouraged the exploration of new methodological approaches to questions of knowledge transfer and dissemination, in particular the use of social network analysis, with essential contributions (on which I rely) by Malcolm Hyman, Roberto Lalli, Matteo Valleriani, and Dirk Wintergrün.[15]

Broader Contexts of Knowledge

Since the history of scientific knowledge can only be understood against the background of other, more basic forms of knowledge, we have also pursued and encouraged studies of intuitive and practical knowledge, Western and otherwise. Katja Bödeker, for instance, pursued comparative field research both in Germany and in Papua New Guinea, where she conducted a study analyzing the development of intuitive conceptions of force, motion, weight, and density.[16] For a well-documented example of an indigenous people and their knowledge (mechanical and otherwise), we have relied on the work of Wulf Schievenhövel and his colleagues on the Eipo, a people living in a remote mountainous area of New Guinea. We have also undertaken studies of practical knowledge based on archeological findings; an interdisciplinary research group directed by Jochen Büttner under the umbrella of the TOPOI Excellence Cluster is, even now, systematically reconstructing weighing technologies of the ancient world. The practical knowledge involved in such technologies has been investigated in fieldwork pursued jointly by Peter Damerow and Matthias Schemmel in Italy and China, including an account of artisanal balance production.[17]

The interpretation of the history of knowledge from an evolutionary perspective can be traced to joint work with Manfred Laubichler, who brought the insights of evolutionary developmental biology to our discussion.[18] Our investigations began to focus on the challenges of the Anthropocene as the result of a major interdisciplinary project dedicated to the subject, co-curated with Katrin Klingan, Christoph Rosol, and Bernd Scherer and based at the Haus der Kulturen der Welt in Berlin.[19] The project involved not only scientists from a broad variety of fields but also artists and representatives of civil society. In this book, we make particular use of the insights of earth scientists such as Peter Haff, Will Steffen, and Jan Zalasiewicz; of an investigation of the concept of the "technosphere" pursued jointly with Sara Nelson, Christoph Rosol, and others; of insights into the transformation of energy systems drawn from the work of Robert Schlögl, Benjamin Steininger, Thomas Turnbull, and Helge Wendt; and of joint work on the history of human interventions in the nitrogen cycle with Benjamin Johnson and Benjamin Steininger.[20]

Toward a Historical Theory of Knowledge

Efforts to integrate specialized research into a larger picture have become rare. If they are undertaken at all, they are often not substantiated by empirical work performed with such a synthesis in mind; or they gloss over detail; or they offer precious few entry points for further research.

Our synthesis focuses on the more general aspects of the historical evolution of knowledge. The emphasis is on ideas and concepts rather than technical details, but specific, key examples are taken from the above-mentioned studies wherever they make for apt illustrations of a general point. Obviously, integrating specialized studies into a larger theoretical framework that can only be partially supported by them is risky business. The resulting edifice must necessarily remain incomplete. It will always be, in part, mere scaffolding. Key concepts of our framework are listed and explained in the glossary at the end of the book.

The combination of a general survey with deep specifics may also be a challenge for readers, who are required to switch repeatedly between a bird's-eye view and a magnifying glass. The reward, we hope, will be a better understanding of the historical role of knowledge in the age of the Anthropocene—an understanding founded on the numerous detailed studies that are brought together here for the first time. It is an invitation to add depth to our vision of the Anthropocene and to use that enhanced vision in addressing the challenges of our predicament.

Darwin's theory of evolution, Marx's political economy, and Freud's psychoanalysis all pursued emancipatory aims by creating more inclusive perspectives on human realities. Darwin's theory may be considered a criticism of the repression of humanity's roots in biological evolution. Marx protested, with his criticism of political economy, against the bourgeoisie's claim to represent the entire species—a claim that denied a dependence on work as the basis for the reproduction of human societies. Freud, in turn, protested against the repression of human beings' needs and drives through their subjugation to civilization.

By the same token, should a history of knowledge not therefore protest against its subjugation to a unilateral conception of science—one that restrains knowledge in the service of formal standards, academic competition, and the interests of profit and power? A conception that separates science from other forms of reflection, thus turning the reflective process itself—the very process by which knowledge becomes scientific knowledge—into an instrument of repression?

In theories of knowledge, the political is often implicit but never absent. In the past, the political dimension was made explicit by externalist views on science, emphasizing the determining role of economic, social, and political structures for science.[21] In an internalist history of ideas, emphasizing the intellectual achievements of heroic scientists, this dimension remained implicit.

In today's discussions, science is often seen either as socially constructed (for example, by "epistemic virtues") or as shaped by "epistemic things." The first position radicalizes a subjectivist view of science and risks marginalizing its objects and contents in favor of a narrative account of shared beliefs and practices within confined communities and cultural contexts. "Social construction" rarely refers, in this context, to the larger economic and political forces (e.g., of capitalism conditioning science as social practice) but rather to its locally situated cultural resources. The second position radicalizes the role of the still unclearly defined objects of research, at the risk of marginalizing its subjects, their intentionality, and their cognition. It also does not really provide a framework that could deal explicitly with the broader societal contexts of science. A third alternative is to downplay the distinction between human agency and the agency of "things," which leads either to depoliticizing and even dehumanizing action or to ascribing human qualities to the natural world, as when the earth is mystified to "Gaia," conceived as a geohistorical agent.

So where does the future of the history of science lie? I believe it lies, in any case, beyond its own specific concerns. Historians of science have developed a comprehensive repertoire of methods and approaches that allow them to analyze many different aspects of the historical development of science. But the history of science has also become somewhat scholastic, concerned more with its internal affairs and its connections to closely related fields in the humanities rather than the world of science and its impact on the human predicament.

While scientific and technical knowledge dominates our daily life, and while the survival of humanity in the Anthropocene depends on a thoughtful application of science-based solutions, the current mainstream of history of science rarely contributes to these discussions. How can we change that? What kind of approach could do justice to a conception of science as a human practice involving, in an irreducible way, mental, material, and social dimensions? How can one conceive knowledge as being constrained, but not determined, by local and by larger political and economic structures? And what historical and political epistemology could help to restitute moral responsibility to the quest of science for knowledge?

In this book, I cannot offer definitive answers to these questions, but I try to look for them beyond the entrenched positions that characterize the current discussion. I am deeply convinced that we need to become experimental again: we should not be content with deconstructing traditional narratives; we should move beyond isolated case studies; we need to forge new alliances with scientists; we should seek new methodologies, including more comparative and systemic perspectives. All of this we cannot simply do as a reflective exercise in the safety of an ivory tower. We must descend into the machine room of science and take part in the daily struggles to turn the Anthropocene into a livable environment for humanity.

Acknowledgments

The integration of specialized studies into a larger picture has required closer-than-usual collaboration across many disciplines, a practice not typically favored by the current, competitive academic system. The special conditions of the Max Planck Society and the collaborative support of my colleagues at the Max Planck Institute for the History of Science have offered me, on the other hand, a unique opportunity to entertain such wide-ranging cooperation for many years. Much of this book was written during my time as chair of the Humanities Section of the Max Planck Society, and I am grateful to my colleagues in the society for the many stimulating discussions during this period. By writing this book, I have tried to show my gratitude to the society, to my mentors, and also to the many colleagues and friends who have shared this experience of collaboration with me.

The discussion of the broader societal role of abstractions is indebted to studies of the history and philosophy of religions conducted by Klaus Heinrich, whose lectures I attended as a student. The analysis of reflective thinking—central, though in different ways, to the work of both Peter Damerow and Yehuda Elkana—stands in a broadly conceived Piagetian tradition that was, for both Peter and myself, represented in an exemplary way by Wolfgang Edelstein and his group at the Max Planck Institute for Human Development.

I would like to extend my special thanks to those colleagues and friends who have supported the writing of this preliminary synthesis with massive investments of their time. The first pages were written during a stay as Francis Bacon visiting professor at the California Institute of Technology at the invitation of Diana and Jed Buchwald, who have lent the project their encouragement from the beginning. It has since gone through numerous revisions. Each new version was critically read and carefully edited by Lindy Divarci, the editorial manager of Department 1 of the Max Planck Institute for the History of Science. Without her loyalty and substantial help, this book would not have come into being. During the second half of the project, Lindy was substantially supported by Manon Gumpert, who was responsible for checking references, facts, and quotations. Manon also supplemented and revised the bibliography, and assisted in editing the final manuscript. The final editing was performed by Zachary Gresham, whose sharp eye, great care, and astonishing persistence and inventiveness helped to transform my Germanic English into American English. His input in the glossary at the end of this volume was invaluable.

A brief comment on the numerous images complementing the text: Images have played a key role in the history of human thinking. They are an important example of its manifold material embodiments, which I designate here as "external representations" of thinking in order to distinguish them from the "internal representations" of the human mind. But this somewhat technical terminology should not fool one into conceiving of these embodiments as passive. On the contrary, images and other external representations lead their own lives, shaping human thinking as much as they are shaped by it, and thus presenting as much as they represent and concealing as much as they reveal. In this book, images often serve not just as illustrations subordinated to the text but convey their own messages, both in dialogue with and sometimes even in contrast to the written content. In a tradition going back to Aby Warburg, Erwin Panofsky, and Horst Bredekamp, I respect the autonomy of the images as much as I am grateful to those who contributed them, in particular to Laurent Taudin, who enriched the book with his charming and sometimes provocative drawings, but also to Lindy Divarci, who carefully chose and arranged all the visual materials.

The path to this final stage was long. It could not have been traversed without the engagement, generous help, and reliable support of my colleagues in Department 1, as well as that of the guests and friends of the department. None of this would have occurred without the unfailing support of my secretary, Petra Schröter, who has been at my side since the founding of the Institute. I also thank Urs Schoepflin and Esther Chen, former and current heads of the Institute's library, and the library staff and digitization group for their invaluable support in obtaining literature and many of the illustrations in this book.

My warmest thanks go to my colleagues Massimilano Badino, Antonio Becchi, Alexander Blum, Sonja Brentjes, Jochen Büttner, Robert K. Englund, Rivka Feldhay, Gideon Freudenthal, Sascha Freyberg, Hanoch Gutfreund, Margaret Haines, Svend Hansen, Julia Mariko Jacoby, Ursula Klein, Jürgen Kocka, Roberto Lalli, Manfred Laubichler, Mark Lawrence, Ariane Leendertz, Wolfgang Lefèvre, Stephen Levinson, Robert Middeke-Conlin, Gabriel Motzkin, Pietro D. Omodeo, Naomi Oreskes, Daniel Potts, Carsten Reinhardt, Giulia Rispoli, Christoph Rosol, Matthias Schemmel, Robert Schlögl, Florian Schmaltz, Urs Schoepflin, Matteo Valleriani, Helge Wendt, and Dirk Wintergrün. With all of them I discussed earlier versions of the book or parts of it at great length, often during departmental meetings, resulting in thorough revisions based on the detailed commentaries, criticism, and suggestions that they generously shared with me. Over the years, I have had many intense discussions and collaborated with numerous other colleagues, including also Giuseppe Castagnetti, Benjamin Johnson, Sara Nelson, Marcus Popplow, Simone

Rieger, Bernd Scherer, Benjamin Steininger, Thomas Turnbull, and Milena Wazeck. With many of them I have published jointly, and I am grateful for their agreement to rework passages of our shared publications for the purposes of the current work. I also thank Paolo Galluzzi, Kostas Gavroglu, Gerd Graßhoff, and Patrizia Nanz for many valuable discussions that are also reflected in this book. Special thanks go to the members of the Scientific Advisory Board of the Max Planck Institute for the History of Science for their reliable support and encouragement over the years, especially also to its current chair and vice-chair, Fabio Bevilacqua and Ana Simões. The voices of my colleagues and friends will thus be heard throughout the following pages. The specific sources on which my text is based are cited in the relevant endnotes. In addition, I have made use of Wikipedia, in particular for some biographical information and for the captions of images where not indicated otherwise. I would like to express my gratitude to all those who have contributed to this wonderful source of knowledge, available as a common good. Finally, I am grateful to two anonymous reviewers for Princeton University Press for their valuable suggestions and helpful criticism of an earlier version of the text, as well as to Al Bertrand (formerly at the Press) and Eric Crahan (currently at the Press) for their encouragement and helpful advice throughout the entire genesis of this book.

The errors and misunderstandings that remain are all mine, but what may be valuable is certainly a joint achievement.

PART 1

WHAT IS SCIENCE? WHAT IS KNOWLEDGE?

Chapter 1

History of Science in the Anthropocene

Anybody who does not earnestly wish that all humanity is well, therefore abuses it. But he is not even a true friend of himself, if he wishes to live as a healthy man among the sick, as a wise man among the dumb, as a good man among the bad, or as a happy man among the miserable.
　　　　　—JOHANN AMOS COMENIUS, *Pampaedia—Allerziehung*

Wherever the problem of knowledge does not appear at the *beginning* of the consideration, it has already been robbed of its true force. The decisive achievement of modern philosophy is that it no longer regards knowledge as a question among others, one that can be treated and resolved incidentally on the basis of other systematic presuppositions, that it has learnt to understand knowledge as the creative fundamental force in the construction of the totality of intellectual and ethical culture.
　　　　　—ERNST CASSIRER, "From the Introduction to the First Edition of
　　　　　The Problem of Knowledge in Modern Philosophy and Science"

Stormy Weather

Humans have changed the planet.[1] Actually, humans have drastically changed the planet, with dramatic consequences. Nearly no untouched nature remains.[2] A large part of the earth's surface not covered by ice has been transformed. The polar ice is melting, the water level of the oceans is rising, and coastal and marine habitats are undergoing massive changes. More than half of the planet's freshwater is being exploited by humans. Oceans are being acidified and contaminated by aquacultures. Agricultural soil is being degraded. Beneath the surface, the earth is being altered by mines and drilling. The construction of thousands of dams and extensive deforestation massively affect water circulation and erosion rates and thus the evolution and geographical spread of numerous species. The loss of biodiversity is greater by orders of magnitude than it would be without human intervention. On average, at least every third nitrogen atom in our bodies has been once processed by the fertilizer industry. Most of the biomass of all living mammals is constituted by humans and domesticated animals.

Through energy-intensive chemical processes, humans use and create functional materials (which are rare under natural conditions), and brought them into wide circulation. Among these are elemental aluminum, lead, cadmium, and mercury, fly ash residues from the high-temperature combustion of coal and oil, and also concrete, plastic, and other man-made materials, many of them displaying properties alien to the natural world. Plutonium from atmospheric nuclear testing will persist in the sedimentary record for the next several hundred thousand years while it decays into uranium and then into lead. We are directly measuring the highest atmospheric concentrations of the greenhouse gases carbon dioxide and methane in at least eight hundred thousand years, and indirectly, in at least four million years. The figures are rising steeply. Even if the use of fossil energy resources were to stop immediately, it would take thousands of years before that concentration sank to preindustrial levels.

Some changes have occurred at a much brisker pace than natural processes. The present concentration of carbon dioxide has been reached at a rate at least ten, and possibly one hundred times faster than increases at any time during the previous 420,000 years. Simultaneously, new diseases have spread via carriers with rapid life cycles that allow them to adapt quickly to the new conditions. How quickly will human societies be able to adapt to the same? The ongoing changes will affect, in any case, different parts of the globe in different ways, and the global nature of these changes will not always be easy to recognize for those who suffer from them. With floods increasingly menacing low-lying cities near large bodies of water, new forms of gentrification emerge, raising the prices of dry and secure locations and displacing the poor. Previously fertile farmlands dry out because of drought, inciting allocation battles and migrations to richer nations. Developed countries may actually appear to benefit from climate change, while developing countries suffer—but ultimately everyone will lose. There will be no escape, not even for the rich.

In short, the planet is changing, with irreversible consequences. The human species has mushroomed all over the planet. Humans have massively intervened in various Earth system cycles, such as the carbon cycle, causing climate change, as well as the water, nitrogen, phosphorus, and sulphur cycles, all of which are fundamental to life on Earth. Humanity has affected the energy balance at the earth's surface, resulting in the transition of our planet into a new stage. Humanity does not act against the backdrop of an unchangeable nature; it is deeply woven into its very fabric, shaping both its imminent and distant future. For all of humanity's interventions in planetary cycles, it is still part of the biosphere, with no power to transcend it. We are not outside observers!

The fundamental revision of our understanding of the state of this planet may only be compared to the upheaval of our physical conceptions of space and time in the wake of Einstein's theories of relativity. In classical physics, space and time seemed to be the rigid stage on which world events were taking place. According to Einstein's theory, in contrast, this stage is no immutable framework but is itself part of the drama; there is no absolute distinction between the actors and the scenery. Space and time do not remain in the background of physical processes but rather take part in them. The new reality of the planet confronts us with a similarly radical need to rethink our situation: we are not living in a stable environment that simply serves as a stage and resource for our actions; rather, we are

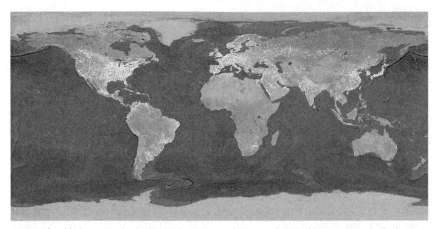

1.1. Artificial light on Earth. The brightest areas are the most urbanized but not necessarily the most populated. Some remain unlit, such as the jungles of Africa and South America or the deserts of Arabia and Mongolia, even though lights begin to appear there as well. Created in 2006 with data from the US Defense Meteorological Satellite Program's Operational Linescan System. Wikimedia Commons.

all actors in a comprehensive drama in which humans and the nonhuman world take part equally.

In the year 2000, Nobel Prize winner and discoverer of the mechanism responsible for the hole in the ozone layer Paul Crutzen felt uneasy with the official account of the state of the planet, according to which we are presently living in the "Holocene" epoch. Geologists have a sophisticated system to organize the enormous time span of the earth's history into intervals. *Holocene* means "entirely recent" and is the second epoch of the so-called Quaternary period, after the Pleistocene. However strange it may sound, the Quaternary is actually an Ice Age that began 2.6 million years ago; more precisely, the Quaternary is characterized by a back-and-forth of polar ice. The Holocene is an interglacial period in which the ice retreats. It began 11,700 years before 2,000 CE, and its climatic conditions have been unusually stable ever since.[3] Crutzen was participating in a conference on Earth system science outside Mexico City when he was struck by a sudden dislike for the Holocene description, which seemed to utterly belittle the human impact on the Earth system. He told the delegates to stop using the term "Holocene" and, while speaking, searched for a better one: "We're not in the Holocene anymore. We're in the . . . the . . . the Anthropocene!"[4]

As it turned out, the term had been used by the limnologist Eugene F. Stoermer since the 1980s.[5] Similar terms have been introduced independently by several scientists. In particular, the concept of a "noosphere" had been introduced and developed by Vladimir Vernadsky, Édouard Le Roy, and Teilhard de Chardin. They conceptualized, albeit in very different terms, humanity as a powerful geologic force and also considered the ethical implications of this assessment.[6] The roots of the idea of humanity as a planetary force go back even to the eighteenth century, when the French naturalist Comte de Buffon (Georges-Louis Leclerc) remarked that "the entire face of the Earth bears the imprint of human power."[7] Its current centrality for a broad discourse on planetary changes and humanity's role in them is, however, connected with that crucial moment in the year 2000. Crutzen himself and scholars of many disciplines have since controversially discussed the Anthropocene and its meaning in understanding humanity's predicament, but also its viability as a geologic time period.[8] The stratigraphic existence and onset is currently being examined by the Anthropocene Working Group (AWG), an interdisciplinary body of geoscientists, which will submit a formal proposal to the Subcommission on Quaternary Stratigraphy. This subcommission, in turn, reports to the International Commission on Stratigraphy (itself responsible to the International Union of Geological Sciences), where it will be ratified by its Executive Committee. To date, the AWG's recommendation is that a new epoch exists, functionally and stratigraphically distinct from the Holocene, and that its boundary layer should be placed around the mid-twentieth century.[9] Whatever the final decision of the geological experts may be (at each step of the formal procedure a majority is required with at least 60 percent of the votes in favor), the concept of the Anthropocene has opened our eyes to a fundamentally altered global environment and the fact that humanity has changed the planet to a degree comparable to geologic forces.

Given the massive impact of human intervention on the planetary environment, the traditional line between nature and culture has become problematic. We are living in an "anthropological nature" resulting from our own interventions.[10] Furthermore, the timescale of human history has become intrinsically meshed with the geologic timescale. Our economic metabolism feeds on fossil energy, consuming within a time span of hundreds of years resources that have been created over hundreds of millions of years. Just as geologic time is turned into historical time, our impact as a geologic force turns human history into a significant part of geologic history.[11]

Who Is Destroying Our Planet?

The question of whether and when the Anthropocene began is still debated. What is clear is that the transformative power of humanity is based on knowledge, accumulated and implemented over generations and ever more quickly since the Scientific Revolution, the Industrial Revolution, and the so-called Great Acceleration beginning in the 1950s.[12] What has also accumulated, however, are the unintended consequences of human activity. Nobody seriously intended to destroy the planet, though many have nevertheless taken bold risks! Some consciously decided not to act against evident dangers and simply continued with their destructive profit-making ventures. Specifically, science and technology have contributed much to bring us into this situation, and were not unaware of the consequences. Global capitalism, industrialization, traffic, and population growth would not be possible without the advances of science and technology.[13] They have catapulted us from the age of horse-drawn plows and carriages to the age of industrialized agriculture and self-driving cars, and they have blessed us with the unequally distributed benefits of modern medicine. But along with their progress, science and technology have generated unintended consequences such as uncontrolled growth, the ruthless exploitation of natural resources, and a rapid increase of the greenhouse emissions that are changing the global climate.[14] Humans are now able to send missions into interplanetary space, but they have not yet found a way to protect billions from poverty or starvation, to contain wars, or to cope with all the other challenges of the Anthropocene for humanity as a whole. True, these are political and economic questions, not just questions of knowledge, but they are also questions of knowledge and of science—they are, as I shall say, "epistemic questions."

Take, for example, the transformation of global energy systems from fossil and nuclear resources to renewable energy. This transformation will be crucial if climate change is to be contained within limits that appear to be manageable today.[15] Its success will depend on a lifestyle change, but also on solutions to many unresolved scientific and technical questions, such as problems of energy storage and networks. New knowledge will also be required to deal with the social and political processes involved in this transformation. Past experience has shown that technical solutions that have proven applicable to a specific geographic site under specific circumstances may not be easily transferable to another.

All future energy supply solutions should be designed and verified in light of their impact on both local and global geologic, physico-chemical, biological, and societal schemes. This will require competency in natural science, technology, the social sciences, and the humanities, as well as local insights on an unprecedented scale. It is as yet unclear by which political and societal processes an energy infrastructure can be transformed in harmony or in conflict with political and economic interests. This will likely require the active participation of a broad, well-informed public, but also new knowledge about the societal processes associated with such changes. Energy transformation is thus a good example of how we must rethink technical transformations as societal transformations and societal transformations as knowledge transformations. However, what matters in the end is that we stop burning fossil fuels, rapidly enough to perhaps mitigate the most catastrophic consequences of further climate change.

In view of this precarious situation, global inter- and transdisciplinary cooperation in science and education have become more urgent than ever. But the international competitiveness inherent in globalization may also have problematic consequences for the organization of research and education—in particular the fragmentation, mainstreaming, conformity, and commercialization of knowledge. In general, the way science and technology are employed depends on the structure of a society, on the way it makes or does not make use of knowledge, on the relation between power and knowledge therein, and on the question of whether and to what extent unintended consequences are taken into account.

Every society has its own "knowledge economy." It comprises the ensemble of its social institutions and processes producing and reproducing the knowledge at its disposal, and, in particular, the knowledge on which its reproduction as a society relies. The action potential of a society (for instance, its reaction to external challenges) depends on its knowledge economy and, in particular, on the societal structures that enable and limit its further exploration of knowledge. While knowledge enables individuals to plan their actions and consider the results, a society cannot "think" but can only anticipate the consequences of its actions within its economy of knowledge. The ability of global human society to cope with the challenges of the Anthropocene will therefore critically depend on the future development of its knowledge economy.

At the beginning of the twenty-first century, the prevailing mechanisms for governing science encourage the production of more and more publications representing ever-smaller units of information. The shattering of scientific knowledge into small contributions has led to a growing fragmentation of science, possibly to the exclusion of insights that could be relevant to addressing the challenges of the Anthropocene—challenges that, by their very nature, cannot be subdivided into disciplinary siloes.

Driven by economic globalization processes, national science policies are increasingly oriented toward international competitiveness, potentially limiting the scope of curiosity-driven research and running the risk of overlooking opportunities that are off the beaten track. No major society today can seemingly permit itself not to foster and regulate its science and education systems according to globalized models. Global competition forces science to cope with economic globalization and its consequences, for example, by feeding into national innovation systems (public and private), but also by conforming to globalized models of the knowledge economy, including both education and research. Incentives are being introduced both on the individual level (i.e., by the management of research according to contractually specified objectives) and on the institutional level (i.e., by implementing quasi-markets through an increasing competition for third-party funding, and through a shift from long-term institutional financial support to short- and mid-term program-oriented financial support).

The dynamics of international competition strengthen globalized models of science and education and the tendency toward a fragmentation of knowledge. The ensuing globalization of science tends to replace reflection with competitiveness and to downplay the role of specific contexts and local knowledge in favor of principles of science organization that are assumed to be of global and even universal validity. Yet it is through this perspective that most societies have come to view their problems, often disregarding the potential inherent in their own particular traditions or in opportunities for adapting those principles—opportunities that sometimes only come with a decoupling from global trends and adapting science policy to local conditions.

The fragmentation of scientific knowledge that is accompanying its globalization becomes particularly problematic when considering the challenges of the Anthropocene. Dealing with these challenges, scientific fields including atmospheric science, seismology, oceanography, environmental science, epidemiology, and space-bound science, as well as sociology, political science, economics, computer science, history, and psychology need to cooperate in forms that transcend not only disciplinary boundaries but perhaps also traditional forms of scientific organization, knowledge production, and education. Given the unpredictability and inevitable serendipity of innovation, it would be shortsighted to bend the practice of science primarily toward the challenges at hand. It would be equally risky, on the other hand, to maintain the self-inflicted fragmentation characteristic of the current knowledge economy, as driven by the competition for real and "symbolic" capital.[16]

Humans intervene in the Earth system without realizing how that system's innate character may influence the consequences of the interventions. In order to understand these consequences, one also has to take into account the entanglement of human and earth history, as well as the potential, whether creative or destructive, of human thinking. Necessary to addressing the global challenges of the Anthropocene and the inseparability of their natural and cultural components is therefore an integrative perspective on knowledge that includes not only the physical science of the Earth system but also the interpretative and critical disciplines of the humanities.

Unwisely, we have left the astonishingly stable state of the Holocene, which has shaped our culture and our ways of thinking. The Anthropocene may be a contrary state of the Earth system, but it is not necessarily the demise of humanity—we simply do not yet know what it holds in store for us. By popping our "holocenic bubble," we are not withdrawing from a static system; we are intervening within a system that is itself highly dynamic. We are performing a global experiment on a system that is already changing itself; our interventions therefore introduce second-order changes. As a consequence, we make ourselves ever more dependent on our understanding of this complex dynamic system and our interactions with it. This understanding is itself not static but subject to a dynamical evolution.

Understanding the dynamics of knowledge is therefore crucial for our future in the Anthropocene. Both knowledge and changes in the environment accumulate across generations in long-term processes—and not necessarily in such a way that the survival of human culture in any recognizable sense is guaranteed.

The World as a Problem of Knowledge

But what is knowledge? Individual knowledge is based on encoding experiences, enabling individuals to solve problems as part of their adaptive behavior. It is rooted in the ability to anticipate actions and their results, and can be corrected in response to consequences, since we can think or "reflect" on our experiences. Due to the dependence of knowledge on prior experience, its predictive power is in principle limited. On the other hand, knowledge can be mentally stored in the form of cognitive structures and repurposed for new goals.

Knowledge has not only mental, but also social and material dimensions. It can be stored, shared, and passed on from individual to individual and across generations with the help of "external representations," such as writing or symbol systems, which are part of the material culture of a society. Material culture not only determines horizons of possible action and forms of social organization, but also a horizon of thinking. The emergence of the concept of energy, for instance, only became possible once actual transformations of motive power (e.g., the replacement of human force by wind or waterpower, and later by the steam engine) emerged historically as material practices. Similarly, twentieth-century cybernetics and control theory were preceded by practical experiences with such feedback mechanisms as James Watt's centrifugal governor, which regulated his pioneering steam engine.[17]

Many scientists would argue that knowledge is philosophically neutral, in that it can be put to good or bad use. This is at least the more traditional position, which shifts the responsibility for the impact of science away from the experts producing new knowledge. By this reckoning, it is not the scientists' responsibility if, for instance, a new chemical substance they develop is employed against civilians in war.[18]

But can we really let scientists (and anybody else who produces knowledge) off the hook so easily, allowing them to shirk all responsibility? Is there even perhaps a perspective from which we may consider the abuse of knowledge to be an expression of ignorance? This point of view certainly requires a broader concept of knowledge than that which is usually employed in academic discourse.

Such a concept of knowledge would also have to contribute to an understanding of what is or is not morally just under particular circumstances, thus informing ethical decisions and political actions. Is it conceivable to arrive at such an encompassing notion of knowledge, a notion that would also facilitate, for instance, the insights that enabled a Martin Luther King Jr. or Nelson Mandela to change the world for the better? This would then constitute a radical answer to the radical neoliberal ideology claiming that when problems cannot be resolved with the help of market forces, the answer is not to limit them but to demand even fewer market constraints. In contrast, I argue here that we should embrace the possibility of rethinking all of our challenges as challenges of knowledge, and that when our knowledge does not suffice, we require more and perhaps different knowledge (e.g., about the functioning of markets).

As a society, we may locally and temporarily establish whatever values and norms we like, and then produce, share, and consume the kind of knowledge that our knowledge economy is capable of generating. Ultimately, however, with growing global connectivity and the planetary impact of our collective actions in the Anthropocene, the totality of our accumulated experiences will determine the fate of the human species, as it does already for many other ones.[19] Some seemingly self-evident or apparently desirable social, economic, and political structures, or even the established norms for social behavior and knowledge production may eventually lead to the demise of human culture as we know it; these would then be unmasked as unsuitable societal structures and imprudent moral and epistemic standards. This perspective suggests that a justification of universal values and knowledge need not involve any form of transcendence, only acceptance of the principle that the highest value is the survival and thriving of the human race, perhaps combined with the sobering but liberating insight that human life is ultimately nothing but a purpose unto itself.

Let us return to our question: According to a tradition going back to the German philosopher Immanuel Kant, the real is not given to us, but put to us by way of a riddle.[20] What does the world look like if one considers its problems as problems of knowledge, and how does one have to conceive of knowledge in order to make this perspective possible? In our individual lives, we experience our ability to change things. We also learn to anticipate behavior and that we are often wrong. Yet we cannot imagine our lives without setting goals, planning, and thinking about our actions, and the experience and the knowledge this provides. Our thinking and our knowledge are major factors in determining what we do with our lives. Our collective lives, the histories of human societies, are also unimaginable without human drives and thinking, without collective experiences, beliefs, feelings, and knowledge.

Any historical account helpful in addressing the challenges of the Anthropocene should therefore do justice to the evident fact that humans are actors whose actions are not just determined by their natural, social, and cultural environments, by their economic, political, or religious interests, or by their drives and passions, but also by their thinking, and in particular by what they actually know about the world and themselves, and by how they know and share it, as well as by the way in which they make use of their knowledge. The presupposition of such an account, however, cannot be that humans own this force as a property naturally (or divinely) granted to them; the challenge is rather to understand how their autonomy as actors could and can be gained through knowledge, if at all. In this way, one might also gain a better understanding of what is usually referred to as human freedom, which is in fact inseparable from our incurably limited and precarious human capacity to understand and judge our predicament, and to conceive of actions to change and even improve it—in short, from the human capacity to think and to use knowledge.

There is a long history of attempts to define knowledge. A review of this discussion would merit a book of its own. For a history of knowledge, and for a general history to which knowledge is central, one might begin with the categories used by the historical actors themselves. Actors' categories certainly offer important clues about the role they ascribed to knowledge in their historical contexts, but these categories do not necessarily cover even their own practices. Such categories also make it difficult to compare different actors and periods, and they hardly live up to the standards that we require today from analytical concepts that allow us to understand historical processes and their dynamics.

Definitions of knowledge that have come from nonhistorical studies, on the other hand, such as the understanding of knowledge in philosophy or the cognitive sciences, may lead to anachronisms, since these intellectual pursuits have no empirical basis on which to judge how knowledge may change in history. Investigating knowledge in history can thus be neither a journey on which one embarks without any conceptual equipment, hoping to pick up whatever the historical actors have left behind, nor a voyage undertaken with one-size-fits-all rigging. A history of knowledge, with knowledge itself as an analytical category, will rather be an exploratory venture that promises new insights not only into historical developments but also into the nature of knowledge itself.

Between the History of Science and the History of Knowledge

One key question is whether and how knowledge evolves in history. Clearly, there is some degree of accumulation through the transmission of knowledge from one generation to the next, but there are also immense losses of knowledge, major failures, and profound transformations of knowledge systems, even when they are not "revolutions" in the sense of sudden ruptures. Investigations limited to specific historical case studies and categories of actors suggest a kaleidoscopic picture in which variety is the only recognizable overall pattern. There is, however, one strand in the history of knowledge to which a developmental logic has been traditionally ascribed: the history of science, which has, for a long time, been conceived as being governed by a logic of progress, occasionally interrupted by relapses and errors.[21] But when considering the history of science within the broader context of a history of knowledge, one wonders whether this notion of its development is an exception or an illusion.[22] This question is, in any case, closely related to the questions of whether society at large advances with the progress of science and whether science depends on cultural contexts.

The self-image of science as a paradigm of progress has accompanied modern science since its inception. For Francis Bacon, only the progress of the sciences, the *progressus scientiarium*, is temporally unlimited, while political improvements are locally and historically confined, and mostly involve violence and chaos. Inventions, on the other hand, bring happiness without inflicting injustice or suffering.[23] During the Enlightenment, the mathematician Marquis de Condorcet linked scientific progress programmatically with social emancipation, aiming at the elimination of inequality through knowledge dissemination and education.[24] But even Condorcet's ideas threatened to end up in a comprehensive rationalization of all of life and thence in a technocratically engineered society. Alexander von Humboldt and his fellow scientists were also convinced that technological innovation, promoted by the sciences, improved the common good.[25] But in the age of the Industrial Revolution, it became ever more evident that the advancements of science and technology do not automatically lead to the progress of society as a whole, as the fruits of technological advances were clearly being distributed unevenly in the emerging capitalist societies, while machinery was used to extract ever greater labor value from workers.

Yet the hope that a link between scientific, technological, and societal progress could nevertheless be established was not abandoned. In the wake of the triumph of Darwin's theory of evolution, some thinkers even declared progress to be a natural law governing both biological and societal evolution.[26] Such hopes were challenged, however, by the catastrophes of the twentieth century.

Whatever relation they perceive between scientific and societal progress, most scientists distinguish their activity from all other cultural expressions of our species primarily by one property—the "cumulative" character of science. Almost all scientists are personally convinced that they can see farther than their predecessors because they can build on their achievements. Scientific progress is similarly taken for granted in traditional studies of the history of science. The underlying image is a kind of relay race of titans, passing the baton of ingenious ideas one to the next—a very undeveloped idea of an economy of knowledge indeed. The history of science thus becomes a chronicle of success, a who, what, when, and where of progress.

In actuality, these questions are better suited to investigating professional sports than science. They do not take into account the fact that the various "sports" of science (i.e., the fields of investigation) have themselves always been subject to redefinition. The history of science is traditionally written from the present backward. Thus, whatever is required to tell a success story of how the present came into being belongs to the history of science; cases that are seen as embarrassing, such as astrology or alchemy, belong to the "prehistory" of science.

Current studies in the history of science tend to question the scientific claim to progress, because such a stance seems incompatible with the extent to which science shares the fallibility of other human endeavors. As a result, science no longer appears distinguishable from other cultural practices. It has ceased to be a paradigm of universal rationality and is presented instead as one more object of study for cultural history or social anthropology. Even the most fundamental aspects of the classical image of science—proof, experimentation, data, objectivity, rationality—have proven to be deeply historical in their nature.[27] On the one hand, this insight has turned out to be liberating, at least for the historiography of science, which now more than ever has begun to take into account the cultural contexts of the scientific enterprise. On the other hand, science no longer offers, as a consequence of this view, a model of rationality that could be applied to other domains of human life.[28]

These more recent studies have opened up a new perspective on the study of the history of science, which is increasingly turning into a history of culture that includes science among other forms of knowledge. These other forms of knowledge include not only academic practices, but also the production and reproduction of knowledge far away from traditional academic settings—in artisanal and artistic practices, for instance, or even in the household and the family.

In traditional terms, the Scientific Revolution of the early modern period was seen as giving rise to modern science not only through specific discoveries but also by establishing a general scientific method, consisting in the formulation of hypotheses that are then tested by experimentation or observation. Modern science and the scientific method were allegedly developed in Europe, first in astronomy and physics, and from there conquered the world of knowledge, as well as the geographical world. Even the traditional account, however, concedes that some of this expansion was achieved only by force, by trying to impose the laws of mechanics on all science, for instance, or by the colonial expansion of Western science, often accompanied by the violent suppression of other forms of thinking.

The traditional argument was that scientific knowledge, wherever it came from, had a quality unique from all previous forms of knowledge. Today, however, some historians of science do not acknowledge a distinction between the validity of scientific knowledge and its historical origins. They no longer see the Scientific Revolution, for example, as a historical breakthrough that has fundamentally changed the practice of knowledge generation and led to the establishment, once and for all, of a scientific method.

Much of the knowledge that became relevant during the Scientific Revolution was the practical knowledge of artisans, engineers, physicians, and alchemists. It was by studying and transforming this kind of knowledge (which dealt, for instance, with the motion of projectiles in ballistics or with the transformation of materials in metallurgy) that contemporary scientists such as Galileo made their great discoveries.[29]

Thus, Galileo opens his final major publication, the *Discorsi* of 1638,[30] which lays the foundations for classical mechanics, with a eulogy to the artisans of the Venetian Arsenal, one of the greatest dockyards of his time.[31] The book is written in dialogue form and begins with a statement by the author's spokesman, Salviati, praising the expertise of the master builders:

SALVIATI: Frequent experience of your famous arsenal, my Venetian friends, seems to me to open a large field to speculative minds for philosophizing, and particularly in that area which is called mechanics, inasmuch as every sort of instrument and machine is continually put in operation there. And among its great number of artisans there must be some who, through observations handed down by their predecessors as well as those which they attentively and continually make for themselves, are truly expert and whose reasoning is of the finest.

To which the other, like-minded interlocutor Sagredo responds, pointing out how much he himself has learned from these experts:

SAGREDO: You are quite right. And since I am by nature curious, I frequent the place for my own diversion and to watch the activity of those whom we call "key men" [*Proti*] by reason of a certain preëminence that they have over the rest of the workmen. Talking with them has helped me many times in the investigation of the reason for effects that are not only remarkable, but also abstruse, and almost unthinkable.[32]

In short, other forms of knowledge, such as the practical knowledge of these artisans, have served as an important but traditionally neglected basis for scientific knowledge, so much so that one cannot truly appreciate the dynamics of the Scientific Revolution without taking them into account. Scientific knowledge is quite obviously connected with other fields of knowledge, and not only with the knowledge housed in theoretical traditions such as philosophy, but also with the practical knowledge of craftsmen and the intuitive knowledge that each of us must acquire in his or her individual development in order to cope with the material nature of the world.

Perhaps even more important, when broadening the vista to include other forms of knowledge, non-Western ways of dealing with knowledge come into view without being immediately gauged against the standards of established Western science. "On their own terms" is the slogan under which Chinese, Indian, and Islamic science are now being analyzed.[33] Similarly, the worldwide circulation of knowledge is now considered not just as a one-sided colonial or postcolonial diffusion process from a center to the periphery, but as an exchange of knowledge in which every side is active and in which knowledge is as much shaped by dissemination as by an active appropriation on the side of the "receivers."

In summary, such an inclusive perspective on knowledge has opened the door to a new understanding of the worldwide dynamics and history of scientific knowledge. It may even seem as though scientific knowledge has lost its place of privilege among other forms of knowledge. But this conclusion is premature. Clearly, there is knowledge outside the sciences. It is equally clear that scientific knowledge is not independent from the knowledge of other areas (nor from other factors, such as technology). It is indeed hardly possible to distinguish scientific knowledge from other forms of knowledge by epistemological criteria alone, or even with the help of a theory of science.

Nevertheless, from a historical perspective, it is generally possible to recognize science in various cultures and periods as a special form of knowledge whose character may, however, change with the historical context. Scientific knowledge involves not only theories but cultural practices consciously directed at the coinage of knowledge that can be transmitted from generation to generation. Scientific knowledge is accumulated and transmitted by "epistemic communities" dealing—often within dedicated educational institutions—with the preservation, improvement, and production of knowledge. Such knowledge is typically encoded using specific external representations, such as texts and instruments.

Scientific practice comprises forms of initiation, education, exploration, discourse, and transmission that are subject to historical change. Historically variable argumentative standards, control structures, and practices for the validation of knowledge all shape its accumulation and corrigibility, thus extending the learning and self-correcting aspects of individual knowledge to a societal institution. A lasting merit of the philosopher Karl Popper is having placed corrigibility at the center of a conception of science as an ever-incomplete quest for knowledge.[34] The specific form that scientific knowledge takes depends on the role a society assigns to knowledge, its "image of knowledge."[35]

Scientific knowledge first emerged in complex societies that created social spaces for exploring knowledge independently from immediate practical purposes. We may therefore speak of "science" whenever the potential inherent in the material or symbolic culture of a society is being explored for the primary sake of knowledge generation.[36] Nobody could have anticipated that human societies would eventually become dependent on such knowledge, a challenge made even greater by the intrinsically uncertain nature of scientific knowledge.

Science as a Golem

A cultural history of science that primarily focuses on specific case studies, while being unable to account for its long-term development, inevitably creates a highly fragmented picture. In this view, science dissolves into a plethora of localized and contextualized activities that are no longer distinguishable from other cultural practices.[37]

This picture can hardly do justice to the overwhelming societal, economic, and cultural significance of science in a globalized world. It has become a mark of political correctness to "provincialize" European or Western science as representing just one among many, equally justified points of view within global culture.[38] But well-meaning political correctness on the part of historians and philosophers can hardly compensate for the destruction of indigenous cultures, for the crimes and the genocides—in short, for the immense damage and abuses that have been committed in world history with the help of science or in the name of Western rationality.

Science may be compared to the golem of Jewish folklore, a being created from inanimate matter, then magically activated to perform useful services, albeit with the risk of becoming independent from, and even hostile to, its creator. As Harry Collins and Trevor Pinch put it, "Science is a golem. A golem is a creature of Jewish mythology. It is a humanoid made by man from clay and water, with incantations and spells. It is powerful. It grows a little more powerful every day. It will follow orders, do your work, and protect you from the ever threatening enemy. But it is clumsy and dangerous. Without control, a golem may destroy its masters with its flailing vigour."[39] The golem of science cannot, in any case, be tamed by underes-

1.2. Rabbi Loew brings the golem to life. Drawing by Laurent Taudin.

timating it, let alone by overestimating our own influence as its creators, witnesses, or critics. There cannot be any doubt: since the nineteenth century, science has dramatically changed the human condition in terms of energy provision and food production, through the introduction of new materials and new forms of transportation and communication, and with new pharmaceuticals and advances in medical care. Now the very survival of our culture in the Anthropocene may depend on the production of the appropriate scientific and technological knowledge.

Considered in this light, any doubts about the cumulative, self-accelerating character of the development of science may seem to be merely a matter of esoteric academic debate. Both believers in scientific progress and skeptics ultimately tend to be united on one point: scientific development is like a mighty, forward-striding golem whose pace establishes, for good or evil, the rhythm of modern industrial and postindustrial societies. Denying the substantial effects of science and technology on modern society amounts to reopening the debate on whether the earth is flat. It is, however, another matter to reconcile the striking impact of science with the complex and often difficult relationship between science and society. Equally challenging is the fact that scientific progress is hardly an automatism or necessity but merely a serendipity of human history.

How Does Knowledge Evolve?

How can radical change in scientific thought be reconciled with the retention and gradual expansion of the knowledge that has preceded it? And how are we to assess the evident existence *and* limits of scientific rationality—as well as its failure to serve as a model for societal progress at large? In this book, I shall show that none of this can be understood without taking into account that science never works in isolation but always as part of larger systems of knowledge, that these systems may profoundly change their structure in the course of history, and that such systems are part of the encompassing knowledge economy of any given society.

One example of a system of knowledge is the curriculum of the medieval university, with its faculties of theology, medicine, and jurisprudence, prepared for by prior study of the seven liberal arts. Another example is the modern ensemble of scientific disciplines. But systems of knowledge do not have to be rigidly organized conceptual systems or intellectual practices; in fact, they do not have to be very systematic at all. The knowledge needed to build a house or to tend a garden may also be conceived as a system of knowledge, composed of many different elements. These elements are not held together by a strict organizational principle, but rather constitute a heterogeneous "package of knowledge." The relations among the components of a system of knowledge may be semantic, as in a scientific theory; they may be institutional, as in a curriculum; or they may be practical, as in the example of a building project.

How can a history of knowledge systems be written that goes beyond a merely descriptive account and yet avoids the pitfalls of forcing this history into a logic of inevitable progress or reducing it to a mere chain of chance events without explanatory value? Here a comparison with natural history may be useful, not in the sense of striving for a narrative of Big History (with a capital *B*) but in the hope of learning from explanatory approaches developed in other fields.

Evolutionary explanations thrived in the nineteenth century, when scientists and philosophers such as Charles Darwin, Ernst Haeckel, Karl Marx, Ernst Mach, Ludwig Boltzmann, Pierre Duhem, Wilhelm Wundt, and many others did not hesitate to seek for connections linking the evolution of life with the evolution of human culture and thinking. In Marx's *Capital*, for instance, we read: "Darwin has interested us in the history of Nature's Technology, *i.e.*, in the formation of the organs of plants and animals, which organs serve as instruments of production for sustaining life. Does not the history of the productive organs of man, of organs that are the material basis of all social organization, deserve equal attention?"[40] In his *Panorama of the Nineteenth Century*, Dolf Sternberger characterized "evolution" as the magic word of the nineteenth century.[41] The Darwinian synthesis of biology and the modern synthesis of evolutionary biology remain evident in and central to today's scientific concerns. This is not the case for the history of science, in which discussions about the evolutionary character of science, knowledge, and culture scarcely play a role.[42]

My point is not to reduce history to biology or to identify a survival of the fittest in the history of knowledge. I rather suggest that we learn from the capacity of evolutionary theory to explain both the historical continuity and the unceasing innovation of life forms, which it does by integrating numerous subdisciplines of biology (from genetics and physiology to paleontology and ecology) within a single historical theory of development, while transforming these subdisciplines in the process. Can a similarly overarching, integrative, and explanatory framework be found for the history of knowledge as an integral part of cultural evolution?

Such an attempt cannot succeed by simply mimicking the biological scheme. Just as the biological theory of evolution was founded on the basis of specific insights into the mechanisms of biological change, an evolutionary account of knowledge has to start from a detailed analysis of the mechanisms of historical change in knowledge and their relation to culture and society. A historical theory of knowledge must similarly rely on a broad array of disciplines to be integrated under a novel perspective and thus to become vulnerable to profound reinterpretations themselves.

Since cultural evolution is ultimately grounded in biology, its greatest selective force is human survival. This ultimate selective force is, of course, mediated through—and buffered by—many layers of culture and society that themselves impose diverse proximate forces of selection on knowledge systems and cultural evolution. Such layers can hardly be anticipated from biological considerations alone. But cultural or social evolution has also been considered an evolutionary process in its own right. This idea goes back to the nineteenth century, to the time right after the appearance of Darwin's *Origin of Species*,[43] with thinkers such as William James and Ernst Mach. It has been revived since the 1980s, when authors such as Richard Dawkins, Luigi Cavalli-Sforza, Robert Boyd, and Peter Richerson began to exploit the mature formal apparatus that evolutionary theory had meanwhile developed (including sophisticated population genetics) to explain cultural phenomena in analogy to biological developments.[44]

In general, these attempts do not reduce culture to biology but rather emphasize parallels—such as the analogy between biological and cultural inheritance through learning processes—and then adopt methods and models from evolutionary theory, as well as from statistics and game theory, to explain cultural change or phylogenetic lineages (e.g., in the evolution of languages). Cultural selectionists thus assume two parallel inheritance systems, one genetic and one cultural. One way in which they are intertwined is "niche construction," which essentially provides a third inheritance system that may be characterized as ecological inheritance. Living beings change their environments, either through phenotypical or through cultural traits, while a changed environment in turn reshapes selective pressures. All three inheritance systems are coupled by feedback loops.[45]

The Heuristic Role of Evolution

It may be tempting to generalize this approach to an evolutionary account of knowledge by identifying different mechanisms of transmission and variability and then seeing how far biological analogies and tools can reach in providing insights into the "population dynamics" of knowledge. This is, however, not the approach followed here. I take biological evolution neither as an overarching process that includes and governs culture and its dynamics, nor as delivering, by analogy with biology, a theoretical framework for the analysis of cultural history that may claim a similarly integrative function for the humanities and social sciences as that which Darwin's theory claims for the life sciences. The former would disregard the autonomy of culture; the latter, the autonomy of cultural studies—both amounting to some form of reductionism.

Rather, I consider the evolutionary theory of biology as a standard of comparison for any historical theory coping with the long-term development of complex adaptive systems like human culture—beginning, however, with its own genuine insights gained through centuries of research within numerous disciplinary traditions. In other words, instead of matching concepts from the realm of cultural analysis (e.g., institution and power, memory and repression, learning and reflection) to an evolutionary framework that is by-and-large borrowed from biology, I take a bottom-up approach. Such an approach begins with concepts, theories, and in-depth investigations from the humanities and social sciences and tries to build an explanatory framework that captures the riches harbored by these investigations, nevertheless keeping in mind that an explanatory framework for human history should comply with some basic lessons learned from evolutionary theory in the life sciences.[46]

Among these lessons is the temporal directedness of the overall process and the asynchrony of particular developments, that is, the lack of a global uniformity of evolution. Evolutionary accounts do not imply "progress" in any traditional sense, and, typically, their outcomes are neither determined by their initial conditions nor by some final goal to be eventually reached—one might say they are neither deterministic nor teleological. In fact, modern biology has long renounced any idea of evolution as the triumphant progress toward the most "highly" developed forms of life, with humans as the crown of creation. Evolutionary theory insists, instead, on the global connectivity of the entire process in which life has unfolded on Earth, often hidden beneath the dazzling variety of local forms. In the history of culture and knowledge we are still far from such a global account.

Another lesson is that evolutionary processes not only allow for chance events to occur but allow them to have long-term effects. Evolutionary processes are path dependent in the sense that current developments depend on past events, even though past circumstances may no longer be relevant. Nevertheless, the unpredictability of future developments, the dependency of later developments on earlier ones, and the role of contingency in such processes in no way force us to resign ourselves to a merely descriptive or taxonomic account, nor to one that is simply a collection of local narratives. Evolutionary accounts do have explanatory potential, beginning with a realization of the sheer complexity that mechanisms for ensuring continuity may give rise to when combined with possibilities for variation and selection.

Evolutionary processes do not just react to external conditions but may also shape their own environments, thus becoming self-referential—this feature is addressed under the label of "niche construction" in evolutionary biology (e.g., beavers constructing their dams). This is obviously also characteristic of cultural evolution. Another striking feature of evolutionary theory that invites a comparison with the history of knowledge is the insight that forms of life that emerge later in evolution do not necessarily eliminate earlier forms of life. To the contrary, simple life forms such as bacteria are by far the most successful models that evolution has ever generated. In a sense, the same holds for the history of knowledge; complex forms of knowledge such as higher forms of mathematics hardly ever completely replace earlier forms such as simple counting techniques.

Evolutionary processes may give rise to convergence (as when eyes developed independently in multiple species)—a phenomenon familiar, for instance, from parallel discoveries in science. Biological evolution generally works with modular components, which are reshaped into building materials for new life forms, as when certain organs are repurposed for adaptation to a new environment. The repurposing of the material environment is a similarly important aspect of cultural evolution. The "horizon of possibilities" shaped by historically bestowed material conditions is always broader than the possibility actually realized at any given moment.[47]

Biological evolution involves genes and their expression in phenotypes, the latter being a subject of developmental biology. Without implying that one can simply transfer this structure by analogy to the realm of culture, it still seems that cultural evolution without a concept of knowledge and a theory of knowledge development would be rather like biological evolution without genes and developmental biology. Indeed, some even define culture in terms of knowledge acquired and socially transmitted in the context of a constructed niche.[48] The transmission of human culture in any case goes beyond the ecological inheritance system associated with niche construction mentioned above, and it involves social learning and the transmission of material artifacts and signs that are detached from their immediate contexts of usage and organized within holistic systems of knowledge.

An Alternative to *The Structure of Scientific Revolutions*

In the following, I sketch such a theory based on detailed historical studies pursued with some of these questions in mind. These studies also cover, in particular, some of the major so-called scientific revolutions: the emergence of classical mechanics in the early modern period, the so-called chemical revolution of the eighteenth century, and the relativity and quantum revolutions of the twentieth century. The basic mechanisms of knowledge evolution that have been identified cannot be reduced to analogies of mutation and selection in biological evolution. They are in fact much more context-dependent and are themselves subject to change in the course of history. There is no general scheme according to which a scientific transformation takes place, for example, according to the sequence of normal science, crisis, and paradigm shift hypothesized by Thomas S. Kuhn in his epochal book *The Structure of Scientific Revolutions*.[49] Still, Kuhn's scheme may serve as a useful foil for identifying the features of an evolutionary account of the history of knowledge by contrasting them with what has meanwhile become a widely popularized—and misleading—idea of the radical breaks associated with scientific transformations.

In short, I argue that major changes in systems of knowledge happen, but that these changes are typically long-term, protracted processes. They cannot be adequately understood without taking into account that knowledge has a layered structure involving the different types of knowledge that are shared within the knowledge economy of a society. It is because of this layered structure of scientific knowledge (which also comprises intuitive, practical, and technical knowledge) that untranslatability or "incommensurability" between scientific worldviews or "paradigms" is a less serious problem in the actual workings of science than is assumed in philosophical discussions. In fact, the notion of incommensurability belongs to the sense of theoretical concepts,[50] whereas the reference of individual concepts (instruments, phenomena, etc.) may remain unaltered irrespective of theoretical changes, thus enabling communication on the basis of other, more pragmatically oriented layers of knowledge.

External representations and embodiments of knowledge such as texts, instruments, or infrastructures serve as the backbone of the transmission of systems of knowledge, ensuring their long-term continuity. Systems of knowledge and their external embodiments are applied and explored by practitioners of knowledge such as the members of a scientific community, either within institutions dedicated to the generation of knowledge or in practical contexts. This exploration leads to the enrichment, extension, and gradual change of systems of knowledge. Furthermore, shared knowledge is always carried by individuals and is thus intrinsically variable. A body of knowledge is never uniquely defined and therefore requires interpretation; different individuals or groups may view it in different ways at times. These variations may become a source of controversy, which is itself a means to conceptual development.

The exploration of the inherent potential of the historically specific means for gaining knowledge thus gives rise to a variety of alternatives within a knowledge system, becoming a source of novelty. In an advanced state of the development of a knowledge system, these variations typically lead to internal tensions and contradictions, which may become the starting point for the reorganization of a system of knowledge or the branching off of a new one. Some of the most crucial steps in the growth of knowledge were indeed not based on the acquisition of new knowledge but rather on developing new ways of using what was already known.[51]

A Global Learning Process?

On this basis, I claim, one can build a history of science as part of a global history of knowledge—without forcing it into a logic of progress and without abandoning the attempt to account for the long-term accumulation of earth-changing knowledge, the accompanying losses and deficiencies, and the dependence of scientific rationality on chance constellations. But how could such a history contribute to answering the questions that we have begun with, in particular the question of what knowledge is required to address the challenges of the Anthropocene? For starters, it would demonstrate in what sense science is just one aspect of a highly fragmented but nevertheless inexorable global learning process in which humanity as a whole, over time, assembles knowledge with the potential for shaping the world. It would also illustrate the ways in which this potential is actually used and would finally demonstrate that science derives its power from being one late result of this global learning process.

When I speak of a global learning process, I again have the comparison to biological evolution in mind. The process as a whole displays the features of individual learning, like the functional adaptation of life forms to their environment—without, however, presupposing an intelligent subject, and without any assurance that such adaptations may not eventually lead to the demise of a species. (Indeed, there are many examples of runaway selection in evolutionary biology, leaving species dependent on a particular ecological niche that may then disappear.) Similarly, human history is evidently not guided by some form of global, collective subjectivity but rather by processes that operate primarily within local settings, albeit with ever more global entanglements and consequences.

Faced with the global challenges resulting from these consequences—such as the changes in the Earth system, of which climate change is perhaps the most visible—we might wish for such a collective subjectivity to emerge and facilitate rational solutions to global problems. Indeed, some advocates of sustainability policies argue in this way, favoring, for instance, measures of geoengineering managed by international expert communities or authorities standing in for an always rationally acting world government. While geoengineering may even become unavoidable as a last resort, the latter hope will likely remain as illusory as the hope for an explanation of the history of life by "intelligent design." Investigating the evolution of knowledge could help us, on the other hand, to conceive of more realistic options for addressing these challenges; it might teach us how new solutions can emerge bottom-up rather than top-down, from the global machinery of knowledge production.[52]

Chapter 2

Elements of a Historical Theory of Human Knowledge

In science, one can learn the most by studying what seems the least.
—MARVIN MINSKY, *The Society of Mind*

[Historical] [s]ources are neither open windows, as the positivists believe, nor fences obstructing vision, as the skeptics hold: if anything, we could compare them to distorting mirrors. The analysis of the specific distortion of every specific source already implies a constructive element. But construction, as I attempt to demonstrate . . . is not incompatible with proof; the projection of desire, without which there is no research, is not incompatible with the refutations inflicted by the principle of reality. Knowledge (even historical knowledge) is possible.
—CARLO GINZBURG, *History, Rhetoric, and Proof*

We still need to learn new and better ways to think, to apply our minds—especially to be able to really get our minds around such massive issues as climate change in the larger context of the Anthropocene. This may require taking a serious step back, and becoming more reflective about how our own thoughts work. If we can learn to do this, then not only will we be able to forecast a safe Anthropocene, but perhaps even more importantly: a beautiful Anthropocene.
—PAUL CRUTZEN, "Foreword: Transition to a Safe Anthropocene"

The Structure of the Book

Rethinking science for the Anthropocene means asking a number of fundamental questions about science and knowledge, as well as about their historical evolution. In this book, I proceed in five steps, which correspond to its principal parts. In part 1, I have already begun to discuss the double character of knowledge crucial for our entry into the Anthropocene, the empowerment it provides, and the unintended consequences it leads to. In part 2, I deal with the historical nature of thinking and ask how knowledge structures change (chapters 3–7). In part 3, dedicated to the "economy of knowledge," I investigate how knowledge structures affect society and how society affects knowledge structures (chapters 8–10). In part 4, I analyze diffusion and globalization processes of knowledge, asking how knowledge spreads (chapters 11–13). Finally, in part 5, I come back to the theme of the Anthropocene, discussing its emergence against the background of evolutionary considerations and asking on what knowledge our future depends (chapters 14–17). In the following, I outline the main line of argument of the remaining four parts in order to offer readers a choice of reading selectively.[1]

How Knowledge Structures Change

Part 2 is dedicated to some remarkable aspects of the cognitive structures of human thinking that help us to understand historical change. One of these aspects is that the structure and content of knowledge are not as independent from each other as a long tradition separating logic from meaning has held. Any concept (e.g., the concept of a tree) may specifically refer to a concrete experience and, at the same time, act as a general cognitive structure to which this experience can be assimilated, connecting it to a web of other experiences. If one thinks about it, every concept is a little theory because it is semantically connected to other concepts.

As I shall argue in chapters 3 and 4, cognitive structures are active in the sense that they change with every implementation. If I see a tree I have never seen before, I may change my concept of a tree and thus the entire network of semantic relations in which this concept is embedded. This is a fundamental feature of cognitive development that also accounts for how scientific knowledge may change in response to novel experiences. In chapter 6, I deal with a particular variety of experience that is hard to digest within a given system of knowledge, a variety that I refer to as "challenging objects." For example, when one expects that all physical interactions require direct contact between bodies, the invisible forces moving the needle of a compass may appear miraculous, as they did for the young Einstein, who remembered this encounter all his life as a challenging experience.

Cognitive structures allow for thinking processes even under circumstances of incomplete (or a glut of) information, which is our typical predicament. As I shall discuss in detail in chapter 4, informational lacunae in parts of the structure may be supplemented by information gleaned from prior experience in the form of "default assumptions," such as when I see only the upper part of a tree and conclude that it must have roots because trees normally do. If it turns out, however, that the root has been cut, I do not have to abandon my conviction that I am dealing with a tree. Art often plays with our default expectations. The artist Herbert Bayer, for example, baffles us with a photomontage showing himself looking into a mirror and finding a slice of his upper arm detached. Where we—and he, of course—would, by default, expect to see a human limb that could only be severed at the price of injury, Bayer discovers to his amazement that his body, as reflected in the mirror, actually behaves as if it were a mannequin from which parts can be removed at will. The marriage of two distinct "mental models" with partly incompatible default expectations (that of a living, intact human body and of a modular mannequin) thus provokes a dissonant

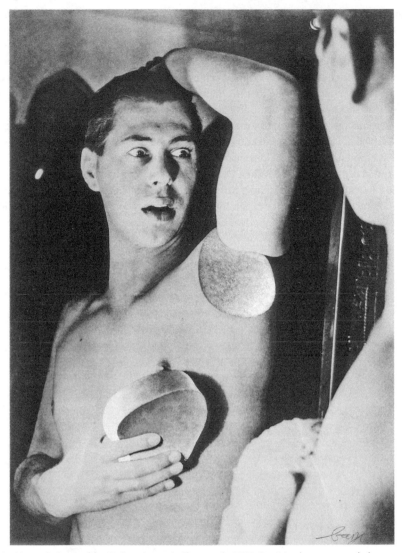

2.1. *Humanly Impossible*. Herbert Bayer (self-portrait), 1932. Bayer's photomontage belongs to a tradition of surrealist experiments with mannequins that take on a life of their own. In a sense, they turned the ancient myth of Pygmalion upside down. According to the Latin poet Ovid, Pygmalion was a sculptor who carved a woman out of ivory. Fulfilling his hidden desire, the goddess of love, Aphrodite, transformed the statue into real woman whom Pygmalion married. In contrast, a surrealist artist would think of himself as a Pygmalion who transformed his vis-à-vis into an artifact. In his photomontage, Bayer transformed his own mirror image into a mannequin that reveals hidden desires, fears, intuitions, and insights—for example, into the mechanization of the world and its resulting frailty. At the same time, art thus demonstrates the power of artifacts to achieve the otherwise "humanly impossible," manifesting and generating as "external representations" such mental states. The generative ambiguity of external representations in the history of knowledge is a central theme of this book. For its role in the history of art, see Bredekamp (2010). © bpk / Los Angeles County Museum of Art / Art Resource, NY / Herbert Bayer.

shock of which many examples can be found in art, as well as in science. This effect
has been regarded as revelatory by the likes of Walter Benjamin and Albert Einstein,
among others.[2]

How do we deal with the remarkable fact that, contrary to the claim of philosophers of science such as Karl Popper, scientific theories are in general not given up
in the face of contrary evidence? In chapter 4, I explain this through the availability
of default settings and the flexibility in replacing them while maintaining the essential argumentative structure of a conceptual framework. Scientific theories are always
embedded in larger systems of knowledge. Incomplete scientific information is typically complemented by background knowledge consisting of default assumptions
rooted in prior knowledge. When Galileo, for instance, saw a blurry picture of the
moon in his self-made telescope, he could identify its irregularities as craters and
mountains owing to his prior experience with landscape images and his expectation
of seeing another Earth-like body.

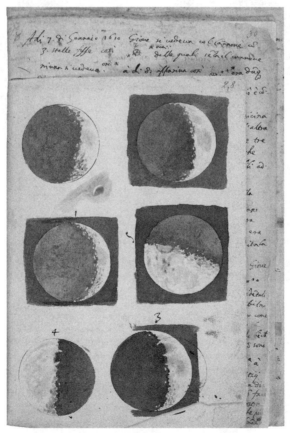

2.2. Manuscript of Galileo Galilei's *Sidereus nuncius*, showing the moon as seen by Galileo through
his telescope. Ms. Gal. 48, c. 28r. By concession of the Ministerio per i Beni e le Attività Culturali—Italy /
Biblioteca Nazionale Centrale di Firenze. See Bredekamp (2019, IV).

The components of a cognitive structure may be derived from other cognitive structures, as when the concept of a house is explicated in terms of its basement, floor, roof, walls, windows, doors, and so forth, with each of these concepts in turn being expandable in terms of its components, such as a door's knob, hinges, lock, wood, paint, and so on. Cognitive structures are thus somewhat like Russian matryoshka dolls, with a nested sequence of other cognitive structures inside. I refer to these complex interdependencies as the "architecture of knowledge."

But, in contrast to a real matryoshka doll, conceptual hierarchies are not necessarily fixed once and for all—one could also start from "inside the box," as it were, expanding it in another direction. For instance, when starting from the concept of a wood panel one might find inside its "Russian doll sequence" the example of a door made out of wooden panels, among many other things. In other terms, we can speak of semantic networks that may be arranged in different hierarchies. What role does the possibility of altering such hierarchies play for conceptual change in science?

In order to address such questions, we have to investigate the architecture of knowledge and the mechanisms of its change in greater detail. As chapter 6 makes clear with examples taken from the history of mechanical knowledge, the architecture of knowledge has a layered structure because of the fact that some experiences, such as those of our moving around in the material world, are rather basic and persistent: lifting heavy objects requires effort, rigid walls are hard to penetrate, and so on. These underlying layers of "intuitive knowledge" often provide default settings for expectations when entering new terrain or when dealing with less concrete matters, where they often reappear as metaphors.

Another feature of the architecture of knowledge that will become relevant when discussing intercultural knowledge transmission in chapter 11, is that people strive for a mental representation of their world that seems rather complete. To some extent, of course, all of us are aware of our blind spots, but we nevertheless tend to avoid an "epistemic vacuum," and thus complement our partial knowledge of the world with default assumptions, striving for a more-or-less holistic view of the world around us. How such views can be challenged is a key question of the Anthropocene, since we will have to overcome much of our preconceived understanding of the world.

Beginning with chapter 3, I discuss the crucial role of external representations such as writing or symbol systems for the transmission and transformation of systems of knowledge. They may serve as "paper tools"[3] to support "reflection" in the sense of thinking on thinking: reflecting on actions with the help of such external representations (e.g., on symbols representing objects to be counted) may engender "higher-order forms of knowledge," (e.g., an abstract concept of number) a process we refer to as "reflective abstraction."[4] These higher-order forms of knowledge may seem to be decoupled from the primary objects of experience, but they actually remain related to them by historically specific knowledge transformations. The resulting abstract concepts, such as the concepts of number, energy, chemical element, or gene, serve to "integrate" numerous experiences within a conceptual system (e.g., the conceptual system of thermodynamics or of modern biology). All of their concepts remain, of course, liable to further change.

Scientific systems of knowledge are indeed subject to persistent change in long-term processes of accumulation, loss, and transformation, which may occasionally give rise to far-reaching reorganizations of both the architecture of knowledge and the relevant epistemic communities. Some of these transformation processes are traditionally called "scientific revolutions." As I have argued above, however, they do not result from "paradigm shifts" but from transformative processes, such as the exploration of an existing system of knowledge to its limits, which may give rise to a "matrix" for the emergence of a new system.

Chapter 7 deals with several examples of scientific revolutions, here conceived as transformations of systems of knowledge. Discussing one important transformative mechanism in detail, we shall see that, under appropriate circumstances, an originally marginal concept within a given system may eventually find itself at the center of a new conceptual system. For example, inertial motion had long been noted when a projectile leaves the hurling motion of a slingshot along a straight line, but it became an explicitly defined anchor point of a conceptual system only with classical mechanics.

Such reorganizations of knowledge systems open up new perspectives, as well as possibilities for disruptive societal change that cannot be predicted by any "linear model" of scientific or technological progress. For this reason alone, focusing all resources on applied research dealing with evident global challenges would make little sense. Grand challenges do not automatically generate grand solutions, and changes of perspective and disruptive innovations can hardly be anticipated.

How Knowledge Structures Affect Society and Vice Versa

Part 3 deals more systematically with the relation between knowledge and society, conceived as two aspects of a common dynamic shaping human existence. A key mechanism of their interdependence is an "entrapment of mutual limitation," a notion introduced in chapter 8. The mental and cognitive development of individuals is limited by the experiential horizon a society offers to them, while societal regulations are constrained by a limited range of individual mental capabilities and perspectives. This interdependence also shapes the capability of individuals and societies to maintain complex forms of mental and social organization, that is, keeping an internal balance and withstanding the ever-present temptation of self-destruction. Societal changes therefore necessarily involve a coevolution of social and mental structures.

The concept of a knowledge economy is at the center of the discussion in chapter 8, and is illustrated by historical examples in chapters 9 and 10. Chapter 9 analyzes the knowledge economy of building in some detail as an example for the historical development of practical knowledge. Chapter 10 offers a broad review of some of the major knowledge economies in history. An important issue in all three chapters of part 3 is how the external representations available to a society—especially its "knowledge representation technologies," be they papyrus, parchment, paper, or the Web—shape its knowledge economy. Chapter 10 pays particular attention to the emergence of scientific knowledge as part of the economy of knowledge. How could it be that the production of scientific knowledge is regulated by a knowledge economy separate from the immediate needs of societal reproduction, and yet scientific knowledge production has become an existential condition of global society? Evidently, the internal regulation of this separate knowledge economy as well as the interfaces shaping the societal implementation of scientific knowledge have become critical factors in humanity's ability to cope with the Anthropocene.

Apart from the focus on knowledge economy, part 3 also deals with the role played by abstract concepts in the regulatory structures of society. Can their historical origin be understood in parallel with that of the scientific concepts analyzed in chapter 3? I suggest that, just as the technology of material practices allows for the formation of scientific abstractions like the concept of energy, societal practices like administrations or markets may give rise to "cultural abstractions," such as the concepts of time and economic value. External representations of these cultural abstractions, such as clocks or money, may in turn become a starting point for new social regulations and developments, such as the rise of capitalism. With the increasing societal role of information technologies, data have become a key cultural abstraction steering societal and economic processes. Cultural abstractions shape our thinking and our social existence. They are not easy to modify because they are the result of the evolution of complex systems that are highly path dependent. Yet, as the history of science shows, they can be changed after all.

While all abstractions tend to be opaque with regard to the historically specific experiences from which they emerge, they may be directly challenged by new experiences. In the history of science, such challenges have triggered major upheavals of fundamental concepts such as space, time, and matter. In societal development, the experiences and challenges of the Anthropocene may also warrant new cultural abstractions as regulatory mechanisms for coping with problems of planetary boundaries and limited resources.

How Knowledge Spreads

Part 4 deals with the spread of knowledge and the mechanisms of knowledge transfer. What role does knowledge play in past and present processes of globalization? Is there an overall tendency to create a flat world in which knowledge will become ever more uniform? In chapter 11, I show that globalization in the sense of the potentially global spread of means of social cohesion (ranging from economy to culture, knowledge, and science) and an enhancement of interdependencies is a layered process. In the course of history, globalization processes have created "sediment" such as the global spread of basic technologies (e.g., food production or writing) that change the conditions for subsequent globalizations. I argue that globalization processes do not necessarily lead to a homogenization of the world because of the crucial role of knowledge, in particular its role in shaping the identity of the actors in such processes. Knowledge transfer always involves an active appropriation on the "receiving" end, associated with the generation and further spread of new knowledge, including repercussions on the "senders."

The transfer of knowledge between societies is shaped by their respective knowledge economies, with the consequence that knowledge transfer generally amounts to a knowledge transformation, perhaps through its modification to the needs of the receiving knowledge economy or a hybridization of existing and new knowledge. As an example, I discuss in chapter 12 the transfer of European scientific knowledge to China in the early modern period and its assimilation into the Chinese economy of knowledge. Because of the embedding of knowledge within systems of knowledge, knowledge transfer may develop self-reinforcing dynamics as elements of the transferred knowledge point to parts still missing. This may be observed, for instance, in the various translation movements shaping the global history of knowledge, also discussed in chapter 11.

The recontextualization associated with knowledge transfer may lead to a "cultural refraction" in the sense of an explication of structures of knowledge implicit in the source knowledge economy. In the context of the disputation culture of ancient Greece, for instance, structures of practical knowledge of Babylonian and Egyptian origin were encoded in writing and made the object of explicit reflections. Cultural refraction may thus foster the emergence of higher-order knowledge with greater independence from local contexts and conditions. Modern science is the result of a long-term global history of knowledge, owing to the formation of sediment of globalization processes as well as the cultural refractions just described. I claim that so-called Western science is actually global science in two senses: First, it is a product of global history. Second, global history now depends on it.

According to the analysis in chapter 12, the "globalization of knowledge" in history is driven by an intertwinement of intrinsic and extrinsic dynamics. The phrase "extrinsic dynamics" implies the spread of knowledge as a "fellow traveler" among other transfer processes, such as commerce, conquest, or missionary activity. Intrinsic dynamics arise as a result of the development of knowledge systems with an ever-broader experiential basis, such as those of modern science, empowering societies to extend their dominance because of their technological, economic, and military superiority. This intertwinement accounts for the path dependency of the globalization of knowledge, in which accidental circumstances become ineluctable conditions of its further development. Even the nature of the most abstract and seemingly universal knowledge available to us is, in other words, shaped by a contingent history of globalization.

In chapter 13, I deepen the discussion of the spread of knowledge by making use of the concept of "epistemic networks" involving social, material, and semantic layers. These networks make it possible to account for self-organizing dynamics in the coevolution of epistemic communities and systems of knowledge. The spread of knowledge over social networks (e.g., about a new experimental method) may lead to the proliferation of that knowledge and its enrichment within an initially loosely connected epistemic community. Novelties often come from outside an established system of knowledge. The community may in turn react to the enriched knowledge that is manifested by the network of corresponding external representations (e.g., a set of scientific papers), integrating the accumulated knowledge into a more strongly organized system of knowledge (e.g., a new scientific discipline) and reorganizing itself around it. It is through an iteration of this cycle of proliferation and integration that distinctive systems of knowledge and epistemic communities organized around them may emerge from weakly connected semantic and social networks.

This also describes the socio-epistemic dynamics underlying some of the so-called scientific revolutions characterized not only by intellectual upheavals but also by a reorganization of scientific communities. This coevolution is not necessarily a fast process, however, as may be illustrated by the emergence of a scientific community focused on Einstein's general theory of relativity, a process that took half a century. The effectiveness of this dynamic evidently depends on the systems of knowledge involved, as well as on the specific properties of the external representations and social networks. These considerations are of relevance when considering the possibility of refocusing the attention of scientific communities on the current challenges of the Anthropocene.

On What Knowledge Our Future Depends

Part 5 returns to the relation between the Anthropocene and the evolution of knowledge. What distinguishes humanity from the rest of the biosphere is cultural evolution, a unique layer of metabolism and learning on top of biological evolution. With their rapidly evolving culture, humans have created an "ergosphere" as a new global component of the Earth system, in addition to the lithosphere, the hydrosphere, the atmosphere, and the biosphere, thus changing the overall dynamics of the system. The ergosphere, a concept introduced in chapter 16, refers to the transformative power of human work and material culture as a new form of metabolizing the planet. It may be on its way to becoming a "technosphere" in which technological and other global infrastructures created by humans assume a self-organizing, quasi-autonomous character.

In chapter 15, I argue that the entry into a new state of the Earth system is not due to a single cause and cannot be tied to a particular event in human history. It can best be described in terms of a cascade of evolutionary processes, as a transition from cultural to epistemic evolution. In cultural evolution, human societies have entered what Marx refers to as "relations of production" dependent on their material culture. In epistemic evolution, human societies' interactions with the Earth system have become dependent on science-based technologies, such as the use of fossil fuels, nuclear power, artificial fertilizers, and genetic engineering. Without the empowerment of the means of production through scientific knowledge, humanity would not have entered the Great Acceleration of the 1950s that is now being discussed in the geological sense as the beginning of the Anthropocene. Once the knowledge economy of science had entered a positive feedback loop with the capitalist economy, a runaway effect eventually transformed cultural evolution into a process ever more dependent on science-based technologies. What stone tools, hunting, gathering, and later food production, clothing, and the building of shelters were for the Holocene, science and technology are now for the Anthropocene: essential conditions of human life as we know it.

The transition from cultural to epistemic evolution suggests the development of knowledge as an important aspect of cultural evolution, a perspective developed in chapter 14. In the transition from biological to cultural evolution, the role of "niche construction" has been transformed from one among several aspects of biological evolution into an essential feature of cultural evolution, as the role of material culture and tool use for the very emergence of modern humans illustrates. In the transition from cultural to epistemic evolution, the role of scientific knowledge has been similarly turned from an aspect into a characteristic feature of novel evolutionary dynamics.

But what role does knowledge play in cultural evolution? The direction in which science and technology develop is not fixed and is certainly not governed by a rigid logic of progress. Their development is rather subject to an interplay between cognitive and contextual factors that includes other forms of knowledge as well. The experiment we are performing with our planet generates both massive changes of the environment and new knowledge. Therefore, we will need to better understand how this coevolution works if we want to be able judge our predicament in the Anthropocene and adjust our behaviors, including our knowledge economies, accordingly. An attempt in this direction is made in chapter 14.

The interplay between regulatory structures of human behaviors and their material environment shows remarkable characteristics: First, among any historically given conditions, societies are capable of reproducing some of them but not others. For instance, in the course of the Neolithic Revolution, also discussed in chapter 14, humans learned to reproduce environmental conditions, thus allowing them to produce their own food. In this way, accidental external conditions, such as the local availability of plants and animals capable of being domesticated, could become indispensable features of further global development, accounting for the path dependence of the process.

Second, a given physical environment and material culture open up specific possibilities for the development of regulative structures. As I have emphasized, the horizon of possibilities shaped by these material conditions may leave room for their exploration. In particular, as the material conditions and external manifestations of human social and behavioral structures play a key role in their regulation, such an exploration will affect these societal structures in turn.

As many examples from the history of science illustrate, exploring the horizon offered by the material means and external representations of a knowledge system may lead to a transformation of the system. Exploring the possibilities of social organization offered by a particular material culture may similarly engender societal transformations. Needless to say, neither kind of transformation is ever simple or straightforward, but both are worthy of deeper investigation from a common perspective. Like biological evolution, cultural evolution is an irreversible, singular, and unrepeatable process whose outcomes cannot be predicted.

Historically, the accessibility of fossil energy resources was an accidental, local condition fostering the emergence of industrialized societies. In the long run, however, it will not be possible to maintain the conditions it offered for societal development, let alone to reproduce them. The seemingly cheap and limitless availability of fossil resources has, for a time, decoupled industrialized societies from local environmental conditions. The abundance of energy has decontextualized their regulatory structures, thus suggesting that they have a universal character; that is, a character that can be superimposed on arbitrary local conditions and projected into an indefinite future. Now the time for such a "local universalism" is over. The consequences of industrialization restore a regulatory function to the environment on a global scale that will affect human behavioral and societal structures, albeit in locally diverse ways, as I discuss in chapter 16.

The inevitable transformation of fossil-fuel based societies will have to be an exploratory process as well, although the transformation to a nonfossil global economy will have to take place under great time pressure. Because local variables will again become increasingly relevant, there will not be a one-size-fits-all solution. Future solutions to energy problems will no longer be able to deal with technical systems independently from local and global environmental and societal contexts. A new knowledge economy will have to be based on "global contextualism," paying attention to both locally varying contexts and global consequences of any human intervention in the Earth system.[5] Therefore, a new knowledge economy will have to deal with the "borderline problems" at the interfaces between the human ergosphere and other spheres of the Earth system. And it will have to provide for modes of integrating the legacy of global human knowledge with the new local knowledge emerging at these interfaces.

We need, in particular, more knowledge about the dynamics of the Earth system and the role of human interventions in its dynamics. But we may also need commonly understandable cultural abstractions referring, for example, to our ecological footprints, that help societies to shape these interventions by locally representing their effects in the lives and minds of human actors.[6] In cultural evolution, the internalization and reproduction of given external conditions within societal development was largely a matter of circumstance. In epistemic evolution, it will have to become more and more a matter of knowledge. The sense of urgency driven by the awareness that we are living in the Anthropocene should thus concern not only politics and economies but also the quest for more knowledge that may trigger cross-scale effects on social behavior.

The fragmentation of scientific knowledge attributable to the current knowledge economy may, however, hinder opportunities for knowledge integration and implementation, blocking urgently needed innovations. In chapter 16, I argue that the Internet offers, at least in principle, a knowledge representation technology with the potential to optimize the current knowledge economy toward a global coproduction of knowledge and new forms of organizing, integrating, locally adapting, and implementing scientific knowledge. A future Web of Knowledge or an "Epistemic Web" would help to balance asymmetries in the ownership and control of knowledge and allow users to become "prosumers," for instance by replacing browsers with interfaces optimized for interacting with global human knowledge as represented on the Web. Not only content, but also the network of links would have to become an openly accessible public good.

Finally, in chapter 17, I briefly review the history of science by comparing it with the history of religion in order to examine the potential of science to offer orientation and guidance to humanity across diverse periods and circumstances. Asking questions about what today's science achieves or misses in addressing questions concerning humanity's survival in the Anthropocene, I demand a new alliance between scientists and civil society in engaging with the challenges of humanity. This alliance will involve rethinking the nature of science itself in ways to which this book may, I hope, contribute.

PART 2

HOW KNOWLEDGE STRUCTURES CHANGE

Chapter 3

The Historical Nature of Abstraction and Representation

What is the meaning of the problem of knowledge? What is its meaning, not simply for reflective philosophy or in terms of epistemology itself, but what is its meaning in the historical movement of humanity and as a part of a larger and more comprehensive experience? My thesis is perhaps sufficiently indicated in the mere taking of this point of view. It implies that the abstractness of the discussion of knowledge, its remoteness from everyday experience, is one of form, rather than of reality. It implies that the problem of knowledge is not a problem which has its origin, its value, or its destiny within itself. The problem is one which social life, the organized practice of mankind, has had to face. The seemingly technical and abstruse discussion of the philosophers results from the formulation and stating of the question, rather than the question itself. I suggest that the problem of the possibility of knowledge is but an aspect of the question of the relation of knowing to acting, of theory to practice.
—JOHN DEWEY, *The Significance of the Problem of Knowledge*

This chapter deals with the significance of abstraction in a historical theory of thinking.[1] After reminding readers of the powerful role of abstraction in human thinking, some major philosophical attempts to explain this role are discussed. I then turn from philosophy to psychology, focusing on the investigations of Jean Piaget into the emergence of abstraction in child development. In order to make investigations of his "genetic epistemology"[2] fruitful for a historical theory of thinking, I deal with the relation between abstractions and social practices. Conceiving these social practices as involving historically specific actions in the material world, I sketch the foundations of a "historical epistemology" that benefits from both psychological and historical insights. (The role of material embodiments and representations of knowledge is key to such a historical epistemology and is discussed in more detail in subsequent chapters.) I conclude this chapter with a discussion of the three dimensions of knowledge emerging from the previous analysis, that is, its mental, material, and social dimensions.

The Power of Abstraction

Any historical theory of knowledge has to account for the emergence, role, and historical transformation of abstract concepts. The critical role of abstraction is evident for the sciences if one thinks, for example, of mathematics. Abstract concepts such as number, space, time, force, and matter are central to scientific knowledge in general and have often been used to underline its privileged status. Abstract concepts may indeed encapsulate vast amounts of experience. But one encounters abstract concepts in other domains of knowledge as well (in philosophy and religion, for example) as well as in daily life. Philosophers have always been fascinated by abstract concepts as keys to understanding the world. Abstract concepts—be they mathematical constructs such as triangles or prime numbers, or physical concepts like matter and forces—appear to retain a timeless validity. They are therefore often held to be independent of the history that led humankind to discover or compose them, and should therefore be accessible even to nonhuman intelligence. Recall for example the suggestion that mathematical laws be employed for communication with extraterrestrials—an idea contemporaries already ascribed to the nineteenth-century mathematician Carl Friedrich Gauss,[3] and that was popularized in more recent times in the novel *Contact* by the American astronomer Carl Sagan.[4]

The pivotal role of abstract ideas for cognition and the mastery of life and nature is, however, no privilege of recent science; it has a very long philosophical tradition going back to ancient Greece. Ancient philosophers dealt with abstract concepts such as being, matter, form, beauty, goodness, and truth. Like later scientific concepts, these came with the claim that they allowed one to capture, master, and even transcend concrete individual experiences. In the case of philosophy, this mastery of reality typically included the capacity to immunize the philosopher against the vicissitudes of life and also to display of some form of technological superiority in managing the practical world. This is illustrated by what is reported in the sources about Thales of Miletus, one of the earliest Greek philosophers, born in the late sixth century BCE and counted among the seven sages of Greece.[5]

Aristotle credits Thales as having founded a school of philosophy that seeks the principles of all things in abstract properties of matter: "Thales, the founder of this school of philosophy, says the permanent entity is water (which is why he also propounded that the earth floats on water). Presumably he derived this assumption from seeing that the nutriment of everything is moist, and that heat itself is generated from moisture and depends upon it for its existence (and that from which a thing is generated is always its first principle). He derived his assumption, then, from this; and also from the fact that the seeds of everything have a moist nature, whereas water is the first principle of the nature of moist things."[6] Thales is also renowned for his astronomical interests, while being simultaneously disinterested in mundane concerns. This aloofness from ordinary experiences is at the center of an anecdote recounted by Plato, in which Socrates says:

> Well, here's an instance: they say Thales was studying the stars, Theodorus,
> and gazing aloft, when he fell into a well; and a witty and amusing Thracian

servant-girl made fun of him because, she said, he was wild to know about what was up in the sky but failed to see what was in front of him and under his feet. The same joke applies to all who spend their lives in philosophy. It really is true that the philosopher fails to see his next-door neighbor; he not only doesn't notice what he is doing; he scarcely knows whether he is a man or some other kind of creature. The question he asks is, What is Man? What actions and passions properly belong to human nature and distinguish it from all other beings? This is what he wants to know and concerns himself to investigate. You see what I mean, Theodorus, don't you?[7]

Yet, the same abstract knowledge that marked the distance of the philosopher from ordinary experiences could also serve him to demonstrate technological superiority, as another anecdote about Thales shows:

> All these methods are serviceable for those who value wealth-getting, for example the plan of Thales of Miletus, which is a device for the business of getting wealth, but which, though it is attributed to him because of his wisdom, is really of universal application. Thales, so the story goes, because of his poverty was taunted with the uselessness of philosophy; but from his knowledge of astronomy he had observed while it was still winter that there was going to be a large crop of olives, so he raised a small sum of money and paid round deposits for the whole of the olive-presses in Miletus and Chios, which he hired at a low rent as nobody was running him up; and when the season arrived, there was a sudden demand for a number of presses at the same time, and by letting them out on what terms he liked he realized a large sum of money, so proving that it is easy for philosophers to be rich if they choose, but this is not what they care about. Thales then is reported to have thus displayed his wisdom, but as a matter of fact this device of taking an opportunity to secure a monopoly is a universal principle of business; hence even some states have recourse to this plan as a method of raising revenue when short of funds: they introduce a monopoly of marketable goods.[8]

Philosophical Struggles

In Plato's dialogues, Socrates demonstrates the primacy of philosophical thinking by challenging ordinary experiences in much the same vein, arguing for the superiority of abstract thinking over concrete experience. He questions, in particular, any claims that artisans know their own techniques, pointing out that any qualification of these techniques or their results as "good" or "beautiful" requires prior knowledge of abstract notions of goodness and beauty. Human knowledge only makes sense if it is ultimately derived from a world of abstract ideas transcending concrete experience.

Plato thus postulated a realm of transcendental ideas, supposedly existing independently of their worldly material manifestations. According to him, the cognition of truth is nothing other than the remembrance of these archetypical ideas, of which all that we concretely experience are mere shadows. Plato described the relation between the world of ideas and the sensible world in his famous allegory of the cave. He compares our sensible world to a cave and the sense impressions we receive from this world to mere shadows. People who see nothing but these shadows throughout their lives will inevitably mistake them for real things. But once freed from the confines of the cave, they will realize that they had seen nothing but a faint reflection of true, stable, and unchangeable reality. The same results will be achieved when we overcome the inadequacies of our senses and finally recognize the world of ideas as true reality, either through philosophical reflection or scientific investigation. Plato, while extolling the fundamental role of abstract ideas such as that of goodness, thus concludes: "In the knowable realm, the form of the good is the last thing to be seen, and it is reached only with difficulty. Once one has seen it, however, one must conclude that it is the cause of all that is correct and beautiful in anything, that it produces both light and its source in the visible realm, and that in the intelligible realm it controls and provides truth and understanding, so that anyone who is to act sensibly in private and public must see it."[9] In this sense, the path to the truth can be conceived quite literally as an "enlightenment." It is the path from darkness into the light of truth, although this truth may be revealed only gradually, and perhaps never in its entirety.

Historically, the most influential and still commonly accepted notion of abstraction goes back to Aristotle.[10] His concept begins with sense experiences as the foundation, which are then decomposed into individual qualities, some of which may then be omitted. In his *Metaphysics* we thus read: "And just as the mathematician makes a study of abstractions (for in his investigations he first abstracts everything that is sensible, such as weight and lightness, hardness and its contrary, and also heat and cold and all other sensible contrarieties, leaving only quantity and continuity—sometimes in one, sometimes in two and sometimes in three dimensions—and their affections *qua* quantitative and continuous, and does not study them with respect to any other thing; . . .)."[11] This concept, however, raises a problem: it seems to be completely arbitrary which qualities are omitted and which are considered essential. Fresh meat, cherries, and blood, for instance, are all red, but this is an arbitrary collection of non-essential similarities that cannot be the basis of science. For Plato, in contrast, abstractions are founded in his transcendental world of ideas. Individuals may thus be guided in discerning which qualities are to be omitted or retained by their recollection of this world of ideas. However, since the world of ideas is not directly accessible, Plato's theory has no advantage over Aristotle's in this respect.

In early modern philosophy, the question of abstraction again became an important issue, this time in connection with legitimizing the role of mathematics for the new sciences. Empiricists like David Hume strictly distinguished between the realm of sense perceptions and that of mathematical propositions. For him, "All inferences from experience, therefore, are effects of custom, not of reasoning."[12] Mathematical propositions are, on the other hand, "discoverable by the mere operation of thought, without dependence on what is any where existent in the universe."[13] The rationalist philosophy of Immanuel Kant aimed instead at justifying the foundational role of abstract concepts for the new empirical sciences. He argued that the abstract form of concepts does not arise from experience by abstraction, but is constituted by the

3.1. Kant's thinking machine. Drawing by Laurent Taudin.

subject itself in the process of cognition. According to Kant's epistemology, the constitution of logical and mathematical forms of thinking is based only on the reflection of mental activity: "The **I think** must **be able** to accompany all my representations."[14]

Kant's aim was to explain why scientific knowledge may be considered, in spite of our ever-fluctuating experiences, a particularly secure form of knowledge, and how it could serve, more generally, as a model case of rationality. Like Plato, Kant inferred from the apparently unassailable security associated with fundamental statements in mathematics and the natural sciences that the validity of these statements is not to be found in sense experience but in certain forms of cognition given a priori, that is, prior to experience. In accordance with the early modern emphasis on the productive role of the individual in this world, Kant identified not an otherworldly realm of ideas as Plato had, but the human mind as the fountain of this secure knowledge. From this perspective, sense experiences are merely raw material for the machinery of cognition, whose architecture is not affected by them. Like an infallible thinking mechanism—anticipating the production machines of the Industrial Revolution—the human mind allegedly transforms this raw material into stable and rational insights.

Among the unchangeable components of his "thinking machine," Kant included a priori notions of space, time, and causality, which he described according to the state of knowledge in the contemporary sciences. As for spatial cognition, for instance, Kant assumed that its basic structures are given a priori in a way that corresponds to Euclidean geometry. He could not anticipate, of course, that other, non-Euclidean forms of geometry would be found later in the nineteenth century—forms that made his choice look rather arbitrary—to say nothing of the subsequent scientific revolutions in physics, in particular when relativity theory introduced new physical notions of space and time.

How could philosophers in the tradition of Kant react to the further development of natural science? One option was to update or replace the outmoded thinking engine Kant had designed with a more modern one in which current concepts of natural science could take their place as the supposedly universal mechanisms by which thinking functions. Or one might react to the failure of the Kantian program by shifting the question from epistemology, the branch of philosophy dealing with cognition, to methodology, thus attempting to identify a general scientific method that assured both the cognitive advancement of science and the validity of its results.

Many such efforts have been made since the beginning of the early modern era. One attempt is associated with an early-twentieth-century philosophical movement called "logical empiricism," which originated in the so-called Vienna Circle after World War I.[15] Strongly influenced by philosophers such as Ernst Mach, some members of this school attempted to reduce all philosophical questions to a special kind of empirical statement, namely, propositions dealing with immediate sense experience, and then to clarify their relation to other, more general statements with the help of logic. In practice, however, it has turned out to be difficult to reconcile any of these schemes with what historians have taught us about the actual workings and developments of science.

A Psychological Perspective

The emergence and role of abstract categories was a central concern not only of epistemology but also of other intellectual enterprises that had emancipated themselves from philosophy by the beginning of the twentieth century. Many fundamental cognitive structures are not present at the outset of life, but are rather constructed over the course of human development. They change over time and may even hold different meanings in different cultures. While philosophical epistemology had largely neglected these developmental, historical, and anthropological aspects of the formation of abstract categories, these aspects now became the objects of empirical investigation in some newly established fields.

In the early twentieth century, psychology was developing its own experimental paradigms. This led, on the one hand, to empiricist approaches to describing human behavior, that is, by focusing on reactions to external stimuli and essentially renouncing an analysis of internal mental structures. On the other hand, attempts were undertaken to capture thought processes, not just by introspection (in contrast to the traditional, more philosophical approach) but by an experimental methodology including, in particular, interviews with test persons.

One of the founders of gestalt psychology, Max Wertheimer, explored his theoretical ideas about thought processes in the context of the history of science. He even had the good fortune to discuss the creation of special relativity at length with Albert Einstein, one of the case studies in his renowned book *Productive Thinking*. Einstein and Wertheimer may have met as early as 1911 or 1912, when Einstein was a professor at the German University in Prague, Wertheimer's hometown.[16] They became close friends in Berlin in 1916, after Einstein had completed general relativity. In the introduction to his case study, Wertheimer writes: "Those were wonderful days, beginning in 1916, when for hours and hours I was fortunate enough to sit with Einstein, alone in his study, and hear from him the story of the dramatic developments which culminated in the theory of relativity. During those long discussions I questioned Einstein in great detail about the concrete events in his thought. He described them to me, not in generalities, but in a discussion of the genesis of each question."[17] At that time, psychology took the natural sciences as a role model for empirically founded investigation. From a philosophical point of view, the categories of logic had been elevated to a universal core of rational thought processes. As Wertheimer puts it in the introduction to his book: "Some psychologists would hold that a person is able to think, is intelligent, when he can carry out the operations of traditional logic correctly and easily."[18] In order to overcome the traditional emphasis of psychology on introspection, a new empiricist outlook demanded human behavior and thinking be subjected to observational and experimental procedures, conceptualized in particular in terms of stimuli and responses and of associations. To quote Wertheimer again: "Many psychologists would say: ability to think is the working of associative bonds; it can be measured by the number of associations a subject has acquired, by the ease and correctness with which he learns and recalls them."[19]

Wertheimer instead conceived of the cognitive structures he analyzed as wholes that play an active role in cognition and that can neither be reduced to the object of cognition nor be seen as the indirect result of external stimuli. With this view, he became a pioneer of what has been called "constructivist structuralism" in psychology, a tradition to which the work of the Swiss child psychologist Jean Piaget also belongs. In his genetic epistemology, the logico-mathematical forms of thinking are abstracted not from the mental activities of the subject but from his or her actions: "It is agreed that logical and mathematical structures are abstract, whereas physical knowledge— the knowledge based on experience in general—is concrete. But let us ask what logical and mathematical knowledge is abstracted from. . . . In this hypothesis the abstraction is drawn not from the object that is acted upon, but from the action itself. It seems to me that this is the basis of logical and mathematical abstraction."[20]

Piaget's Concept of Abstraction

A fundamental starting point of Piaget's theory is that cognition is an essentially active process, conditioned by experiences in which reality is transformed through behavior and action. Actions and their attendant consequences accordingly shape structures of thinking. If I count pebbles arranged in a certain geometrical figure (e.g., in a row or in a circle) and I find that I always come up with the same number, no matter from where I begin or in which figure the pebbles are displayed, I have come across a mathematical property called commutativity, allowing the change of order of elements subject to an activity such as counting. This is not so much an insight into a property of the pebbles but into the actions of counting I can perform with

3.2. Swiss psychologist Jean Piaget (1896–1980) teaching children in a Washington Heights classroom in New York City. Piaget pioneered the study of cognitive development in children, emphasizing the role of actions in building up cognitive structures. Bill Anderson / Science Source.

them. Piaget claims that fundamental structures of logical and mathematical thinking result from an increasingly sophisticated mental organization and reorganization triggered by experiences and enabling such action coordination in practice.

This process begins with sensorimotor schemata resulting from an internalization of behavioral patterns closely related to sense perception, such as sucking and later following movement. The sensorimotor period extends from birth to the appearance of language. In this period children acquire, among other faculties, the coordination between vision and prehension, as well as between means and goals, which happens roughly within the first year and a half of life. For instance, a child may first produce a noise with a rattle hanging from the cradle as the result of a chance discovery but will then purposefully use the same noise to make his mother stay in the room.[21]

During the sensorimotor phase, children also build up a mental construction of objects as entities independent of the self and its actions, what is called the schema of "object permanence." At first, there is no homogeneous understanding of space, since actions and the spatial aspects of the senses (oral, visual, auditory, tactile) are not integrated. Action schemata under visual control and the perceptual constancy of the shape and size of an object presuppose the coordination between vision and prehension, the ability to grasp what is seen.[22] Changes in the perception of bodies in motion, for instance, can be attributed to changes in perspective rather than as transformations of the object. Still, children first seek an object that was hidden where they last saw it and not where it vanished. This is because the object is conceived as part of a situation and not as an entity independent of the action of the child. Only

3.3. Two-year-old boy building a tower and checking the size of the cups. The child first tries to put the large green cup on top of the small green cup and then again on top of the yellow cup. With knowledge of the transitivity of size, it would follow immediately that the large green cup does not fit on top of the yellow cup because this cup is even smaller than the small green cup which the child had tried initially.

after several stages is the cognitive structure of object permanence fully developed, as a result of the coordination of more elementary action schemes and their integration, mediated by actual experiences with the environment.[23]

Later, children build up action schemes and cognitive structures that Piaget characterizes as "operative." Operations are actions that can—at least mentally—be reversed, such as when I pour water from one vessel into another and then I pour it back. Operations always presuppose the conservation of something, an invariant, in this case the amount of water.[24] Attaining this level of thinking is less evident than it may seem. Children at the "preoperative" stage do not yet have a notion of the conservation of volume and do not realize that the quantity of water is conserved when it is poured, say, from a broad and flat vessel into a narrow and high vessel.[25] Operations are typically related to an entire system of actions subject to general principles like the conservation of volume or commutativity.

Let me explain the idea with reference to the striking example of seriation.[26] This cognitive structure is relevant to the following kind of psychological task: children are asked to sort different sticks by their lengths and to put them in a row starting with the shortest stick and ending with the longest. Young children confront this task empirically, that is, they take a short stick and a long stick, then another short stick and a long stick, and so on, but they do not coordinate these pairs. Typically, they do not succeed in ordering all the sticks by their lengths. Somewhat older children succeed, but only by trial and error. From a certain age onward, however, children proceed differently, in a more systematic way. They start, for instance, from the shortest stick, then search for the next shortest one, and so on. They realize that if stick A is longer than stick B, and B is longer than C, then A is also longer than C. Sticks A and C do not have to be compared directly since their relation can be inferred from a cognitive structure the children have meanwhile built up. This structure is characterized by principles such as reciprocity, which allows the children to infer that when they search for the next shortest stick among all the remaining ones, this stick is longer than all the ones previously taken. In other words, they are able to coordinate the relation "longer than" with the relation "shorter than."

Remarkably, such cognitive structures are built up in the individual development of a child—or as psychologists say in "ontogenesis"—from an interaction between empirical trial actions and cognitive development. Before these structures exist, the task has to be performed without making use of transitivity, comparing each stick with every other. Once such a cognitive structure has been built up, specific pieces of information are assimilated to the cognitive structure that defines a consistent system of actions. This system now allows the actions with the sticks to be performed in a way that allows anticipation of the outcomes of comparisons. In other words, thinking is an exploration of the possibilities opened up by a given means of action within the cognitive structures that have been built up.[27]

The example of the seriation task, which has been extensively investigated in child psychology in the tradition of Piaget and his Geneva school, also illustrates other remarkable features of cognitive structures. Learning involves what Piaget calls the "assimilation" of new contents to existing cognitive structures, but also the modification of these structures, which Piaget calls "accommodation."[28] Every learning process is an active process characterized by the appropriation of new experiences in terms of both assimilation and accommodation. Development begins with a stage in which the cognitive structures are clearly not available. This stage is then followed by an intermediate one in which the goal to be reached can be anticipated and the means for its solution are available, but in which problem solving remains purely ad hoc. Finally, a stable stage is reached in which actions are clearly governed by the existence of the structure (for example, presupposing the transitivity of ordering relation). The assimilation of new experiences into a cognitive structure is a historical process in the sense that the cognitive structure changes each time it occurs.[29] Its extension and enrichment by experiences ultimately provides the material for its accommodation—that is, its transformation into a more developed cognitive structure. A crucial mechanism for such a reorganization is what Piaget calls "reflective abstraction," which refers either generally to thinking about structures of thinking, or more specifically to the possibility of making an action the object of a higher-order cognitive structure, for example, considering it as a reversible operation.[30]

This understanding of ontogenetic development seems to imply, however, a predetermined universal hierarchy of steps leading from actions with concrete objects to ever higher order mental operations. According to Piaget, any child in any culture at any time undergoes these predetermined ontogenetic steps of development. As we have seen, the sequence begins with sensorimotor intelligence, the level of practical intelligence in which sensory data are assimilated to schemes of coordinated, repeatable actions functioning below the level of conscious thinking. What follows is the level of what Piaget calls "preoperational thinking," in which sensorimotor intelligence is supplemented by the symbolic function, allowing objects to be represented by symbols distinguishable from their meanings.[31] The highest level is that of operational thinking, which emerges when internalized actions become reversible mental operations, from which concepts such as quantity, time, space, and other abstract concepts of logical and mathematical thinking are created in a seemingly universal way through reflective abstraction.

Is this then an answer to our question about the origin of abstract concepts? Here we find them not in the lofty realms of science and philosophy but as end results of child development, at least as Piaget sees it. Not all cultures, however, have identical abstract concepts. The question therefore is: How do we account for the emergence of different systems of abstract concepts? A clue comes from the fact that abstract concepts such as time, length, volume, and weight are also important categories of social life that regulate social institutions as different as the organization of labor or the regulation of traffic—concepts that may change in history and differ between cultures.[32] These concepts are abstract in the sense that they cover and connect an enormous range of social experiences, from bodily to planetary dimensions, and capture them within the same categories and standards of measurement. What is the relation between the psychological and the social meanings of abstract concepts?

The Roots of Abstraction in Social Practices

The existence of abstract categories is by no means self-evident and universal. It is rather closely related to historically and culturally varying practices establishing actual connections between different realms of experience. In the ancient world, as well as in some recent nonliterate societies, for example, no abstract concepts of space or time integrated notions as different as the interval between two heartbeats and the length of a year, or the breadth of a finger and the length of a day's journey. One of the best studied recent nonliterate societies are the Eipo, who live in the central highlands of West New Guinea, close to the Eipomek River.[33] Prior to 1974, when intense research on their culture was pursued, the isolated tribes living in this area had experienced few contacts with either Western civilization or the Indonesian culture nearby. Until recently, the Eipo did not use measures of distance; their language even lacked a general term for distance. When they traveled to a distant village for more than two days, they described their journey as a "journey on which I slept twice" without applying the same sort of assessment to, say, distances in their immediate environment. Whereas our abstract concept of space covers all kinds of spatial realities, from the size of our body to the distance traveled on a journey, and integrates them into a unified metric framework, their navigational space is not structured by distances and directions but by a closely knit network of toponyms representing landmarks. We will come back to different systems of spatial thinking in chapter 14.

The historical origin of abstractions in social practices is suggested by the example of another abstract concept, that of the exchange value of commodities represented by money. In the beginning of his work *Das Kapital*, Marx emphasizes the curious character of the abstraction associated with the social practices that generate a commodity: "A commodity appears, at first sight, a very trivial thing, and easily understood. Its analysis shows that it is, in reality, a very queer thing, abounding in metaphysical subtleties and theological niceties. So far as it is a value in use, there is nothing mysterious about it, whether we consider it from the point of view that by its properties it is capable of satisfying human wants, or from the point that those properties are the product of human labour."³⁴ Marx then proceeds to explain that the relation between products as expressed by their identities as commodities possessing an exchange value is actually an expression of the social relationships among their producers: "A commodity is therefore a mysterious thing, simply because in it the social character of men's labour appears to them as an objective character stamped upon the product of that labour; because the relation of the producers to the sum total of their own labour is presented to them as a social relation, existing not between themselves, but between the products of their labour. This is the reason why the products of labour become commodities, social things whose qualities are at the same time perceptible and imperceptible by the senses."³⁵

Commodities may thus be regarded as representations of social relations. Marx sees a close relation between this economic abstraction and the reign of abstractions in the field of religion, drawing attention to the fact that the human origin of such abstractions is typically concealed: "In order, therefore, to find an analogy, we must have recourse to the mist-enveloped regions of the religious world. In that world the productions of the human brain appear as independent beings endowed with life, and entering into relation both with one another and the human race. So it is in the world of commodities with the products of men's hands. This I call the Fetishism which attaches itself to the products of labour, so soon as they are produced as commodities, and which is therefore inseparable from the production of commodities."³⁶ Historically, the origin of exchange value as an economic abstraction can be traced back to the world of the early urban civilizations, for instance in Mesopotamia. There, during the period commonly referred to as Ur III, toward the end of the third millennium BCE, silver coins or ingots started to play the role of universal currency in economic exchanges, in particular with foreign powers.

Cuneiform specialists have demonstrated (on the basis of the rich documentation surrounding state-regulated fisheries) that certain commercial agents were responsible for trading surplus goods both within the Ur III economy and with neighboring societies.³⁷ These economic activities eventually created an expanding system of value equivalents. While this happened at first at the margins of state-regulated economic activities, in particular in foreign trade, it eventually led to an entire economy oriented toward the acceptance of silver as the standard. Simultaneously, working time became integrated into this system of economic commodities, with the workday representing a crucial measure of value, very much in accordance with Marx's statement: "In all states of society, the labour-time that it costs to produce the means of subsistence, must necessarily be an object of interest to mankind, though not of equal interest in different stages of development."³⁸ Value thus became an abstract category, represented by standard measures such as silver weights, within a society passing from a barter economy to commodity exchange.

The historical emergence of abstract categories in the context of social practices connecting different realms of experience seems to be a general pattern. But what are the cognitive mechanisms that give rise to such abstractions? And how do these cognitive mechanisms relate to the role of the social processes that I have just indicated?

Actions as Material Practices

Let us come back to Piaget's "genetic epistemology" as an account of the psychological development of abstract categories in a sequence of well-defined stages. If this understanding is applied to the historical dimension of cognitive structures, as has indeed been attempted by Piaget and others, we similarly end up with a conception of the evolution of knowledge in which its results are universally predetermined. In other words, this line of reasoning would end up in a teleological conception of the evolution of knowledge.[39] By that reckoning, one could hardly relate the emergence of abstractions to the historically specific social processes and the collective experiences they generate as we have begun to discuss them.

At this point, we must tap into other theoretical traditions that have paid more attention to such historically specific social processes. In the 1920s, Lev Vygotsky, Alexander Luria, Aleksei Leontiev, and others formed an influential circle of psychologists in the young Soviet Union.[40] They focused on socially meaningful activities, and in particular on material and symbolic action as an explanatory principle of human cognition. In a famous work titled "Tool and Symbol in Child Development," Vygotsky and Luria stressed the close links between the practical use of tools and symbols in the development of human actions.[41]

In the 1930s, the historians of science Boris Hessen and Henryk Grossmann (also in a Marxist tradition) developed important insights into the technological and economic conditions surrounding the emergence of modern science, insights that also placed material practices in the foreground in the development of certain forms of thinking.[42] According to Hessen, human concepts of nature and mechanics changed in the early modern world (with the predominance of cities and the rise of the capitalist economy) because natural phenomena were continually more often conceived as produced by machines rather than by organisms.

This insight extends Marx's observation that the evolution of machinery in this period triggered insights into the exchangeability of motive forces.[43] Machines came to replace the hand of the laborer, reducing his or her skilled activity to that of a motive force. The function of the driving power was thus gradually isolated from other functions in manual production processes. This made it possible to replace human motive force by animal or natural powers such as wind, water, or gravity. When the engine became a separate entity in the production process, it constituted a basis for identifying and exchanging different motive forces. Grossmann extended this insight into the cognitive realm: the historical transformation of machinery laid the groundwork for the abstract concepts of motive force, work, and motion. The social position of science was simultaneously enhanced by the growing economic importance of technology, which also stimulated an increasing integration of practical and academic knowledge.

I will come back to these historical developments in chapter 10 when discussing the early modern knowledge economy in greater detail. Here I am only interested in the fact that the so-called Hessen-Grossmann thesis aims at identifying the horizon of cognitive possibilities on the basis of historically given material and symbolic means.[44] As for the symbolic means, Ernst Cassirer captured their fundamental role succinctly in his 1944 "An Essay on Man": "Man has, as it were, discovered a new method of adapting himself to his environment. Between the receptor system and the effector system, which are to be found in all animal species, we find in man a third link which we may describe as the *symbolic system*. This new acquisition transforms the whole of human life. As compared with the other animals man lives not merely in a broader reality; he lives, so to speak, in a new *dimension* of reality."[45] The aim is to combine these insights with Piaget's sophisticated theory of cognitive development within a larger framework that takes into account three strands of development: *phylogenetic development*, the biological evolution of the species homo sapiens; *ontogenetic development*, the development of the individual; and *historiogenesis*, economic, social, and cultural development—in short, the cultural evolution of human societies.[46]

The historiogenetic strand of evolution is linked with the other two in several ways. Phylogenetic and historiogenetic factors were intimately connected at the genesis of humanity. Not only was biological evolution the prerequisite for the emergence of human culture, but, as we know, this culture decisively shaped the final steps of anthropogenesis, the biological evolution of modern humans. This is especially obvious when one considers the biological repercussions of social interactions and the use of tools. Finally, the development of the species is realized—both phylogenetically and historiogenetically—by the ontogenesis of the individual. The historiogenesis of cognitive structures, in particular, depends on individuals who acquire the shared knowledge of a society at a certain historical moment in their individual development and participate in the transmission and transformation of this knowledge through their cognitive activities.

One can now reconceptualize the constitutive role of actions and their coordination in the generation of cognitive structures in Piaget's model. Taking into account what one can learn from a tradition going back to Marx, these actions should be endowed with a concrete, historically specific meaning that includes the material means and social constraints of these actions.[47] The understanding of human actions as part of socially embedded material practices creates a bridge between societal developments and the ontogenetic emergence of cognitive structures as described by Piaget and other psychologists. The bridge is constituted, in particular, by means of production, material embodiments, and external representations of knowledge and other material conditions of human action. Examples are instruments, infrastructures, texts, symbol systems, or any other aspects of the material culture of a society that may serve as an encoding of knowledge.

The Role of Material Embodiments of Knowledge

All this hints at the role of material culture as the backbone of an evolution of knowledge. From the history of astronomy, for instance, it becomes particularly evident that its advances are shaped by instrumentation, beginning with Galileo's use of the telescope, to the introduction of spectroscopy in the nineteenth century, to radio astronomy in the twentieth century, and now to instruments allowing us to pursue gravitational wave astronomy. Evidently, the history of science is punctuated by the invention, use, and exploration of such material tools, instruments, and laboratory apparatus, from the simple machines of antiquity to the particle accelerators of high-energy physics. But what if the game-changing role of material culture as a means of cognition also extends to the symbolic means of our thinking, to the number systems of arithmetic, to writing as an external representation of language, to the symbol systems of mathematics, physics, and chemistry, or, more generally, to what one may call "paper tools"?[48]

All material contexts of action, such as the tools or media employed, the action itself, its accompanying gestures and sounds, or the places where it happens, may be considered material embodiments of knowledge. Material embodiments of knowledge may not only take different forms, but one and the same material embodiment may exert different functions depending on the context of usage and the cultural environment. A tool like a hammer is, first of all, a material means of action, but it may also embody different kinds of knowledge. We distinguish, in particular, between the knowledge required to use a tool, to produce it, and to invent it. Even if one has never seen a hammer before, the knowledge necessary to use a hammer may be inferred from its form and physical constitution, or may easily be acquired by manipulating it. It is typically more difficult to extract the knowledge to produce a tool merely from inspecting it. Reconstructing this knowledge from a given artifact is known as "reverse engineering."

Material embodiments comprise "enactive" aspects that are immediately associated with the action itself, as well as communication systems specifically designed to transmit mental contents. They include tools, artifacts, models, rituals, sound, language, music, and images, as well as symbol systems and writing. If a material object or condition is used to represent knowledge, we speak of an external representation of knowledge.

The functioning of external representation essentially depends on what Piaget has called (following Cassirer and Simmel) the semiotic or the symbolic function.[49] The semiotic function is the ability to distinguish events and objects from their meaning. External representations rely on the technologies intentionally developed for knowledge representation, which may range in complexity from notches carved on a stick as a simple tallying mechanism to sophisticated formal symbol systems such as mathematical or chemical formulas and, of course, include computer systems today. Historically, knowledge representation technologies developed not in direct succession but overlapped, and almost all persist today.

The identification of external representations as important players in the evolution of human culture has itself a long history and includes the works of philosophers such as Georg Wilhelm Friedrich Hegel, Karl Marx, Georg Simmel, and Ernst Cassirer. External representations may be treated as signs in the sense of semiotics, a field of inquiry going back to the work of Ferdinand de Saussure, Charles Sanders Peirce, John Dewey, and others.[50] The cultural historian Gary Tomlinson has emphasized the crucial role of semiosis, that is, the capacity of understanding and using signs for the transmission of knowledge: "Knowledge taught, imitated, learned, and passed on to successive generations is always knowledge *about*, just as a sign is always a sign *of*. This is as true of cultural knowledge that appears in behavioral form—a technique for flint knapping passed along by hominins possessing no modern language—as it is for the most abstract ideational construct."[51] Signs can involve iconic, indexical or symptomatic, and symbolic dimensions. One speaks of an "icon" if a sign has some resemblance to the object represented, such as an image or a sound; an index or symptom is physically or causally related to the signified object as smoke is to fire; a symbol does not need to have such direct relations but is typically subject to rule-based usage, as when objects are represented by the words of a language.[52]

External representations often involve more than one of these dimensions, depending on their context of usage. Onomatopoeic words such as "bang" are not just symbols but also carry an iconic dimension, mimicking a particular sound. Symptoms may be interpreted as symbols, as when an astrologer reads signs in the heavens as symbolic messages about human events. In the history of religion and the history of science, we encounter the claim that the world may be read as a text. Famously, Galileo Galilei writes in *Il Saggiatore* (*The Assayer*, 1623): "Philosophy is written in this grand book—I mean the universe—which stands continually open to our gaze, but it cannot be understood unless one first learns to comprehend the language and interpret the characters in which it is written."[53] Symbols only derive their power to refer to external objects by being part of a system and from the capacity of an actor to understand their relation to other symbols of the system. It is the system as a whole that assumes the role of an index referring to sets of objects or processes outside of it.[54]

As Gary Tomlinson has argued, semiosis and the different functions that signs can assume (from icons, via indexes, to symbols) may have been closely related to the phylogenesis of culture, reaching into the animal kingdom and beyond into prehistory and early human history: "Representation appeared in the course of evolutionary history along with semiotic organisms that introduced a new kind of repetition into the world. On top of a repetition by *dis*placement—informational correspondence at the ends of Shannon's channel—there appeared repetition by *re*placement. There is every reason to think that the advent of organisms capable of this reflects a major transition or set of transitions in the history of life."[55] He illustrates his view with an example: "A gull perceiving a shell in the water as an icon of food has enacted semiosis, but not cultural semiosis and certainly not systematic culture. The shell as insignia, on the other hand, enacts all three: it is a representation of an object (it is a sign), it can be displaced and transmitted (it is a cultural sign), and its meaning emerges as a function of its place in the constellation (it is a sign in systematic culture)."[56] The handling of external representations is guided by the cognitive structures represented by them, as when writing a text or drawing a figure is guided by the intention to express specific intellectual contents. An individual triangular figure drawn on paper, for instance, may serve as an iconic representation of a general mathematical triangle (with lines that have no thickness) by similarity, but it takes the cognitive structure of a triangle to know the difference between them. The use of external representations may also be constrained by rules characteristic of their material properties and their employment in a given social or cultural context, such as orthographic rules or stylistic conventions in the case of writing. We refer to such rules as the semiotic rules or regulative structures associated with a given external representation.

External representations can be used to share, store, transmit, or control knowledge, but also to transform it. Individual knowledge generally results from the individual appropriation of shared knowledge through reconstruction from external representations. External representations of knowledge have, as a rule, no uniquely fixed relation to internal, mental representations. This ambiguity may be a disadvantage when it gives rise to misunderstandings, but it is also a crucial factor in innovation, since the knowledge built up in individual cognition from shared external representation may be characterized by variations that in turn may become the starting points for novel insights.

Forgotten Traditions in the History of Science

In the twentieth century, reflection on science developed into distinct branches that represented entirely different perspectives on how science embodies human rationality. Each branch tended to emphasize a single aspect of science—its normative aspects or its ideas, its historical contingency, or its social and economic contexts—forsaking all others. From the perspective of a comprehensive reflection on science that tries to take into account its historicity as well as its claim to a privileged form of rationality, this divergence amounts to a veritable split of rationality. It was engendered by specialization and reinforced by ideological divisions, the world wars, the Holocaust, and the Cold War. As a consequence, valuable insights into the nature of science were lost, fragmented, or marginalized. Three episodes from the 1930s, representing traditions in the history of science that remained obscure for many years, may serve to illustrate this split of rationality.[57]

The Second International Congress on the History of Science and Technology was held in London from June 29 to July 3, 1931. It was attended by a Soviet delegation led by Nikolai Bukharin, a circumstance that attracted public attention because it raised the expectation that the Russians would use the occasion for political propaganda. One of the Russian delegates (the director of the Institute for Physics at Moscow University, Boris Hessen) presented, however, a profound scholarly study with far-reaching implications for the history of science. It was titled "The Social and Economic Roots of Newton's *Principia*."[58] The audience was hardly able to follow his arguments; most of the other papers presented at the conference consisted of "reminiscences from the elderly, and trivia from obscure amateurs." One account suggests that Hessen's talk was drowned out by a ship's bell, rung by the British historian of science Charles Singer in order to disrupt his message and the Soviet contingent more generally.[59]

The left-wing British scientist and historian of science J. D. Bernal later observed that the Russian speakers "had a point of view, right or wrong; the others had never thought it necessary to acquire one."[60] He added, however, that this impressive performance was hardly effective: "The Russians came in a phalanx uniformly armed with Marxian dialectic, but they met no ordered opposition, but instead an undisciplined host, unprepared and armed with ill-assorted individual philosophies. There was no defense, but the victory was unreal."[61] Back in Stalinist Russia, Hessen was later falsely accused of conspiracy and terrorism; he was executed in 1936. His contribution was not forgotten, but distorted: Hessen's claims tended to be misinterpreted by Western historians as crude reductions of science to economic *needs*. They failed to realize that Hessen had stressed instead the *enabling* role of production forces for scientific research as an activity in its own right.

3.4. A delegation of Soviet scientists attending the 1931 Second International Congress on the History of Science and Technology in London. *Left to right*: (*front*), Boris Hessen, Nikolai Bukharin, Abram Ioffe, Ernst Kolman; (*back*), Boris Zavadovsky, Modest Rubinstein, Nikolai Vavilov. See Chilvers (2015, 74). © Hulton-Deutsch Collection / Corbis via Getty Images.

In 1935, the *Zeitschrift für Sozialforschung* published an article titled "The Social Foundation of Mechanistic Philosophy and Manufacture," authored by the Marxist economic theorist Henryk Grossmann.[62] With the Nazi regime in power, the German-language journal meanwhile appeared in French exile. The article, written as a critique of Franz Borkenau's book *The Transition from the Feudal to the Bourgeois World Picture*,[63] was hardly noticed and fell immediately into oblivion. Grossmann himself survived the war in exile, but, like Hessen's contribution, Grossmann's work offered profound insights into the relation between science and the technological and economic conditions of the early modern period.

Taking Marx's concept of labor as their starting point, Hessen and Gross-mann independently developed a view of science in stark contrast to widely spread prejudices on Marxism. Following the analysis of Gideon Freudenthal and Peter McLaughlin, three "Hessen-Grossmann theses" can be identified: First, that technology opens cognitive horizons for science, explaining the material conditions under which certain abstractions become possible. Second, that technology also limits the horizons of science, explaining why certain abstractions are not possible under given conditions. And third, that other social factors—including ideologies, beliefs, and political and philosophical theories—can affect science.[64] Before the 1970s, the Hessen-Grossmann theses had virtually no influence on the history of science.[65]

Early in the twentieth century, philosophers of the so-called Vienna Circle (such as Moritz Schlick) had sought a closer unity of science and philosophy, interacting with prominent scientists such as Albert Einstein on the challenges of contemporary science.[66] But by the mid-1930s, some of them had retreated to a formal analysis of language. The abyss between a philosophical and a historical epistemology of science became strikingly evident when, in 1933, Schlick received a manuscript by the Polish bacteriologist, physician, and historian of science Ludwik Fleck under the title "The Analysis of a Scientific Fact: Outline of a Comparative Epistemology."[67] Fleck hoped that Schlick could help him publish his work, which offered a new perspective on science as a collective enterprise, arguing that the historical development of all knowledge is socially conditioned.

In order to describe this collective enterprise, Fleck introduced such notions as "thought collective" and "thought style."[68] For Fleck, the history of scientific knowledge was related in essential ways to social mechanisms like education and tradition. In his letter to Schlick, he criticized traditional epistemology: "I could never shake the impression that epistemology examines not knowledge as it actually occurs, but its own imagined ideal of knowledge, which lacks all its real properties. . . . The statement that all knowledge originates in sensations is misleading—because the plurality of all human knowledge stems quite simply from textbooks. . . . Finally, the historical development of knowledge shows some remarkable common aspects as well, such as for instance the particular stylistic closeness of the respective systems of knowledge, which demands an epistemological investigation. These considerations prompted me to treat a scientific fact from my area of expertise epistemologically, whereupon the aforementioned manuscript emerged."[69] Schlick's publisher, Springer Verlag, decided, however, not to publish the book. Fleck survived Auschwitz and Buchenwald. His book appeared in 1935 with a Basel publishing house and remained obscure until it was rediscovered in the 1960s.

The relation between material embodiments of knowledge and their use in action and thinking is flexible and context dependent. This intrinsic ambiguity is a major source of novelty in the development of knowledge, as I shall discuss in greater detail on the basis of several examples. The material representation of a cognitive structure may have features that were not inherent in the original structure but can be discovered when using and exploring it. To illustrate this, I refer to the role of mathematical formalisms for representing scientific knowledge about nature. A mathematical formalism can serve as an external representation of certain physical laws, as when the law of falling bodies or the orbits of the planets are described with the help of the differential equations of Newtonian mechanics. Exploring these equations by manipulating them according to the semiotic rules of the relevant symbol system (in this case that of differential calculus), one may arrive at novel concepts or insights, for instance, concerning how this formalism may be used to describe the orbits of comets.

Operating with a symbolic system in sciences like physics or chemistry is not just "blind thought" wherein the symbol system can be substituted for what it stands for, as philosophers such as Gottfried Wilhelm Leibniz or Moses Mendelssohn claimed in the eighteenth century. The symbolic system rather resembles a natural language in which the combination of signs does not only have to satisfy semiotic rules but must also be constantly checked for its semantics.[70] Often the manipulation and exploration of a symbol system may produce expressions that are formally correct but to which no meaning can be ascribed—for instance, because they describe physically nonsensible entities or impossible situations. But such an exploration of the formalism may also give rise to expressions to which, surprisingly, a new meaning can be ascribed, even a meaning that was not anticipated within the conceptual system that the formalism was originally designed to represent.

For instance, I could start building up Newtonian mechanics based only on a law of force, without initially introducing a concept of energy. I can then begin to manipulate my equations and form certain mathematical expressions, for instance, from the symbols for the mass and velocity of a body, which may seem arbitrary at first, but then turn out to represent quantities that are conserved in physical interactions. Perhaps I call them "momentum" or "energy." I may then even reformulate the entire framework by starting with these new expressions, insisting that they are more "fundamental" than the force law from which I originally started. This example is a bit simplified, but not entirely made-up: the question of whether there is something conserved in mechanical interactions did indeed become the subject of the famous *vis viva* controversy in the eighteenth century, to which I shall return below. The example illustrates how intricate the interaction between the exploration of an external representation of a system of knowledge and the development of its conceptual structure may be. Later I shall discuss the impact of the exploration of a formalism on the transformation of a conceptual system in greater detail.

The essential point is that systems of knowledge are never given in their entirety at the outset of an investigation. Their meaning unfolds only gradually, through the exploration of their implications with the help of external representations—for example, by doing calculations, applying a conceptual framework to new circumstances, exposing systems to debate within an epistemic community, and so forth. We therefore speak of the "generative ambiguity" of external representations, which leads to their seemingly paradoxical function in knowledge transmission: on the one hand, they secure and stabilize the transmission of knowledge over generations, thus ensuring its longevity, while they serve, on the other hand, as tools for thinking and thus agents of change and innovation in the transmission of knowledge.

A fundamental property of both material means of action and external representations of knowledge is that the range of possible applications associated with these given means and representations is larger than the goals pursued by any given set of actors. Hegel considered this as a "cunning of reason."[71] The pragmatist philosopher John Dewey described the double function of tools for stabilizing and innovating meaning as follows: "The invention and use of tools have played a large part in consolidating meanings, because a tool is a thing used as a means to consequences, instead of being taken directly and physically. It is intrinsically relational, anticipatory, predictive. Without reference to the absent, or 'transcendence,' nothing is a tool."[72] If one has adopted or invented a certain tool for a specific purpose, it is always possible to come across new applications of that tool, when employing it in a new context, for instance. The material means of actions determine, however, the limited range of what can possibly be achieved by them. This range may be described, adapting a formulation of the philosopher Edmund Husserl, as a "horizon of possibilities" associated with the given means.[73] Of course, this horizon is not only determined by the actions' material means but also by the cognitive structures underlying and coordinating the handling of these means.

In other words, the range of possible results of actions performed with given means may exceed those encoded in the preexisting cognitive structures and become the starting point for the emergence of new cognitive structures. The explorative use of external representations is, therefore, a major driving force in the evolution of knowledge systems. This dynamic must even have played a role in the very emergence of our species, since the use of tools long preceded the appearance of modern humans. Tool use was probably not merely a consequence of the biological development of the brain; it must have involved a two-way causality, since tool use offered opportunities for enhancing cognitive abilities based on regulating actions involving tools, which in turn made an enlargement of the brain an advantageous mutation in the further biological evolution of hominids. The wider role of material representations for human societies is discussed in the box "Representations, Power, and Transcendence" at the end of chapter 8.

From Genetic to Historical Epistemology

A given stage in the evolution of knowledge is distinguished by the specific, historically available external representations of individual cognition. From a historical point of view, the reconstruction of this horizon of possibilities is therefore a crucial task, making it possible to identify a common social field in which actors may achieve similar results, whether they are directly connected to each other or to a common source, or whether they make use of similar potential inherent in the available material means for representing knowledge. This helps to explain, for instance, the occurrence of simultaneous discoveries and inventions, and, more generally, the emergence of common themes, methods, or intellectual achievements that seem to occur more or less independently from each other in a specific historical situation and that come to be associated with it. When dealing with these phenomena, it is often overlooked that mutual influences and common resources depend on the nature of shared knowledge and that external representations carry a shared potential for arriving at similar results. For instance, Galileo and many of his contemporaries developed strikingly similar approaches to the phenomenon of accelerated motion, not because they necessarily exchanged their views with one another but because they all drew on the same intellectual resources of medieval philosophy for analyzing change processes. I will come back to this medieval technique below when discussing diagrams as an example of external representations.

It is useful to distinguish between external representations of different orders. First-order representations of cognitive structures materially represent real objects by symbols or models with which one can essentially perform the same operations one can perform with the objects themselves. These proxy operations can, as a rule, be performed more easily since they are not constrained by the contingencies of the real situations. Think of numbering discrete objects with the help of counters, using one counter for each object. While dealing with counters is a merely symbolic action and obviously cannot substitute for actions with the real objects, it does allow the results of real actions to be anticipated in order to plan and control them.

But operations with first-order external representations not only serve to execute operations according to existing cognitive structures. They also play a constructive or innovative role, since they are constitutive in building up new cognitive structures by the internalization of operations. Thus, the cognitive structures involved in arithmetic may be built up by manipulating simple counting systems. In this respect (in giving rise to new cognitive structures) symbolic actions such as the manipulation of counters can completely substitute for real actions.

The newly arising cognitive structures (as, in our example, those of arithmetic) may in turn be externally represented by a symbolic system, for example, that of decimal numbers. But in order to calculate using such second-order representations, one already has to know arithmetic. While first-order external representations represent real objects, second-order representations represent cognitive structures that no longer need to refer directly to concrete objects. While the simple counters in our example may be manipulated without presupposing an abstract number concept or the knowledge of arithmetic, the addition of decimal numbers requires previously acquired cognitive structures, that is, prior knowledge of arithmetic.

Like first-order representations, second-order representations may also be symbols or material artifacts; the rules for handling and transforming them essentially follow from the cognitive structure being represented. Their adequate use presupposes that real objects and actions have been assimilated into the cognitive structure they represent. Second-order representations are, however, only indirectly related to these objects and actions. They do not play a constructive role with regard to the preexisting cognitive structures they represent, but they may in turn play such a role for higher-order structures emerging from their usage, as when the cognitive structures of algebra are built upon reflections on arithmetical operations.

Material means or external representations generate a cognitive structure by determining a system of results of actions that is then reproduced in the structure's generative rules. Insofar as the means of action are culturally and historically specific material objects, the cognitive structures that emerge from employing them are also

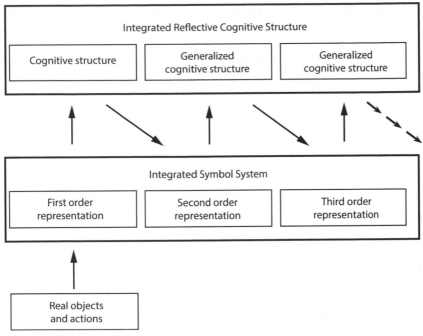

3.5. The reflective structure of abstraction. From Damerow (1996a, 379).

culturally and historically specific. This circumstance is often overlooked when investigations focus on more or less universal features or fail to include a comparative dimension. From this point of view, knowledge takes on both a structural, as well as a historically and empirically concrete quality. Knowledge of some object can be defined as the capability to solve problems involving this object by connecting information about it with cognitive structures relevant to understanding its behaviors. Knowledge, in this sense, is always structured knowledge. But although such structures may appear to be entirely abstract, they actually depend on content because they result from concrete actions with objects and material means external to them.

While material means of action thus play a constitutive role for the genesis of cognitive structures, cognitive structures also shape material actions in turn. This is obvious when one considers the fundamental role they play in coordinating actions in the first place. But cognitive structures may also guide the construction of improved or novel material means that enable actions and applications beyond those encoded in the given structures (structures that have allowed for the construction of these devices in the first place) with the consequence that new cognitive structures may emerge.

Cognitive structures change in the course of their application, becoming more general in the sense that their system of rules apply to ever richer content. At the same time, a cognitive structure is also being connected to other cognitive structures and thus becomes more differentiated, more powerful, and, in a sense, more concrete. When a cognitive structure fails, reflection on the actions on which the cognitive structure is based may guide the construction of a new structure by making use of novel or hitherto unused thinking tools and the possibilities they open up. Reflection, in this context, offers thinking a direction. Even in the simple example of ordering sticks by their length, it is true that no attempt to solve this problem takes place under conditions identical to those of the previous attempt. Each attempt to solve a problem on the basis of a given cognitive structure can in principle be remembered, thus resulting in an enrichment of the structure by accretion. It is as the Greek philosopher Heraclitus claimed: you cannot step twice into the same river.[74]

With these notions, Piaget's concept of reflective abstraction can be turned into a thoroughly historical concept in which material, socially contextualized actions serve as the origin of cognitive structures. In this sense, reflective abstractions, such as those giving rise to the abstract mathematical concept of number, ultimately depend on the material actions from which they originate, such as the concrete actions of counting material objects with the help of number words, clay tokens, or number signs. Reflective abstraction thus becomes a constructive process in which novel cognitive structures are built up by reflecting on operations with specific external representations like language or mathematical symbols. These symbolic means typically represent previously constructed mental structures, so that a potentially infinite chain is created involving a succession of reflective abstractions and representations that we designate as "iterative abstraction."[75] This is the hallmark of a more developed cognitive structure: it allows for a reconstruction of previous positions but cannot itself be expressed in terms of the previous framework. In this way, Piaget's "genetic epistemology" becomes a veritable "historical epistemology."

Social Cognition in the Historical Context

The historical evolution of knowledge is realized through the cognitive activities of individuals. It is, on the other hand, a collective process spanning populations and generations, involving the interaction of individuals whose minds work independently from one another. Human cognition is essentially social cognition, comprising the ability to communicate, to share knowledge, and to learn from each other. It presupposes that humans conceive of each other as intentional beings, and that they are capable of taking someone else's perspective. Individuals grow up under culturally specific conditions that determine the challenges and constraints under which they learn systems of knowledge.

From the point of view of an individual, internal representations are related to individual experiences and to the possible usages of a given means of action that the individual may anticipate. This is the *individual* or *subjective* meaning of external representations of knowledge. From a societal point of view, the external representations embody action potentials that may not be perceived or anticipated by every individual but that are nevertheless realized in societal practice; in fact, the external representations are typically a product of actions earlier performed. That is the *social meaning* of external representations. The social meaning of the external representations of knowledge available to a group or a society constitutes its shared knowledge.

Individual cognitive development may be conceived as a process in which subjective meaning is acquired by an appropriation, in interactions and through communicative processes, of the shared knowledge offered by society. When, in turn, novel insights attained by an individual do not lead to cognitive structures that are transferable in socialization processes, they are lost in transmission and irrelevant to historical development. Individual contributions can only affect the historical transmission of shared knowledge when the results of individual cognition are systematically reproduced and extended by appropriate social institutions, contributing to the perseverance and accumulation of knowledge.

The possibility of intersubjective communication is also based on shared knowledge and is typically mediated by language. Language is itself an external representation of knowledge that is being generated by cognitive structures enabling a speaker to produce formally correct linguistic constructs.[76] Linguistic performance, however, goes beyond linguistic competence in requiring a meaningful use of language in concrete contexts of action. Such contexts also involve nonlinguistic cognitive structures, which are integrated with linguistic structures that thus assume meaning through subjective semantics. In the process of linguistic performance, linguistic constructions generated in such contexts of action become the external representations of the subjective meanings involved.

Different subjective meanings result from different individual developmental trajectories. They find their communality in the underlying social meaning of the means of action and external representations of knowledge. Without such a basis in collective experience and shared knowledge, intersubjective communication would be impossible. Shared knowledge furthermore presupposes that the social meaning of the objects of communication can be matched with the subjective meaning or be reconstructed on the basis of the individual experience.

Understood in this way, the historical development of shared knowledge is a highly path-dependent process involving iterative abstractions that are contingent on a series of concrete historical experiences. The codification of experience in terms of such knowledge structures is one reason for the characteristic recursive blindness of abstract thinking with regard to its own experiential sources—a recursive blindness that also accounts for the seemingly a priori character associated with some of its results.

In subsequent chapters, I will discuss examples of this process. The concepts of space and time in classical physics, for instance, cover a wide range of experiences and constitute what may seem to be a universally applicable framework. Understanding and applying this framework does not presuppose any knowledge about its historical emergence. These circumstances could indeed suggest that the physical concepts of space and time are given prior to and independent of any specific experiences. Yet this seemingly a priori character was challenged when, eventually, certain new experiences made it necessary to reflect on some of the basic operations of measurement from which these concepts originated. The result of this challenge was, as we shall see in chapter 7, a new framework of space and time that again acquired a seemingly universal range of validity.

The Generative Ambiguity of External Representations

Against this background, we are in the position to discuss a characteristic difficulty of the reconstruction of historical forms of thinking. Since the basic structures of cognition reflect, according to Piaget, the coordination of actions and not the actions themselves, it is difficult to reconstruct the level of cognition from an observed behavior alone. Indeed, the same behavior, for instance, moving from one place to another, may result from sensorimotor activities, from the use of a map and a compass, or from employing a GPS system. Only a fuller account of such behaviors in their social and cultural contexts and, in particular, an identification of the historically given means of action make it possible to identify the level of cognition involved. It is also true that external representations of knowledge may serve to express knowledge of different layers of reflexivity, as when, for instance, the equation $5 + 7 = 12$ is used as an example of simple arithmetic, of sophisticated number theory, or of the so-called synthetic a priori in the sense of Kant's philosophy.

The use of external representations for communication, be it in educational or in cross-cultural contexts, creates an inherent uncertainty about which layers of reflexivity are involved.[77] The same problem occurs, of course, when we are confronted with historical examples of external representations, such as calculations in ancient Babylonian or Egyptian mathematics, which tend to be misinterpreted. Such misinterpretations typically result in anachronistic reconstructions which ascribe to the historical actors the same type of reflexivity that characterizes modern thinking, neglecting the fact that the same cognitive tasks can often be accomplished by different means. This ambiguity of external representations is a trace of their original role in generating such different levels of reflection in the first place, as there clearly would be no number theory, for instance, or Kantian epistemology without the external representations of simple arithmetic that historically preceded them. We may therefore refer to the "generative ambiguity" of external representations.

Who Thinks Abstractly?

To conclude, let us return to the question of abstract thinking. In an essay, the philosopher Georg Wilhelm Friedrich Hegel asks: "Who thinks abstractly?" He then surprisingly answers: "The uneducated, not the educated." Hegel illustrates his astonishing answer by colorful examples, describing, for instance, how a market woman accused by a maid to have sold her rotten eggs insulted her in return: "What? My eggs rotten? You may be rotten! You say that about my eggs? You?" She then continued to reduce the entire personality of the maid and even her family to the fact that she had accused her of selling rotten eggs: "Did not lice eat your father on the highways? Didn't your mother run away with the French and didn't your grandmother die in a public hospital?"[78]

What Hegel implies here is that cognition begins by subsuming experiences under abstract concepts, and that only thereafter is it capable of generating ever more concrete representations of the object. Cognition is as much a process of concretization as it is of abstraction because it links abstract concepts in cognitive structures and

3.6. The Hegel story. Drawing by Laurent Taudin.

uses them to interpret reality. The usefulness of abstract concepts is only validated in this process of concretization, in which cognitive structures are enriched by experiences and relations to other cognitive structures.

We have thus, by drawing on a combination of philosophical, psychological, and historical insights, found an answer, albeit a somewhat abstract answer, to our question about the emergence and role of abstract concepts. Certain key abstract concepts of scientific thinking emerge from a reflection of material actions belonging to historically specific practices. But we have learned more: the framework sketched in this chapter not only allows us to understand, at least in principle, the nature of abstract concepts—it also indicates how to reconcile the nondeterministic character of the evolution of knowledge and its dependence on very specific historical circumstances with the possibility of producing a long-term accumulation of knowledge. One should not forget, however, that the accumulation of knowledge is not guaranteed by this process, nor is the nature of the higher-order forms of knowledge predetermined, as we shall see more closely when looking at examples in the following chapters.

Three Dimensions of Knowledge

To summarize, we may distinguish three dimensions of knowledge: a cognitive, mental, or "internal" dimension; a material, embodying, or "external" dimension;

and a social dimension, referring to the societal processes involved in producing, sharing, transmitting, and appropriating knowledge. All of these dimensions are equally relevant to the understanding of the evolution of knowledge as expounded in this book.

The dynamics of knowledge evolution can only be understood when all three dimensions are taken into account. This is obvious in the case of the social dimension, which includes the transmission processes that account for the continuity of this evolution over time. But it also holds for the material dimension, which constitutes the backbone of this transmission in terms of the material culture underlying it.[79] Finally, a historical evolution of knowledge only makes sense when it includes a shared knowledge base and its interactions with individual thinking processes.

Knowledge is a problem-solving potential, that is, the capacity of an individual or a group to solve problems and to mentally anticipate corresponding actions. Knowledge is based on experience and encoded in mental, material, and social structures. It is generated by reflection on environmentally embedded actions and serves as a potential for the anticipation and control of actions. Knowledge is internally represented by cognitive structures that enable the connection between past and current experiences. It is shaped (but not determined) by the material culture and existing social relations, and ultimately arises from experiences accumulated in socially constrained material practices.

The historical continuity, dissemination, and transmission of knowledge relies on its external representations, which serve the knowledge reproduction structures within a society. External representations may become prerequisites for the construction of new knowledge structures, as these new structures may result from actions applied to the external representations. These new knowledge structures are removed from the primary actions, but in ways that are dependent on the historically specific material and social nature of external representations. They can again be encoded through external representations of a higher order, which then, in turn, become the prerequisites for further development.

The long-term evolution of knowledge can only be understood on the basis of the entanglement, via material embodiments, of the ontogenetic and historical developments of cognition. Material embodiments and external representations mediate between socially shared knowledge, which is the object of historical development, and individual knowledge, which is the only true manifestation of this shared knowledge.

Chapter 4

Structural Changes in Systems of Knowledge

Experience gained over several years of working in the venereal disease section of a large city hospital convinced me that it would never occur even to a modern research worker, equipped with a complete intellectual and material armory, to isolate all these multifarious aspects and sequelae of the disease from the totality of the cases he deals with or to segregate them from complications and lump them together. Only through organized cooperative research, supported by popular knowledge and continuing over several generations, might a unified picture emerge, for the development of the disease phenomena requires decades.
—LUDWIK FLECK, *Genesis and Development of a Scientific Fact*

Science, like the Mississippi, begins in a tiny rivulet in the distant forest. Gradually other streams swell its volume. And the roaring river that bursts the dikes is formed from countless sources.
—ABRAHAM FLEXNER, "The Usefulness of Useless Knowledge"

In the preceding chapter, I focused on the emergence of abstract concepts. But a historical theory of knowledge must also account for how reasoning operates in knowledge frameworks that are not mathematized, or that may differ considerably in their conceptual structure from those of scientific knowledge. I have claimed that scientific knowledge is partly based on and includes other forms of knowledge, such as practical or technical knowledge. How can the relation between such different forms of knowledge be understood?[1]

I have also begun to discuss, in general terms, how cognitive structures may change by processes of assimilation and accommodation, by reflective abstraction, and by exploring the horizon of possibilities opened by external representations. So far, I have described these processes in very general terms and have in particular left the question open of how one can understand processes of major historical change that involve a multitude of such cognitive structures, such as when so-called scientific revolutions happen.

In this chapter I therefore deal with complex aggregates of knowledge, as they are also characteristic of science and technology. I refer to them as systems of knowledge, though this term is by no means intended to imply a high degree of organization. In the sense I adopt here, "systems of knowledge" may also refer to loose collections and assemblies of knowledge elements held together by cultural practices. Specifically addressing such collections that have a low degree of organization, I sometimes also refer to "packages of knowledge."

Yet all these complexes are characterized by an epistemic architecture that changes over time. In order to describe this architecture, I make use of the mental, social, and material dimensions of knowledge introduced above. To investigate the dynamics of change, I distinguish, somewhat schematically, between three kinds of knowledge: intuitive, practical, and theoretical. This short list is not meant to be exhaustive or exclusive, but it nonetheless helps us analyze the interplay between different kinds of knowledge in scientific change.

The evolution of cognitive structures is described with the help of concepts from cognitive science such as frames, procedures, mental models, and default settings. I argue that taking insights into cognitive change gained from current observations in cognitive psychology and adapting them to historical investigations is analogous to assimilating observations about biological change to a historical theory of biological evolution. Both cases involve not only a transfer of concepts or knowledge from one domain to another but an adaptation of the concepts themselves to the new domain. In this sense, Darwin and his appropriation of a wide array of biological knowledge may serve as a model for a historical theory of knowledge. Finally, having laid the groundwork by discussing systems of knowledge, mental models, and default assumptions, I sketch characteristic features of the overall dynamics of scientific change, emphasizing the role of exploration processes, triggers of change, and the reorganization of knowledge systems.

Systems of Knowledge and Their Architecture

Knowledge systems are socially shared collections of knowledge elements that are somehow connected. They may be produced and reproduced by the same social group or institution—a workshop or a university, for instance. The heterogeneous knowledge shared in a workshop or a university curriculum (including but not limited to what is taught in courses and learned from textbooks) would, in this sense, constitute a system of knowledge. A knowledge system may also be considered as such on the basis of its elements being cognitively related, for instance, the various aspects of a scientific theory such as relativity theory, including its elaborations, applications, and open problems. We may alternatively emphasize the role of material links between knowledge elements, like the books collected in a library. "Knowledge integration" refers to any process by which new knowledge becomes connected to an existing system of knowledge. The dynamics of knowledge development depend above all on the available means for knowledge integration, which in turn depend on social institutions, means of representation, and, of course, the cognitive features of a knowledge system.

The various paths to constituting a system of knowledge may lead to very different architectures. In particular, knowledge systems and their components may be characterized by degrees of reflexivity (How far is knowledge removed from the primary objects of action?), systematicity (How systematically is knowledge organized?), and distributivity (To what extent is knowledge shared?).

The range of knowledge forms with different degrees of reflexivity includes the categories mentioned above: intuitive knowledge, practical knowledge, and theoretical knowledge. Intuitive knowledge results from the interaction of humans with natural and cultural environments in the process of ontogenesis. Practical knowledge is the specialized knowledge resulting from the experiences of specially trained practitioners. It is characteristic of all kinds of craftsmanship and has been conveyed, for long historical periods, as part of the transmission of professional skills.[2] Theoretical knowledge can typically be appropriated by an individual through texts that are understandable only with prior knowledge and under certain external circumstances. These categories overlap; practical knowledge today may also involve scientific knowledge and vice versa, and knowledge transmitted by texts is not automatically theoretical. Reflexivity is at its lowest in intuitive knowledge, when unaccompanied by conscious reflection and unmediated by symbolic forms; it is higher when the object of knowledge is itself a form of knowledge. In the preceding chapter, I have indicated how different orders of reflexivity may be identified in the context of iterative chains of abstractions.

Knowledge can also be systematized to varying degrees, ranging from chunks of knowledge (via packages of knowledge without close cognitive ties) to more or less coherent systems of knowledge that may, for instance, be deductively structured. The degree of systematicity depends on how closely the components of a knowledge system are interwoven, that is, on how many and which links exist between its components, whether the system as a whole can be represented in such a way that the components and their role in the functioning of the system can be inferred from it (which requires a certain degree of reflexivity), and whether it is possible to reconstruct the system from just some of its components. The distributivity of knowledge may range from knowledge shared by social groups to knowledge that is basically the same for all cultures and historical periods.

In subsequent chapters, several different knowledge systems are discussed in detail. To briefly illustrate the concept here, I will limit myself to a few examples and remarks.

The mechanical knowledge of the Eipo of New Guinea mentioned in the previous chapter belongs to a package of knowledge strung together primarily by certain material tools and their use. Various kinds of sticks are used for specific agricultural purposes such as the digging stick for tilling the soil, the lever for digging out big stones, the planting stick for seeding and planting vegetables and roots, and the harvesting stick for digging out plants with their roots. The tips of these sticks are prepared by pointing them according to their specific purposes using some knowledge about suitable wedge shapes.[3]

The most frequent application of the digging stick in particular is its use as a lever by placing the left hand close to the end serving as the fulcrum and applying the force of the right arm to the other end. A skilled man will hold the stick in a way that he attains, depending on the kind of soil to be prepared, the best possible result, implicitly choosing the distances in accordance with the law of the lever. Clearly the related knowledge is held together by a set of specific practical purposes and the use of similar means. It is not clear, however, whether this system also includes some form of generalized mechanical experience. In any case, there is no obvious evidence of a vocabulary for a generalized mechanical knowledge. The links between components in this system of knowledge are realized in the actual participation in practical activities, rather than being codified independently therefrom.

Practical knowledge is usually described as "implicit" or "tacit" knowledge because its verbal expression represents only a very limited aspect of its transmission. In reality, such knowledge is characterized by a broad array of media and structured information necessary to its communication.[4] Its external representation may be composed of samples, a variety of tools and demonstrations of their usage, verbal explanations (possibly involving technical terminology), drawings or models, and a specific distribution of labor, as well as social and material contexts that may not be made explicit but taken for granted in a particular culture.

As a rule, this renders the transmission of practical and technological knowledge more context dependent than the communication of theoretical knowledge through texts, as witnessed by the difficulties of reverse engineering an artifact. This context dependency—and hence locality—of practical and technological knowledge is often reinforced by the fact that technical solutions, at least until the premodern period, were themselves attuned to specific contexts rather than produced for a global market as is mostly the case today.

Historically, practical knowledge was frequently transmitted by means of family traditions, but also in the context of institutions such as workshops or guilds. Since this knowledge is closely connected with interpersonal traditions, it is often recognizable in the stylistic features of artifacts that owe their creation to it. Formal education supported by institutions during long historical periods has been, by comparison, a rather marginal activity.

The dependence of practical knowledge on multiple forms of external representation, each connected with its own regulative structures, also accounts for the stability of this kind of knowledge, at least as long as the relevant contexts for its transmission do not substantially change. If they do change, however, practical and technological knowledge is much more easily lost than theoretical knowledge. It is also more difficult to reflect on a wide-ranging array of external representations than on the operations of a single device, or on the symbolic representations of theoretical knowledge. This difficulty in challenging existing traditions is one reason for the impressive stability of practical knowledge traditions in the premodern world.

The codification of practical knowledge with the help of external representations (e.g., the emergence of technical manuals in early modern Europe) has been an important driving force of knowledge development, as it may turn knowledge itself into an object of transmission independent from its implementation.[5] A similar emancipation from concrete contexts also takes place in dedicated processes of education. Teaching occasions a reversal between problems and methods. When knowledge is taught in an educational context, the relation between problems and solution strategies may indeed be reversed: Initially, the problem determines the choice of methods to be taught. But in teaching, the methods become central and the problems are chosen in dependence on the methods to be transmitted. A striking example are mathematical problem texts from ancient Mesopotamia specifically designed for the teaching of scribes.[6]

In the historical evolution of knowledge, this "emancipatory reversal" often played an important role, allowing for the creation of exploratory knowledge unbound from specific purposes. It "frees," so to speak, a system of knowledge from concrete contexts of application by focusing on its means rather than on its ends. For example, a very systematic form of knowledge is arguably so owing to the needs of teaching. Thomas Aquinas justified the writing of his *Summa theologiae* with the argument that an "exposition" (i.e., a commentary) necessarily follows the order of the books on which it comments; the training of novices instead requires a systematic presentation of doctrine.[7]

The greater context independence of a theoretical system of knowledge over that of a collection of practical knowledge may be illustrated by the system of knowledge surrounding the law of the lever going back to Archimedes.[8] This system is externally represented by written texts that establish explicit links (for example, in the form of proofs or explanations) between statements about mechanical devices and behaviors, such as the equilibrium of a balance or the functioning of a lever. The same system may, of course, be expressed in very different ways, choosing, for instance, some of the statements as axiomatic bases from which to derive the others.

Abstract concepts such as that of the center of gravity come with a wide range of applications that subsume numerous devices and behaviors under the field of mechanical knowledge; here a system of knowledge determines what its components are and how they function. A system of knowledge with such a high degree of systematicity can typically be reconstructed from some of its components, as may be illustrated historically by the remarkably successful efforts of Renaissance scholars to reconstruct lost or fragmentary scientific and technical writings from antiquity, sometimes basing entire theoretical edifices on just a few extant concepts and statements from ancient authors.

The Role of Cognitive Science and Darwin's Example

In the following, I discuss cognitive science as a reservoir of concepts and examples for a historical theory of knowledge. In order to illustrate its usefulness for historical investigations, I first briefly discuss procedural knowledge as a kind of knowledge that frequently occurs in the context of both practical and theoretical knowledge, and that lends itself particularly well to the introduction of some basic notions of cognitive science such as "chunking" or "slots."[9]

Knowledge bound to specific ends is often transmitted as problem-solving knowledge. It is articulated in examples and procedures involving the steps necessary to solve the problem. A procedure is a relatively stable series of repeatable actions that can be encoded as a set of instructions allowing a person to implement, each time in a new situation, the execution of a number of actions in a definitive sequence.

Such knowledge, when explained in speech or in writing, is often termed "recipe knowledge." It typically has a goal, namely, the accomplishment of some task or the resolution of some problem. As with any cognitive structure, the implementation of a procedure in a new situation or context will have some influence on the procedure itself. More than just practical knowledge can be organized in terms of procedures; resolution techniques for mathematical problems may also take this form. Early abacus texts from the Middle Ages, for instance, typically teach their readers calculation techniques in the form of such procedures.

An elementary example is a cooking recipe. It strings together discrete actions, like cleaning and cutting vegetables, as independent procedural steps. Procedures thus rely on prior knowledge (e.g., about elementary actions such as cleaning or cutting) that they embed in a sequence of actions. Typically, procedures have variables or "slots" to be filled with empirical data, such as observing whether a piece of meat in the oven has assumed a certain color or tenderness.

Procedures may have branching points involving decisions about whether to pursue a particular action or an alternative. They may also invoke subprocedures or may themselves be subprocedures of higher-order procedures. When a procedure calls up a subprocedure, the latter is executed and then returns control, so to speak, to the original procedure. The construction of a house, for instance, may require the preparation of some building materials, such as bricks or wooden frames, and can only proceed after the latter has been accomplished.

Procedures may be more or less integrated. They may consist of a loose string of actions moderated by clues, such as when one is driving a vaguely familiar road, occasionally relying on landmarks or even on asking locals for help. Or they may be rather like an automated sequence of actions that are smoothly executed without much thought, such as when one is driving along a very familiar road.

In the latter case, a procedure is memorized as a single unit (chunking) and is said to have achieved "noun-status." Such procedures often carry labels to support their retrieval from memory. Thus, recipes typically have names. In addition, procedures may be associated with descriptors: a mathematical operation may be remembered as the distributive law because it relates addition to multiplication.

This description of procedural knowledge owes much to the terminology and ideas of cognitive science—for instance, when I characterize a procedure memorized as a single unit in terms of "chunking" or being associated with a "descriptor."[10] Cognitive science, which in turn owes much to computer science, indeed has useful concepts and distinctions to offer to a historical theory of knowledge. We may distinguish, for instance, a situation in which a knowledge representation structure is called up from memory ("retrieval") from a situation in which it is constructed in the process of reasoning ("real-time synthesis"). Real-time synthesis often involves the real-time construction of an external representation of the knowledge structure.

While cognitive science may provide tools for a history of knowledge, these tools are usually not rooted in an understanding of the architecture of socially shared knowledge, and they often do not take into account the cultural and historical diversity of thinking. Researchers in this field are interested primarily in acts of individual cognition and the way certain cognitive structures are built up, identified, or retrieved from memory during these acts. But for an understanding of the evolution of knowledge, such structures are part of a transmitted macrostructure of knowl-

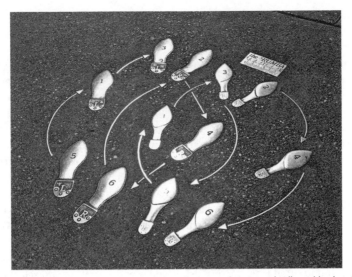

4.1. Dance steps as recipe knowledge. The sidewalks in Seattle's Capitol Hill neighborhood feature embedded bronze footprints outlining dance steps. This one, on Broadway Avenue, shows the rumba. Wikimedia Commons.

edge. They belong to a social, that is, shared, reservoir of knowledge from which individual knowledge draws and to which it contributes. For the study of reasoning processes, the situation is thus somewhat similar to that of biology prior to the work of Charles Darwin.

Just as in the study of cognition today, in which cognitive mechanisms are investigated from diverse perspectives (in particular, on the one hand, from an "internalist" perspective, emphasizing the functioning of the brain; and, on the other hand, from an "externalist" perspective, considering the history of ideas and cultural practices), biology in the first half of the nineteenth century was split into a variety of mutually unrelated subdisciplines, from botany to zoology, and from morphology to paleontology. Darwin's theory of evolution radically altered this situation by making it possible to establish systematic conceptual links between "contextual" factors such as the geographical distribution of species and "internal" factors such as the structures described by morphology. This analogy between pre-Darwinian biology and present-day studies of cognition suggests that the opposition between contextualist and internalist approaches to knowledge may indeed be overcome by a theory of the evolution of knowledge.

As I argue in chapter 14, more recent developments within evolutionary theory provide an even better framework for bridging internalist and contextualist perspectives. Insights into the role of complex regulatory structures from genomes to epigenetic and environmental systems have contributed to a mechanistic understanding of the origins and patterns of phenotypic variation, while at the same time the concept of niche construction has captured the dynamic interactions between organisms and their environment. Taken together, these perspectives link traditionally internalist and contextualist approaches as part of complex regulatory and causal networks governing biological systems on both ontogenetic (developmental) and phylogenetic (evolutionary) timescales.

What conditions enabled Darwin to forge conceptual unity from a bundle of mutually unrelated pursuits?[11] An approach called "actualism," introduced by the geologist Charles Lyell,[12] turned out to be crucial because it suggested interpreting history in terms of currently observable processes. In his autobiography, Darwin later remembered: "After my return to England it appeared to me that by following the example of Lyell in Geology, and by collecting all facts which bore in any way on the variation of animals and plants under domestication and nature, some light might perhaps be thrown on the whole subject. . . . I soon perceived that selection was the keystone of man's success in making useful races of animals and plants. But how selection could be applied to organisms living in a state of nature remained for some time a mystery to me."[13] In other words, the historization of nature that is the hallmark of Darwin's achievement was in part based on observations that did not have an obvious bearing on the question of the historical development of species, namely, the practice of breeding and its attendant practical experiences with variation and selection. This practice supplied Darwin with controllable examples of change in biological life forms, which he was able to exploit for a theory in which not man but nature itself took on the role of the breeder.

As far as the historical evolution of knowledge is concerned, we are in a rather similar situation. There is indeed a comparable nonhistorical experiential foundation that can be made fecund for such a theory in a way similar to Darwin's use of breeding; in this case, the foundation consists of studies of cognition (in the widest possible sense), including neuroscience, psychology, educational research, computer science, and cognitive science.

This large field of study offers both a wealth of empirical knowledge and a pool of theoretical models that can be brought to bear on an evolutionary theory of knowledge. That the full potential of this resource has not yet been realized is probably due to the fact that it is not sufficient to simply transfer the results of laboratory studies into the field of history. As was the case during the emergence of the biological theory of evolution, it is necessary to elaborate a genuinely historical conception— which will necessarily challenge many of the theoretical presuppositions implicit in the laboratory studies. In the previous chapter, we saw how Piaget's genetic epistemology could be transformed into an historical epistemology by "historicizing" some of its fundamental concepts.

Beyond the key insight into the relation between representation and reflection, studies of cognition have provided us with a wealth of further insights, for example, into the working of memory and procedural thinking, and into the frameworks of everyday thinking. These insights are based on psychological experiences (modeling human thinking) and to an increasing extent on neuroscience. By using their borrowed conceptual tools, we can solve a number of the fundamental problems of a historical theory of thinking, such as the role of so-called implicit knowledge, the possibility of correcting inferences, reasoning under conditions of incomplete information, and the complementarity of different forms of knowledge.

Mental Models

In the following, I again tap into studies of individual thinking, adapting some of their findings to my purposes. When taking up concepts from cognitive science such as "mental models," we are primarily concerned with them only insofar as they are part of a historically transmitted architecture of shared knowledge that raises questions not usually posed in cognitive science, in particular about their origin and transformation in the context of historically specific experience and about the mechanisms by which they are shared and appropriated in a given society. Cognitive science does not usually deal with social processes, and its concepts and theories account for them insufficiently.

Cognitive scientists have reconstructed surprisingly coherent, powerful, and yet diversified inferential structures of everyday thinking, in the case of qualitative reasoning about physical processes, for example. Such structures provide the underpinning for thinking about physical processes, even in the presence of a developed theory (such as that of classical mechanics). It is always necessary to relate the abstract constructs of a theory to our handling of the material objects to which the theory must ultimately refer. It is a well-known pitfall of science education that the difficulties of applying an abstract theory to a concrete problem are often due to incongruences between the mental models governing the qualitative thinking about the problem and the conceptual structures of the theory.

Studies in cognitive science have shown that even in early childhood the realization that objects can be moved by exerting force on them leads to the inverse conception that every perceived movement of an object must have been caused by a mover exerting a force upon it. In other words, the experience of forces as causes of motions is converted into an interpretation of perceived motions as being caused by force exerted by a mover. At the same time, it is expected that (under otherwise equal conditions) a greater force must also cause a stronger motion. This set of expectations may be viewed as a typical example of the formation of a mental model—which, in this case, is called the "motion-implies-force model."[14]

The experiences represented by this model are so general that we might venture that the motion-implies-force model is probably acquired by people in each and every culture in the course of their ontogenetic development. It specifies a general mental model of motion whose variables are the mover, the moved object, and the motion performed. Whenever this intuitively acquired model is retrieved by the perception of changes, it requires the identification of the mover, the moved object, and the motion of the object in the given situation. If any of these three variables cannot be instantiated, the application of the model fails and the perceived changes are not recognized as the motions of an object. These variables are therefore also called the "critical variables" of the model.

A Children's Story

The example suggests adopting a conception according to which knowledge obtained from experience is represented by a cognitive structure that comprises a relatively stable network of possible inferences relating inputs that are variable. The term "slots" is used to indicate the nodes in the structure that are filled with variable inputs that have to satisfy specific constraints. The basic idea goes back to an investigation of the question of how children understand stories in Eugene Charniak's 1972 dissertation, with insights later extended and publicized by Marvin Minsky, his thesis adviser, who introduced the notion of "frames" for such knowledge structures.[15]

Mental models or frames offer a different perspective on the same cognitive structures that were discussed in the previous chapter. At first glance, the notion may seem to be oversimplified, suggesting that we are reducing cognition to some kind of mechanical slot machine into which various contents can be inserted without affecting the machine itself. Previously, and in contrast, I have emphasized that cognitive structures change with every implementation. Enriching a concept with a new experience may change the entire semantic network of which it is part. If the concept of force, for instance, is no longer related simply to motion—as in the motion-implies-force mental model—but rather to the concept of acceleration, as in Newtonian mechanics, its semantics are profoundly changed. "Inertia" no longer means a tendency to come to rest but now refers to the preservation of a state of motion. Clearly, this semantic change was a major transformation, achieved in the course of the transition from Aristotelian to classical physics, a process that extended over centuries and to which I shall return below. Nevertheless, as we shall see, the slot machine metaphor associated with concepts such as frames or mental models turns out to be quite useful for understanding some important details of this process. It helps to account, in particular, for the way in which prior knowledge informs how we judge a situation, make a scientific argument, or simply understand a text.

Let us assume that a story in a children's book begins with the following words:

It was Paul's birthday. Jane and Alexander went to get presents.

"Oh, look!" Jane said. "I'll get him a kite."

"No, don't," Alex responded. "He already has one. He'll make you take it back."

Let us presume further that questions like the following were posed to the reader:

Why are Jane and Alex buying presents? Where did Jane and Alex go? What did Alex refer to by the word "it" in his last sentence?

Most people familiar with the tradition of celebrating birthdays will be able to answer these questions spontaneously: Jane and Alexander bought gifts because it was Paul's birthday; to do so they went to a shop; and the word "it" refers to the kite that Jane would have to return. Most people would also be convinced that these answers are contained in the paragraph they read. Both of these facts are actually somewhat surprising. None of the answers is contained directly in the text; they have to be inferred from the information given.

The form this reasoning takes, however, does not correspond to the traditional understanding of deduction; for, in the sense of formal logic, the answers cannot be concluded without introducing additional assumptions. The answers are actually obtained from the information contained in the text by linking it with previous experiences. The example suggests the existence of a mental model of a birthday whose variables must be instantiated using information either from the given situation or from prior knowledge before conclusions can be drawn.

Default Settings

The slot fillers or "settings" may have different origins, such as empirical evidence, reasonable expectations, or a preliminary hypothesis, or they may be implicitly determined by other reasoning processes. In particular, they may also result from prior experience or prior reasoning. In this case, we speak of "default settings" or "default assumptions," referring, for instance, to plausible assumptions that represent previous experiences. Once a mental model is instantiated, that is, if the input information satisfies the constraints of the slots, the reasoning about the object or process is to a large extent determined by the mental model.

Consider again the motion-implies-force model. It may be the case, for instance, that the force causing a motion is not immediately visible, as in the case of a car in which the engine is under the hood. But even if we do not see it, we still believe that a car moving on the street is actually driven by an engine—we simply supply this information from prior experience. The perception of a motion is thus linked with structured knowledge stemming from such previous experiences. This knowledge allows conclusions to be drawn about the given situation that may extend far beyond what was directly perceived.

Scientific communication would be just as unthinkable without such default assumptions, as would be human communication in any other area involving external representations—like the spoken or written word. Default assumptions supplement the always incomplete specification of the objects of scientific investigation, or, for that matter, of that given by any external representation.

4.2. An absurd timetable. Drawing by Laurent Taudin.

4.3. Is this Tweety? Drawing by Laurent Taudin.

Take, for example, the task of ascertaining whether there is a train from Boston to New York at 10 a.m. At Boston Union Station, we find a timetable that lists no such train. We conclude that no train from Boston to New York leaves at that time, although the timetable does not explicitly say so. There is not even a note in the timetable informing us that unscheduled trains do not run. It is simply assumed that the reader has learned to supplement timetable data with the default assumption that schedules do not exist if they are not listed. Knowledge about the world in which we live, a precondition for understanding scientific theories, is of course not confined to such simple default assumptions. In the past it has, however, often been neglected by historians of science, and it is rarely subjected to a deeper analysis in terms of cognitive structures.

Conceptions such as that of a frame or mental model allow us not only to grasp deductions based on incomplete information but also make it possible to conceive changes in conclusions as a consequence of changes to context.

Default assumptions may indeed be easily corrected without questioning the entire framework of reasoning. A popular example is a bird by the name of Tweety.[16] From the information that Tweety is a bird we infer that Tweety can fly, although nothing of the kind is stated explicitly. We just know that normally birds can fly. But if we are also told that Tweety is a penguin, we are quite ready to retract the previous inference without abandoning any of our previous premises. We are still convinced that Tweety is a bird, and that typically birds can fly. The statement that birds can fly was not given explicitly in this example; it was a supplement from our prior experience. It represents a standard assumption about birds that allows inferences to transcend the actually available, incomplete information. Thus, the mental model of a bird inherits certain default assumptions—that it needs food, for example—from the higher-order model of a living being.

Mental Models in Scientific Thinking

This type of everyday inference corresponds to what we frequently observe in scientific reasoning. How does a scientist react when an inferred event does not take place, or when a theoretically predicted effect does not occur in an experiment? Typically, a scientific theory is not abandoned but altered in accordance with the unexpected observation.

Take, for example, the theory of the ether, the basis of optics and electromagnetism in the nineteenth century. The concept of the ether was shaped according to the familiar mental model of a medium carrying waves, analogous to air as the medium for sound waves. If one assumes an ether at rest as the carrier of electromagnetic waves like light, then it ought to be possible to experimentally detect the motion of the earth with respect to this ether. But attempts to actually measure such an "ether wind" failed. At first, however, this did not lead to an abandonment of the underlying mental model, but rather to specific assumptions about the behavior of bodies in the ether—for example, a length contraction—explaining why their motion with regard to the ether remains unnoticeable.

Mental models are typically context specific and not universally valid. They therefore allow for the theoretical description and explanation of object-specific frameworks of reasoning and their changes through history. Mental models bridge various levels of knowledge that represent the same object in various forms, ranging, for instance, from the technical knowledge of practitioners up to scientists' theories. Mental models therefore make it possible to grasp implicit conclusions that are embodied in the logic of practitioners' actions but not recorded explicitly in the form of language or writing.

How do mental models help us to explain changes in systems of knowledge? Just as I have claimed for cognitive structures in general, the assimilation of different objects and processes into a mental model and the outcome of such instantiations are constantly changing the model by enriching its reservoir of experiences. New default settings may become available; different instantiations may become related to each other.

The result is gradual adaptation and modification according to a differentiation of mental models, their tailoring to different contexts, and their occasional integration. The applications of mental models may also become the object of reasoning that produces new knowledge in the sense of higher-order knowledge. This may occur as an immediate accommodation of the model in reaction to insufficient fit in a given situation, but also by its deliberate reorganization as a result of accumulated higher-order knowledge obtained by reflection. When a mental model does not fit, the object of cognition may be assimilated to another model or the model may be tailored to the new experience.

It is possible that more than one mental model is appropriate for application to a specific object or process. In this case, different mental models are linked to each other or, so to say, "integrated" or "networked" by application to the same subject. Thus, originally independent domains of reasoning may become connected through the object to which different mental models are applied. This may result in complex knowledge representations—but it may also lead to insurmountable contradictions. Another source of change is the relation of mental models to their external representations. A material model may serve as an external representation of a mental model. Such a material model supports the use of the corresponding mental model. Exploring a material model may, however, also reveal features that had not been included in its initial conception as the materialization of a mental model. Furthermore, the semiotic rules governing the employment of a material model may run into conflict with the running of the corresponding mental model, provoking challenging problems of interpretation.

Because of all of these characteristics, concepts like that of the mental model borrowed from cognitive science are more suitable to account for transformations of knowledge than are theories based on a strict differentiation between thought processes and empirical experiences. In fact, the conception of frames and mental models was itself developed in opposition to the assumption that human thought could ultimately be traced back to the laws of a universal, formal logic or linguistic theory.

How Do Systems of Knowledge Evolve?

Against the background of this preliminary account of the role of mental models, I want to come back to systems of knowledge at large. Systems of knowledge, being part of the shared knowledge of a group or a society (an epistemic community or a thought collective in the sense of Ludwig Fleck), tend to be simultaneously robust and mercurial. On the one hand, they need to be robust if the life or even the survival of a group or a society depends on them. They are typically stabilized by being embedded in societal institutions but also by the power of accumulated collective experiences; institutional frameworks for the transmission of knowledge may involve correctives and control procedures guaranteeing the stability of the transmitted knowledge system. On the other hand, systems of knowledge also tend to be flexible and capable of adapting to new circumstances, that is, they possess precisely the features identified in mental models that constitute the typical building blocks of systems of knowledge.

How do systems of knowledge change? Systems of knowledge continue to change in a gradual way through the application and exploration of a community of practitioners that is itself constantly changing. Systems of knowledge are never conceived in their entirety, that is, in full flower of their potential conclusions and applications; they actually only unfold with the intellectual practice in which they are embedded. In 1915, for instance, the general theory of relativity was little more than a set of equations established by Einstein. Only its elaboration and exploration through the efforts of a growing community of practitioners created the framework for astrophysics and cosmology that allowed for the recognition of such phenomena as black holes and gravitational waves and the ability to probe the empirical validity of such concepts. I come back to this example in chapters 7 and 13. Eventually, a system of knowledge may become so stable that its structures often overpower any impact from individual contributions by the sheer mass of the experiences it reflects. While individual thinking is governed to a large degree by these shared resources, it may, however, also affect them in turn, amplifying them and occasionally even changing their structures. Otherwise the theory of relativity would have never been created.

The fundamental structures of systems of knowledge may indeed undergo major revisions. Over the course of history, novel concepts such as number, weight, inertia, chemical element, and genetics emerge, while natural motion, ether, or phlogiston disappear, familiar today only to historians of science. The wax and wane of such fundamental concepts, which have often played a pivotal role for entire conceptions of the world, marks the cognitive side of these major architectural changes.

Evidently, such major transformations typically also involve changes in the social organization and external representation of knowledge systems. At least in hindsight, major architectural changes to systems of knowledge often appear as disruptive events. This has been a favorite subject of the history of science. It has been described in terms of scientific revolutions and associated with notions such as breakthrough discoveries or paradigm shifts, and in particular with outstanding scientists whose names are attached to such innovations. The belief that such breakthroughs are primarily based on ingenious insights by single individuals is as widely spread as the suspicion that they are unimaginable without some dose of irrationality. Great scientists themselves have contributed in no small measure to the mythology of discovery, often indulging in recollections of crucial moments—overnight, in the course of a conversation, or during a vacation—when the decisive thought supposedly crystalized. How does this emphasis on disruptive change fit with the perspective on the long-term development of shared knowledge? In order to answer this question, I shall proceed in three steps, first discussing triggers of change, then exploration processes, and finally the reorganization of a knowledge system.

Triggers of Change

Triggers of change may arise in all three dimensions of a knowledge system: the mental, the material, and the social. Humans' interactions with their environment continuously change the material world around them, thus creating new objects of cognition as well as new media and knowledge representations that may affect the architecture of a knowledge system. There is always individual variation when new experiences are assimilated to a system of knowledge.

Transformations of knowledge, and of scientific knowledge in particular, may be triggered by an extension of the range of experience, by the introduction of new knowledge representations, and in particular by conflicts, which may have various origins and often roots going far back into history. But the transformation of knowledge systems does not follow a generic scheme valid for all cultural and historical contexts. To what extent new experiences are simply obliterated by the corrective mechanisms of an institutionalized knowledge system or taken as triggers for its further development depends on the specifics of a given historical situation, and, in particular, on the knowledge economy of a society, a concept that I discuss in greater detail in chapter 8. New knowledge can be a stimulus of change, but how it acts depends on the control structures and images of knowledge in a given knowledge economy. Even when novelty, innovation, or curiosity is negatively valued, new knowledge is nevertheless often implicitly assimilated into a given system of knowledge, while it may at the same time be explicitly rejected, or reclassified as "ancient" or "lost" knowledge.

Nevertheless, it is worthwhile investigating characteristic examples of triggers of change in order to analyze the developmental dynamics of systems of knowledge. One such example is what we have called "challenging objects." Challenging objects are phenomena, artifacts, or other parts of material culture that confront existing theoretical frameworks with explanatory tasks that cannot be accomplished through the available conceptual means, thus triggering their further development and ultimately their transformation. Challenging objects typically embody other forms of knowledge—for instance, the knowledge needed to generate the phenomena in question or the practical knowledge of artisans to invent, produce, or make use of such objects. Challenging objects act on systems of knowledge in specific ways shaped by both their material nature and the concrete contexts in which they occur, so that the materiality and accidental character of these contexts may leave profound traces on the further evolution of knowledge systems.

Challenging objects played, for instance, a key role in the rise of early modern science. These objects had their origin predominantly in the rapidly developing technology of the day.[17] Examples include the pendulum and the flywheel used in machine technology, or the understanding of projectile motion for artillery. They were the focus of scholars and engineer-scientists of the period who attempted to explain them in the context of the available systems of knowledge, such as that of Aristotelian natural philosophy or Archimedean mechanics, both highly developed systems. The close observation of projectile motion raised, for instance, the question of how to explain the continuation of the motion of the projectile after it had lost contact with the original mover, such as the throwing hand or the cannon. From the point of view of Aristotelian physics, this was a tricky question indeed. The relevance of projectile motion to military technology in the early modern period also made it an urgent question, turning projectile motion into a challenging object. I return to this problem at greater length in chapter 6.

The theoretical frameworks employed in the investigation of such challenging objects largely determined the possible theoretical questions and answers. The concepts of the extant theoretical frameworks were probed in their application to these objects, leading to new results as well as internal inconsistencies, driving the transformation of these frameworks into what eventually became classical mechanics. Thus, Galileo's new science of motion, for instance, can be conceived as resulting from a struggle with the challenges represented by the pendulum and the motion of a projectile. Galileo related both to the study of accelerated motions along inclined planes, which thus became another challenging object at the center of his new science of motion.

The transition from Aristotelian natural philosophy and Archimedean mechanics to classical mechanics, triggered by the challenging objects of contemporary technology, suggests that new knowledge systems develop historically out of preexisting knowledge systems. Furthermore, a transition between two systems of knowledge significantly involves several layers of knowledge, in this case not only scientific theories but also practical knowledge about challenging objects, for instance, the knowledge of projectile motion accumulated by practitioners of ballistics.

The second example of a trigger of change is what I call a "borderline problem." In a sense, borderline problems are also challenging objects, but with the additional characteristic that they belong to distinct systems of knowledge that they bring into contact and sometimes conflict with each other, triggering their integration and reorganization. Borderline problems are particularly relevant to understanding changes in highly developed, disciplinary science, since here we often encounter a situation in which an object or problem falls under the domain of application of more than one system of knowledge. An example that I discuss in greater detail in chapter 7 is the problem of heat radiation, which became a trigger of the so-called quantum revolution of modern physics around 1900. This problem simultaneously belonged to two different subdisciplines of contemporary physics: the theory of heat and radiation theory. These subdisciplines had independent conceptual foundations that were brought into contact in their application to this specific problem, which fell under both their domains of application. The further exploration of the problem of heat radiation by scientists such as Max Planck and Albert Einstein then led to concrete insights that could not be easily assimilated into either domain but that were supported by empirical knowledge gained through experimental investigation.[18]

Now we come to the next step, the developmental dynamics of a system of knowledge driven by triggers of change. Explorations of a system of knowledge are rarely random; they typically have a specific direction imposed on them by triggers of change. Triggers and the ensuing exploration of existing systems of knowledge may result in internal tensions within these frameworks, including ambiguities, often associated with the proliferation of alternative solutions, paradoxes, and even contradictions—in short, in a loss of systematicity. The many well-known paradoxes and puzzles of the emerging quantum theory may serve as an illustration of this state of development.

Triggers of change, such as challenging objects or borderline problems, provoke the emergence of alternative perspectives that manifest themselves in intellectual controversies.[19] Gideon Freudenthal has characterized such controversies as persistent disagreements concerning substantial intellectual issues. They are ubiquitous in all historical periods. They typically involve the shared knowledge of the participants and presuppose a common structuring of this knowledge by shared conceptual systems. They often arise because the discussants adopt different interpretations of the shared framework and draw different conclusions therefrom.

This can happen within a single system of knowledge, but it is all the more likely when the relevant knowledge is shared by diverse conceptual systems or alternatives in foundational concepts that serve as starting points for conceptualizing the given phenomenon. In any case, in the course of the exploration of the shared knowledge, partial differences of subjective meaning may arise. The fact that these differences are only partial allows a meaningful exchange over the course of the controversy, while the very existence of these differences makes the controversy itself unavoidable.

One example of a scientific controversy from the early modern period was the so-called *vis viva* controversy toward the end of the seventeenth century. It was concerned with the question of which causal agent—what we call today force, energy, or momentum—produces certain mechanical effects and which is conserved in mechanical interactions. This debate, launched by a critique of Descartes by Leibniz, eventually led to the insight that the traditional term "the force of a body in motion" ambiguously refers to two distinct causal agents (energy and momentum) and that there are conservation laws for both.[20] It is a key achievement of classical mechanics.

In the process of exploring the available means of a system of knowledge, alternatives to an established framework may emerge. They are the more viable the richer their experiential basis, which in turn is largely provided by that established framework. The elaboration of alternatives (with the help of the means the established framework provides) often results not in the abandonment of the established solution but in its reconceptualization in new and different terms, for instance, with the help of a new formulation. Such reformulations have a double function: they serve to assimilate extant but unintegrated experiences into a system of knowledge, thus rooting it even more deeply in experience; and they may also become the starting point for novel experiences and, possibly, for the eventual overcoming of the established system.

Typically, scientific controversies are not resolved by victory, but rather by the development and subsequent transformation of a system of knowledge into something new—something in which the original question has changed or even lost its meaning. But even when no party clearly prevails, one of the opposing positions may have a greater impact on the emergence of a novel system. In any case, both antagonistic positions can be recognized, in hindsight, as alternative interpretations of the same underlying foundation of shared knowledge. As I have emphasized before, this is precisely the hallmark of a more developed system of knowledge: it allows for a reconstruction of earlier positions while it cannot itself be expressed in terms of the previous framework.

From this perspective, a number of features may be recognized as characteristic of a scientific controversy, such as the multiplication of examples in one's favor, attempts to reconstruct the adversary's position from one's own perspective, but also unavoidable instances of misunderstanding and a shift toward a more reflective stance. In sum, these thrusts and parries constitute the exploration of the limits of a system of knowledge.

As I have also emphasized above, such systems are never established in their entirety from the outset, that is, in all of their potential applications and conclusions. The evolution of knowledge proceeds through the exploration of shared knowledge in a community of practitioners. Controversies are one essential form in which this evolution takes place. The effectiveness of these controversies may depend on the specific historical conditions, be they material, social, or intellectual, favoring or impeding their productivity. As a result, some controversies may be resolved, in the sense outlined above, in a very short time, while others may extend over centuries.

The omnipresence of controversies and rivaling theories in science may, however, suggest a greater variability than a closer examination of its developmental character reveals. In fact, however, the development of knowledge is much more constrained than a history of ideas can account for. It is largely shaped by a historically granted set of means, which accounts for the path dependency of the development of a system of knowledge. This path dependency is in part due to a winner-takes-all logic according to which any established solution—"established" both in an intellectual and an institutional sense, for example, Newtonian classical mechanics—is typically stabilized and extended through the assimilation of a maximal range of experiences, thereby gaining an advantage over conceivable alternatives (e.g., Leibnizian mechanics), which will never be granted a similar chance of being implemented and institutionalized.

In the course of the exploration of a system of knowledge, whether driven by an individual investigation or by exchanges within a community of practitioners, the degree of systematization may vary over the various parts and domains covered by the system. If we conceive a system of knowledge as involving a network of epistemic operations (arguments and inferences, applications to specific problems, hypothesis building, constructions and calculations, experimental practices, etc.), then some parts of it may be more closely interwoven than others. Systems of knowledge typically have a tight-knit core of mental models, conceptual frameworks, patterns of argumentation, practices, instruments, applications, and results that show a high systematicity and remain stable for a longer time. In Aristotelian physics, for instance, the classification of different causes, the distinction between celestial and terrestrial motions, and the subdivision of the latter into natural and violent motions formed such a core, which remained stable over centuries.

In the course of the exploration of a system of knowledge, new clusters of insights, results, or activities may develop at a certain distance from this core, in the sense of being only loosely coupled to it, by nonstandard applications, by unfamiliar fillings of the slots of mental models, by weak or daring hypotheses, or by arguments to which there are plausible alternatives. These "epistemic islands" may, however, develop their own systematicity. Such an epistemic island was constituted, for instance, by Galileo's theory of motion, which (as I discuss in greater detail in the following chapters) emerged from a transformation of Aristotelian physics triggered by challenging objects such as the pendulum and the motion of projectiles. Initially, this epistemic island was nothing more than an insight into the parabolic shape of projectile trajectory, the law of fall, as well as a few theorems about motion along inclined planes. From the perspective of Aristotelian natural philosophy, these insights could be considered as being of marginal significance—however problematical it may have been to assimilate them to its framework. The discrete insights were, on the other hand, related to one another, but did not yet constitute a theoretical system with a comparable range as that of Aristotelian natural philosophy. Only as a consequence of the further accumulation of knowledge around this epistemic island did a broader new system of knowledge arise, in this case in the form of classical mechanics.

The Reorganization of Knowledge

This brings me to the third step, the establishment of a new system of knowledge by reorganizing accumulated knowledge. Such a reorganization may take many different pathways. Here I will focus on the special but important case in which an epistemic island serves as the seedbed or matrix for a new conceptual framework. The "matrix" refers to the set of results gleaned from exploring a given system of knowledge from which the new system eventually emerges. Although an epistemic island acting as a matrix does not yet have the internal systematicity and general applicability of a fully developed system of knowledge, it does have the potential to integrate further knowledge. The result of this integration is what in hindsight may appear to be an intermediate construction, the scaffolding for the final result of a knowledge transformation, serving as the actual matrix for the new conceptual framework.[21] The very existence of such intermediate stages makes it clear that structural changes in systems of knowledge are neither disruptive breaks nor merely cumulative processes, but protracted reorganizations of knowledge. The preclassical mechanics of Galileo's time, Hendrik Antoon Lorentz's electron theory, the so-called old quantum theory of Niels Bohr and Arnold Sommerfeld, as well as the preliminary gravitational theory of Einstein and Marcel Grossmann are typical examples of such intermediate stages in the history of physics.[22]

How does an epistemic matrix engender the transformation of a system of knowledge? Because it is epistemically isolated from the core of the existing knowledge system, it is liable to a reinterpretation that is not confined to the overarching principles of the original system. The very need for such a reinterpretation may be suggested by this distance from the center of a conceptual framework. Consider, for instance, the introduction of a "length contraction" as a problematic ad hoc hypothesis in Lorentz's electron theory (another example is investigated in greater detail in chapter 7). The possible reinterpretation of the matrix, on the other hand, is suggested by the external representations of the knowledge encapsulated by the matrix, for example, the famous Lorentz transformations mathematically describing length contraction. A reflective abstraction induced by operating with such external representations may result in cognitive structures different from the ones at the core of the preexisting system of knowledge, in this case the new concepts of space and time in Einstein's special relativity. The relevant external representations themselves typically result from pushing a given cognitive framework to its limits, in this case those of classical electrodynamics.

These limits are made evident by the previously mentioned signs of crumbling systematicity and increasing distance from a well-functioning core. The external representations of the island knowledge may then act as a starting point for the construction of the new conceptual framework. The matrix is thus the ultimate and often problematic terminus of the old framework; it constitutes, at the same time, the core of the new one. We thus recognize once more the functioning of external representations as a bridge between the old and the new, ensuring the continuity of knowledge evolution—even if this continuity is here linked with a transformation of cognitive architecture that is itself grounded in external representations.

The transformation of a system of knowledge is also a matter of perspective. A new system of knowledge emerges from existing knowledge to which it remains related in many ways. An emergent system of knowledge—as it is embodied in what I have called the "matrix"—can therefore be seen from two angles: as an advanced post of the old system or as the nucleus of something new. Which of these two

4.4. The starting point for a new system of knowledge is often found within an existing conceptual network when it is being picked up at a different point, as when Einstein reinterpreted Planck's work from a new perspective. Drawing by Laurent Taudin.

perspectives is taken by a historical actor may be a generational question, as has often been emphasized. But it may also be a question of centrality or marginality within a field of investigation, or a question of general overview versus focused specialization. Different perspectives are not simply presented but emerge in the course of the exploration of a system of knowledge. Since such an exploration involves individually differentiated learning processes, a multiplicity of perspectives may evolve as the exploration runs its course.

The transition from an old to a new system of knowledge is typically attended by a shift of the intellectual center of gravity. We may call this process "re-centering." What was initially peripheral, whether as a domain of application, as an argument, or as auxiliary concepts or constructions, may now come to the center and emerge as part of the constitution and core of a new knowledge system. Such re-centering processes are often accompanied by a reversal of the direction of inferences: what had been a tediously achieved result in the old system may turn into a premise in the new system.

The transformation of knowledge systems may have surprisingly far-reaching consequences, well beyond what the historical actors expected. Some transformations may start from a rather specific, local problem and then touch on fundamental issues of amazing generality, while others stay within the same level of generality as the original problem. This diversity is a consequence of the different layers of knowledge involved in such transitions. In some way, all problems somehow involve the most general frameworks, which act as background knowledge for their investigation, such as our understanding of abstract concepts like space, time, bodies, matter, causation, as so forth.

Mostly, however, this knowledge remains just that, background knowledge that does not itself become an issue. Under certain, rather specific conditions, however, emerging knowledge systems will resonate with these general background structures and lead to the transformation of abstract concepts such as space and time.[23] For an emerging knowledge system to impact general background structures, it must adhere to a description of such abstract concepts on the level of its external representations that is both consistent and distinct from the preexisting one. As we shall see in chapter 7, this was the case, for instance, in the transition from classical to relativistic physics.

In a sense, we have now come full circle. We started out, in the beginning of the last chapter, with the question of the nature and development of abstract concepts. We have now seen how much it actually takes to change existing abstract concepts (of space and time, for instance) and for the changes to become part of a new comprehensive framework.

Systems of knowledge do not in fact change in one fell swoop. In particular, there are no scientific revolutions, properly speaking. Only after long periods of exploration, in the course of the assimilation of new experiences and owing to an ongoing internal reorganization, do systems of knowledge reach a certain stability and universal applicability. For instance, the classical mechanics ascribed to Galileo and Newton reached such a state only after its reformulation in the eighteenth century by Joseph-Louis de Lagrange, Leonard Euler, and other representatives of "analytical mechanics"; Darwin's theory of evolution only after the "new synthesis" of the early twentieth century; and Einstein's theory of general relativity only after its "renaissance" in the second half of the twentieth century (an example to which I return in chapter 13).

Although all of these cases are quite different, we have begun to identify generic aspects of the developmental dynamics of systems of knowledge. In particular, I have pointed to the role of triggers of change, to the role of individual and societal exploration processes of the given systems of knowledge and their external representations, as well as to the extended processes of knowledge reorganization. These processes require robust and flexible mental models as important components of systems of knowledge, and they depend on those models' layered structure of intuitive, practical, and theoretical knowledge. History does not repeat itself, and not all transformations of knowledge systems follow the same pattern. Yet, these specific features may hint at more general aspects of the evolution of knowledge—not in the sense of a generic scheme of transformation, but in the sense of revealing mechanisms and criteria that may be relevant to other cases as well and may therefore be used as part of a tool kit for analyzing them. To which extent this is indeed the case, I examine in the next three chapters.

Chapter 5

External Representations at Work

External representations are tools of thought that mediate between the material and imaginary worlds and also create a reality of their own.[1] In this chapter, I consider four important examples from the history of knowledge that illustrate the pivotal role of external representations for the transformation of knowledge systems.[2] These examples will also serve as illustrations of different types of external representations, from symbolic systems to diagrams. The first two examples concern the emergence of arithmetic and writing in ancient Mesopotamia. The third example deals with the invention of chemical formulas. The fourth example illustrates the role of diagrams in shaping a conceptual system for quantifying processes of change. In all four cases, the transformations made possible by the respective external representations had a lasting impact on the history of knowledge. Even today, their influence can be detected in the way that we write and calculate, describe chemical reactions, or understand processes of change.

The Emergence of Arithmetic

The research of Peter Damerow and his colleagues on the emergence of writing and arithmetic in ancient Mesopotamia has been paradigmatic for understanding the role of external representations in the historical genesis of abstract concepts.[3] I therefore begin my survey of historical examples with the emergence of arithmetic and writing. I intend to illustrate the role of external representations in the genesis of an abstract number concept and in the evolution of a writing system suitable to represent spoken language from earlier context-dependent forms of accounting and administration.

Indeed, when writing and arithmetic emerged during the urban revolution of the fourth millennium BCE, they did not immediately give rise to abstract numbers or a general-purpose writing system but rather to techniques appropriate to specific contexts and mental models of administrative activities. The emergence of these mental models was closely tied to the development of symbolic systems serving to

Timeline of Early Writing[5]		
BCE	8500	Simple tokens in the Middle East
	3500	Complex tokens and clay envelopes
	3400	Numerical tablets
	3300–3200	Earliest writing in Mesopotamia and Egypt
	3200–3000	Egyptian Hieratic script
	3100	Proto-Elamite script
	2500	Adaptation of cuneiform to write Semitic languages in Mesopotamia and Syria
	1850	Proto-Sinaitic alphabetic texts
	1650	Hittite cuneiform
	1600	Earliest Proto-Canaanite alphabetic inscription
	1400–700	Anatolian hieroglyphic script in use
	1250	Ugaritic alphabet
	1200	Oracle-bone inscriptions, China
	1200–600	Development of Olmec writing
	1000	Phoenician alphabet
	900–600	Old Aramaic inscriptions
	800	First Greek inscriptions
	800	South Arabian script
	650	Egyptian Demotic script
	600–200	Zapotec writing at Monte Alban
	400–200	Earliest Mayan writing
	250	Jewish square script, used for Hebrew and Aramaic
	100	Spread of Mayan writing
CE	75	Last dated Assyro-Babylonian cuneiform text
	200–300	Coptic script appears
	394	Last dated hieroglyphic inscription
	452	Last dated Egyptian Demotic graffito
	600–800	Last classic Mayan writing

coordinate collective human actions in complex societies. As we shall see, the emergence of an abstract concept of numbers and a generic writing system followed the sequence familiar from the preceding chapter: triggers of change (in this case the expansion of the administrative economy), the exploration of possibilities offered by existing external representations, and the reorganization of a knowledge system originally centered exclusively on accounting practices.[4]

Let us therefore consider this important example in greater detail, making use of the theoretical toolbox presented in the preceding two chapters. Identifying a concrete object with a name, a sign, or a counter creates a first-order representation of that object. These representations can then be handled in much the same way as the objects themselves; for instance, they can be joined or separated, or be arranged in an order. Historically, the earliest forms of arithmetical activities are distinguished by the use of standard sets of concrete objects or symbols as first-order representations of objects with which a one-to-one relation is established. The quantity of a set of objects can then be represented by a repetition of the counters. When a standardized sequence of words is assigned, in a temporal sequence, to the elements of a given set in a fixed order, a counting sequence emerges. It represents what is called the "ordinal" structure of the given set.

At this earliest level of thinking, which may be characterized as "protoarithmetical," there are no symbolic transformations that correspond to arithmetical operations such as addition or multiplication. Symbolic transformations only apply to the representations of the objects themselves, not to representations of higher-order cognitive structures, which embody a more abstract concept of number. Protoarithmetical forms of thinking are found in many surviving nonliterate cultures. Historically, they are harder to attest but probably go back to the beginning of sedentary agricultural cultivation and animal husbandry.

But on the Mesopotamian plain and in the Persian highlands, clay tokens have been found that go back to the beginning of the eighth millennium. It is conceivable that these clay tokens of various shapes may have served as first-order representations of objects and been used for representing and controlling their quantities in the context of protoarithmetical counting and accounting.[7]

In the second half of the fourth millennium, a transformation and repurposing of older forms of external representation took place. The modest accounting techniques that had been developed in the context of the rural economy of Mesopotamia were extensively exploited in the administration of the emerging city-states. The earliest verifiable counters contained bundled units of higher order (six, ten, sixty) that themselves indicated a release from a one token–one item relationship. Among the traditional accounting techniques were seals representing certain administrative acts, such as documenting the labor of workers or their provision with food. The exploration of these external representations of administrative knowledge in the context of an expanding economy eventually led to a transformation of traditional symbolic culture. The potential of tools of symbolic representation was exploited to its limits—a characteristic feature of the evolution of knowledge, discussed above. This is evident, in particular, from a proliferation of accounting practices, which had played only a minor role in the context of rural communities.

Timeline of Numerical Concepts and Arithmetic Thought[6]

Stage	Dates	Periods	Activity	Characterization
0	Up to 10000 BCE	Approximately until the end of the Mesolithic period	Pre-arithmetical quantification	No arithmetical activities. All judgments about quantities are based on direct comparisons of amounts and sizes. Communication and transmission only by transmittable techniques of comparison and by comparative expressions of language.
1	10000–3000 BCE	Neolithic period and Early Bronze Age	Protoarithmetic	Quantities are precisely identified by one-to-one correspondences. Communication and transmission with the aid of conventionalized counting sequences and tallying systems.
2a	3000–2000 BCE	Period of the early city cultures (until the invention of the sexagesimal place value system in the ancient Near East)	Symbol-based arithmetic with context-dependent symbol systems	Quantities are structured by metrological systems. Communication and transmission of these systems and of the corresponding mental constructs through complex symbol systems and developed techniques for the transformation of symbol configurations.
2b	2000–500 BCE	Period of developed city cultures (in the ancient Near East)	Symbol-based arithmetic with context-independent symbol systems	Quantities are structured by abstract numerical systems with object-independent arithmetical operations. Communication and transmission of these systems by unified, context-independent, but culture-specific symbol systems for the representation of arbitrary quantities, including abstract "rules of calculation." Emergence of first forms of "preclassical mathematics" that are abstract but dependent on culture-specific symbol systems.

3a	500 BCE–late 19th century CE	Classical antiquity, late antiquity, Middle Ages, and early modern era (until the emergence of analytical mathematics)	Concept-based arithmetic with deductions in natural language	Abstract number concept with "a priori" provable properties. Communication and transmission with the aid of a written representation of "propositions" about abstract numbers and their mathematical properties. Propositions are logically ordered and systematically arranged by deductive theories according to the model of Euclid's *Elements*.
3b	Since late 19th century CE	The modern mathematical tradition	Concept-based arithmetic with formal deductions	Formal understanding of arithmetical structures and expansion of the number concept by construction of new arithmetical structures. Communication and transmission with the aid of formal language systems

The symbols used for accounting remained, for the time being, essentially first-order representations of relevant quantities and administrative or economic practices. They now became, however, part of developed symbol systems and served more complex administrative purposes. They did not represent abstract numerical values; rather their meaning and numerical value depended on the context, that is, what they were supposed to count. Simultaneously, symbol transformations were performed that did not represent real actions; they merely illustrated the potential of the symbol system to produce knowledge relevant to the context of accounting.

A critical turning point was reached when the two main elements of traditional accounting techniques—the counters used for keeping track of the quantities of administered objects and the seal impressions documenting the relevant administrative acts—came to be represented within a single medium. The seals, in particular, carried information about the administrative and societal context that determined their meaning. They testified to property, legal acts, or socially correct behavior. The two elements were initially integrated in the form of sealed hollow clay balls (*bullae*) that conveyed certain combinations of clay counters. Occasionally, the combinations of tokens inside were represented by marks on the bullae's surface. In principle, sealed clay tablets served the same function but were easier to handle than the bullae. In any case, two initially separate accounting techniques—the counters and the seals— thus became integrated into a new form of external representation, whose enormous potential could be (and was) explored thereafter.

Representing an early stage of protowriting and protoarithmetic, these sealed tablets became the starting point for the exploration of new forms of information storage and processing in the archaic period of Mesopotamian society. One of the reasons that I focus here on Mesopotamian history rather than on Egyptian arithmetic and writing, for instance, is indeed this rich evidence of intermediate developmental steps eventually leading up to an abstract number concept and a mature writing system.

The tablets could hold more information than the earlier administrative techniques, and this information could be structured more flexibly and efficiently. For instance, it was now easy to invent new signs to denote new semantic categories. Conversely, the existing economic and administrative activities were shaped by these new techniques of representation. As a result of this development, so-called protocuneiform administrative texts became the external representations of a mental model of the accumulation and distribution of resources and products by the administrative officials. This mental model was in turn generated by a reflection on the specific practices of this administration.

Second-order representations of quantities and arithmetical activities resulted in turn from reflection on these protoarithmetical mental models and their symbolic representations. While these quantities and arithmetical activities may have initially retained their context dependency, their meaning was no longer primarily determined by handling the concrete objects themselves but by actions performed with their first-order external representations. Accordingly, the second-order representations are characterized by well-defined systems of numerical symbols and semiotic rules for their application. These formal rules apply directly to the symbols. They may be either implied or explicitly stated. In any case, their application is strict and no longer de-

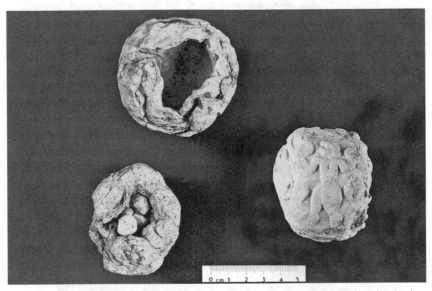

5.1. Depiction of hollow clay bullae, containing small tokens used for counting. They date back to approximately 3500–3300 BCE. Deutsches Archäologisches Museum, Foto: Irmgard Wagner.

pendent on the contingent conditions of the concrete world. In the Mesopotamian context, one encounters a clearly marked stage (corresponding to a period that extends from the first emergence of writing around 3200 BCE until the invention of the sexagesimal place value system by the latest at the end of the Ur III period around 2000 BCE) in which the second-order representations of symbol-based arithmetic and their meanings vary with context.[8] In this period, a generalized system of notation that is standardized and covers all domains is still lacking.

In a subsequent stage, such standardization did occur, leading to a notation into which all context-specific representations could be transformed. Such a generalized representation now allowed for formal operations without any reference to specific applications. In the Mesopotamian context, this stage was reached with the invention of the sexagesimal place value system (by ca. 2050 BCE), which unified all prior forms of context-dependent notation for counting discrete objects. This happened in the course of the unification of the city-states by the empire of the third dynasty of Ur and the corresponding emergence of a centralized administration. As a consequence of the establishment of the sexagesimal place value system, arithmetical techniques were also unified. Reflecting on arithmetical operations in this new context gave rise to technical terms, problem types, and solution strategies that now became the core of what came to be known as Babylonian mathematics.

In summary, arithmetic emerged as a system of rules for handling external representations. No abstract concept of number has to be presupposed as a starting point for its historical development. Similar developments also took place in other cultures. A symbol-based arithmetic emerged in cultures as different as the Chinese,

5.2. Protocuneiform tablet from Uruk (ca. 3000 BCE) showing a calculation for the amounts of ingredients needed in the production of dry cereal products and beer. See Nissen, Damerow, and Englund (1993, 42–43). Courtesy of Cuneiform Digital Library Initiative.

the Egyptian, the Indian, the Central American, and the Mesopotamian cultures.[9] In China, for instance, *The Nine Chapters on the Mathematical Art* goes back to the first millennium BCE.[10] The work refers to operations involving counting rods on a reckoning board.[11]

Such a reckoning board served as the material model of a particular mental model of arithmetical operations; rod numbers mimicked the composition and decomposition of sets of real objects. In Egyptian arithmetic, this role was played by the written hieroglyphic numerals.[12]

How the Practical Roots of Mathematical Knowledge Were Suppressed

Greek philosophers tended to suppress the practical roots of their abstract knowledge. Plato writes in his *Republic* about the relation between theoretical and practical mathematics, belittling the latter:

> Then it would be appropriate, Glaucon, to legislate this subject for those who are going to share in the highest offices in the city and to

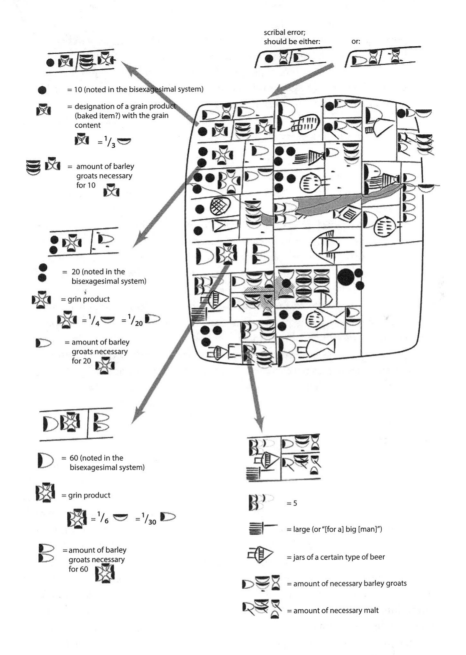

scribal error;
should be either: or:

● = 10 (noted in the bisexagesimal system)

⊠ = designation of a grain product
(baked item?) with the grain
content

⊠ = $\frac{1}{3}$ ◡

⊠ = amount of barley
groats necessary
for 10 ⊠

●● = 20 (noted in the
bisexagesimal system)

⊠ = grin product

⊠ = $\frac{1}{4}$ ◡ = $\frac{1}{20}$ ◗

◗ = amount of barley
groats necessary
for 20 ⊠

◗⊠ = 60 (noted in the
bisexagesimal system)

⊠ = grin product

⊠ = $\frac{1}{6}$ ◡ = $\frac{1}{30}$ ◗

⊟ = amount of barley
groats necessary
for 60 ⊠

= 5

= large (or "[for a] big [man]")

= jars of a certain type of beer

= amount of necessary barley groats

= amount of necessary malt

persuade them to turn to calculation and take it up, not as laymen do, but staying with it until they reach the study of the natures of the numbers by means of understanding itself, nor like tradesmen and retailers, for the sake of buying and selling, but for the sake of war and for ease in turning the soul around, away from becoming and towards truth and being. . . . Then, for all these reasons, this subject isn't to be neglected, and the best natures must be educated in it.[13]

When discussing the roots of writing and arithmetic in the large administrations of early state societies, one should also not forget the exploitation and suffering imposed by the power structures these administrations helped to maintain.[14] The state of the Ur III period (ca. 2100–2000 BCE) was centrally organized and imposed strict controls on all of its resources, including masses of state-owned workers. Children were forced to enter the work process as soon as they could be exploited; elderly laborers had to work until they were incapacitated. Termination of state employment was either by flight into an uncertain existence or by death. The sophisticated administration, which gave rise to so many innovations important to subsequent human history, maximized the exploitation of this workforce. By way of example, consider the recording of day-to-day activities on small tablets—in many ways presaging the time cards and credit memos of modern enterprises— and then the ingenious act of assembling this collected information into larger economic control structures.

The Emergence of Writing

In Babylonia, the development of writing closely paralleled that of arithmetic.[15] The use of the protocuneiform writing system mentioned above was at first entirely governed by its function within the Mesopotamian administration and its growing sophistication. But that very sophistication also created contexts in which new applications of the system could arise and be considered from a new perspective, at some distance from the sphere of primary applications. One such context was education. Such recontextualization is another generic feature of the evolution of knowledge that I have addressed above as "emancipatory reversal." The growing complexity of the system required institutional support for its transmission from generation to generation. But schooling implies a separation of the cognitive means of administration from their immediate context of application. It thus opens up a perspective in which the potential of these cognitive means can be explored independently of the constraints of their application to concrete administrative problems.

This next step in the development of writing was thus shaped by the fundamental property of external representations emphasized in chapter 3: that the range of their possible applications is larger than the specific goals for which they had initially been introduced. In fact, the potential of protocuneiform texts to represent mental constructions reached far beyond the limited field of application in Mesopotamian administration. In its most evolved form—reached by around 2600 BCE—it may already have included the possibility of representing spoken language.[16]

Written language in turn became an important external representation of many other forms of knowledge. I will return to the issue in chapter 10 when discussing the knowledge economy of early literate societies. Here I limit myself to a particular example, related to the further development of mathematics: Writing could be used to encode the results of reflection on arithmetical operations, thus leading, for instance, to explicit linguistic statements on the nature of the concept of number as we find in the works of Greek philosophers such as Plato. The linguistic encoding of mathematical operations led, more generally, to systems of arithmetical propositions that could be ordered in a linear and hierarchical way within a so-called deductive structure.

An example is the definition of and propositions about properties of even and odd numbers, a doctrine from ancient Greece probably going back to the Pythagoreans of the fifth century BCE, and later integrated into Euclid's *Elements*.[17] By the linguistic representation of operations with symbols representing even and odd numbers, the independence of such concepts from the specific properties of a symbol-based arithmetic was substantially enhanced. The conceptual meaning of even and odd numbers was now no longer directly constituted by the arithmetical techniques that had originally given rise to them, but by the application of linguistic techniques. An example of such techniques is the determination of meaning with the help of a definition, for instance, when the concept of prime number is defined with respect to the possibility or impossibility of certain arithmetical operations.

Ever since the transmission of knowledge could rely on writing as an external representation, writing has comprised the structured and often systematized arrangement of terms, problems, recipes, and propositions native to specific areas of knowledge. After the spread of writing in Mesopotamia, new forms of written knowledge emerged, such as grammatical texts, divination texts, lists on various subjects, historiographical texts, healing texts, astronomical texts, and so on. Later, some of this knowledge diffused gradually into the Greek world and became the subject of new forms of written representation that in turn acted as influential models for representing knowledge—including elaborate treatises on philosophical and scientific subjects. The fact that ancient knowledge was encoded in writing encouraged the transmission of complex systems of knowledge to other cultures, such as the Syrian, Persian, and Aramaic worlds and the Latin West, stimulating the creation of new knowledge and the further elaboration of such models.[18]

Writing as a means for representing language has since become an essential feature of most subsequent societies. The invention of writing shows how more or less accidental consequences of historical processes may turn into the indispensable precondition for the functioning and stability of a society, and in particular for its further development. The evolution of knowledge is a highly path-dependent process in which the dynamics at any given stage depend not only on the outcome of the stage preceding, but on the entire evolutionary trajectory back to some of its initial biological and ecological conditions.

The Emergence of Chemical Formulas

Another striking case—analyzed by Ursula Klein—in which a reflection on external representations gave rise to new structures of knowledge is the case of chemical formulas as they were invented in the early nineteenth century by the Swedish chemist Jacob Berzelius.[19] Berzelius introduced formulas such H_2O for water that simultaneously represented basic systematized experimental knowledge in inorganic chemistry and theoretical models of the constitution and behaviors of organic chemical substances. As we shall see (in a way similar to what we have observed in the emergence of arithmetic) these theoretical models were in part the result of a reflection on symbolic operations performed with these chemical formulas used as "paper tools." Such paper tools serve to codify and thus to preserve and transmit certain aspects of knowledge in the wider context of written external representations, without being confined to written language.

A central issue of research at the time was what Berzelius called the "laws of chemical proportion."[20] Based on the earlier work of Joseph-Louis Proust and John Dalton,[21] it had become clear that the elements in inorganic chemical compounds always occur in the same proportion, and that if two elements can form more than one compound, the mass ratios of one element, which combine with a fixed mass of the other element, will always be ratios of small whole numbers. Although these insights gave a boost to chemical atomism, it remained difficult to connect the empirical insights gained in the field of inorganic chemistry with organic chemistry and contemporary physical ideas about the nature of atoms. It took more than a century before a reliable basis for reconciling chemical and physical atomism was found in modern quantum theory. In the meantime, the further evolution of chemical knowledge was shaped by symbol systems more closely related to the realities of chemical experimentation, such as the chemical formulas of Berzelius.

The key entities in a theory of chemical proportions were actually not atoms but rather scale-independent portions among chemical substances.[22] Chemical substances were investigated with regard to their chemical properties and behaviors, independent of their mechanical qualities such as shape, orientation, or size. What therefore mattered quantitatively were scale-independent portions of such substances, which is exactly what Berzelius's formulas represented. This representation simultaneously shaped and stabilized the meaning of such portions and thus contributed to their extension to organic chemistry, independent of contemporary atomistic notions.

While these formulas thus served as first-order external representations of certain chemical manipulations involving discrete, scale-independent portions, they were also the result of reflection on the limits of representing chemical portions by ordinary language and by Daltonian representations of atoms and molecules. Moreover, they were embedded in a wider semantic network comprising notions of chemical elements and compounds, as well as the measurable relations between these elementary constituents, among other things. Their usage in organic chemistry played a constructive role in building up mental models of organic chemical constitution and reaction. The formulas became part of developed symbol systems and served more complex purposes of theory formation. Symbol transformations could now be performed with them that no longer represented real actions but models of action and constitution that could elicit predictions. In this context, they served both as symbols of semantic relations and as icons. In fact, the iconic properties of the Berzelian formulas—for example, the possibility of connecting letters that represent chemical elements with a plus sign—mirror aspects of the mental model represented, for instance, by the constitution of a chemical compound from building blocks of more elementary constituents.

Graphical Representations of Change

A striking example of the role of diagrams used as paper tools, ensuring both continuity and innovation, is provided by a medieval tradition of representing the change of qualities with the help of diagrams. The revival and exploration of this tradition in early modern science helped lay the groundwork for the understanding of motion in classical physics and its conceptualization with the help of infinitesimal calculus.[23]

The method goes back to the fourteenth-century scholar Nicolas Oresme, who introduced it to graphically represent scholastic ideas about the quantitative change of qualities.[24] A quality can be a sensory quality such as whiteness, heat, or motion, but also a moral quality such as virtue or the grace of God. Such qualities may change in time or space over a certain interval. One example might be heat variation along the length of an iron rod. Another would be a motion whose velocity changes over time. Either space or time is considered the extension of the quality. In the case of the rod, the extension would be its length. In the case of a motion varying in speed over time, the extension would be the time.

The varying intensity of the quality is represented by a straight line erected orthogonally at each point of its extension. The length of the vertical line represents the "degree" of the quality at that point of its extension (a common characteristic in this tradition). Connecting all the upper endpoints of the lines of intension one obtains a summit line. Together with the line of extension, the summit line, and the first and the last degree of the changing quality, a two-dimensional figure emerges that represents the change. Its area represents the "quantity of the quality." The figure can now be used to characterize the change of qualities. In the case of a uniform motion over a certain interval of time, for instance, one obtains a rectangular figure representing a "uniform quality." A triangular figure represents a "uniformly difform quality" for which the intension uniformly varies over the extension.[25]

In the medieval context, this method of graphical representation was part of a sophisticated philosophical discourse about change. It was employed in order to analyze sophisms, to clarify theological questions, or to discuss fictional situations—such as what happens when Socrates continuously increases his pace. The graphical aspects of the method bear some similarity to modern representations of integration with the help of two-dimensional diagrams. Consider, for instance, a motion that is described by its velocity as a function of time. In the case of a constant velocity over a certain interval of time, the integral of this function over time would give the

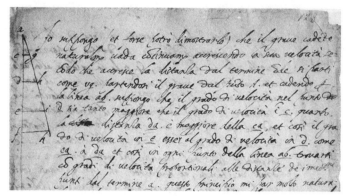

5.3. Section of a manuscript page by Galileo Galilei from ca. 1604 with the sketch of a demonstration of the law of fall based on the use of a medieval representation of change. Ms. Gal. 72, folio 128r, by concession of the Ministerio per i Beni e le Attività Culturali—Italy / Biblioteca Nazionale Centrale di Firenze.

distance traversed by this motion in the given time interval. This integral is represented by the area of a rectangular surface that is determined by the time interval and the horizontal line representing the constant velocity. From a medieval perspective, we would have a uniform quality. In the case of a uniformly accelerated motion, starting from rest at time zero, the velocity function would be given by a straight line through the origin; the corresponding surface would be triangular. From a medieval perspective, we would now have a uniformly difform quality.

Conceptually, however, the medieval method is very different from modern calculus or classical mechanics. It involves neither a concept of function nor a concept of integration, while the medieval notions of extension, intension, and degree of a quality are alien to the modern framework. Yet, in the early modern period, protagonists of preclassical science such as Galileo Galilei, René Descartes, Isaac Beeckman, or Thomas Harriot all relied on the medieval tradition of representing qualities in order to tackle the same challenging problems that would be solved by calculus in later classical mechanics.[26] An example is the derivation of the law of fall, that is, the quadratic increase of the distances traversed by a falling body with the time elapsed, from the assumption of uniform acceleration. How can the emergence of modern concepts of motion (involving other concepts like momentary and average velocity) be explained, and what was the role of the medieval tradition for that emergence?

The answer is provided by the principle, discussed above, that external representations, as tools of thinking, have a wider horizon of applicability than that for which they had originally been introduced. Early modern scientists repurposed the existing external representations to address challenges emerging from a new context. As discussed in the previous chapter, this context was characterized by an increasing entanglement of practical and theoretical knowledge, mobilized in order to treat the challenging objects of contemporary technology such as projectile motion, the motion of a pendulum, or the effect of machines. Here we encounter medieval external

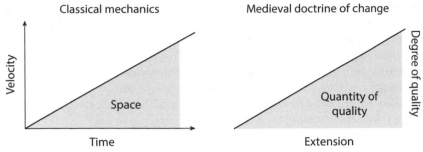

Classical mechanics

Medieval doctrine of change

5.4. Different representations of change. In classical mechanics (*left*) the increase of velocity in a uniformly accelerated motion of fall can be represented by a diagram showing velocity as a linear function of time and the space traversed as an area representing the integral of that function. In the medieval doctrine of changing qualities (*right*), velocity can be represented as a "uniformly difform" or "triangular" quality with its extension (time) and its changing degrees. While the two conceptualizations are different, the diagrams are similar, so the graphical representation could act as a bridge between them.

representations serving as a conduit from one conceptual system to another, while themselves being reinterpreted in the process.

The medieval conceptualization of change indeed helped to prepare (through novel applications by the protagonists of early modern science) the groundwork for the emergence of classical mechanics and modern calculus. One element of continuity between the two conceptual systems is the iconic quality of the medieval diagram. In the words of Charles Sanders Peirce, a diagram is "in the main an Icon of the forms of relations in the constitution of its Object, the appropriateness of it for the representation of necessary inference is easily seen."[27] The diagram representing, for instance, uniform acceleration captures, through its geometrical arrangement, the essential features of such a motion, whatever the conceptual milieu in which it is embedded. As a consequence of what I have called above the generative ambiguity of external representations, the extended application of such diagrams in the early modern period, in new contexts and for new purposes, fostered the emergence of new conceptual relations.

The study of motion was indeed central to early modern science, and different arrangements were considered in real experiments and thought experiments alike. For instance, an accelerated motion of fall could be converted into a uniform motion by deflecting it into the horizontal so that the motion continued along a horizontal plane. This horizontal motion would then continue with a constant velocity corresponding to what in the medieval framework would be designated as the "last degree" of the accelerated motion. In this way, the obscure medieval notion of "degree" could be interpreted in terms of the well-defined constant velocity of a uniform motion. It thus received a direct operational meaning resembling that of the momentary velocity of classical physics, which characterizes motion in an infinitesimally small moment of time.

This is but one example from a much richer exploration in which various interpretations of the existing means of external representation were investigated and either validated or excluded. In the course of this exploration, the medieval framework was enriched with new experiences, extended, modified, and eventually reinterpreted and transformed into part of the conceptual apparatus of infinitesimal calculus and the treatment of motion in classical physics.

This episode belongs to the long-term history of mechanical knowledge, whose investigation has helped to identify many of the general features of the development of a knowledge system discussed above. In what follows, I will therefore use this history to illustrate some of these features with concrete examples.

Chapter 6

Mental Models at Work

Hitherto, what has been boasted of is what production owes to science, but science owes infinitely more to production.
—FRIEDRICH ENGELS, "Dialectics of Nature"

Default assumptions are more than mere conveniences; they constitute our most productive way to make generalizations. Although such assumptions are frequently wrong, they usually do little harm because they are automatically displaced when more specific information becomes available. However, they can do incalculable harm when they are held too rigidly.
—MARVIN MINSKY, *The Society of Mind*

A Periodization of Mechanical Knowledge

In the following, I will illustrate the origin, function, and transformation of mental models in human thinking about nature in the domain of mechanical knowledge.[1] Mechanical knowledge concerns material bodies in time and space, their motions, and the forces that cause or resist such motions. Mechanical knowledge allows us to predict how bodies will change their position over time if we know their current state and the forces acting on them. A rough survey suggests that the long-term history of mechanical knowledge can be divided into six more or less coherent periods.

The first period may simply be called the "prehistory of mechanics"; it comprises the long period of time in which human cultures accumulated practical mechanical knowledge without documenting it in written form and without developing theories about it. Although the origin of sciences such as mathematics and astronomy can be traced back to the ancient urban civilizations of Babylonia and Egypt, this is, surprisingly, not the case for mechanics. In fact, although there are numerous sources testifying to the large construction projects of these civilizations, there is no document referring to the mechanical knowledge that must have been brought to bear on these endeavors.

The next period is that which properly merits the label "the origin of mechanics." It saw, in particular, the formulation and proof of the law of the lever. More generally, it is characterized by the appearance of the first written treatises dedicated to mechanics and to physics, associated in the West with names such as Aristotle, Euclid, Archimedes, and Heron, and in China with the so-called *Mohist Canon*

墨經, dating from about 300 BCE. The works of the Western authors had an enormous impact on subsequent development, while the earliest Chinese writings on mechanics exerted an influence only much later.

The beginning of the third period is characterized by the transformation of mechanics into a "science of balances and weights" in which the law of the lever again played a key role. This period covers the Arabic and Latin Middle Ages, which saw the production of an extensive mechanical literature, albeit focused on a relatively narrow range of subjects.

The fourth period is that of preclassical mechanics, ranging from the sketches of Renaissance engineers such as Leonardo da Vinci to the mature works of Galileo Galilei. In contrast to the preceding period, it deals with an increasingly large number of subjects, among them the inclined plane, the pendulum, the stability of matter, the spring, and so forth. Nevertheless, the law of the lever continued to play an important role.

The fifth period saw the rise of a mechanistic worldview. It extends from the first comprehensive visions of a mechanical cosmos (such as that of Descartes, via the establishment of classical and later analytical mechanics by figures such as Newton, Euler, and Lagrange) to the attempts of nineteenth-century scientists to build physics on an entirely mechanical basis.

The sixth period comprises the decline of the mechanical worldview and the disintegration of mechanics from the turn of the nineteenth to the twentieth century. This period is associated with the emergence of modern physics and its conceptual revolutions, as represented by quantum theory and the theory of relativity.

This overview of the long-term development of mechanics raises a number of puzzling questions. For example: How did theoretical mechanics originate in the ancient world, and why did it not happen earlier? What kind of knowledge made the formulation of the law of the lever possible, and what knowledge was required for its proof? What accounts for the remarkable differences in the development of Western and Chinese mechanical knowledge, and what accounts for the difference between the medieval science of weights in the Arabic and Latin worlds, on the one hand, and early modern preclassical mechanics, on the other? What kind of empirical knowledge made the emergence of classical mechanics possible, and what accounts for its remarkable stability over the more than two hundred years of classical physics? What explains the even greater stability of Aristotelian physics over more than two thousand years? How can one explain the disintegration of mechanical concepts around the turn of the past century, along with the emergence of revolutionary theories such as the theory of general relativity—a theory that proved to be foundational to knowledge unavailable at the time of its creation? Finally, how did basic principles like the law of the lever survive all these changes?

In view of the remarkable continuities and discontinuities in the development of mechanical knowledge, it may be tempting to look for contingent reasons that shaped its history and that are unrelated to the intrinsic nature of mechanical knowledge. Take the long dominance of Aristotelian physics, which even extended up to the period of preclassical mechanics and the so-called Scientific Revolution with its widespread anti-Aristotelian attitude. Is it really reasonable to explain this dominance by external factors like the adoption of Aristotelian philosophy as the official doctrine of the Catholic Church? Such explanations only sound plausible as long as one assumes that scientific knowledge is primarily encoded in scientific ideas and theories.

But if one takes all types and layers of knowledge into account, such as the intuitive knowledge governing thinking and behavior in our natural environment, it rather seems (as is argued in chapter 4) that certain aspects of Aristotelian physics must have been as convincing for medieval and early modern scholars as they are for children and even high school students today because it was structured by commonly shared mental models. In short, an understanding of the long-term development of mechanical knowledge must take into account, in addition to the theoretical knowledge usually considered in the history of science, intuitive physics and practical mechanical knowledge. We thus return to the three types of knowledge discussed in chapter 4, but this time I will tailor them to the case of mechanical knowledge.

Types of Mechanical Knowledge

Intuitive physics is based on experiences acquired through universal or nearly universal human activities. Experiences relevant to intuitive mechanical knowledge include, for instance, the perception of material bodies and their relative permanence, their impenetrability, and their physical behavior. Intuitive physics also underlies the arguments of scientific theories of mechanics. In proofs of the law of the lever, it is, for instance, usually assumed tacitly and without any need for justification that if one arm of the balance goes up, the other one cannot go up as well but necessarily must go down. Among the basic models of intuitive physics is the motion-implies-force model described in chapter 4, providing an interpretation of perceived motions as caused by the force exerted by a mover. Another example is a mental model of stability that leads us to deduce that a heavy object, not supported or held by another body, falls downward, but at the same time, that a body tending to fall can be stabilized by another body. Conversely, we can conclude from this mental model that a heavy object, which does not fall downward, must be supported by another one, even if the support is not directly visible or identifiable.

A second kind of mechanical knowledge, which predates any systematic theoretical treatment of mechanics, is practical mechanical knowledge—the knowledge achieved by dealing with mechanical tools such as the lever. Such tools serve to increase mechanical forces beyond biologically limited human capabilities. In contrast to intuitive mechanical knowledge, this type of knowledge is no longer universally shared by every human being; it is culture specific and, within a culture, not usually shared by all its members. It is closely linked to the production and use of such tools by professionalized groups of people.

Once the requisite tools for gaining such experience and building up such knowledge have been invented, the knowledge of how to produce and use them adequately is communicated among individuals and transmitted from one generation to the next by material tools, by shared practices, and by oral communication using esoteric terminology.

Since ancient times, a foundational example of practitioners' knowledge has been determining the weight of a body and the force required to lift it. This is prototypically embodied in a real model, namely, that of the balance with equal arms. The force that keeps the balance in equilibrium is equal to the weight of the body. Hence, we call this model of compensation between force and weight the "equilibrium model."

The practical knowledge of the ancient technicians and engineers also involved other basic experiences, in particular how to free oneself from the constraint of the equivalence between weight and force. In fact, the art of the mechanician consisted precisely in overcoming the natural course of things with the help of instruments such as the lever. According to this understanding, a mechanical instrument serves to achieve, with a given force, an "unnatural" effect that could not have been achieved without the instrument.

We therefore call the model underlying this understanding the "mechanae model"—according to the Greek word *mechanae*, which means both device and trick, and is at the origin of the word "mechanics." A more specific version of the mechanae model is the "lever model," which recognizes the lever as such a force-saving device. Since ancient times, the lever has existed in the form of tools (like the digging stick), as an element in the construction of wine and oil presses, and as an aid in moving and positioning heavy loads.

Another example from the area of practical knowledge is the mental model of the arch. This is based on the practical experience that an opening in a building can be spanned by an arch, which only begins to bear its own weight when the keystone is inserted. Arch technologies were already known from prehistoric times. Initially, however, a method called "corbel technology" was used, which involved vertically distributing the arch's weight over the supporting walls, while the keystone had no static significance. In other words, the arch model was already a result of learning processes, probably based on the widespread use of vaults for construction both below and above ground.

The fact that human mechanical knowledge (in contrast to the mechanical knowledge inherent in the behavior of animals) has its roots not only in biologically determined activities but, in addition, in the use of historically transmitted mechanical tools implies that the development of tool-based mechanical knowledge is determined by social contexts. The skillful use of mechanical tools requires accumulated experience, which can best be acquired by a specialization in the handling of these tools. Hence, the development of tool-based mechanical knowledge is to a great extent the result of a social division of labor leading to the emergence of professions and the transmission of professional knowledge from one generation to the next, including the ever-present danger of its loss.

This transmission is predominantly realized through participation in the process of work during an apprenticeship. Tool-based mechanical knowledge comprises the mechanical tools themselves and the techniques of handling them; its development, application, and historical transmission does not in principle require any symbolic representation in written descriptions or graphical depictions. This development may, however, leave traces in the semantic topologies of languages, in particular in the

6.1. Corbeled vault in the tomb of the kings of the Third Dynasty of Ur (twenty-second to twenty-first centuries BCE). © Bildarchiv Foto Marburg / Albert Hirmer / Irmgard Ernstmeier-Hirmer.

emergence of professional terminology—a vocabulary derived from the tools, skills, and practices to which they are related.

One consequence of the social preconditions for the historical occurrence of mechanical tools is that these tools are developed predominantly in areas that are essential to the sustainability and reproduction of a community. Thus, civilizations in early stages of human history developed mechanical tools for hunting and gathering, food processing, and tribal warfare. In such areas, mechanical tools turned out to be favorable for the community, so incentives emerged for their systematic production and their transmission to the ensuing generations.

In contrast to intuitive and practical mechanical knowledge, which depends on familiarity with the use of specific mechanical tools and techniques, theoretical knowledge representing mechanical experience is based on generalized concepts encoded in language. It typically comes either in the form of explicit mechanical theories based on reasoning codified in natural language or in the form of deductive theories based on the representation of reasoning in rule-based systems of formalized, typically mathematical language. An example is the mental model of the center of gravity. If you suspend a body at this point (the center of gravity), it will not turn, no matter in which position you suspend it. This model is derived from generalizing the mental model of a balance in the context of a mechanical theory according to which every body can, in principle, be considered as a balance. It can thus be assigned a virtual fulcrum—an abstract point from which the body is suspended in an equilibrium that will not falter no matter the position.

Theoretical knowledge has its origin not so much in the immediate processing of sense data, as philosophers such as David Hume claimed, but in reflection on mental models, in particular on the mental models of intuitive and practical mechanical knowledge. This reflection leads to intuitive inferences in the medium of mechanical rules and statements. It furthermore paves the way toward an integration of such rules and statements into higher-order systems of generalized concepts that represent (in theoretical form) experiences embodied in intuitive and practical mechanical knowledge.

Philosophers of Greek antiquity, such as Aristotle, explicated elementary models of intuitive thinking (like the motion-implies-force model) conceptually and in written form. Aristotle made this model the core of his theory of locomotion, which shaped philosophers' theories of motion well into the late Middle Ages and the early modern period.[2] In doing so, philosophers availed themselves of a metalanguage that no longer relates to the perceived reality itself, but rather to the thinking directed toward this reality. Implicit mental models of intuitive reasoning thus became generalized affirmations of an explicit natural philosophy with a claim to universal validity. In this context, mental models of theoretical thinking are communicated by an explicit description of their structure and of the conditions for their application.

This provides an answer to the question raised above concerning the relation between different forms of knowledge, in particular with regard to the relation between practical and theoretical knowledge. Theoretical mechanical knowledge emerges with vocabulary related to specific forms of mechanical tools and techniques, which are then used to elaborate general relations between them by discursive descriptions and, finally, by deductive systems based on implicit or explicit rules of operating with formalized language representations. These terms and relations directly or indirectly reflect the integration of mental models that originated as structures of intuitive and practical mechanical knowledge. The models now ensure the applicability of the generalized mechanical concepts to the specific contexts of mechanical experiences from which they were originally derived, which thereby become examples of general theoretical statements on mechanical experiences.

Challenges of Theoretical Mechanical Knowledge

The shift from a model of intuitive thinking to a theory of motion fixed in written language had a number of consequences for the further development of thought. The expanded possibilities for reflecting on conclusions and their results became, in particular, the point of departure for a further differentiation of the model. A first form of such a differentiation results from processing experiences that formally fall under the generalized written conceptions of the model but that cannot be assimilated to the intuitive model on which it is based.

Specifically, in Aristotle's *Physics*, the failure of the intuitive model of the causation of motion in areas where it was difficult to identify a mover led to the definition and exclusion of particular types of motion. Aristotle excluded two types of motion from the motion-implies-force model: first, the motion of celestial bodies; second, the case of heavy objects falling toward the center of the earth, as well as the ascension of light objects in the opposite direction. The two excluded kinds of motion were demarcated as "natural" motions, which require no causal force, as opposed to the "violent" motions, which are the sole subject of his theory of the causation of local motions by forces.

According to his theory, the motion of celestial bodies comes about through the natural circular motion of the fifth element, the so-called ether, while the natural, rectilinear motion of the heavy and light objects is generated by their inbuilt tendency to reach their natural place. The consideration of these natural motions thus triggered a conceptual differentiation achieved by theoretically excluding them from the domain to which the motion-implies-force model could be applied.

Another observation that caused an internal differentiation of the model through theoretical reflection was that of the phenomenon of inertial motion, that is, of motion that proceeds without a driving force. When a projectile is thrown from the hand of the thrower, it continues its motion without being driven by any detectable cause. However, for the motion-implies-force model to be applicable, it requires the critical variable *mover* to be in direct contact with the moved object in order to exert a force on it.

Philosophers of antiquity, including Aristotle, presented various subtle (and sometimes overly subtle) arguments trying to assign a causal force to an inertial motion, such as the flight of a projectile, though the presence of such a force was not obvious. Such arguments generally proceed from the assumption that the throwing hand also induces motions in the medium, for example, motions of the air that, in turn, then cause the continuation of the movement through direct contact with the moved object—a rather counterintuitive explanation. Yet such artificial explanations, especially those of Aristotle himself, were handed down, owing not so much to their explanatory power but rather to a kind of theoretical dogmatism. In contrast to the intuitively plausible mental models underlying other claims of Aristotelian physics—such as the distinction between natural and forced motions—the passing on of such contrived attempts to repair the Aristotelian theory of local motion was linked to specific social conditions, and in particular to the existence of learned institutions, which guaranteed that they could be established as dogmas and their validity enforced.

Another opportunity for adapting a mental model to a challenging problem is the introduction of new mental models into the extant model's free slots. For instance, as it later turned out, the generic variable *motion* could be refined to include the characteristic property of *acceleration*. In this way, the motion-implies-force model of Aristotelian natural philosophy changes into the acceleration-implies-force model of classical mechanics. According to this model, it is not motion that requires a force, but the change of motion, that is, acceleration. The acceleration-implies-force model eventually became a cornerstone of classical physics. The new model is still rooted in intuitive physics, as is suggested by its anthropomorphic concept of force and also by its genesis as a modification of the motion-implies-force model. But the acceleration-implies-force model is no longer part of intuitive physics or of practitioners' knowledge, as is evident by its counterintuitive consequences, such as the implication that a force-free motion will never come to rest. It therefore took considerable effort and time before this modification was historically accepted.

Adapting a Mental Model of Intuitive Physics

While this brief account shows how such a momentous transition like that from the explanation of motion in Aristotelian physics to that of classical physics could be accomplished in principle, it does not explain how this transition was engendered by the transformation of an entire system of knowledge. The acceleration-implies-force model was no ad hoc invention but rather resulted from a long process of knowledge accumulation and transformation that eventually gave it the stability and power to integrate this knowledge within the emerging framework of classical mechanics.

One point of departure in the transformation of the motion-implies-force model was also grounded in intuitive physics, namely, in the everyday experience of inertial motion, as studies in developmental psychology have confirmed.[3] An object that is set in motion by a force generally continues to move for a while after the force has ceased to affect it, as if something had been transferred from the force to the object. In late antiquity, this experience became the starting point for an adaptation of the motion-implies-force model that included the explanation of inertial phenomena that were otherwise difficult to integrate. This adaptation was achieved by transforming the conception of force. The force that causes a motion was no longer conceived as the power of a mover directly affecting the object but rather as an entity that can be transferred from the mover to the object moved. This transferred force was typically designated as an "impressed force" (*vis impressa*) or as "impetus."[4]

In the early modern period, such expanded explanatory possibilities constituted an important point of departure for overcoming Aristotelian patterns of explanation without having to abandon basic Aristotelian assumptions from the outset. For instance, if one assumes that the impetus impressed on a projectile is gradually used up, then one obtains an explanation for why its movement slows down. If one assumes, by contrast, that the impressed impetus can be used up only by counteracting forces (like the resistance of a medium), then one arrives at the conclusion that the object remains in motion as long as no such forces affect it. In this case, the motion-implies-force model thus leads to conclusions that correspond to those implied by the principle of inertia in later classical physics, thus effectively scaffolding the development of an inertial physics.

Aristotelianism as Shared Knowledge

As we have seen, Aristotelian physics captures many common experiences, which is one reason for its persistence over the centuries. Another was its importance to institutionalized education. In the early modern period, when Aristotelian physics was basic knowledge taught at the universities, any attempt to create an equally general theory of nature had to start from this basic body of knowledge, even if its goal was to revise the predominant Aristotelian system.

This explains why Galileo and his contemporaries often combined an anti-Aristotelian attitude with an adherence to basic Aristotelian assumptions. They also relied on some of the sophisticated elaborations of Aristotelian physics in the medieval scholastic tradition that I have touched on in the previous chapter when discussing the diagrammatical representations introduced by Oresme in the fourteenth century. This knowledge was not rooted in intuitive physics; it was part of the shared knowledge of the time because it was transmitted as a canon of theoretical knowledge within the university tradition.

None of the assumptions of this theoretical tradition were as self-evident and seemingly irrefutable as the elements of Aristotelian theory rooted in intuitive physics; some of them were subjects of debates extending over centuries. This strand of shared theoretical knowledge thus depended on a continuity embodied either in writings or teaching traditions. It explains a remarkable communality of many contemporary attempts to deal with the phenomenon of acceleration by figures such as Galileo, Descartes, Beeckman, Harriot, and many others.[5]

They all conceptualized, in particular, accelerated motion as a quality that has an extension as well as an intension, and that was characterized by changing degrees. This conceptualization was in fact closely related to the tradition of representing change that I have discussed as part of a medieval Aristotelian tradition.

The Challenging Objects of Early Modern Technology

But why was the study of accelerated motion so important to early modern thinkers, and why did they pay so much more attention to it than had Aristotelian philosophers before them? Part of the answer has already been encountered in the discussion of the triggers that drive change in a knowledge system. In this case, it was the challenging objects of contemporary technology, in particular mechanical devices or ballistics, that suggested the importance of a better understanding of accelerated motion. In dealing with these challenging objects, resources of theoretical knowledge (such as the Aristotelian distinction between natural and violent motion) were brought together with the new experiences of practitioners.

Any serious theory of projectile motion advanced at that time had to take into account the common knowledge of the practitioners of ballistics as was passed down not only by participation and oral transmission, but also by numerous published and unpublished military treatises. The knowledge of the artillerists based on their professional experience included the fact that the speed of a projectile increases with the force the exploding powder exerts on it, that more projectile weight requires more force to reach the same distance, that the distance of the shot depends on the angle to the horizon in which the cannon is directed, that there is an angle at which this distance reaches a maximum and that there are angles at which flat and steep shots reach the same distance with different effects.[6]

Galileo, for instance, became interested in challenging subjects like the motion of a pendulum and the curve described by a projectile.[7] He convinced himself that the trajectory of a projectile could be likened to the form of a hanging chain, both being close to a parabola. He also compared the motion of a pendulum to the motion of a body rolling up and down within a spherical bowl—thus forming networks of mental models. The motion of a body rolling up and down within a spherical bowl could in turn be related to the motion up and down inclined planes inscribed onto the sphere. This encouraged Galileo to intensely study falling motion along inclined

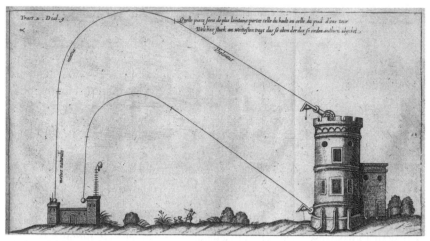

6.2. An elaboration of the Aristotelian interpretation of the motion of a projectile in terms of "violent," "mixed," and "natural" motion. From Ufano (1628). Courtesy of the Library of the Max Planck Institute for the History of Science.

planes. He soon developed a small set of theorems with which he attempted to build a new theory of motion. This set of insights formed what I have called in chapter 4 an "epistemic island." This island was still embedded within a larger Aristotelian system, but it began to develop its own systematicity.

The Emergence of Classical Physics from a Reorganization of Knowledge

The comparison of projectile trajectory with the curve of a hanging chain helped Galileo to see its symmetry as an essential pattern and suggested identifying both curves with a parabola.[8] Other students of artillery (and the young Galileo himself) had argued that the trajectory is first governed by the violent motion conveyed to the projectile by the cannon, and hence directed along the line of the shot, but is ultimately dominated by the natural motion of the heavy projectile, thus transforming into a vertical fall downward.[9] The trajectory should therefore not be symmetrical, but should rather consist of three different parts: an oblique and rather straight initial part, a curved middle part, and a vertical part at the end.

In 1592, however, Galileo performed a simple, qualitative experiment with his patron Guidobaldo del Monte, rolling inked balls over the surface of an inclined plane. In contrast to airborne projectiles, these balls left a visible trace of their trajectory. Galileo and Guidobaldo observed that this trace bore a striking similarity to the trajectory of a hanging chain (turned upside down). This suggested to them that the two curves are generated by similar forces (which according to classical physics is actually not the case). This similarity convinced Galileo that the symmetry of the trajectory is an important feature of projectile motion. He further conjectured that the shape of the trajectory must be a parabola. In the following years, he fleshed out these and other consequences of his experimental insights but without giving up the Aristotelian dynamical framework.[10]

Constructing a parabola from the composition of two motions, a horizontal and a vertical, made it plausible to assume first that the horizontal motion is uniform, and second that the vertical motion uniformly accelerated, which suggested something like a principle of inertia (at least for horizontal motions) and a specific form for the law of free fall. The assumption that horizontal motion is neither accelerated nor decelerated fitted well into Galileo's emergent theory of inclined planes, which still operated within the framework of Aristotelian physics. A motion of descent was conceived as an accelerated "natural" motion, and a motion of ascent as a decelerated "violent" motion, so Galileo could classify a uniform motion along a horizontal plane as "neutral."[11]

This set of insights into projectile motion and motion along inclined planes forms the core of Galileo's emerging theory of motion. While this theory was still embedded in an Aristotelian understanding of motion with its distinction between natural and violent motion, it started to display remarkable deviations, such as emphasis on the phenomenon of acceleration, which played no role in Aristotelian physics, or the claim that there is something like a uniform neutral motion, or the hypothesis that the projectile trajectory is symmetrical.[12]

However, from principles in essential accordance with the Aristotelian framework, Galileo could only unambiguously derive the trajectory of horizontal shots. One component was the uniform horizontal motion familiar from the case of the neutral motion along a horizontal plane. The other was the accelerated motion of free fall. Taken together, these two motions generate the parabolic trajectory of a horizontally projected body.[13]

But in the case of oblique shots, the motion was difficult to anticipate. Oblique motion in an upward direction along an inclined plane could be clearly recognized as a decelerated motion. Does this imply, in the case of oblique projection, a decelerated motion along the line of the shot, and an accelerated motion of a downward fall?

This possibility, which appears strange to us because it conflicts with classical mechanics, was indeed seriously entertained by Galileo and his contemporaries.[14] The assumption that motion along an oblique trajectory is decelerated was simply a default assumption following from applying the mental model of motion along inclined planes to the analysis of projectile motion. In the case of horizontal shots, it suggested that the horizontal component of projectile motion is uniform, just like motion along a horizontal plane. In the case of oblique projection, it suggested that the oblique component of projectile motion is decelerated, just like upward motion along an inclined plane. The problem was, however, that the resulting shape of the trajectory failed to exhibit the expected parity between horizontal and oblique shots.

In contrast, according to classical physics, the motion along the line of the shot actually constitutes an inertial motion that will proceed uniformly along a straight line as long as no other forces intervene. The parabolic shape of the trajectory of a projectile can be derived, for any direction of a shot, from this principle of inertia and the vertical acceleration of free fall.

But even without possessing a general principle of inertia as part of the foundation of his physics, Galileo was aware that the trajectory would likely be symmetrical, as in horizontal projection, even in the case of oblique projection.[15] He had thus achieved a result that corresponded exactly to what one would derive from classical

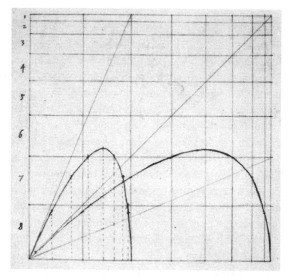

6.3. Point-by-point construction of projectile trajectories under the assumption that the component of the motion along the oblique direction of the shots is decelerated, as if taking place along an inclined plane. Thomas Harriot, Add. MS 6789 fol. 5r. © The British Library Board.

physics but without having its principles at his disposal. On the contrary, given the insight that the shape of a projectile trajectory is parabolic in general, the principle of inertia now became a plausible assumption from which this result could be derived, and vice versa: inertial motion could now be conceived as that motion which a projectile on a parabolic path would follow if gravity were not acting.

The assumptions of preclassical mechanics thus acted as a matrix in the sense introduced in chapter 4 because they could be reinterpreted as linchpins of a newly constructed system of knowledge: the system of classical mechanics. Toward the end of the seventeenth century, Isaac Newton, in his famous *Philosophiae naturalis principia mathematica*,[16] turned the principle of inertia into a general principle of a deductive theory of the motion of bodies. He applied a modified explanatory scheme for the causation of motions by forces to an explanation of celestial motion, a subject hitherto exempted from such an explanation according to Aristotelian natural philosophy. Newton was thus able to mathematically derive Kepler's laws of planetary motion.

From the perspective of classical physics, this step appears to constitute the ultimate victory over the traditional Aristotelian theory of motion. It seems that Aristotle's incorrect assumption about the causation of motions had been simply replaced by the correct one. In fact, however, Newton, like his predecessors, still regarded the motion of a body as being caused by the impression of forces.[17] Newton's reluctance to fully abandon a conception of motion still reminiscent of the theory of impetus may appear as an irrational and superfluous insistence on traditional ideas—at least if his crucial step is reduced to a simple correction of an incorrect theoretical assumption. Newton's definitions in terms of traditional notions, however, hint at the constitutive conditions under which the change from Aristotelian to classical physics actually

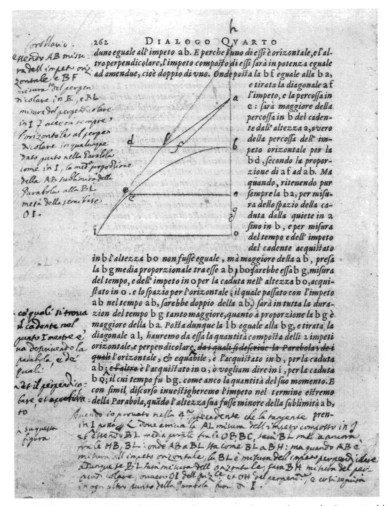

6.4. Emended diagram in Galileo's hand copy of his major work on mechanics, the *Discorsi* published in 1638. The lines added to the diagram and the related handwritten notes amount to a formulation of the principle of inertia, as an afterthought to Galileo's published argument still rooted in Aristotelian dynamics. Ms. Gal. 79, c.157v, by concession of the Ministerio per i Beni e le Attività Culturali—Italy / Biblioteca Nazionale Centrale di Firenze.

took place, namely, as the historical transformation of a traditional mental model of the causation of motion (the motion-implies-force model) into an acceleration-implies-force model.

The radical change in physical explanations, which the Newtonian model implies when compared with explanations based on the motion-implies-force model, was the *result* rather than the *presupposition* of a comprehensive process of knowledge integration that included Galileo's studies of accelerated motion and Johannes Kepler's investigations of planetary motion. Thus, from our theoretical perspective, processes of knowledge transformation no longer appear as corrections of existing theories on the basis of new experiences. Rather, new structures of knowledge typically emerge as a result of development processes that take place within the framework of existing structures, which are themselves enriched by the accretion of new experiences.

La querelle des anciens et des modernes

According to the historian and philosopher Arnold J. Toynbee and the anthropologist Jack Goody, renaissances are not an exclusively European specialty: they are a recurrent phenomenon in all written cultures.[18] Renaissances may serve to legitimize innovation by constructing an idealized past and, in doing so, construct themselves.[19] Reference to an idealized past can make radical innovation appear to be the restitution of a lost golden age—an age temporarily obliterated by loss or neglect, but capable (or so it is claimed) of being restored in all its glory through the reintroduction of something ancient in contemporary disguise. Often, however, the past is simply exploited as a resource for the struggles of the present. In any case, such constructions are subject to reinterpretation, and their practice has not left the present unaffected.

These processes have also shaped the emergence of "modernity" in culture and science. They may be illustrated by the so-called *Querelle des anciens et des modernes*, or "Battle of the Books," as it was called in English. This controversy extended over a considerable period of time, from roughly the mid-sixteenth to the mid-eighteenth century, mainly in France and England.[20] It began as a debate about aesthetic forms and criteria but widened into a debate about how to position contemporary achievements with respect to those of antiquity. Of course, the debate took place against the backdrop of the European Renaissance, which already constituted a tradition (reaching back, by this time, more than two centuries) of referencing contemporary culture in relation to classical antiquity. In this sense, the *Querelle* was a kind of second-order Renaissance and played an important role in the self-consciousness of modernity. A central concern of the *Querelle* was therefore also to challenge the perceived need for and normative relevance of constant references to antiquity in view of new achievements that signaled the increasing independence of modernity from past models—including imagined and idealized ones.

Similar processes also took place in the history of science, as may be illustrated by the case of Isaac Newton, whose work was even invoked as testimony in the *Querelle*—surprisingly, however, not on the side of modern science. His *Philosophiae naturalis principia mathematica*, published in 1687, was simultaneously the capstone of preclassical mechanics, the cornerstone of classical mechanics, and the harbinger of the new analytical mechanics of Jean le Rond d'Alembert, Leonhard Euler, and Joseph-Louis Lagrange. Newton himself, however, privileged Greek geometry, rejecting innovations such as the new symbolic algebra of François Viète and Descartes's analytical geometry.[21]

Even when Newton himself contributed an evident innovation—such as the calculus of fluxions (modern infinitesimal calculus)—he attempted to conceal its novelty by rephrasing it in terms that made it appear as a continuation of something old. In the method of first and last ratios, for example, he conceived of differential quotients as limits to the geometric proportions of Euclidean geometry. Rather than placing the fundamental role of the infinitesimal calculus in the foreground of the new mechanics, Newton thus rephrased his derivations in the *Principia* in terms closer to classical geometry. In contrast, his competitor Gottfried Wilhelm Leibniz made vigorous use of the new symbolic means.[22] There was, however, also good reason to write the *Principia* using more familiar geometrical language; Newton's readers otherwise would have found it much harder to understand his physical theories.

Nevertheless, for some of Newton's readers, particularly in France, the *Principia* eventually became the foundation of a new science that needed no grounding in ancient geometric methods, nor in geometric intuition. In his *Mechanique analytique*, Lagrange in fact took pride in having renounced all illustrations.[23] For him and his peers, the power lay in a universally applicable symbolic formalism, whereas Newton had cherished and praised the possibility of phrasing (or rephrasing) his arguments in intuitively clear, natural language. Thus, for the sciences, the counterpart of the *Querelle*, raising the issue of modernity, was also a dispute about formalism versus intuition, about synthetic versus analytical methods—in short, about the viability of the so-called analytical turn of science.[24]

As it turns out, however, the relation between formalism and intuition is not one of mutually exclusive, alternative approaches to science. Rather, as discussed in chapter 3, abstraction and reflection are grounded in an interplay between formalisms acting as external representations of thought and cognitive structures. While it is true that analytical mechanics no longer relied on mental models of intuitive and practical knowledge, it did not, however, renounce mental models altogether. Rather, it created its own intuitively plausible forms of thought, which developed a force of persuasion equal to those mental models of ancient science from which its leading representatives polemically distanced themselves in the *Querelle*. One example is the mental model of a function, represented by the unspecified analytical expression $y = f(x)$. Another is the so-called principle of minimal action, which is the core of a powerful new formalism of great relevance to physics even today. At the same time, the principle of minimal action lent itself to an intuitive interpretation regarding nature's tendency to accomplish something with minimal effort.

In short, the *Querelle* is just one prominent example of another recurrent feature in the evolution of knowledge: a dispute about the primacy of formalism versus natural language, of symbols versus semantics.

Chapter 7

The Nature of Scientific Revolutions

Concepts can never be derived logically from experience and be above criticism. But for didactic and also heuristic purposes such a procedure is inevitable. Moral: Unless one sins against logic, one generally gets nowhere; or, one cannot build a house or construct a bridge without using a scaffold which is really not one of its basic parts.
　　　　—ALBERT EINSTEIN, letter to Maurice Solovine, 1953

Paradigms are not corrigible by normal science at all. Instead, as we have already seen, normal science ultimately leads only to recognition of anomalies and to crises. And these are terminated, not by deliberation and interpretation, but by a relatively sudden and unstructured event like the gestalt switch.
　　　　—THOMAS S. KUHN, *The Structure of Scientific Revolutions*

On closer inspection, not only was the Scientific Revolution of the early modern period the result of long-term developments of knowledge, but so were other major changes in the architecture of scientific knowledge. As I have argued above, innovation is often born of the exploration of the horizon of an existing system of knowledge and its concurrent transformation by reorganizing the accumulated knowledge therein. In the following I review the so-called Copernican, chemical, and Darwinian revolutions and will detail some of the basic features of Einstein's two "relativity revolutions": the first that led to special relativity in 1905, the second that led to general relativity and a new theory of gravitation in 1915. At the end of this chapter I make a very brief excursion into the quantum revolution, a topic dear to Thomas Kuhn. In chapter 10, I add a further example, the geological revolution of the twentieth century.[1]

I stress a number of common themes, taking up the concepts introduced in chapter 4. None of these achievements are creations out of nothing. They result from the transformations of preexisting systems of knowledge. These transformations are typically protracted processes that may extend over generations.[2] External and internal triggers of change (such as challenging objects or borderline problems that arise when two different conceptual schemes are applied to the same case) may act as catalysts of these transformations. New structures of knowledge may emerge from the preexisting ones by an exploration of the potential inherent in their external representations and by the formation of epistemic islands somewhat decoupled from the core of the preexisting system of knowledge. Such islands may then act as matrices or seedbeds for building up a new system.

Kuhn versus Fleck

Kuhn's analysis of the structure of scientific revolutions was shaped by the ideological controversies of the Cold War, in particular by an opposition to Marxist views.[3] Kuhn deemed economic structures, technical abilities, and practical conditions, for instance, to be largely irrelevant to the history of science proper; by his account, discoveries are almost mystical individual achievements, whereas the community of those practicing "normal science" is conceived of as being an elitist, albeit conservative, majority.[4] In contrast to Kuhn, Ludvik Fleck neither confined thought collectives to elitist scientific communities (as Kuhn did) nor did he reduce science entirely to a social and intellectual activity, as is often assumed.[5] He rather emphasized the role of the "resistance" experienced from the outside world in shaping a scientific thought style, similar to the manner in which I have described the role of challenging objects in chapter 4.

The difference in "thought style" is also evident from Kuhn's ambivalent reaction to Fleck's book: "I don't think I learned much from reading that book, I might have learned more if the Polish German hadn't been so very difficult. But I certainly got a lot of important reinforcement. There was somebody who was, in a number of respects, thinking about things the way I was, thinking about the historical material the way I was. I never felt at all comfortable and I still don't with [Fleck's] 'thought collective.' "[6] Fleck's text is actually written in elegant Standard German. In the preface to the English translation of Fleck's book, Kuhn is even more explicit in his critique of Fleck: "But the position it elaborates is not free of fundamental problems, and for me these cluster, as they did on first reading, around the notion of a thought collective. . . . Rather I find the notion intrinsically misleading and a source of recurrent tensions in Fleck's text."[7]

Kuhn's views on science changed over the course of his career. He later turned away from a conception of paradigm shifts in terms of gestalt psychology and became interested in the philosophy of language as a means to express his ideas on science. In this context, he even abandoned the use of the concept of paradigm, replacing it with "lexicon."[8] But he continued to stress the character of science as a communal enterprise.

In 1969, shortly before he died, Fleck summarized his ideas in a manuscript that, for a long time, remained unpublished. He evidently found his view on the social character of science confirmed by its recent development: "In the present day, in the era of team cooperation, of articles published by several coauthors, of so many journals, reviews, conferences, symposia, committees, governing bodies, societies and congresses, the communal nature of scientific knowledge becomes evident."[9]

The Copernican Revolution

Commonplace accounts of the Copernican revolution are a good case in point, illustrating a widespread understanding of paradigm shifts in science. In past reconstructions of the Scientific Revolution (most prominently by Alexandre Koyré and Thomas Kuhn), the heliocentric shift was often seen as a sort of "paradigm of paradigms." Copernican astronomy was considered to be the main instigator of the epistemological break out of which modern science eventually emerged.[10]

Copernicus's *De revolutionibus orbium coelestium* (1543) placed a formerly peripheral celestial body, the Sun, at the center of the planetary architecture and launched a new astronomy on the basis of geometrical considerations. The new astronomy cast doubt on the crucial assumptions of a well-established geocentric tradition that rested on an integrated system of natural, philosophical, metaphysical, and theological doctrines.[11]

From the viewpoint of the Aristotelian tradition, the heliocentric system amounted to a fundamental change in the construction of the world system—a change that was unacceptable for many reasons, in particular because of its apparent physical absurdity. From the viewpoint of established natural philosophy (*philosophia naturalis*), the motion of Earth raised more problems than it offered solutions. To be sure, Copernicus simplified planetary theory by offering geometrical explanations for a series of phenomena that had forced Ptolemaic astronomers to develop bizarre ad hoc devices. For instance, the very limited elongations of Mercury and Venus, and the retrograde motion of all the planets, could be treated as necessary consequences of a theory in which Earth and the other planets move around the Sun.[12]

Still, the new cosmology infringed on fundamental assumptions of Aristotelian physics (or "natural philosophy"), beginning with the essential distinction between the sublunary realm of corruption and rectilinear elemental motions on the one hand, and the realm of superlunary perfection and circular ethereal motions on the other. As discussed in chapter 6, such a framework made an explanation of the different characteristics of terrestrial and heavenly motions possible. In fact, heavenly motions were accounted for by "celestial spheres," deputed to transport the planets.

Copernicus had no articulated physical alternative to offer. Rather, he was guided by an inversion of priority between physics and mathematics in favor of basic issues of mathematical astronomy. By taking the revolution of Earth around the Sun as the means to establish planetary distances, Copernicus made it possible to determine the order of the planets, which had been a controversial issue. Previously, that order had been established according to the natural philosophical postulate that distances and periods of revolution are proportional—but this was problematic for geocentric astronomers, especially with respect to the periods of the inferior planets. The periods of the superior planets were estimated one by one without any systematic interconnection. In light of the Copernican theory, planetary order became the physical consequence of geometrical reasoning and, to the great satisfaction of Copernicus, could be brought into agreement with the distance-period relationship.

7.1. Nicolaus Copernicus (1473–1543), mathematician, astronomer, and founder of the heliocentric worldview. Anonymous painting, sixteenth century. © bpk-Bildagentur.

Copernicus's reversal of the dependency of mathematics on physics raised methodological and epistemological doubts by questioning disciplinary divides. Nevertheless, his "revolution" was less of a radical break with tradition than it may at first appear. It solved a series of problems and offered challenging vistas of inquiry, raising fundamental new problems, particularly for physics.

But, to a large extent, Copernican thought was based on an exploration of the limits of traditional systems of knowledge. Its geometrical models were taken from Ptolemy and the later tradition of Arabic astronomical writings.[13] Rather than starting with a *tabula rasa*, Copernicus took over a complex mechanism of planetary astronomy that had been worked out in a long astronomical tradition. Geometry was the technology of representation whose exploration made Copernicus's achievement possible.

Still, in order to make sense of the heliocentric proposal, it was necessary to re-engage an age-old debate on the possibility of harmonizing geometrical modeling with physical causal explanations. This problematic assumption (that it was in fact possible) had constituted the core of considerations on heavenly physics from late antiquity to the Islamic and Latin Middle Ages. Inquiry into the physical causes of heavenly motions can be traced back to the seminal work by Alexander of Aphrodisias, a Peripatetic philosopher who lived in Athens at the beginning of the third century. Its later systematization in the tradition of the twelfth-century Andalusian

Muslim philosopher Ibn Rushd (Latin: Averroes) led to several attempts to develop planetary theories in accordance with the basic philosophical and metaphysical postulates of Aristotelian philosophy (foremost the principle of the perfect circular uniformity of celestial motions).[14] And Renaissance scholastic astronomers still faced the problem of the relation of mathematics to physics in Copernicus's time.

The most important aspect of Copernicus's geometrical approach and conceptual tools can be found in the works of his immediate predecessors, figures such as Johannes (Müller) Regiomontanus, Georg von Peuerbach, and Albert Brudzewo.[15] In this case as well, mathematical modeling can be seen as the elaboration of external representations that formed a bridge between a traditional system of knowledge and an emerging new system. At the same time, this elaboration highlighted the tensions inherent in the traditional system, most importantly the problematic match between geometrical representation and physical explanation. Peuerbach's *Theorica planetarum*,[16] for instance, illuminated the tension between geometrical modeling and the scholastic belief that the celestial orbs were carrying along the planets, thus providing motivation to search for better solutions.[17] With the emergence of the Copernican heliocentric system, these tensions were, however, not resolved. On the contrary, the exploration of both possible mathematical models of the heavens and attempts at their physical interpretation continued for generations.

It was particularly in the wake of the parallax measurement of cometary trajectories and the establishment of their superlunary location that the existence of the heavenly spheres deputed to transport the planets was cast into doubt. From the 1580s onward, this new evidence opened up heated debates about the accountability of planetary motions and the necessity of a new physics. From the perspective of the history of celestial physics, one can say that the Copernican system offered a planetary model that could not convince the broader scholarly community until it became clear that it could be supported on natural and physical grounds. In fact, Giordano Bruno's vitalistic physics, Johannes Kepler's celestial physics of forces, and René Descartes's corpuscular mechanics were all attempts to transform the Copernican theory into a physical revolution.[18]

The Chemical Revolution

Another example of the protracted transformation of a system of knowledge is the so-called chemical revolution of the eighteenth century, traditionally associated with Antoine-Laurent Lavoisier. Ursula Klein has aptly characterized this transformation as a revolution "that never happened."[19] Supposedly, Lavoisier challenged the basic understanding of chemical substances, including the concept of elements, and even rewrote the causal explanation of chemical transformations. Among other issues, Lavoisier's "chemical revolution" is commonly regarded as a rupture with the ancient concept of chemical "principles" and its replacement with the modern concept of chemical elements, in the sense of simple substances that cannot be further decomposed through chemical analysis.

However, long before Lavoisier, chemists had already argued that the building blocks of common natural substances are more basic substances that resist further

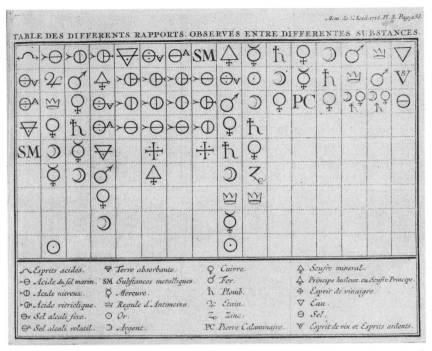

7.2. Table of affinity relations between different substances. From Geoffroy (1777, plate 8, 268). Courtesy of the Library of the Max Planck Institute for the History of Science.

decomposition in chemical analysis. Moreover, the contemporary chemical community, including Lavoisier, neither dismissed the ancient term "principle" nor the related idea that the simplest chemical "principles" (also known at the time as "elements") conferred particular properties. For example, Lavoisier derived the term "oxygen"[20] from the Greek word *oxys* (sharp, acid) in order to convey the idea that oxygen was the component (or "principle") of chemical compounds that conferred the property of acidity to them; he thus designated oxygen as an "acidifying principle."[21] Likewise, he argued that heat or "caloric" was a principle that conferred the property of elasticity to the gases.[22] What's more, caloric shared many features with the hypothetical phlogiston: it was imponderable, penetrated all bodies, and could not be stored in vessels.

Lavoisier's work was in fact based on century-long studies of chemical transformations, summarized in particular in so-called affinity tables.

They represented a large part of the shared chemical knowledge of the time and constituted a common point of reference for both Lavoisier and his adversaries, the so-called phlogistonists. The latter believed that a substance called phlogiston is contained within combustible bodies and released during combustion, whereas Lavoisier could show, in the 1780s, that a ponderable gas (oxygen) combined with the imponderable matter of heat is required for combustion. In his experiments, Lavoisier used refined quantitative measurements and systematically pursued theoretical questions.

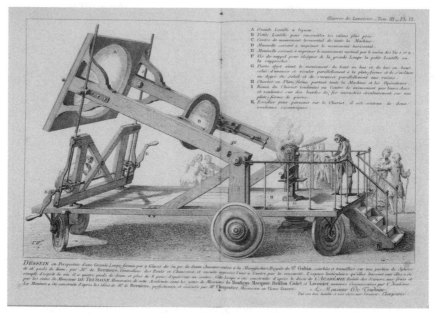

7.3. Antoine Lavoisier (wearing goggles) with his solar furnace. From Lavoisier (1862–93, plate 9). Courtesy of the Library of the Max Planck Institute for the History of Science.

But, as Ursula Klein and Wolfgang Lefèvre have argued, both his methods and his theoretical analysis relied on a long tradition that Lavoisier continued.[23] By applying a new chemical nomenclature and taxonomic system to theoretical questions such as the understanding of combustion, permanent gases, or acidity, he was able to draw novel conclusions by extending and systematizing the accumulated shared knowledge of his time. He did not, however, create a new world of chemical substances that was in some sense incommensurable with the world in which he had begun. Lavoisier proposed a coherent theoretical system, but this proposal was limited to the most ordered subfields of chemistry, thus creating an epistemic island, which became the point of departure for further structural changes of chemical knowledge.

The Darwinian Revolution

The emergence of the biological theory of evolution is yet another example of the protracted transformation of a system of knowledge, but it also highlights other characteristics that have been emphasized in my theoretical framework.[24] First of all, the development of evolutionary theory was indeed a long-term process. Evolutionary thinking goes back at least to the eighteenth century. Over the first half of the nineteenth century, evolutionary conjecture built up a more and more significant undercurrent in several biological disciplines, such as comparative anatomy, systematics, and biogeography. In these disciplines, findings and conclusions from within entirely nonevolutionary conceptual frameworks came to be increasingly at odds with the established dogma of the constancy of species.

It was these irritating findings and conclusions, not explicitly evolutionary conjectures, that Darwin resorted to when coming to grips with his own evolutionary agenda. In order to capitalize on these results and insights, Darwin had to discover their evolutionary potential, detach them from the conceptual frameworks in which they had been achieved, and finally transform them into evidence and arguments for his own theory of biological evolution. In Darwin's own words: ". . . innumerable well-observed facts were stored in the minds of naturalists, ready to take their proper places as soon as any theory which would receive them was sufficiently explained."[25]

The publication of Darwin's *The Origin of Species* in 1859 was certainly the decisive feat that transformed this evolutionary undercurrent into an assumption held openly by more and more biologists. But it was not Darwin's explanation of the evolution of biological species by variation and selection that was widely accepted by biologists of evolutionary convictions. Rather, the majority of these scientists developed other evolutionary theories that conformed to more familiar models of evolutionary processes, such as ontogenetical development or the belief in historical progress.

In contrast, Darwin's theory centered not on inner-organismic developmental tendencies but on adaptation to environmental changes. It also included coincident alterations. This made it a revolutionary and totally unacceptable theory for his contemporaries. Consequently, by the end of the century Darwin's theory had been almost completely "eclipsed" (as Julian Huxley described it) by more familiar theories (orthogenetic theories of evolution assuming an innate tendency of organisms to evolve in a definite direction and, above all, several varieties of neo-Lamarckism, assuming an organism can pass on acquired characteristics to its offspring).[26] This situation changed around 1930, when classical genetics (and particularly population genetics) destroyed the assumption of the inheritance of acquired characteristics—an assumption also held by Darwin, though in contrast to Lamarckian theories, not one on which his theory stands or falls.[27] Darwin's theory was then given, as Wolfgang Lefèvre puts it, a "second chance."[28]

In this history, we see some features common to other episodes of the evolution of knowledge. New systems of knowledge emerge by extended processes of knowledge accumulation and integration. Only eventually do they reach a certain stability and universal applicability. But what the history of the theory of biological evolution also illustrates is how this process of knowledge integration actually operates on the accumulated knowledge.[29] This knowledge comprised (as mentioned in chapter 4) the practical experiences of domestication but also morphological, taxonomic, paleontological, geological, and other forms of knowledge that had been acquired before Darwin, often under assumptions in conflict with each other and with Darwin's emerging theory.

These preexisting, heterogeneous domains of knowledge thus generated borderline problems in the sense discussed in chapter 4, which turned them into triggers of change. The preexisting elements of knowledge constituted a historically contingent set of premises that Darwin eventually transformed into internal presuppositions of his theory. As a result of his work, these domains of knowledge, whether taxonomies or morphological traits, were themselves reinterpreted, acquiring an "evolutionary" meaning that they did not possess before. Similarly, the transformation of preclassical mechanics triggered by the challenging objects of early modern technology had turned contingent boundary conditions into essential building blocks of classical mechanics.

Even after the so-called modern synthesis of the 1940s, developmental and evolutionary biology remained separate epistemic traditions, each with its own theoretical and experimental practices, as the biologist and historian of biology Manfred Laubichler has observed.[30] The search for an integrative evolutionary framework, perhaps one that also includes cultural evolution, thus goes on to this day. I come back to the challenges of cultural evolution for such a framework in chapter 14.

Borderline Problems of Classical Physics

The role of borderline problems in triggering a change in the architecture of previously accumulated knowledge is particularly evident in the transformation of classical physics into modern physics in the early twentieth century. At the turn of the twentieth century, classical physics was divided into the subfields of mechanics, electrodynamics, and thermodynamics. Scientists at the time even discussed these domains as candidates for alternative foundations for the entire conceptual edifice of physics, in terms of a mechanistic worldview, an electrodynamical worldview, and another worldview that was to be built on basic concepts of thermodynamics such as energy.[31] However, hopes for the realization of such encompassing views based on a set of classical concepts soon turned out to be futile. The rapid development of specialization into separate subdisciplines brought forth knowledge relevant to more than one of these subdisciplines, thus inevitably leading to overlaps between distinct, highly structured bodies of knowledge. And these overlaps (the borderline problems of classical physics) proved to be a crucial driving force in further developments, acting as challenges from inside.

One such borderline problem was that of heat radiation in equilibrium, which was briefly introduced in chapter 4. This borderline problem arose in the overlap of the theories of heat (thermodynamics) and radiation (part of electrodynamics). There it was found that radiation, enclosed in a cavity at thermal equilibrium, has thermal properties that are independent of all specific features, such as the properties of the cavity's material. The universal energy distribution of this radiation could be determined experimentally by precise measurements and described exactly by the famous radiation formula proposed by Max Planck in 1900.[32] Planck's radiation formula solved this special problem at the borderline between electrodynamics and thermodynamics in a form that is still valid today. However, this success veiled a foundational crisis: the classical picture of a continuum of waves of all possible energies could not be reconciled with Planck's radiation formula. Instead, it turned out that

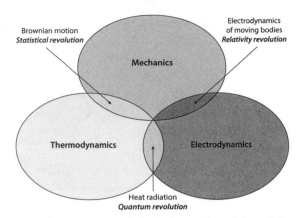

7.4. Scientific "revolutions" at the borderline problems of classical physics.

totally new, nonclassical concepts were necessary to find a physical interpretation of the energy distribution of radiation in thermal equilibrium, as described by this formula. The conceptual transformation triggered by this crisis eventually led to quantum theory.

Another borderline problem was that of the so-called electrodynamics of moving bodies, which arose at the crossroads between mechanics and electromagnetism. This became the birthplace of relativity theory.[33] It originated in the question of how optical and mechanical phenomena proceed in a moving frame of reference. The relativity principle of classical mechanics suggested that physical laws should be the same in reference systems moving with respect to each other in rectilinear and uniform motion. Classical optics suggested, on the other hand, that light was carried by an essentially immobile ether—a favored reference system at absolute rest. The Dutch physicist Hendrik Antoon Lorentz made increasingly successful but ever more contrived efforts to develop a theory that would do justice to the empirical knowledge about the problems in this border region, combining insights from optics, electromagnetism, and mechanics.[34]

Exploring the Horizon of Classical Physics

Eventually, Albert Einstein succeeded in harmonizing the seemingly contradictory principles of electrodynamics and mechanics by introducing new, nonclassical concepts of space and time. He was, however, not the only scientist to address these problems, and not even the only one to treat them as catalysts for rethinking classical physics. Einstein had found it curious that the interaction between a magnet and a conductor in relative motion is explained in different ways in electrodynamics, although the resulting effect is independent of whether the magnet or the conductor is considered to move.[35] In both cases, the resulting effect is indeed an electric current in the conductor, but according to classical electrodynamics, the explanation

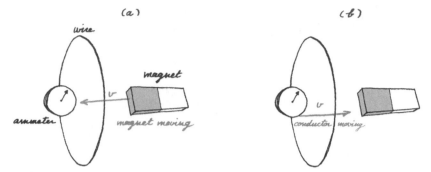

7.5. The magnet-conductor thought experiment. From Janssen (2014, 178). Drawing by Laurent Taudin.

varies with the choice of the system of reference: When the conductor is considered to move while the magnet is at rest, the charges in the conductor experience a magnetic force. But when the conductor is considered to be at rest while the magnet moves, the conductor experiences an electric force.

The independence of the interaction between magnet and conductor from the state of the observer would be an immediate consequence of the principle of relativity in mechanics . . . if only this principle were also valid for electrodynamics. But this possibility seemed to be excluded, as the hypothetical ether constituted the system of reference on which contemporary electrodynamics was based.

Motivated by such puzzles, Einstein explored the horizon of possibilities offered by contemporary theoretical and experimental means and the paper tools available to him. He even attempted to elaborate an alternative theory of electrodynamics in which the familiar laws of mechanics apply, the principle of relativity in particular.[36] Borderline problems are typically the stimulus for the proliferation of alternative theories. But Einstein's youthful attempt hardly stood a chance against the impressive theoretical framework elaborated by the Dutch physicist Hendrik Antoon Lorentz on the basis of Maxwell's theory. Although built on the notion of a stationary ether, Lorentz's theory was quite capable of coping with electromagnetic phenomena in moving frames. Lorentz first wrote down approximate transformation equations to derive phenomena in a moving frame of reference from the known laws for a rest system in 1895,[37] and then in their exact form in 1899.[38]

By 1904, Lorentz had a comprehensive and systematic theory.[39] With his transformations he could, in principle, explain all phenomena of the electrodynamics of moving bodies. The French mathematician Henri Poincaré called these transformations "Lorentz transformations";[40] they would become a core feature of the later theory of special relativity. In its well-developed form, Lorentz's theory already encompassed many of the remarkable phenomena for which the theory of relativity is known today: length contraction, as well as the dilation of time for processes observed from different frames of reference, and also the increase of a body's mass with its speed.

However, these aspects had a different meaning in Lorentzian theory. To explain the absence of any observable effect of the motion of a terrestrial laboratory through the stationary ether, Lorentz saw himself forced to extend his otherwise successful theory of electrodynamics by additional assumptions about the behavior of bodies moving through the ether. In 1892, he had introduced a new time coordinate for moving systems, which served as an auxiliary variable and allowed him to explain why experiments with terrestrial light sources show no effects of the order v/c owing to motion in the ether, where v is the velocity of the system and c the speed of light.[41]

Later, Lorentz introduced the assumption of a length contraction in moving systems[42] in order to be able to explain the famous Michelson-Morley experiment[43] as well, an experiment that can detect such effects to higher order in v/c. The introduction of special auxiliary quantities for lengths and times in moving systems extended the formalism of Lorentzian electrodynamics by elements whose physical interpretation was problematic from the perspective of classical physics. These addenda loosened the bonds of Lorentz's theory from the conceptual foundations of classical physics, turning the problem of the electrodynamics of moving bodies into an epistemic island.

The Emergence of an Epistemic Island

By the beginning of 1905, at the latest, Einstein clearly saw that there was no alternative to the electrodynamics of Maxwell and Lorentz and abandoned his attempts to search for an alternative theory. Yet, he had good reasons to doubt the fundamental conceptual asset of Lorentz's theory, the concept of a stationary ether, not only because it violated the relativity principle to which he was committed, but also because this notion had become physically problematic in other contexts in which Einstein was working at the same time, in particular in examining the thermal properties of the ether. He was also investigating (following the work of Wilhelm Wien and Max Planck[44]) the problem of heat radiation. This investigation had led him to the conclusion that Lorentz's concept of ether failed to yield an acceptable thermal equilibrium for radiation because the ether would consume all the heat, and the concept was hence unacceptable.

Thus, as a result of his broad overview of contemporary physics and familiarity with a number of its borderline problems, Einstein perceived the electrodynamics of moving bodies as being a peculiar scientific field that was sharply separated from the rest of classical physics. This "epistemic island" displayed its own internal logic ever more clearly, while its connections to the main body of classical physics were severely weakened. As we shall see below, it was by exploring the implications of this internal logic that Einstein eventually arrived at his insight into the need for new concepts of space and time to solve this problem.

The role of Einstein's specific perspective may seem to emphasize individual creativity to the point of undermining the attempt to account for such transformations in more general terms. One should not forget, however, that the emergence of multiple perspectives is part of the exploration of a knowledge system. It naturally involves individually differentiated learning processes, which, in the presence of triggers of change, tend to diverge even more widely than usual.

Lorentz's theory had emerged as the only viable solution to the problems of the electrodynamics of moving bodies. Yet, from Einstein's perspective, once he had questioned the concept of the ether, the entire theory became unacceptable, because without an ether there was no longer a basis for the crucial assumption of Lorentz's theory: that the speed of light in the ether is constant. Moreover, without an ether there was no physical mechanism for explaining the contraction of lengths, which Lorentz had introduced to account for the result of the Michelson-Morley experiment.

The Emergence of Special Relativity

Einstein's conclusion that all these critical concepts belonged to kinematics suggested to him that it was misleading to search for an overarching meaning at the concrete level of the theory of electromagnetism.[45] He instead focused on precisely those elements that held the key for the eventual solution: for one, the auxiliary assumptions introduced by Lorentz about times and distances in moving systems of reference, and for another, the constancy of the speed of light and the principle of relativity. Einstein therefore introduced the principle of a constant speed of light independent of a theory of ether, then placed this nonclassical principle alongside the classical principle of relativity as an axiom of a new theory. As it turns out, the Lorentz transformations can then be derived from fundamental principles and cease to represent an auxiliary construct.

If the space and time coordinates that occur in the Lorentz transformations are now interpreted as physically meaningful quantities, then these transformations imply that identically constructed measuring rods and clocks in a system that moves relative to an observer define units of length and time that are different from those given by identical rods and clocks stationary to the observer. This behavior obviously required an explanation, not at the level of electrodynamics (as in Lorentz's problematic attempt at a physical justification) but at the level of kinematics. At this point, further reflection became necessary—namely, a reflective abstraction incorporating another level of knowledge, not the level of specialized physical theories such as electrodynamics, but of practical operations with rods and clocks.

The first question that posed itself was whether there was any way of verifying the curious behavior of bodies and processes in a moving frame of reference. How do measuring rods and clocks behave in such a system? What does it mean to say "these events take place simultaneously," and how does one check that? Such questions made Einstein recognize the problem of ascertaining simultaneity in relatively moving systems of reference as the critical step for solving his problem. This resonated with his philosophical reading, particularly with the writings of David Hume and Ernst Mach.[46] Against this background, Einstein realized that the concept of time is not simply given, but represents a rather complicated construct, and that to ascertain simultaneity at different locations one needs a definition that must be based on a practical method. It is at this point that Einstein's insight into the nature of simultaneity becomes a plausible consequence of the preceding chain of reasoning,

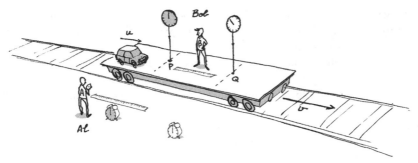

7.6. Example of a measurement with clocks and rods in a moving system. According to Al, Bob over-estimates the velocity of the remote control car with respect to the railroad car because, according to Al, Bob is measuring that velocity, the ratio of distance and time, with rods that are contracted and with clocks that are out of sync and run slow. See Janssen and Lehner (2014, appendix A, sec. 1.5). Drawing by Laurent Taudin.

and, at the same time, takes on its significance as the crucial step toward the solution of his problems.

Einstein's method for ascertaining simultaneity at different locations—the synchronization of spatially distant clocks by light signals that propagate with a finite speed—at first sight has nothing to do with the complicated physical problems he was trying to solve. It does not even seem to go beyond classical physics. Instead, this procedure is quite consistent with our everyday ideas about time measurement; it was even used in contemporary technical practice, as Einstein knew from reading popular scientific texts and his experience as a patent clerk.

Because of the inextricable connection between the concepts of velocity and time (which Einstein recognized through his synchronization procedure) the arbitrariness of the relation between definitions of time in different frames of reference could now be removed in two ways.

One way was to introduce the hypothesis that Einstein's method of establishing simultaneity should lead to the same results regardless of the state of motion, and to therefore conclude that there is indeed an absolute time, as assumed in classical physics. The second hypothesis was that the speed of light (as opposed to time) should remain constant independent of the state of motion—a hypothesis that Einstein naturally preferred in view of the success of Lorentzian electrodynamics, in spite of its nonintuitive consequences.

If the second hypothesis is chosen, then simultaneity is relative and depends on the state of motion—and one immediately obtains all the startling kinematics of the special theory of relativity, from the slowing of clock rates to the contraction of measuring rods in a moving system of reference.

Developing a Matrix for a New Theory

We thus see that Einstein's crucial last step connected two levels of knowledge—a theoretical and a practical—in a new way. His reflection on the foundation of the concept of time interlinked a simple method of time measurements at different locations with theoretical knowledge about the propagation of light from contemporary electrodynamics. Only through this connection could specialist studies on electrodynamics affect the fundamental concepts of time and space with consequences reaching well beyond the specific problem of the electrodynamics of moving bodies.

This connection also explains the specific historical place of special relativity, in the sense that it could not have emerged from philosophical reflection on time measurements performed at any prior time. In fact, the postulate of the constancy of the speed of light, which actually determined Einstein's new concept of time, was the result of the long-term development of the system of knowledge we call classical physics; it represents the quintessence of nineteenth-century electrodynamics and its problems at the borderline with mechanics.

The formalism developed by Lorentz (and in particular his treatment of the Maxwell equations of electrodynamics in a moving system of reference) acted as a bridge between the old, ether-based theory and the framework of special relativity. Einstein could essentially use the same formalism but give it a new interpretation. Lorentz's framework thus acted as a matrix for generating the theory that later became known as special relativity. Its new kinematics followed in essence from Lorentz's results by reversing the direction of inference in his arguments. What Lorentz had tediously achieved with the help of auxiliary assumptions—the compliance of electrodynamics with the relativity principle and the constancy of the speed of light—now became the starting point for Einstein's straightforward derivations.

Galileo's followers had similarly formulated the principle of inertia by working backward from Galileo's most advanced findings on parabolic trajectory, which also served as a matrix to develop a new approach to the theory of motion. As we saw in chapter 6, Galileo had achieved the insight into the parabolic and symmetric shape of trajectory by stretching and twisting the conceptual framework at his disposal. The assumption that a projectile shot in an oblique direction would continue its course along a straight line in uniform motion, if gravity ceased to act on it was a daring and problematic implication. Galileo's disciples and followers could see, however, that by starting from such an assumption (i.e., the principle of inertia) the parabolic shape of trajectory could be derived in an easy and straightforward manner. In the same way, Einstein could accomplish the final step in developing the special theory of relativity by inverting the problematical arguments that had led to the Lorentz transformations.

In the beginning, the new theory of special relativity (which Einstein published in 1905) was itself an epistemic island because it was only explicitly validated in a restricted domain of experience. As we have seen, it emerged from an investigation into the special problem of the electrodynamics of moving bodies, but it established a new framework of space and time that claimed validity for all of physics. It remained to be seen how the rest of physical knowledge could be assimilated into this new framework. This did not happen by a sudden switch from one view of the world to another, as Thomas Kuhn's concept of scientific revolution implies, but involved the work of an entire scientific community extending over many years.

This integration of the accumulated knowledge of classical physics with a new understanding of space and time was itself a conflict-laden process. One obvious task within this research program was the inclusion of the force of gravitation into the new framework. A natural way to pursue this task was to create a field theory of gravitation analogous to electromagnetic field theory and in agreement with special relativity. Accordingly, such a theory would have to comprise a gravitational potential analogous to electromagnetic potential and field equations analogous to Maxwell's equations to ensure that gravitational interactions do not propagate faster than light. In other words, the example of electrodynamics offered a mental model for how such a gravitational field theory should look, with an appropriate adaptation of its default settings, of course.[47]

Understanding the role of this theory as a mental model for Einstein's heuristics has indeed turned out to be crucial for reconstructing his search for a gravitational field theory. The basic idea of this mental model is that matter acts as a source, that a source creates a field, and that this field then prescribes the motion of matter. This qualitative model does not yet involve any mathematical framework and is hence extremely malleable, which is its strength. At the same time, it can be flexibly enriched with more specific information. The way the source creates the field can be described by a field equation, while the way the field prescribes the motion of matter can be captured by an equation of motion. At each step in the further specification of this model, additional information can be taken either from prior experience (in particular from experience with electromagnetic field theory) or from newly identified information, obtained, for instance, through real-time constructions using the paper tools at hand. Let us now take a close look at Einstein's investigative pathway.

Special Relativity

Einstein based his special theory of relativity on two postulates: the relativity postulate and the light postulate. The relativity postulate says that the same laws of physics hold for all observers moving with respect to one another at a constant velocity. The light postulate says that in the frame of reference of any such observer, light always has the same velocity c regardless of the velocity of its source.

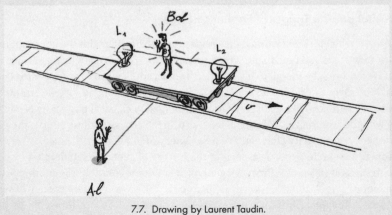

7.7. Drawing by Laurent Taudin.

It is a direct consequence of these two postulates that whether or not two events at different places happen at the same time depends on the state of motion of the observer. This is called the relativity of simultaneity. It is illustrated in the figure here. Bob is standing at the exact center of a railroad car moving to the right at constant velocity v with respect to Al, who is standing by the tracks. According to the relativity postulate, the laws of physics should be the same for both. The light postulate is one of those laws. Two light bulbs, L1 and L2, are attached to the ends of the railroad car. Each light bulb flashes once. Suppose these flashes reach Bob at the same time; on this Al and Bob will agree. But they will disagree about the simultaneity of events happening at different places—namely, the flashing of L1 and L2.

Did these two light bulbs flash at the same time? Bob will say that they did. The light postulate tells him that both light flashes had velocity c with respect to him (they just moved in opposite directions). Since Bob is standing in the middle of the railroad car, both flashes had to cover the same distance to reach him. If they reached him at the same time, they must have left the light bulbs at the same time. For Bob, therefore, the flashing of L1 and the flashing of L2 are simultaneous events.

Al disagrees. The light postulate tells Al (as it told Bob!) that both light flashes had velocity c with respect to him, Al (again, they just moved in opposite directions). From Al's point of view, Bob was rushing away from the flash coming from L1 and toward the flash coming from L2. Hence, according to Al, the flash coming from L1 had a longer distance to cover to get to Bob than the flash coming from L2. Yet the two flashes reached Bob at the same time. It follows, Al concludes, that L1 must have flashed before L2.

Whether the flashing of L1 and the flashing of L2 happened simultaneously or not thus depends on whether we ask Al or Bob. In general, the simultaneity of two events occurring in different places must be dependent on the state of motion of the observer if the velocity of light is independent thereof.

Gravitation as a Trigger of Further Change

In 1907,[48] when Einstein worked on a review article dealing with the consequences of the new space and time framework introduced by special relativity for various domains of physics, he found, to his surprise, that its application to gravity raised new puzzles.[49] Over a period of several years, extending from 1907 to 1915, he attempted to reconcile the knowledge of gravitation embodied in classical physics by Newton's law with the requirements of special relativity. In the course of these efforts, the general theory of relativity emerged as a new theory of gravitation.[50]

There are two remarkable aspects of the history of general relativity that I would like to stress at this point: First, the general relativity revolution was another "slow revolution." It took about half a century before the theory was mature enough to serve as a generally applicable framework for treating problems of space-time physics. This only happened in connection with the emergence of a broader epistemic community exploring the consequences of the equations established by Einstein in 1915. I return to this issue in chapter 13 when examining the coevolution of systems of knowledge and epistemic communities as another aspect of scientific revolutions.

The other aspect I would like to emphasize concerns the astonishing robustness of Einstein's original achievement, in spite of all the subsequent transformations of the theory of general relativity. In the century following the publication of his gravitational field equations in 1915,[51] the new theory was indeed spectacularly confirmed.[52] It predicted the bending of light and a redshift of its color in a gravitational field; it could explain the expansion of the universe, black holes, and gravitational waves. None of these effects were known when Einstein formulated the theory. The only empirical clues suggesting that a modification of Newton's law of gravitation might actually make sense were certain, very minor deviations of the planetary orbits from Kepler's laws. We are thus led to a new question: What gave general relativity this impressive predictive power and stability? A plausible answer to this fundamental question cannot be given as long as one sees general relativity as the brainchild of one particularly ingenious scientist. General relativity was rather, as I argue below, not a replacement but a reorganization of the knowledge of classical physics, which continued to underpin and fuel the theory's explanatory power.

Einstein's Principle of Equivalence

This is perhaps most evident from a thought experiment that Einstein used when trying to solve the puzzles he encountered when first attempting to fit gravitation into the scheme of special relativity. One of these puzzles was that such a special relativistic theory of gravitation seemed to imply a violation of Galileo's principle that all bodies fall with the same acceleration, whatever their constitution. Bringing together the classical understanding of gravitation with special relativity thus amounted to identifying a new borderline problem.

Einstein then had the idea that Galileo's principle could be preserved if one is allowed to simulate gravity by acceleration. In a uniformly accelerated box somewhere in outer space, an observer would feel a force dragging bodies to the floor, just as if the box were placed on the firm ground of a planet like Earth, subject to a real gravitational field. In such a box, all bodies would fall to the floor with the same acceleration owing to the so-called inertial forces acting within it. Considering such forces of acceleration to be equivalent to "real" gravitational forces and treating them according to special relativity gave Einstein a powerful instrument with which to infer the properties of a relativistic theory of gravitation, long before he was able to actually formulate such a theory.

Einstein called the insights of his thought experiment the "principle of equivalence." The principle of equivalence states that all physical processes in a uniform and homogeneous gravitational field are equivalent to those that occur in a uniformly accelerated system of reference without a gravitational field. It can be conceived as a combination of two mental models comprising two laboratories in different states of motion. In the first case, the laboratory is at rest and a gravitational field is present. In the second case, there is no gravitational field, but a constant external force

7.8. Einstein on his way to the equivalence principle. Drawing by Laurent Taudin.

accelerates the laboratory. By way of this thought experiment, Einstein connected the two mental models, each of which embodies shared knowledge of classical and special relativistic physics. The combination of the two models allowed him to consider this knowledge from a new perspective that opened up the gateway to general relativity by allowing him to simulate gravitational effects by studying accelerated systems of reference.

Mathematical Representation versus Physical Interpretation

At some point, it became clear to Einstein that the gravitational potential in his new theory of gravitation should be mathematically represented by the so-called metric tensor, a complicated mathematical object known from non-Euclidean geometry as it had been developed in the mid-nineteenth century.[53] The key problem of the new field theory of gravitation was to find a field equation relating this gravitational potential to a given distribution of mass and energy (the sources of the field). Through such a field equation, mass and energy would then determine the gravitational field, just as electric charges and currents generate an electromagnetic field.

The main challenge of finding a gravitational field equation was to relate the sophisticated mathematical formalism of which the metric tensor was part to the equally sophisticated physical demands on such a field equation. Among these physical demands was, in particular, the requirement that the new field equation reproduce the familiar and well-established Newtonian law of gravitation under appropriate circumstances. After all, this law was, apart from the minor deviations mentioned above, confirmed by a wealth of empirical observations.

It was, at first, not at all clear to Einstein how this physical demand could be fulfilled within the new mathematical formalism of the metric tensor. That formalism came, of course, with its own rules—rules that suggested candidates, none of which however seemed to fulfill the physical criteria that had to be imposed on them. We thus recognize again the heuristic role and epistemically productive character of paper tools and their semiotic rules.

In reaction to this situation, Einstein (together with his mathematician friend Marcel Grossmann) developed a double strategy.[54] On the one hand, they explored sophisticated mathematical paper tools, attempting to interpret them as external representations of physical concepts such as energy and gravitational force. That was their mathematical strategy. On the other hand, they investigated mathematical representations that could be directly interpreted in familiar physical terms because they were so similar to Newton's well-known law. Then they attempted to elaborate on their mathematical consequences, seeking contact with the sophisticated formalism. That was their physical strategy.

The Metric Tensor

The concept of a metric is essentially a generalization of the concept of distance. In 1907, Hermann Minkowski, Einstein's former teacher at the ETH (Eidgenössische Technische Hochschule) in Zurich, developed a mathematical formalism to represent physical events in space and time and also the relations between these events as implied by Einstein's special theory of relativity. This formalism combines space and time into one entity—space-time—and is equipped with a "metric" instruction that is employed to measure the geometric distance between any two physical events that occur at different positions and different times. The mathematical formalism consists of a four-dimensional space-time in which coordinate systems representing physical frames of reference are used to characterize each physical event by four numbers: its three space coordinates and one time coordinate. The time coordinate is multiplied by the constant speed of light, and thus also acquires the dimension of a spatial coordinate, so that Minkowski's four-dimensional world becomes even more similar to the three-dimensional Euclidean world of classical physics.

In three-dimensional Euclidean space, the familiar metric instruction is to measure the distance between two points with the help of the Pythagorean theorem by forming the sum the squares of their Cartesian coordinate separations. In Minkowski's space-time, the square of this distance is instead the square of the time separation between the two events *minus* the square of their spatial separation. It is essentially an extension of the Pythagorean theorem to four dimensions, adapted to the special nature of the time coordinate. Observers moving at constant velocity with respect to each other may compute this value using their respective positions and time measurements and will obtain the same result.

Whereas a flat surface is characterized by a metric that behaves in the same way everywhere on the surface, the geometric properties of a curved surface must be described by a variable metric. Such a metric associates different actual distances with a given coordinate distance at different locations on the surface. This variable metric turned out to be a suitable representation of the gravitational potential. In a curved four-dimensional space-time, ten numbers are needed to calculate the distance from one point to any neighboring point. It is convenient to present these numbers in a four-by-four matrix array. This array is the "metric tensor," which reflects the geometrical properties of space-time in the chosen coordinate system.

A Transitional Synthesis

Only after several years of exploring these complementary lines of research did Einstein eventually find a field theory of gravitation in which a mathematical representation could be seamlessly matched with the physical interpretation of the new theory. An important step on this path was the elaboration (again, with Grossmann) of a provisional theory that acted as a transitional synthesis and eventually as a matrix for formulating the final version of general relativity. In hindsight, we can see it as the scaffolding with the help of which the final theory was built.

This provisional theory thus played a similar role as a transitional synthesis for the formulation of general relativity as did Lorentz's theory for the emergence of special relativity, and as preclassical mechanics did for classical mechanics. It became a key instrument for bringing together various relevant components of knowledge: Newton's law of gravitation, field theory, mathematical formalisms, and planetary astronomy were suddenly housed within a single framework. Like the various heterogeneous elements that fed into Darwin's theory of evolution, these components were eventually turned from contingent premises into intrinsic elements and supportive evidence of Einstein's theory.

It was the development of the provisional theory that eventually allowed Einstein and his collaborators to build a bridge between the physical and the mathematical requirements on his emerging theory of gravitation. In the end, the final theory could essentially be gleaned from the provisional theory by reinterpreting some key components of the formalism developed for it. The physical interpretation of the new theory was now significantly different from the concepts that had formed Einstein's starting point. The new interpretation was the result of adapting the meaning of prior physical concepts to the implications of the mathematical formalism explored along Einstein and Grossmann's double strategy. The exploration of this external representation thus guided a process of concept transformation in which its implications eventually became constituents of new physical insights, such as the understanding of gravitation as a geometric property of space-time.

The Protracted Process of Knowledge Integration

The integration of knowledge at the roots of general relativity was the result of a laborious and conflict-laden process to which not only Einstein but also many other scientists contributed. This process was in no way concluded when Einstein published the gravitational field equations in November 1915; it lasted at least until the 1950s. Just as we saw in the case of Darwin's theory of evolution, general relativity was, at the moment of its birth, hardly a broadly accepted and generally applicable theoretical framework.

Again, we see the *longue durée* and incremental phases in the transformation of a system of knowledge, beginning with a stage of conceptual reorganization involving intermediate constructions that in hindsight may appear as scaffolding for the final result—a result in this case strongly associated with Einstein as the single pivotal figure.[55] Such a singular association is, however, not necessarily the case for all such transformations, as the emergence of quantum physics, with its many key players illustrates.

This first stage is followed by one in which the new system is cleansed of traces of the transitional phase (the scaffolding being removed). At the same time, paradigmatic domains of application are established, and the immediate implications of the external representations and their interpretation are further explored, often with conflicting consequences. In the case of general relativity, these were its formative years,[56] with the establishment of the first exact solutions of the theory and its first application to cosmology (soon to become one of its paradigmatic domains).

Finally, we may see a stage in which the limitations of such paradigmatic applications are overcome, while the new system of knowledge becomes a stable, broadly applicable framework in which its formalism and its conceptual interpretation are finally balanced. In the case of general relativity, this only happened after its so-called low water mark period (as described by Jean Eisenstaedt),[57] during which it was hardly taken seriously as a conceptual system in its own right, only to come into its "renaissance" in the late 1950s (Clifford Will's term),[58] a theme to which I return in chapter 13.

This is, of course, by no means a necessary sequence of phases. As I have emphasized before, triggers of change may historically change (e.g., from challenging objects to borderline problems); so too may mechanisms of knowledge integration. Most important, the evolution of knowledge is no autonomous process in its own right, shaped as it were by the intrinsic structures of systems of knowledge. Whether or not a system of knowledge is actually explored in a given historical situation, thus initiating a dynamic of change, depends on broader societal conditions. In order to understand the evolution of systems of knowledge, we must therefore consider these societal conditions in greater depth.

Heisenberg's Matrix Mechanics as a Transitional Synthesis

In 1925, Werner Heisenberg published a seminal paper titled "A Quantum Theoretical Reinterpretation of Kinematical and Mechanical Relations."[59] It effectively founded quantum mechanics as we know it.[60] Heisenberg supposedly came across the key idea in a single night in the summer of that year, during a short vacation to the island of Heligoland. In truth, the so-called Quantum Revolution was another protracted process of knowledge transformation. This process unfolded over more than a century, from the enigma of spectral lines discovered in the nineteenth century to the mature formulation of quantum mechanics in the early 1930s and on through the interpretational debates that continue today. One of the turning points was Planck's formulation of his famous radiation law in 1900;[61] another was the atomic model which Niels Bohr proposed in 1913.[62]

The period between 1913 and Heisenberg's 1925 paper is often described as the reign of the "old quantum theory." It resembles the era of preclassical mechanics in that a group of scientists attacked a patchwork of loosely related problems, seeking a unifying conceptual system that would supersede an older framework and offer a coherent solution. In both cases, the new system emerged from the exploration of a variety of paths still stemming from the old framework; this helped bring together the relevant knowledge, albeit in a preliminary form.

Heisenberg's paper constitutes just such a transitional synthesis. It acted as a seedbed for the new theory in the sense that it offered a generalizable formalism that could subsequently be extended and interpreted as a new semantics. Heisenberg's formalism could be extended to a calculus of matrices and interpreted as a quantum translation of the dynamical equations of analytical mechanics, as was soon realized by Max Born and Pascal Jordan.[63]

Still, Heisenberg's seminal paper was merely a transitional synthesis. This was not only the case because he did not yet realize that his formalism corresponded to matrix calculus, but also because the specific physical problem from which his proposal had originally emerged—the calculation of spectral intensities—restricted his semantics. In his paper, Heisenberg therefore emphasized that his formalism only involved such observable quantities, a circumstance to which he ascribed a programmatic role.[64]

This narrow interpretation was only overcome when Born and Jordan completed Heisenberg's approach to a full-blown formalism, which could now be seen as a universal framework based on classical analytical mechanics. However, the semantics of the new quantum mechanics remained a key issue in the interpretational debate following the establishment of its formalism. The history of quantum mechanics thus offers another example of both the role of knowledge integration through transitional synthesis and the importance of the interplay between external representations and their interpretation in the emergence of a new system of knowledge.

Before we turn to the societal dimension of the evolution of knowledge, one final remark may be appropriate on epistemology, or more precisely, on the role of the architecture of knowledge for different outcomes of knowledge developments. At the end of chapter 4, we investigated the conditions under which changes in knowledge systems lead to the transformation of a fundamental framework involving changes in such abstract concepts as space and time. I have pointed to the role of deeper levels of knowledge such as those concerned with an elementary understanding of measurements of space and time using rods and clocks. When an emerging new formalism—for instance, a mathematical formalism—can still be interpreted in relation to such practical knowledge, the formalism may in turn assume a concrete physical meaning and at the same time induce a change in the abstract concepts related to the practical knowledge.

These were essential prerequisites for the creation of Einstein's theory of special relativity (and of general relativity, but for this argument and for reasons of simplicity I remain with special relativity). As discussed, special relativity emerged from an attempt to solve the concrete problems of the electrodynamics of moving bodies, which eventually became a matrix for building up a new, general framework for space and time. Intermediate stages functioned as scaffolding for this new framework, sustaining the construction of a new formalism that could be interpreted in terms of space and time concepts by linking elements of the new formalism to familiar measurement operations. Subsequently, a new mathematical formalism was set up: Minkowski's four-dimensional space-time framework. This served as the general platform for formulating specific physical theories, automatically fulfilling the requirements of relativity theory.

The case of quantum theory is similar but somewhat more complicated. It also emerged from the treatment of specific problems that gave rise to a transitional synthesis: Heisenberg's 1925 paper. This paper (along with Erwin Schrödinger's wave mechanics) acted as the seedbed or matrix for the emergence of a new kind of mechanics by first establishing a general formalism and then by interpreting it as a quantum version of classical mechanics (see box above). As in the case of relativity, the mathematical formalism serving as the general platform for formulating specific quantum theories was elaborated only later in 1927 by John von Neumann, who formulated quantum mechanics as a theory of observables represented by operators on an abstract Hilbert space of quantum states.[65]

But in contrast to relativity theory, it turned out to be much harder to find a convincing operational interpretation of this framework by establishing contact with a more familiar layer of practical knowledge such as that of the space and time measurements in relativity. The much discussed "measurement problem" of quantum physics and the ongoing interpretational debate of quantum theory are consequences of this difficulty. While this debate is still open, the need to establish *some* relation of the theory to layers of practical knowledge has always been broadly acknowledged, beginning with Niels Bohr's contentious ideas on the "complementarity" of the macro- and the micro-world of quantum physics.

Unlike relativity theory, which can be characterized as offering a new understanding of the familiar concepts of space and time, it is not so obvious to say what is actually the more elementary layer of knowledge that is revised by quantum theory. What is the theory really about? Is it the quantization of energy, the wavelike behavior of particles, or the structure of information, embodied in new sorts of probabilistic correlations in nature?[66] The fact that quantum theory has no similarly familiar counterpart in more elementary layers of knowledge (as is the case for relativity theory) makes it no less convincing but much harder to digest. The Nobel Prize–winning physicist Richard Feynman went so far as to claim that nobody understands quantum mechanics:

> There was a time when the newspapers said that only twelve men understood the theory of relativity. I do not believe there ever was such a time. There might have been a time when only one man did, because he was the only guy who caught on, before he wrote his paper. But after people read the paper a lot of people understood the theory of relativity in some way or other, certainly more than twelve. On the other hand, I think I can safely say that nobody understands quantum mechanics. . . . I am going to tell you what nature behaves like. If you will simply admit that maybe she does behave like this, you will find her a delightful entrancing thing. Do not keep saying to yourself, if you can possibly avoid it, "But how can it be like that?" because you will get "down the drain," into a blind alley from which nobody has yet escaped. Nobody knows how it can be like that.[67]

The deeper reason for this incomprehensibility of quantum theory lies in its somewhat skewed architecture of knowledge, that is, the lack of a simple resonance between its abstract theoretical concepts and those of a more familiar "ground level" of physical knowledge. The epistemological parallels displayed by the genesis of relativity and of quantum theory are nevertheless striking: In both cases, the emergence of a new fundamental framework required the construction of a new formalism, its operational interpretation, and its transformation into a platform with universal applicability.[68] These parallels hint at a more general significance of the concepts used for describing them, and this is indeed what I have in mind. Like the concept of scaffolding (introduced into this context by Michel Janssen), the concept of platform is also used here (following Manfred Laubichler) in an evolutionary sense.[69] In this more general sense, a platform can be understood as an environment with a regulatory function that opens up and determines a specific space of possibilities, similar, for instance, to the role of an operating system that enables various apps to run on a smartphone, or of a cell in which genes can execute their functions. In chapter 14, these recent ideas in evolutionary theory and their application to cultural evolution are discussed in greater detail.

PART 3

HOW KNOWLEDGE STRUCTURES AFFECT SOCIETY AND VICE VERSA

Chapter 8

The Economy of Knowledge

We see how the history of *industry* and the established *objective* existence of industry are the *open book of man's essential powers*, the exposure to the senses of human *psychology*. Hitherto this was not conceived in its inseparable connection with man's *essential being*, but only in an external relation of utility, because, moving in the realm of estrangement, people could only think of man's general mode of being—religion or history in its abstract-general character as politics, art, literature, etc.—as the reality of man's essential powers and *man's species activity*.

—KARL MARX, *Economic and Philosophic Manuscripts of 1844*

In this chapter,[1] I analyze the relation between knowledge and society with the help of a general concept of a knowledge economy as the ensemble of practices and institutions by which societies produce and reproduce knowledge.[2] I begin with a few remarks on the concept of practice as treated in current social and behavioral sciences and then formulate some fundamental assumptions to serve as the background for my analysis of institutions (the "anthropological gamut," as I call it). I consider institutions as regulatory structures of human societies that involve knowledge in an essential way. On this basis, I then develop the concept of a knowledge economy in greater detail, discussing knowledge transfer, the role of material embodiments of institutions, the character of normative systems, and mechanisms by which knowledge economies change. Just as chapter 4 introduced some basic notions for the analysis of systems of knowledge, illustrated with concrete examples in subsequent chapters, I here mainly discuss general aspects of knowledge economies, an analysis that is then put to work in chapters 9 and 10; in part 4, I investigate the dissemination of knowledge more generally.

Knowledge in Society

I have argued in the beginning that our predicament in the Anthropocene is also the result of the knowledge that human societies have accumulated over the course of millennia. In particular, human societies have developed structures for fostering this accumulation of knowledge and have become, in turn, ever more dependent on the knowledge produced and the ways in which it is distributed, shared, lost, or suppressed. But what exactly is the role of societal structures in the production and evolution of knowledge, and how do societal institutions distribute and rely on knowledge in return? The relation between the two is, in any case, not one-sided. Social structures such as institutions of learning evidently condition the development of knowledge, but knowledge may also become essential to societal stability and change—as when the generation of new knowledge becomes a matter of economic productivity in capitalist societies. Knowledge may be the answer to problems of social order (as has been claimed)[3] just as social order is shaped by the systems of knowledge available to a given society. Sheila Jasanoff refers to this interdependency as "coproduction": "Briefly stated, coproduction is shorthand for the proposition that the ways in which we know and represent the world (both nature and society) are inseparable from the ways in which we choose to live in it. Knowledge and its material embodiments are at once products of social work and constitutive of forms of social life; society cannot function without knowledge any more than knowledge can exist without appropriate social supports."[4]

How can one conceptualize this interdependency in general terms? In the following pages, I introduce the notion of a "knowledge economy" to describe and analyze the interconnection between society and knowledge. I thus use this term much more generally than do current economists and sociologists when they refer, for instance, to the transition of the global economy from a labor-intensive economy to a knowledge economy. In my usage here, a knowledge economy is the ensemble of all societal practices, including institutions, concerned with the production, reproduction, transfer, distribution, sharing, and appropriation of knowledge in a given society.

This perspective requires a reexamination of some basic concepts of social science—an undertaking I can pursue here in only a preliminary way. In analyzing societal structures, I will limit myself to a broadly conceived concept of societal institutions and how they involve knowledge. I argue that societal development can ultimately be understood as a coevolution of institutions and knowledge as two among several regulatory structures of human practice (individual and collective). In part 5, I return to this use of evolutionary terminology. The starting point of my investigation here is the concept of practice.

Practice as a Borderline Problem

As I have argued in the preceding chapters, when studying the history of knowledge, understanding the nature and role of actions is central. Knowledge results from experiences acquired in actions, it accompanies human practices, and it constitutes a potential for mentally anticipating actions. Here I introduce the wider concept of practice as referring to ensembles of actions within a societal framework, that is, actions that are not just performed by an individual ad hoc but that are part of societal structures, material culture, and knowledge traditions.

In the late nineteenth century, questions about the nature and role of practice became central to pragmatism, a philosophical school launched in the United States by Charles Sanders Peirce and William James, and then developed by John Dewey,

George Herbert Mead, and others.[5] The work of these thinkers made it evident how far-reaching an investigation this subject could and must be, including not only the classical themes of philosophy, but also sociology, psychology, and history. Many studies in the social sciences today are influenced by what has been called the "practice turn" in social theory, which ranges from organizational research via the history of science and technology, and from gender studies to media and lifestyle studies.[6]

Different social theories center on different conceptions of the social. They either place the emphasis on societal structures, on actors guided by some form of rationality, on actors bound by institutionalized norms, or on culture as a material and interpretative matrix from which the social is constructed. At first sight, it is surprising that these approaches should come as alternative, rather than fundamental possibilities. Human societies are evidently governed by structures that are not necessarily directly accessible to individuals—nor need individuals even be aware of them, such as those related to societal or even global divisions of labor, as emphasized by Émile Durkheim.[7] Individual actors are clearly guided by some form of rationality and self-interest, although these can hardly be taken as universal, as was claimed by rational choice theory. Individual actions often follow normative rules corresponding to social expectations and roles, as Talcott Parsons, following Durkheim, has observed.[8] Finally, actors are enabled to act within the world they have access to through material means, which they can interpret in terms of shared symbol systems, cultural codes, and orders of knowledge.[9]

While the former has been emphasized in Marxist theories, the latter has been stressed by philosophers such as Ernst Cassirer, and then again in the cultural turn of the 1970s. In their concrete investigations, sociologists and historians are often pragmatic in making use of these various approaches, as well as of analyses on the micro, meso, or macro levels, in eclectic ways. The challenge is to integrate these approaches into a theoretical framework that also includes the evolution of knowledge. This appears to be all the more necessary since, as I discuss in part 5, our entry into the Anthropocene was the result of cumulative and often contingent development involving all social dimensions.

Meanwhile, all of these domains have undergone substantial independent developments, including accumulation of empirical knowledge and a diversification of theoretical directions. Investigations into the nature and role of actions and practice are being pursued by many disciplines (e.g., sociology, history, psychology), and pragmatism is only one among many possible theoretical approaches. Yet, the study of actions and practice remains particularly challenging because it now constitutes, against the background of the unfolding of different disciplinary perspectives, a borderline problem in the sense discussed in chapter 4.

As we have seen, borderline problems occur when objects of investigation cause different disciplinary perspectives to intersect, generating clashes and novel insights. These insights are not only triggered by tensions between the various conceptual systems and disciplinary practices that converge in a particular area of investigation; they are also engendered by the inevitable confrontation with a shared problem that is independent of these disciplines and can be concretely explored. Actions and practices are currently explored within the social and behavioral sciences, in the context of historical investigations and even from biological and neurological perspectives.[10] Bringing all these perspectives to bear on a common problem should therefore provide a source of major innovation, giving rise to an integration of knowledge that may well challenge some of the fundamental assets of the disciplines concerned.

The Anthropological Gamut

I suggest that such an integration of knowledge should begin from what I would like to call, for lack of a better term, the "anthropological gamut": Humans are animals. They share with other life forms the need to metabolize with their environment, to sexually reproduce, and to die. This biological constitution induces needs, desires, and fears. Furthermore, humans share with other animals the capacity to anticipate some of the consequences of their behavior and to interact with their environment using tools.

What makes humans somewhat special (albeit not exclusively so) is that they are social beings who live in communities and maintain themselves through socially transmitted interactions with the environment, some of which are traditionally designated as labor. These interactions comprise, in particular, sets of socially shared patterns of individual or collective actions involving a variety of tools and knowledge. These actions are neither predetermined by biology nor by the environment but can be directed at solving problems that individuals or groups encounter in their environ-

8.1. Peter Paul Rubens, *Massacre of the Innocents*, ca. 1609/11, at a time when massacres were a reality in Antwerp. Self-destruction is an always imminent danger in human societies, even in the presence of sophisticated culture. Wikimedia Commons.

8.2. Quarry at Rano Raraku, Easter Island, used by the Rapa Nui people to supply volcanic stone for sculpting their gigantic statues, the moai. The reasons for the demographic collapse of the indigenous population remains an issue of scholarly debate; contact with Europeans has had, in any case, the most devastating effect on the local population. See Hunt and Lipo (2012); Stevenson, Puleston, Vitousek, et al. (2015); Middleton (2017). Wikimedia Commons.

ments. Practices are appropriated in ontogenesis and through individual experiences and are transmitted by participation in collective activities. These learning processes cannot be accounted for without taking into account human thinking, communication, and external representation of thinking, as have been investigated in the vast field of psychology. Human culture is transmitted in terms of holistic systems of knowledge.

The reproduction of human societies relies on a variety of goods, their availability to satisfy human needs and desires, and the tools for their production. It depends on the transmission of a material culture that comprises these tools, on the transmission of the social relations in which these tools are employed, and on the transmission of the knowledge to use these tools and to act in accordance with existing social conventions. In any given human society, the available means of production are employed within particular divisions of labor, which, from a certain historical stage onward, creates a separation of intellectual and material labor. The material means and contexts of action regulate and constrain practices, and occasionally open up unanticipated opportunities and difficulties.

While there is a wide range of conceivable human practices beyond those strictly involved in societal labor, human societies will perish if the sum of these practices does not accomplish their physical survival, if it uses the available resources in non-sustainable ways, or if it otherwise leads to self-destruction. Similarly, practices of drinking, eating, sleeping, giving birth, and so on may be almost infinitely malleable but cannot be arbitrarily suspended without the risk of death or extinction. Death also imposes specific constraints and challenges on human practices, such as their transmission to the next generation. All in all, we see the necessity for iteration and the potential for conflict, for example, over limited resources and opportunities, opportunities to learn and forget, the accumulation or loss of material culture, and so on. These necessities remain as persistent to human history as they are left unspecified by these critical anthropological facts.

Social Institutions

Human societies are typically structured by an ensemble of institutions that, taken together, must ensure their reproduction.[11] But institutions are not the most general form of social interaction. Social interactions may form, for instance, networks that have only a spontaneous or temporary nature. I discuss such networks in greater detail in chapter 13. What distinguishes institutions is the regular and more permanent character of the social interactions involved in them. Their regulative structures are transmitted from one set of actors to another and, in particular, across generations by learning processes. This regularity may be articulated in various forms, for instance, in rituals or rules that become means for the transmission of an institution independent from the concrete contexts of social interactions.

Social institutions regulate human interactions, enabling human societies to cope with certain regularly recurring problems, from those related to mastering collective challenges by cooperation, to the division of labor and the power structures related to it, to the acquisition and redistribution of resources, and to the resolution of societal conflicts. This occurs in the context of biological and economic reproduction, which constitute the origins of all social institutions, but also by ritualizing collective behavior, by exerting power, by explicit agreements, by the market, or by material arrangements such as architecture or traffic systems. Institutions are constrained but not determined by the material culture of a society. They possess a self-referential quality that is characteristic of human societies, that is, individual actions are simultaneously constitutive of and determined by the totality of interactions maintaining an institution or a society. Hence, the enforcement of institutional regularities may take various forms but typically has a systemic character. Their maintenance (e.g., through mechanisms of social control) involves costs, that is, a certain amount of societal labor.

With the division of labor, the formation of social hierarchies, and the further differentiation of societies, institutions gain relative autonomy and enhance both their interdependence and their self-referentiality as arenas of coordinated human interaction. In other words, institutions may become their own purposes. Nevertheless, the collection of institutions existing in a given society is confronted with the challenge to reproduce the society as a whole, ensuring the survival of its members. This global challenge consists primarily in maintaining a suitable division of labor but may involve adaptation to new environmental conditions. How such global challenges are translated into individual and collective reactions is mediated by the ensemble of a society's institutions and by the knowledge transmitted through them.

Institutions and Knowledge

Just as knowledge relies on institutions, institutions rely on knowledge. They embody and transmit knowledge, in particular the capacity of individuals to anticipate actions, as well as knowledge on social cooperation, control, and conflict resolution. Institutions form the bases of knowledge systems, which then in turn become the condition for those institutions' stability and further development. Institutions, however, do not think.[12] The knowledge they embody or enable is typically intersubjectively shared knowledge, but it is ultimately always the knowledge of individuals. Since institutions mediate cooperative actions, they have to rely on this distributed knowledge and engender collective thinking processes. A critical condition for the stability of institutions when faced with challenges is the capacity to break down solutions into individual actions. The failure to do so may result in the modification of existing institutions or the creation of new ones.

I have adopted here a broad concept of institutions, not requiring, for instance, that rules be formalized or enforced from outside.[13] Yet only when there is some form of external, material representation of institutional regulations (as well as an internal representation in the minds of individuals) can they structure collective actions distinguishable from random behaviors. Institutions represent the potential of a society or group to coordinate the actions of individuals and to interact with their environment. In the most general sense, institutions can be conceived as encoded collective experience, resulting in sets of shared behaviors connected by cognitive, social, and material links, including those of the symbolic world of external representations, paper tools, and so forth.

As "action potentials," institutions are a regulative structure of human practice similar to knowledge. But there are also important differences. There is no knowledge without the possibility of a *mental* anticipation of actions—but institutions must regulate collective behavior without mental anticipation in cooperative actions. The regulation is rather realized by the constraints and material conditions inherent in such cooperative actions, by behavioral norms and belief systems, and, in particular, by what we call here the economy of knowledge. Thus, for example, "the police"—as an institution—does not think and a tax authority cannot *mentally* anticipate its actions (e.g., the ramifications of raising the income tax by 1 percent). Thinking is the privilege of individuals; for an institution, the anticipation of the consequences of its actions is a matter of mediating between individual and shared knowledge. The extent to which institutions are capable of anticipating the consequences of their actions always depends on the ways in which they store prior experiences, make them available to individual actors, and coordinate individual thought processes—in other words, on their knowledge economy. The incapability of societies to "think" is an epistemic dilemma that we may try to alleviate by improving its knowledge economy, but from which we can hardly escape.

What Is an Economy of Knowledge?

Let me summarize: Every society needs to deal with processes of production, integration, transmission, distribution, appropriation, and reproduction of knowledge.

Cooperative practices require prior knowledge, as well as a coordination or parallelization of thought processes, thus involving communication as an unavoidable correlate to individual thinking. Communication, preservation, and accumulation of knowledge require external representations. As discussed above, practical knowledge, for instance, can be represented by instruments and rules of trade, and is usually transmitted by direct participation in work processes and by the oral communication of rules.

All institutions rely on specific knowledge, which they generate or presuppose. They thus depend on the broader societal knowledge economy, and their activities may include the generation, transfer, transformation, and appropriation processes of knowledge involving different kinds of external representations. Material and knowledge economies are interdependent. For any material product of a society (a particular machine, for example) we may distinguish between the knowledge required to invent it, to produce it, and to use it. The knowledge economy of a society has to generate, allocate, and maintain the corresponding systems of knowledge. And it has to be capable of integrating new knowledge into existing or newly created knowledge systems.

Individual actors acquire historically specific cognitive and normative judgments through acculturation under given social conditions. Education, learning, and training in the contexts of family, apprenticeship, and schooling (more generally speaking, processes of socialization) are therefore important aspects of a knowledge economy. The knowledge economy may comprise not only institutions and social practices like organized educational processes, but also certain second-order knowledge about such practices, which we call "images of knowledge."[14] Second-order concepts that serve to categorize knowledge and epistemic practices (such as "fact," "evidence," "proof," and "objectivity," for instance) denote shared control structures of a knowledge economy. These second-order concepts are typically related to institutionalized practices of science, establishing its supposed universality, albeit in specific, historically contingent ways.[15]

Just like the material economy, the knowledge economy as a whole rarely becomes the subject of conscious reflection by its participants. From the perspective of individual actors, its systemic characteristics are primarily experienced as conditions of their own actions, which they generally must submit to.

With the separation of intellectual and manual labor in early urban societies, the production and transmission of some knowledge became part of a societal division of labor, for example, with priests or scribes as the privileged owners of certain kinds of knowledge. A separate, higher-order knowledge economy with its own epistemic (e.g., educational) institutions emerged that had to be sustained by the material economy and integrated into a society's other institutional frameworks. The carriers of this separate knowledge economy formed an intellectual or "epistemic" community. The rise of epistemic institutions, such as philosophical and scientific schools, creates new challenges for dealing with resource allocation and other generic problems of the institutional coordination and control of societal practices.

These challenges also include, more specifically, the need to handle intrasocietal knowledge transmission and access regulations, plus the corresponding transaction costs.[16] Since epistemic institutions are, by definition, to some extent decoupled from the immediate reproductive activities of a society, they also harbor a potential for creating knowledge for its own sake, or rather, for the sake of purposes defined self-referentially by the institution itself. Such knowledge may generate important spillover or subversive effects as well as feedback loops with other societal domains. In chapter 4, I have referred to this dynamic as an "emancipatory reversal" characteristic of epistemic institutions.

Knowledge Transmission Processes

Knowledge transmission processes may involve the co-transmission of knowledge and practice as a singular unit, the spreading of hearsay information, and the institutionalized transmission of knowledge by ritualized oral communication, apprenticeship, schooling, or dissemination by market-driven processes (as with commercial media). In the past, knowledge spread, for instance, by manuscript copying and circulation, and subsequently by print media. Knowledge also spread by the communication of ideas and by its reconstruction, adaptation, and accommodation through reverse engineering, that is, by extracting knowledge from an artifact with the purpose of reproducing it. Knowledge transmission is practiced in institutions such as workshops, laboratories, factories, academies, universities, and research organizations. It may be regulated by policies and is inevitably affected by changing knowledge representation technologies, such as the invention of printing, a theme to which I return in chapter 10.

Transmission and transfer are used here as generic terms for the movement of elements of knowledge from one place to the other. When referring to the overall process, typically involving systems or packages of knowledge, we may speak of the "spread" and sometimes also of the "diffusion" of knowledge without, however, implying unidirectionality in the sense of a one-sided transmission, for example, from a "center" to a "periphery, ignoring the repercussions of such transfer on the transmitters."[17] When the overall process of knowledge spread is intentional, we generally call this "dissemination." When knowledge exchange is part of a structured knowledge economy, we speak of "circulation." A transfer process of knowledge is designated as "globalization" when the knowledge transferred is no longer bound to local contexts and hence, in principle, capable of global spread. All of these processes involve an active role of the recipients of knowledge, and transmission is usually associated with transformation of knowledge.[18]

Knowledge transmission processes vary along three basic dimensions. The first is *mediation*: Is the knowledge transmitted through direct personal contact or through external representations? In personal contact (immediate transfer), the principal external representations are ephemeral—speech and action. Its two main constituents are instruction and imitation. In mediated transfers, stable external representations play a constitutive role, even though they are not explicitly designed for this purpose. Stimulus transfer, in the sense of Alfred Louis Kroeber and Cyril Stanley Smith,[19] is a paradigmatic case of mediated transmission via a representation not explicitly designed to represent knowledge; transmission by writing is a paradigmatic instance of

8.3. A printing press, ca. 1600. Plate from a print series, *Nova reperta* (New inventions of modern times), engraved by Jan Collaert I, after Stradanus. Antwerp: Theodoor Galle. Metropolitan Museum of Art: Harris Brisbane Dick Fund, 1934.

the other case. The second dimension is *directness*: Is the knowledge transmitted in a direct, continuous process, or are there relays? The transmission of orally encapsulated knowledge through time and space is an instance of transmission by relay. The third dimension is *intentionality*: Is the knowledge transmitted intentionally or accidentally? In the case of oral transmission, both instruction and imitation play a role.

Receivers of knowledge should not be conceived of as passive receptacles, since they may resist the transmitted knowledge or appropriate and adapt it to knowledge of their own. In short, receivers are typically involved in a process of coproduction of knowledge. Knowledge transmission involves complex dynamics of cognitive processes on the one hand, and of the preservation and transmission of established bodies of shared knowledge on the other. All these processes are shaped by the relevant economies of knowledge and in particular by the diverse media of knowledge transfer; by products, tools and technologies, shared experiences, and oral communication; as well as by symbol and information processing systems.

External Representations and Knowledge Economy

In part 1, I stressed the productive and heuristic function of external representations such as paper tools. But external representations also serve, so to speak, as the commodity (or even the currency) of a knowledge economy. Knowledge may travel with external representations and with people. These are its main vehicles. Various vehicles possess their own peculiar characteristics, such as speed of transmission, reliability of transmission, or the mobility of actors—be they individuals, social groups, or societies.

As I have argued above with respect to paper tools, the form of representation has implications both for the structure of knowledge and for the operations that can be performed on the represented knowledge, as well as for its potential for transmission. Knowledge may be so embedded in culturally specific external representations that it is difficult to extract and hence difficult to transmit. Or the processes of extraction may so radically change the structural relationship between the knowledge and other items of knowledge that the knowledge extracted is transformed into new knowledge. Transmission processes are not simply successful or not but always involve selection and transformation; thus writing, for example, becomes a selective force in the transmission of knowledge, as what is not written is usually lost.

Particular knowledge representation technologies shape the knowledge economy in different ways, since they vary along a set of economic dimensions in addition to and beyond the traditional concept of transaction costs.[20] One of these dimensions is *portability*, characterized by the ease with which a representation can travel. Radio and television broadcasts propagate very quickly, whereas inscribed monoliths generally don't move at all but are commonly transmitted in the form of copies. Another dimension is *durability*. Cuneiform tablets have endured for thousands of years, and before the invention of recording devices, spoken language vanished without a trace. Further critical questions concern the *ownership* of knowledge representation technologies: Who has access to the means of production? How easily can this access be controlled? Another economic dimension is *rivalry*: Does an individual's use of a representation decrease the value of that representation for others? Only one person can read a manuscript at a time, but many people can listen to a storyteller or watch a television program. A further issue is *reproducibility*, that is, at what cost a representation can be copied. Books were more expensive before the invention of printing with movable type; now the cost of a digital copy approaches zero.

From the perspective of a digital world, we realize that *interactivity* is another important economic dimension. How flexibly can a representation be accessed? A monologue can only be listened to from beginning to end; parts of a book can be skipped or reread; an electronic text can be searched in more powerful ways than its analog counterpart. I have emphasized the cumulative and iterative character of the evolution of knowledge. From this perspective, *recursiveness* is a critical economic dimension of knowledge representation technologies: Can higher-order knowledge about a representation, resulting from reflection on its usage, be in turn represented and integrated with the existing representation? Books can be annotated in the margins, but electronic texts can be annotated more extensively and easily; a spoken monologue, on the other hand, cannot be annotated at all. Finally, we have to pay attention to *connectivity*: To what degree and how explicitly is a representation connected to other knowledge? An epic poem may contain allusions to other literature, but these are less direct connections than the footnotes in a scholarly article or (a fortiori) the hyperlinks in a Web document. I return to these dimensions of

knowledge representation technologies when I analyze historical knowledge economies in chapter 10 and again when I discuss the idea of an Epistemic Web at the end
of chapter 16.

Marx's Value Theory and Its Counterpart in the Information Society

At the beginning of *Das Kapital*, Marx analyzes the relation between use
value, commodity, exchange value, money, and capital.[21] A commodity has
use value for an individual user; it assumes an exchange value only when it
is exchanged for other goods—on the market or by bartering, for example.
It is, as discussed in chapter 3, through this real exchange that exchange
value as an abstract concept comes into being. As a consequence, commodities assume the dual aspect of representing use value and exchange value
at the same time. The exchange value that commodities assume in an economy based on distributed labor ultimately represents (according to Marx's
"labor value theory") the amount of societal labor needed to produce a
good; prices on the market fluctuate around this value.

Money emerged as a specific commodity (represented, for example, by
silver) that came to represent exchange value. It is typically exchangeable
with all other commodities and easily storable. Thus, in a market economy,
a commodity is sold for money, which in turn buys another commodity. In
a capitalist economy, the cycle of commodity-money-commodity turns into
another cycle by which money becomes capital: money-commodity-more
money. Marx writes: "The simple circulation of commodities—selling in order
to buy—is a means of carrying out a purpose unconnected with circulation,
namely, the appropriation of use-values, the satisfaction of wants. The circulation of money as capital is, on the contrary, an end in itself, for the expansion of value takes place only within this constantly renewed movement. The
circulation of capital has therefore no limits."[22]

In this way, Marx explains the emergence of the capitalist economy
against the background of a broader vision of society according to which
the market is not the only conceivable form of organizing societal labor. On
the contrary, the emergence of the market, of money, and of the capitalist
mode of production is the result of contingent historical processes. It can only
be understood by focusing on developments of societal reproduction as a
whole, rather than ahistorically postulating individual economic behaviors
as a starting point—such as the greed and "rationality" of a fictitious *Homo
oeconomicus.*[23]

Knowledge must be similarly conceived of as both a presupposition and
a result of societal labor, rather than as the possession of some lonely Robin-
son Crusoe who then shared it with his fellow humans. In this way, the relation
between such concepts as meaning, external representation, information,
data, and Big Data (or "data capital") can be interpreted in a societal con-

text and in analogy to the series of concepts discussed above (use value, commodity, exchange value, money, and capital).

Knowledge has meaning for individuals who appropriate the shared knowledge of a society. This is its use value. Within a knowledge economy, information is the exchange value of knowledge; it is knowledge encoded for transmission. The modern concept of information goes back to Claude Shannon's information theory, where it is conceived in relation to the communication of messages involving a source, a transmitter, a channel with a certain capacity, a receiver, and a destination for the message: "The fundamental problem of communication is that of reproducing at one point either exactly or approximately a message selected at another point."[24] The concept emerged against the background of practical and technical knowledge about such messaging and became a widely used cultural abstraction in societies in which the related information and communication technologies play a key role for their knowledge economies.

External representations unite meaning and information because they serve as both representations of the meaning of knowledge for its individual appropriation and as carriers of information from sender to receiver. Data may be considered as the monetary form of information—a specific but universally applicable external representation (encoded in symbolic language and typically housed and transmitted today in an electronic medium) that can serve as its universal standard and measure.[25] Big Data is thus the capital form of information; it is data whose accumulation has become a purpose in itself, transitioning from the cycle information-data-information to the cycle data-information-more data. To give an example, a quest for information using Google makes sense because of the Big Data the firm has already accumulated; that quest will in turn augment that data and hence the power and wealth that comes with its accumulation. I return to data capitalism in chapter 16.

In other words, data and information are to be construed from knowledge, contrary to the notion of conceiving data as atoms from which information and then knowledge can be derived or constructed. Data only become the universal standard of information once an external representation of knowledge can be symbolically coded, brought into the same format, and stored and transmitted in the same medium. Only in the age of electronic data and its transmission via electronic networks has this actually occurred.

All knowledge is originally shared knowledge, a common good, as I discuss in chapter 16. Different external representations lend themselves to privatization via different paths. While the electronic medium offers, in principle, new opportunities for returning knowledge to the state of a common good, data capitalism leads to an ever more comprehensive subjugation of knowledge under private or state interests.

Normative Systems

Institutions hold knowledge about their own functioning, rationale, and meaning for the actors involved. This knowledge allows individuals to act in accordance with the regulations and forms of cooperation germane to an institution. I call this knowledge "normative knowledge." More generally, I refer to normative systems as conglomerates of values, beliefs, and knowledge concerned with the functioning, rationale, and meaning of institutions for the actors involved. The enforcement of institutional regulations and the power structures behind them never merely comprise violence, military, political, or economic dominance, or authority. They also rely on the support of normative systems that regulate social order beyond the exertion of crude force. The coordination of individual actions mediated by institutions accordingly presupposes behavioral norms and practices, as part of normative systems like religion, law, morality, or ideology. The knowledge involved here is social knowledge. It comprises not only specific contents incorporated in such systems (e.g., knowledge about certain traditions and mythologies) but, above all, the capacity to act in compliance with certain behavioral norms implicit in such systems.

Institutions need to be psychologically viable in the sense of being capable of dealing with individual needs, drives, and desires that otherwise would destroy social cohesion. Human beings exist as social entities, and their social existence constantly confronts them with the challenge to maintain a balance between their individual needs, drives, perspectives, and actions and the requirements imposed by the functioning of society. Institutions can mediate between these requirements only if they coordinate not just human actions, but also their motivational and cognitive presuppositions thus offering appropriate processes of socialization and equilibration.

Normative systems may, of course, have other functions as well, such as representing knowledge about the natural world or about emotional needs and experiences. In the course of history, we see a differentiation between these various functions, which will, however, hardly ever be complete. Typically, even the understanding of scientific objectivity within a scientific community includes both social norms and role models. Normative thinking and knowledge cannot be separated in any absolute way: maintaining social cohesion requires problem solving and hence knowledge, while problem solving presupposes cooperation and hence moral norms and practices.

Normative systems allow individuals to interpret and control their own behavior and that of others in the framework of the societal group to which they belong, forming the basis of normative judgments and their legitimization. Individual behavior thus becomes "meaningful" by being practically and mentally related to group behavior. Religions, in particular, offer resources for articulating collective self-awareness and self-reflection, including reflection on unresolved tensions whose virulence they both maintain and assuage.

An Entrapment of Mutual Limitation

Institutions are subject to continuous change. They constitute dynamic equilibria in societal interactions. Variations in institutional regulations occur constantly as a consequence of the fact that individuals are different and hence embody and enact societal structures in ways that often differ in time and space, from occasion to occasion, and in particular from generation to generation. Just as cognitive structures change every time they are implemented, the same holds true for the implementation of an institutional structure in a concrete situation.

The appropriation of normative behaviors, in particular, also occurs in ways that may vary from individual to individual; it depends on the concrete experiences involved in the appropriation and implementation of an institutional framework. As a consequence, there is always a lot of play between the regulative structures of an institution on the one hand, and its externalization in concrete behaviors on the other. Concrete experiences generate not only knowledge about what is being regulated but also about these regulations, producing feedback on the regulations themselves. More generally, every social experience may become the germ of new regulations.

The interactions of an individual with others, as mediated by an institution and interpreted within a normative system, are relevant to the constitution of both an individual's identity and of its relation to a communal identity. In other words, institutions shape what one may call role models or "institutional personae."[26] A society typically comprises several institutions, which may exist in relative separation from one another, or which may be superimposed or in conflict. The superposition of institutions therefore creates, on the side of the individuals, the opportunity and the need to form multiple identities. These are in turn embedded in different normative systems, often independent of each other, which are more or less capable of sustaining such societal as well as cognitive and emotional complexity.

With the division of labor and the emergence of multiple institutions in one society, individuals will, furthermore, acquire different shares of societal wealth, different positions of power, and different perspectives on a society and its institutions. Privileged positions allow individuals or groups to control cooperative actions as well as access to resources—and hence to exert power. Individuals in underprivileged positions conversely become the subjects of power, deprivation, and exploitation.

"Power" is understood here as the (often institutionalized) potential of an individual or a group to direct and control the actions of other individuals or groups, typically by controlling the conditions of their actions. For various reasons, power can never be absolute. This is because it primarily affects action and thinking only in a mediated way, but also because it is usually conditioned by the need to reproduce a society, as well as the material conditions of its existence. Power is subject to the constraints of various economies, political, epistemic, moral, and ecological. In any given historical situation, "power" depends on the institutional structure of a society and also, in particular, on its knowledge economy and thus on the available representations that critically shape this economy. But power also depends more directly on the problem-solving capacity of knowledge, with regard to the control of natural and social processes.

The societal positions created by the division of labor also correspond to specific role models or institutional personae constituting what one may describe as the "objective interests" of the individuals filling these positions, though they are internally realized in individually diverse forms as "subjective interests." Clashes of interest, together with other social and epistemic contrasts and differences, drive societal dynamics in a way similar to the manner in which the triggers of change discussed in previous chapters foster cognitive development. They constitute challenges and problems that do not determine the outcome of societal development, which instead depends as much on the society's available internal resources and dynamics.

Challenges may specifically concern the knowledge economy of a society, in particular when they are recognized and acknowledged as problems of knowledge. Both cognitive and societal innovations often begin with individual deviations from transmitted knowledge and practices. Their broader adoption within a society and their integration into the continuing tradition of shared knowledge or institutional regulations depends on the specifics of a society's knowledge economy. In loosely structured societies, innovations can be taken up relatively quickly and easily but can be lost again just as quickly—the lack of institutionalization also means that there are scarcely any selective instruments that allow deviant knowledge or collective practices to be preserved.

Socialization processes deal with actual individuals and their needs, desires, and cognitive abilities, with real living situations and their practical challenges embedded in a given material culture. They also deal with unforeseen events. From the perspective of individuals, socialization processes are decisively shaped by the local interactions they experience. The values and mental models of the institutions they develop are fostered and limited by these experiences. For instance, in ontogenetic development, normative concepts such as justice are shaped by children's experiences with cooperation and the opportunity to interchange perspectives.[27] The actual outcome of such developmental processes is always conditioned by the tangible experiential spaces available in a given historical and social situation. Institutions must in turn function under the premise that the behavior of individuals is regulated by the limited internal representations in these experiential spaces.

The result is entrapment, a feedback loop of mutual limitation in which individual capabilities and perspectives are blunted by the confined experiential horizon a society offers them, while institutional regulations are themselves narrowed by the bounded mental capabilities and perspectives of the individuals for whom they exist. This is why authoritarian societies tend to produce authoritarian characters—who in turn prefer authoritarian societies. The market, to give another example, is a mechanism that tends to distill societal interactions into relations between buyers and sellers, thus creating specifically limited experiential spaces. It demands institutional personae according to the model of *homo oeconomicus* with the hypothetical ability to make rational choices among a range of options. In real life, however, this demand is hard to fulfill in view of the "bounded rationality" of human choices and decisions.

The intertwinement and mutual reinforcement of institutional personae and the regulatory mechanisms of institutions is an important mechanism in stabilizing a society and its institutions, accounting in large part for societies' conservative and resilient character. This mechanism never works perfectly, however—for one, because human beings never perfectly live out the institutional personae they are supposed to embody, and because the reality of an individual life experience never perfectly fits the coordinative behavioral patterns of an institution.

The entrapment of mutual limitation imposes severe boundaries on the capacity of a society to adapt to new situations—for instance, when dealing with the challenge of integrating immigrants who do not fit the expectations embodied in the mental models of the "autochthonous" members of a society. A knowledge economy mediates between processes of socialization: some in which individuals are shaped by society, and some in which society is shaped by individuals through the implementation of knowledge. The capacity of a society to deal with its challenges (e.g., integrating immigrants by overcoming prejudices against them) will therefore crucially depend on the capacity of its knowledge economy to collectively share, weigh, and reflect on the entire range of experiences induced by such challenges and to adapt existing normative systems to them. Changes in the knowledge economy of a society constitute important points of political intervention, since they may open up new experiential horizons by offering novel mechanisms for the equilibration of conflicting perspectives.

Material Embodiments of Societal Institutions

I have thus far neglected the material dimension of institutions as an important mediator between individual actions and societal practices, though we have already seen its crucial role in the discussion of knowledge systems. I refer to the fact that societal practices are fundamentally shaped by material conditions and various forms of embodying institutions, such as productive forces, infrastructures, symbol systems, or humans playing institutional roles.

Means of production forces (such as machinery, plants) or infrastructures (such as architecture or technological networks) solve problems of the material world and simultaneously regulate collective behaviors by creating a material environment suited to the cooperation and interaction of individuals. Large technological systems typically develop "technological momentum" in the sense that societies may initially be able to control the use and scope of the underlying technical innovations, while the economic dynamics they unfold over time tends to increase their power over society.[28] Material embodiments of institutions may also take on a symbolic meaning, acting as clues for triggering individual behaviors in accordance with relevant institutional regulations. They thus function as external representations of the knowledge systems associated with these regulations, in particular of normative systems.

When material embodiments of institutions are employed in order to implement institutional regulations, they act as "social technologies." They represent the social relations regulated by an institution or society, just as a technological artifact may represent the working of certain physical forces, for instance, when a lever is used to save force. Operating a lever does not require knowledge of the physical law of the lever, but it may well serve as an external representation of such knowledge. While a technological artifact addresses problems of the physical world, material embodiments of institutions help to solve problems of collective interaction.

We may thus distinguish between the operative function of material embodiments and their role as external representations of institutional regulations. The externalization of institutional regulations contributes to solving societal problems because the coordination of individual interactions can be charged (in part) to these external representations, which then serve as social technologies—for example, following a command chain, dealing with paperwork in an administration, applying written law to a violation of norms, or exchanging goods for money on the market. The external representations reduce the effort required to solve problems of collective interaction because individuals need only know how to handle the external representations in order to participate. Individuals' remarkably unconscious use of external representations is succinctly expressed in the words of the economist Friedrich Hayek: "We must look at the price system as such a mechanism for communicating information if we want to understand its real function—a function which, of course, it fulfills less perfectly as prices grow more rigid. . . . The most significant fact about this system is the economy of knowledge with which it operates, or how little the individual participants need to know in order to be able to take the right action."[29]

Material embodiments of institutions are essential for their transmission from one generation to the next. This always involves learning processes, that is, the transmission of knowledge. Material embodiments guarantee the long-term transmission of the regulation of cooperative action and thus form a kind of institutional memory. Since reflection on institutionalized interactions is mediated by their external representations, these representations shape the internal, mental representations of an institution, which may of course differ from individual to individual. The external representations are in turn charged with a new "transcendental" meaning that reflects, albeit in an opaque way, their societal significance. Persons, places, or artifacts may take on new meaning in the form of rulers, gods, sacred places, or objects that now represent a normative social order, defining a field of actions compatible with the regulations of an institution.[30]

Both in the case of knowledge and in that of social order, external representations can become themselves the objects and means of actions constituting higher forms of cognitive or societal organization. Money, for instance, does not only represent the exchange value of goods on the market but can also become a means of social control or of financial speculation by enterprises using money as capital. Similarly, the documentation of commercial transactions can become the starting point for building up a tax system. The emergence of new institutions that rely on external representations in order to steer existing institutions or practices constitutes a form of "societal reflexivity," understood as a mechanism of hierarchy formation. (Evidently, such societal reflexivity does not, unlike individual reflexivity, necessarily increase the self-awareness of societies.)

External representations can thus become the starting point for the creation of "second-order" institutions, which enable the functioning of existing institutions or represent supervisory authorities that impose a superordinate logic on individuals and alter the conditions of cooperative action; in our examples, this phenomenon is represented by the emergence of tax authorities or capitalist enterprises. Second-order institutions are generally based on the possibilities of externally representing

8.4. "Time is money." A coal worker checks out from the Inland Steel Company in Floyd County, Kentucky, ca. 1946/47. US Government Archives.

the regulation of cooperative action and use these possibilities implicitly or explicitly in order to control whether individual actions correspond to the demands of complex cooperative endeavors or societal processes.

Administrations are second-order institutions of this kind. Their task consists in perpetuating the functioning of existing institutions. As we shall see in chapters 9 and 10, administrations typically emerge only in societies with a differentiated division of labor and in connection with complex cooperative projects. Because of their complexity, particular forms of institutionalized action are only possible through administration, especially when the challenge is to direct the actions of a large number of people toward a common task, a communal endeavor of longer duration such as the irrigation projects of the ancient river cultures in Babylonia and China.[31]

As a result of the externalization of institutional regulations, abstract concepts (as discussed in chapter 3) may take on important social functions as "cultural abstractions" that are not just relevant in the context of systems of knowledge but also in the realm of social technologies. Cultural abstractions belong, in other words, to knowledge systems whose representations also function as social technologies. Consider what are perhaps the two best-known examples, time and money.

In modern societies, certain aspects of the coordination of interpersonal interactions—to put it simply, making an appointment or organizing a work flow, for instance—are governed by an abstract "time" externally represented by clocks.[32] Regulating one's actions with the help of a clock thus becomes an efficient substitute for the coordination of actions among the members of a complex society by other, more direct forms of communication. This is the operational function of clocks: as an external representation of a specific coordination of social practices. But no knowledge about these coordinative mechanisms is required to adequately use clocks for their operational function. I have earlier emphasized that it is useful to distinguish between the knowledge required for the invention, the production, and the usage of an artifact. A short history of time as an abstract concept can be found in chapter 14.

Money similarly coordinates economic interactions as an external representation of value (i.e., of a certain system of social relations, as discussed in chapter 3). More precisely, as an external representation of social relations, money represents the co-ordinative mechanisms that determine its value, for instance, the costs needed to produce a certain commodity. As part of a system of social knowledge, on the other hand, money may simply represent the options of its owner to purchase commodities with an equivalent exchange value. In order to adequately use money for its operational function, no knowledge about the coordinative mechanisms that determine its value is required.

The externalization of institutional regulations contributes to the opacity of institutions from the perspective of individuals (screening the "machine-room" aspects of an institution) because it decouples actions with external representations from the concrete interactions at lower levels of societal reflexivity. This also holds for related cultural abstractions, such as time and money, which are no longer conspicuously connected to the specific experiences from which they historically emerged (in this case, the cycles of nature and the facilitation of human cooperation). The noise of the machine room of a society can, on the other hand, hardly be completely suppressed. Even if it were possible, for instance, to conceptualize all human and natural resources in economic terms with the help of cultural abstractions such as "human capital," "natural capital," or "ecosystem services," there is no guarantee that the underlying reality can be reliably mastered by the corresponding social technologies.

Symptomatic Consequences

Let me turn to another aspect of the materiality of societies and their institutions, which in the Anthropocene is becoming ever more important: material embodiments of societies and institutions are not just a precondition of their functioning but may also be a conspicuous *consequence* of human activity. (In chapter 14, I come back to these consequences under the label of "niche constructions," adopting a term from evolutionary biology.) One such consequence is the creation of new external representations of the knowledge systems associated with human practices and societal institutions—not only the ones I have discussed so far in the context of social technologies but also more broadly the creation of "cultural representations," as manifested in works of art, for example.

Cultural representations capturing and preserving conflictual experiences (from religious manifestations to works of art and science) possess an ambivalence that may be described as "symptomatic." Freud suggested that neurotic symptoms are not just the expression of unresolved conflicts in an individual mind but also the result of a

8.5. Banks of the Ciliwung River with anthropogenic sediments in the area of the district of Kampung Bukit Duri, Jakarta, Indonesia, 2011. © Jörg Rekittke.

societal need to regulate, control, and repress needs, drives, and conflicts—an enterprise that never works perfectly. Such symptoms may therefore serve to simultaneously conceal and reveal subliminal or suppressed conflicts and traumata, as well as this imperfect control by society.[33] As James Agee points out: "it needs to be remembered that a neurosis can be valuable; also that 'adjustment' to a sick and insane environment is of itself not 'health' but sickness and insanity."[34]

Similarly, cultural manifestations may be combined with a rationalization of substitute behaviors, which then attain a ritual character. On the other hand, they may be culturally liberating. As representations of unresolved problems (and hence as a form of communication about them) the sediment of our past activities may become the starting point for dealing with unresolved conflicts at a more conscious and communicative level. In other words, in this symptomatic aspect of cultural representation we recognize not only a form of collective memory but also another instance of what I have called the "generative ambiguity" of external representations.

Societies are also confronted with the ambivalent consequences of their past actions by the material traces they have left in their environment. These traces are also the result of a process of externalization, which may be described as a formation of sediment.[35] The traces are ambiguous because they may constitute *action plateaus*, in the sense of opportunities for future action, planning, and innovation on the basis of new material means and preconditions. We call them "platforms" (as in "platform economy") when they function as an infrastructure with regulatory functions determining a specific space of possibilities. Action plateaus may also become effective because sediments of earlier societies can often be applied to entirely new perspectives. But the traces of the past may, on the other hand, constitute a liability and a severe burden for the future—for example, the environmental changes that are catapulting us into the Anthropocene.

While such sediment could traditionally be "externalized"—in the narrow economic sense of indirect costs or of waste being discharged to a third party, and in particular to weaker societies—this has become ever more difficult in the global world of the Anthropocene, where there will ultimately be no third party left upon which to offload the unintended consequences of our actions. We are therefore increasingly confronted with the challenge to see and understand the formation of sediment as a limiting condition for the functioning of global society. It is here that the consequences of human practices turn into the challenge of understanding the material sediment of our actions also as cultural symptoms of our global predicament.

A possible answer to this challenge is the development of appropriate external representations of human behaviors in the sense of social technologies establishing novel feedback mechanisms between human societies and their environment. One currently discussed possibility is, for instance, to put a price on carbon emissions in the hope of making those who are the most responsible for climate change pay for the sediment of their actions. But it is hard to imagine that any such measures alone will suffice to transform an expansive economy into a subsystem of a closed ecology, at least without enhancing the role of knowledge as a further regulative mechanism of this transformation. Such an enhancement will crucially depend both on cultural representations that allow us to reflect on our predicament and the further development of the current knowledge economy, including new forms of active participation in this economy (a theme to which I return in the final chapter).

Representations, Power, and Transcendence

Representation is a crucial consideration for understanding the peculiar metabolism that humans have developed with their environment. Human livelihood is ultimately based on work as a transformation of this environment through material means. The resulting changes and the means themselves thus become representations or material embodiments of these human interventions, often persisting even after the interventions themselves are terminated. Representations objectively reflect these interventions, but they have also come to play a fundamental role in the capacity of human cognition to create and understand signs that carry meaning beyond their immediate existence as material conditions or objects. This capacity is at the origin of a characteristically human symbolic—or, as some would say, *transcendent*—world in which, in principle, everything can acquire a meaning pointing beyond itself. What is actually being represented is variable but not arbitrary. The representation of those factors that enable collective survival—the common ground—is important to all human societies and the subject of shared cultural practices. However, what is understood as common ground may vary, as may the modes of its representation.

Representations invoke the presence of something that is absent, but they may also intensify what is already present by doubling it. As we have seen in chapter 3, representations therefore support the function of thinking in a double way by enhancing the permanence of its objects and by allowing thinking to act on itself (what we call *reflection*), thus altering its subject. In a similarly double way, representations also support the function of institutions:[36] they lend immediate violence a mediated but long-distance effectivity as power (making violence present even if it is absent); and they turn institutions into social regulations constituted by such representations (thus affecting their identity), as when, according to the legend, Louis XIV famously expressed the essence of absolute monarchy by saying, "*l'état, c'est moi.*" I return to the nature of the absolute state in chapter 10.

The two realms of representation are not independent from each other because the representations supporting the functioning of an institution (and, in a sense, constituting it) must simultaneously become the objects of perception, feeling, and thinking. The thinking triggered by these representations may serve to legitimate the exertion of the indirect power they mediate, but the representations may also undermine this legitimacy when they are perceived as symptoms of the malaise the underlying power relations create.

Struggles about representation are thus a form in which societal conflicts may articulate themselves. What does it mean, for instance, when religious traditions forbid the representation of God? It could mean that the common ground is not to be constrained or confined in human terms, thus delimiting its power. All that may be possible is to strike an alliance with this unpredictable power, keeping the future open for changes. Or the prohibition to form an image may elevate the common ground to a transcendent world beyond any form of mediation between its unlimited power and concrete human existence. Theological debates can often be read as debates about representation, while philosophical debates about representation may acquire a religious and political connotation in the sense that they also negotiate the relation between the individual and the common ground of societal existence.

What about the much-debated philosophical question of the relation between essence and appearance, for instance, which can also be put as the question of whether appearances represent essence?[37] It is the essence that appears, but the appearance is never the essence. Philosophers have used this seeming contradiction to set up doctrines proclaiming that appearances must be transcended in order to get at the essence, and that only the essence is essential. But this entails sacrificing the world of appearances and abandoning the opportunity to resolve its tensions. Isolating the appearances, on the other hand, from a common ground and accepting them just as they are (and all on the same footing, as phenomenological traditions have argued) ultimately favors indifference—at its most extreme, even with regard to life and death.

8.6. Layers of external representation. What makes the King? William Makepeace Thackeray, *The Paris Sketchbook* (1840, 302). Internet Archive.

However, representations do not necessarily reduce the individual to a mere derivative of the common ground nor, alternatively, do they liberate the individual from this origin at the price of indifference. Representations can rather be a middle ground in which the individual may be taken just as seriously as the common ground from which it emerges. Thus, representations become a means of negotiating the relation between the common ground and the individual. Representations, rather than any form of transcendence, which is, after all, only derived from them, are the unique medium of human existence.

Darwin's theory of evolution challenged the aspiration of human reason to represent divine reason by questioning the boundary between the realm of humans as rational beings and the realm of nature. He insisted that both are connected through one natural history in which the same forces affect both realms.

Similarly, a history of knowledge should insist not on transcendence as a uniquely human condition but on the role of material embodiments and external representations in systems of knowledge in order to anchor the actions and thinking of human societies in a continuity of natural history.

We have considered societies as ensembles of institutions with the capacity to reproduce themselves in all aspects of the anthropological gamut. I am referring to societies here not as absolutely distinct entities; the ensemble of institutions forming a particular society is simply more closely interconnected by internal structures and interactions than the society as a whole is to another society. This classification may not, however, coincide with the self-perception of a society as a distinct entity mediated by its knowledge economy.

I have stressed that the knowledge economy is critical to the capacity of a society to react to external challenges. It also determines whether and how new experiences are turned into the development of knowledge systems. Among these knowledge systems are those belonging to normative systems governing compliant behaviors. In the case of the knowledge economy of a society, such normative systems also comprise the images of knowledge that shape the roles and values a society ascribes to particular kinds of knowledge. Owing to their systemic reproduction as parts of the institutional ensemble of a society, normative systems tend to be locked in by mutual limitation, with the effect of a mutual stabilization and reinforcement of institutions and normative systems.

This also holds for the knowledge economy governing specialized disciplinary science in the modern world and its corresponding images of knowledge. In the Anthropocene, we are confronted with the question of whether such a highly distributive knowledge economy as that of disciplinary science has become too path dependent and inflexible to assure the production of the knowledge necessary for the continued existence of human civilization as we know it.

Finally, we have seen that institutions involve material embodiments, including those related to material means, social technologies, and external representations. I have emphasized that the available knowledge representation technologies open up a specific horizon of possibilities for a knowledge economy. The material embodiments of an institution similarly determine a specific horizon of possibilities for social interactions. Today, for instance, algorithmic institutions have emerged, such as Web-based arbitration or dating sites, making use of the computational capabilities and interactivity offered by digital media.[38]

Critical interfaces of knowledge and society include the economy of knowledge, social technologies and their corresponding political structures, and the possibilities offered by cultural manifestations for reflecting on and intervening in the mechanisms shaping our predicament. The increasing significance of sediment formation and global society's shrinking latitude for maneuvering within the constraints of the planetary environment make it necessary to reexamine and question these interfaces in light of the current challenges. But before we come back to planetary constraints and their implications in part 5, we first have to understand the historical development of these interfaces and consider some examples of knowledge economies in greater detail. We will begin this more concrete discussion of knowledge economies with one of the most basic infrastructures of human societies, which existed long before the advent of science, ever since humans began to construct their own shelters.

Chapter 9

An Economy of Practical Knowledge

When one wants to build something, one needs to know one's purpose—for example, this provides strength and this is why it is necessary. But study and thought are necessary in order to understand one's purpose: just as the archer would not hit the jug if he did not aim, the Maker would not achieve his purpose if he focused on something else instead.

—DANIELE BARBARO, *I dieci libri dell'architettura di M. Vitruvio*

We presuppose labour in a form that stamps it as exclusively human. A spider conducts operations that resemble those of a weaver, and a bee puts to shame many an architect in the construction of her cells. But what distinguishes the worst architect from the best of bees is this, that the architect raises his structure in imagination before he erects it in reality.

—KARL MARX, *Capital, a Critical Analysis of Capitalist Production*

In the following I illustrate some of the general features of knowledge economies introduced in the preceding chapter in the context of a largely practical system of knowledge that is of fundamental importance to human existence: the history of building and architecture.[1] This knowledge economy involves material embodiments, namely, the buildings themselves, which often simultaneously carry institutional, cultural, and epistemic meaning. An epistemic history of architecture focuses on the kinds of knowledge (and the mechanisms of its transmission and implementation) that have enabled the architectural achievements of the past.[2] Here I focus, in particular, on the remarkable and often monumental achievements that preceded the major intervention of science and engineering in building beginning in the sixteenth century.[3]

After discussing the knowledge economy of building and its representations in general, three aspects of the evolution of this knowledge economy become the center of attention: I first consider the emergence of second-order institutions (such as building administrations) and of architecture as a new profession at the interface between the building process and its supervising authorities. Next we will take a closer look at plateaus and the formation of sediment, as well as the role of canonical buildings as an especially influential form of representing and transmitting building knowledge. Against this background, I finally explain some of the conditions under which an exploration of building knowledge and various experiments became possible, even prior to the advent of modern science and technology. Although the limitation of the following account to a largely European perspective is merely a reflection of my own knowledge of the subject, the following examples show under which conditions an economy of practical knowledge could open up spaces for innovation processes, including interfaces with scientific and technological knowledge.

The Knowledge Economy of Building

The history of building until far into the modern era was based on the practical knowledge of craftsmen, master builders, and architects. The design of buildings, the knowledge of materials and building techniques, and the organization of logistic procedures was transmitted largely by oral instruction and participation in the work process.[4] This knowledge economy enabled the construction of the impressive buildings of the early civilizations of Mesopotamia and Egypt; the exemplary architecture and infrastructure technology of Greek and Roman antiquity;[5] the great architectural achievements in ancient China, India, and Japan; the monumental buildings of the ancient American civilizations; the sacred and defensive buildings of the Middle Ages; and also the risky, innovative building projects of the Renaissance.

Building, along with food supply, has been one of the fundamental subsystems of human practice since the early period of human history. It fulfills basic human needs through transforming the environment and contributes to the regulation of coexistence; it is therefore manifest in all cultural traditions. Building is usually a cooperative activity that presupposes shared knowledge within a given knowledge economy.[6] Since there is hardly any other way, a direct personal involvement in building is necessary for understanding the manual aspect of the use of tools, the knowledge about which method makes sense in which situation, and the knowledge about the character and quality of building materials.[7] But even in early history, there were cases in which building knowledge was also transmitted through teaching exercises (such as those found on Mesopotamian cuneiform tablets, for example) and through targeted training.[8]

The knowledge economy of building was and is shaped by a society's economy of resources, the available materials and technologies, practical knowledge about them, and the institutions concerned with their administration and the transmission of relevant knowledge. Such institutions shape stability and continuity. A knowledge economy also depends on the organization of labor within a society. The knowledge economy of building specifically depends on the level of professionalism on the building site, the related forms of training, and the available planning instruments. The acquisition of building knowledge belonged, for a long time, to more general processes of socialization. It was part of a broadly shared practical knowledge of how to build shelters and houses and was consequently tied to the societal structures that were primarily responsible for these processes.[9]

The stability of building knowledge over longer periods is therefore inherent in the conditions and processes broadly occurring within a society. On the one hand,

9.1. Oral transmission of practitioners' knowledge: construction, without centering, of a Nubian barrel vault using sun-dried bricks, Elephantine, Egypt, 2001. Photo by Dietmar Kurapkat.

this stability is supplied by the material buildings themselves, provided they survive over time. On the other hand, it is conditioned by the embedding of building knowledge in social structures. The transmission of knowledge about management and control across generations greatly depends on the continuing survival of particular social formations—but practical building knowledge may survive even beyond the collapse of such formations.[10]

This is due, first, to the characteristic social stratum of the carriers of this knowledge, a stratum that was often capable of surviving political breaks as well. Second, as mentioned above, this unusual continuity of practical knowledge also lies in its material representations, which also give it some independence from political constellations. The oral and practical communication of this knowledge, on the other hand, might also limit its continuity, since it is easily lost through lack of use.[11] All in all, this results in the historical evolution of building knowledge, an evolution that occurs more or less along two different timescales: within existing social formations and over longer periods that include political, cultural, and social breaks.

Representations of Building Knowledge

Building knowledge can be represented in many different forms: through instruments, objects, practices, models, drawings, plans, and accompanying written or spoken information. It can, in particular, also be represented by the buildings themselves, though they usually offer no direct information about the methods of their construction.[12] Although the monumental constructions of ancient cultures often retained their impact over millennia, each time the corresponding social order collapsed, the transmission of the building knowledge on which they were based was at risk of disappearing as well.[13] However, transmitting knowledge by means of the constructions themselves generally took the form of a "stimulus diffusion,"[14] a process in which the shell of a system is received by parties who must fill it with content and explanation themselves.

After breaks, building techniques had to be either reconstructed or reinvented. Building techniques, whose transmission by direct contact with workers was disrupted and could not be reconstructed on a one-to-one basis by later generations of builders, were therefore substituted by more or less new techniques—as was the case, for instance, when Renaissance architects tried to emulate buildings of Roman or Greek antiquity, an example I return to below.[15]

In premodern societies, the transmission of written knowledge about building was rather limited.[16] Texts concerning the technical and formal aspects of building, which were written, for instance, by Greek architects as early as the end of the sixth century BCE, are no longer extant.[17] The practice of transmitting written information was only resumed later. The famous *Ten Books on Architecture* by the Roman architect and military engineer Marcus Vitruvius Pollo[18] (first century BCE) was one of the first books on architectural theory and became enormously influential in the Renaissance. Owing to the lack of basic schooling in medieval times, the masters of the magnificent constructions of the Middle Ages were typically unable to write. Leaving aside the occasional mention of architecture in texts from late antiquity and the Middle Ages (and of course Villard de Honnecourt's famous sketchbook) building knowledge only reemerged as a subject for literary discussion in the treatises of the Renaissance.[19]

Administration and Representation

Large public building projects, such as dam constructions or other irrigation measures, or the erection of temples, palaces, and tombs require considerable societal resources, which means that achieving them is closely related to the available steering and control mechanisms for these resources. In early literate cultures, these mechanisms were largely defined by institutions relevant to the whole of society.[20] But adaptations to the particular practical requirements of building also occurred repeatedly, for instance, in the assignment of personal or institutional responsibility for "extra-large" building projects, that is, those with outsized requirements of building materials and manpower, requiring the optimization and control of complex implementation processes on the building site and the corresponding need for special legitimization strategies in view of the huge societal effort involved. For this reason, building projects tended to have repercussions on more general mechanisms of social control, through the emergence of second-order institutions such as administrations, for example, but also through the codification of knowledge in writing.[21]

In the course of history, the development of building administrations went hand in hand with the written documentation of some of its aspects. This increase in written administrative records enabled better command and control of logistics, particularly when it involved transport and communication over larger distances and time spans.

9.2. Life-size statue of Gudea, Prince of Lagash, dedicated to the god Ningirsu, called "Architect with Plans," ca. 2120 BCE. The statue depicts Gudea as the architect of his temple, Eninnu, which is probably shown on the tablet laid on his knees engraved with architect's drawings. The plan on the tablet follows the conventions of Mesopotamian clay and brick architecture. © bpk / RMN—Grand Palais / René-Gabriel Ojéda.

The codification of such control processes also created new possibilities for reflecting on these processes, thus creating opportunities for their optimization.

As mentioned above, the structures of building administration in early history coincided with the generic power and administrative structures of society as a whole.[22] Later, however, independent institutions emerged, such as mason's guilds and building enterprises, in which logistic and technical knowledge was accumulated, implemented, and transmitted to future generations.[23] In fact, the decoupling of task-specific institutions (such as building guilds) from more generic societal structures opened up new possibilities for hoarding material and technical knowledge.[24] This emancipation of specialized institutions gave rise to new types of knowledge economies—as can be seen, for example, in the emergence of the architect's profession in ancient Greece, and even more clearly in the role of the architect in relation to the specialized administration of large building projects in the early modern period.[25]

The Emergence of Architecture as a Profession

In classical Greece, as in Mesopotamia and Egypt, building initiatives were run by authorities that were not exclusively responsible for building.[26] In classical Greece, it was the people's assembly of the *polis* that initiated public building contracts.[27] Starting from the mid-fifth century BCE, major building projects were run by construction commissions deployed specifically for this purpose. Such commissions achieved a high degree of transparency and controllability, but, owing to their members' short service periods and the fact that expertise was not a precondition for membership, they provided no guarantees of continuity or the successful completion of a building project.[28] What emerged here was a new interface between areas of activity that had been regulated by a comprehensive hierarchy in earlier societies—areas that now had to be connected by structures of a new knowledge economy. As a result, the actual responsibility for executing a building project largely devolved onto the architects and craftsmen who worked on the building site.[29]

The political context of the Greek *polis* played a key role in shaping the architect's expertise and profession as an institutional persona. The architect was the thread of continuity in the building process that state hierarchies had provided in earlier societies. The necessity of making a distinct profession responsible for this continuity emerged against the background of a rapid change of personnel in political posts and also in light of the growing importance of private enterprise. In this context (and that of competition between urban centers) the architect attained an independent position as an expert with respect to designing, planning, and shaping a building.[30]

In the early modern period, architects assumed an even more creative role within a knowledge economy of building, which was now characterized by elaborate administrative structures ensuring the transmission and evaluation of building knowledge. I return to this point below, in connection with the emergence of a more consciously experimental approach to building. But let me first discuss the general processes by which an accumulation of building knowledge was achieved.

The Formation of Plateaus

The historical knowledge economy of building was shaped by the formation of plateaus through the sedimentation of prior results. The scope of a building plan, for instance, is determined by the number of steps that can reasonably be anticipated. The potential for more in-depth planning depends on accumulated building knowledge and its representation. The need for planning depends, on the other hand, on the preconditions for action—the plateaus—that provide the basis for a building project. If prefabricated construction elements, for example, are available on the building site, no further effort is required to obtain them.

In a building project, the question may therefore arise as to whether all the necessary materials should be obtained at the beginning of the project or gradually. In the first case, the building project could be achieved more quickly (on the condition that a larger number of workers are available), whereas in the latter case, the project might use resources more economically. In the Neolithic period, given the new dimensions of communal constructions, optimization was realized by dividing building preparations from the actual construction process.[31]

This division was evidently closely related to the introduction of standardized building components such as bricks.[32] The transformation of clay from a complementary building material for covering and sealing wood into a primary material in its own right influenced every later type of construction. Clay was initially applied to wood in damp layers that were left to dry for several days before the application of more. This method demanded a minimum of predictive planning but made construction time-consuming and difficult. Furthermore, large projects required large numbers of workers.

9.3. A square room with rounded corners at the early Neolithic ruins at Göbekli Tepe, Turkey. Photo by Dietmar Kurapkat.

As early as the ninth millennium BCE, walls were built from handmade bricks molded by hand and dried in the open air. The advantage of this prefabrication was that the material result of one work phase became the means of starting the next. The preparatory building phase could thus be separated from the actual construction phase, which considerably accelerated the entire process. However, this method also placed heavier demands on preplanning logistics, since the required amount of material had to be estimated in relation to the building to be erected. Beginning in the eighth millennium, regular rectangular clay bricks were made, evidently using molds. As a result, proper brick bonds could be made without necessitating the use of large quantities of grout.

The standardization of building components achieved in this manner reinforced the tendency toward right-angled construction forms and improved the stability of walls through the covering of mortar joints.[33] This example shows that challenges of coordination from the distant past may have far-reaching consequences through their material traces—consequences that in this case still have an impact on present-day building, which, after all, continues to favor bricks and right-angled construction forms.

Generally speaking, it can prove useful to break up a complex process into smaller acts to enable optimal exploitation of extant plateaus like natural transportation (such

9.4. Mud bricks being manufactured from Nile sludge in Sheikh Hamed, ancient Athribis, near Sohag in Middle Egypt. The drying procedure takes several days and thus requires a large area. Photo by Ulrike Fauerbach.

as rivers) or sources of materials (such as quarries or forests, or any of a given society's recyclable materials).[34] Against this background, it becomes understandable why the extensive building activities of the early empires prioritized an effective infrastructure that allowed the planning of a building project to rely on standardized elements and more or less permanent resources.[35]

The creation of more or less permanent infrastructures and the emergence of standardized building elements illustrate one way in which building knowledge could be accumulated and transmitted from one generation to the next through the formation of sediments from prior actions. A specific but very influential way of facilitating the transmission and spread of building knowledge was the introduction of canonical building forms, a theme that I take up next.

Canonical Buildings as External Representations

Buildings can be understood as external representations of societal institutions. They enable and enforce cooperative actions and, moreover, embody an awareness of what it means to be part of the community that is constituted by its institutions. One can scarcely understand the perception of buildings without taking account of basic human needs (the need for shelter as well as the need for communal representations), which are continually reinterpreted through buildings and thus incorporated into other social experiences.

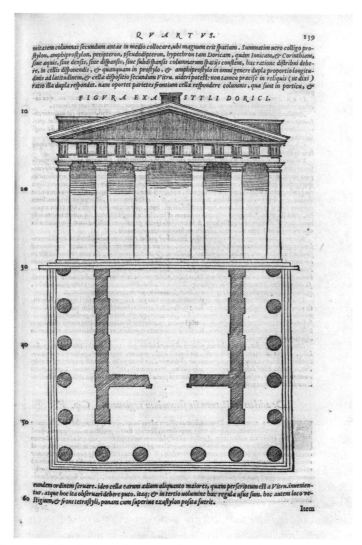

9.5. Elevation and plan of an exastyle Doric temple. Vitruvius, *De architectura libri decem*, ed. Daniele Barbaro (1567, book 4). Courtesy of the Library of the Max Planck Institute for the History of Science.

Canonical buildings such as Greek or Roman temples served as external representations of shared mental models of both social and practical knowledge. The establishment of canonical buildings changed the knowledge economy of building. For one thing, planning knowledge was structurally simplified. In the case of canonical buildings with well-defined proportions among its elements, the essential challenge for the planning process consisted in calculating the necessary resources and organizing the individual construction phases. Conversely, the available resources determined the dimensions of the building to be erected. Once it was clear what type of building was planned, the determination of its scale was all that was needed to define it completely. Canonization and modularization also simplified the process of planning and communication on the building site. Hence, brief written information was sufficient to determine the main components of a building to be erected.[36]

Knowledge of canonical forms could have a long-term historical effect, while the transmission of practical building knowledge presupposed "action by contact" to some extent. In contrast to the knowledge of stylistic elements, practical knowledge about the building process—conditioned by the regional availability of building materials—had a limited mobility that greatly depended on the historical circumstances. Practical knowledge tended to be regionally specific and persisted partly because it was nonetheless adaptable to new requirements of form. In other words, it was extendable. Craftsmen on the ground were regularly left to choose the technique with which to implement a building plan. The concurrence of local knowledge with new building forms could thus become a source of innovation.[37] As we shall see below, this was the case during the Renaissance, when new aesthetic requirements met with traditional practical knowledge.

The extent to which the transmission of canonical building forms involved explicit and reproducible knowledge (together with the longevity of surviving building records) determined the independence of these forms from the specific building knowledge of each receiving society. This is particularly clear in relation to the long-term impact of the canonical building forms of Greek and Roman architecture. Beyond major historical and cultural breaks, the effect of explicit knowledge transfer through individuals was, as a rule, much more limited than the cultural influence of canonical models of building forms. This only began to change in the Renaissance, with the increasing explication and scientification of building techniques through the incorporation of scientific knowledge, in particular concerning stability and material characteristics.[38] During this period, the printing press, through the reproduction of architectural treatises, became the main driver of the spread of the canonical building forms of classical antiquity in Europe and beyond.

Buildings as Collective Experiments

Major building projects like the construction of a Gothic cathedral can be seen as collective experiments in which experience could be gained, represented, and taken into account later on. The effectiveness of such experiments depended on the forms in which the building knowledge involved was represented, as shaped by the available media through which the evaluation of buildings could be expressed. Medieval and Renaissance construction workshops, for instance, boasted a knowledge economy that encouraged the integration and exchange of building knowledge among different projects and its transmission across generations.[39]

Beginning in the cities and on the major building sites of the late Middle Ages, institutional structures emerged that helped to steer such experiments and to provide the institutional memory that gave them an enduring impact. The increasing use of written records in management and administration processes played a key role in this transformation.[40] The rapid economic development in Italy and certain parts of northern and central Europe that began in the late Middle Ages led to a considerable

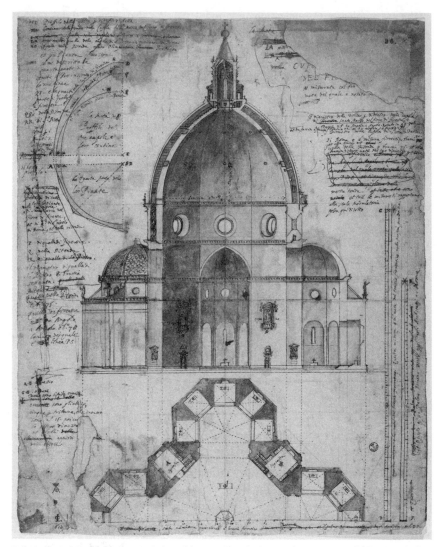

9.6. Lodovico Cigoli, section and plan of the cupola and tribunes of Santa Maria del Fiore, including comparison with the profiles and plans of the domes of the Pantheon and St. Peters, Rome, 1601. Foto Scala, Firenze—su concessione Ministerio Beni e Attività Culturali e del Turismo.

expansion of building activities. The administrative organization of late medieval building projects was influenced by previous construction experiences on the one hand, and defined by a highly developed mercantile culture on the other.[41]

The process by which the planning activities of architects became independent of their executive activities began in the early Middle Ages and continued in the Renaissance era.[42] The professional profile of an architect was reinforced by an educational canon and an institutionalization of an architect's education and training. The canon was based particularly on antiquity as the model of Renaissance architecture and thus on the importance of the only surviving antique architectural treatise, that of Vitruvius, as mentioned above.[43]

In terms of construction techniques, on the other hand, Renaissance builders essentially continued the traditions of the Gothic period but adapted a new approach to form and stylistic sensitivity primarily influenced by models of antiquity as they were found in Italy. The rejection of Gothic building forms on aesthetic grounds led to new construction tasks, such as crowning buildings with domes instead of crossbeam rib vaults. As we shall see in detail below, in Florence, this led to the erection of the largest dome of its time, actually an octagonal cloister vault in form although realized with technical innovations that allowed its self-supported construction with some of the static characteristics of a rotational dome.[44]

Historically we can think of the octagonal cloister vault as a step along the evolution toward the ideal circular-plan dome. This kind of vaulting over the crossing of late medieval basilicas was present on a smaller scale in earlier Tuscan cathedrals, such as Siena and Pisa. These proto-domes must have been constructed like all Gothic vaults with the support of wood armature, still possible on the scale and elevation of their structures. In Florence, even as the projected scale of the whole building was vastly increased in the revised planning of 1366–68, traditional wood centering must have been foreseen as the means for the construction of the cupola. This is supported by the record of the decision taken only in 1431, when the dome was nearly finished, to demolish the superseded 1368 model next to the Campanile, *except* the armature of its cupola, which was to be carefully conserved for future reference.[45]

Working within the given economic parameters, architects transformed their clients' wishes into concrete building plans, simultaneously taking responsibility for directing and executing the project. This created a bridge that, in principle, made it possible to use the accumulated experience gained in the course of construction work for planning. This type of feedback could result in building projects that amounted to veritable experiments in which planning was deliberately targeted toward innovations that could be immediately validated (or not). But these experiences would have remained ephemeral without institutional memory. As mentioned above, since the late Middle Ages, administrative structures had existed that could serve as repositories of institutional memory and thus as components of a new knowledge economy.[46]

As developments in fourteenth- and fifteenth-century northern Italy show in a particularly instructive way, administrative responsibility for a building project was frequently assigned to individuals or organizations with commercial knowledge.[47] A large construction site could require its own building management, which, in practice, completely took over the tasks involved in the construction, expanding its own purview at the expense of the supervisory authority. The growing autonomy of this kind of building administration provided the basis for new forms of interaction between a building project and the society in which it was realized.

Notably, the building administration set up expert commissions to discuss the specific challenges of a building project. It could also prepare and present options for referenda or decision making by superior authorities. This gave the building administration a new role beyond the task of organizing and controlling the construction process: it also organized the knowledge management in a way that, first, incorporated and made optimal use of the society's shared knowledge resources and, second, contributed to the project's social legitimation.[48] The new kind of knowledge economy that emerged from this transformation stimulated the development of technical and economic practices that increasingly also made use of other knowledge resources, including scientific and engineering knowledge. In this way, the new knowledge economy involving experts transformed societal needs and challenges into triggers for the development of knowledge. In the following, I illustrate these processes with one of the most prominent examples of Renaissance architecture, the construction of the cupola of the cathedral in Florence.

The Magnificent Cupola of the Florentine Cathedral

The cupola of Santa Maria del Fiore, the cathedral of Florence, represents, with its external diameter of circa fifty-five meters and elevation of circa thirty-three meters, the greatest masonry dome ever erected without sustaining armature.[49] It was built between 1420 and 1436 according to the plans of Filippo Brunelleschi (1377–1446), a master goldsmith who became one of the first modern architects, combining the roles of designer, planner, engineer, and construction supervisor.[50]

The cathedral shapes the urban landscape of Florence well beyond the borders of the city. Its cupola forms a kind of man-made mountain visible from afar in the surrounding hills. The cathedral and its magnificent cupola served not only as a symbol of pride and prosperity for a thriving community, but in many senses also represented a civic achievement, the product of a special economy of shared knowledge. The cupola, erected without the support of a wooden centering, represents a unique engineering feat. How was it possible? Was it, as Brunelleschi's contemporary Leon Battista Alberti suggested, the feat of one single man who besides being a goldsmith, was also a clockmaker and genius of all kinds of mechanical appliances? In his *De pictura*, Alberti writes: "Who could ever be hard or envious enough to fail to praise Pippo the architect on seeing here such a large structure, rising above the skies, ample to cover with its shadow all the Tuscan people, and constructed without the aid of centering or great quantity of wood?"[51] To erect the cupola an estimated 37,000 tons of materials, including more than 4 million bricks, had to be hoisted.[52] To make this possible and to place heavy stones with great precision on the rising cupola, Brunelleschi invented special cranes and complex hoisting machinery that were later studied with admiration by, among others, Leonardo da Vinci, who at the time was an apprentice in the workshop of Verocchio, which was charged with preparing the bronze sphere to be placed at the top of the lantern.[53]

The cathedral was a challenging object from the very beginning, when in 1294 the construction of the church was approved by the city council. Its design by the architect Arnolfo di Cambio probably already included a smaller dome over the crossing. In 1355–57 and again in 1366–68, the project underwent important transformations, raising the height of the nave with soaring cross vaults and vastly expanding the dome and east end configuration. The community decided in favor of an ambitious model with self-supporting cupola rising high on an octagonal tambour over the massive piers of the crossing area, whose robust structure guaranteed stability without the extensive

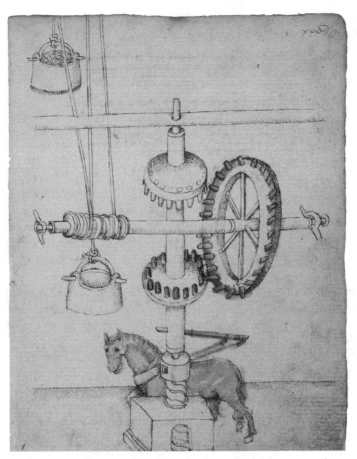

9.7. Drawing by Mariano di Jacopo detto il Taccola of Brunelleschi's hoisting winch, moved by a horse. Unlike Brunelleschi's machine, Taccola's version does not allow for variations in speed. Ms. Palatino 766, c. 10r, by concession of the Ministerio per i Beni e le Attività Culturali—Italy / Biblioteca Nazionale Centrale di Firenze.

exterior buttressing commonly used in Gothic churches to stabilize lofty architectural structures. The bold project elaborated by a committee of painters and masons, perhaps already with specification of a double cupola, with external and internal shells, was, after long deliberations, preferred over a more traditional proposal.[54]

The decision was taken knowing that the decisive constructive questions were as yet unresolved. In the further history of the building project, this daring choice repeatedly presented challenges and controversies regarding the feasibility of such a vast, freestanding cupola. But it was backed by strong political and cultural motives, which included not only the wish to distinguish Florence from enemies in the north, such as Milan, whose strongly Gothicizing cathedral was begun only in the 1380s, but also to underline its ambitions as a republican community that was renewing the traditions of classical antiquity.[55]

The architectural knowledge for building a cupola could not yet rely on scientific calculations, but it was not simply based on individual intuition either. It was rather practical knowledge, organized in mental models rooted in a long tradition of building knowledge. The problem in applying these mental models to the construction of a specific building was, in a way, similar to the task of a surveyor who tries to measure the surface of uneven terrain by covering it with familiar geometrical shapes such as triangles or rectangles whose area can easily be calculated. Here, the skill of the architect lay in the deployment and combination of familiar mental models that captured the traditional knowledge of statics when confronted with singularly challenging building tasks.

One of the mental models relevant to building a dome was that of hoop tension. The most familiar material model for hoop tension and how to deal with it is a barrel surrounded by hoops, which prevent its breaking under the stress exerted by its load. The model suggested that the lateral stress produced by the weight of a dome on its wall could be contained by constructing chains around it or within its walls. Another relevant mental model is that of a self-sustaining arch. When this model is generalized to a rotational symmetric dome, it becomes clear that such a structure can indeed support itself. But this does not yet provide a clue as to how it could be erected without scaffolding. In order to approach this task for a masonry dome, one has to imagine a rotational symmetric dome being decomposed into rings that form successive layers of bricks.

But how can these bricks be prevented from falling down, especially when the walls of the dome are ever more steeply inclined at the top? Here, another mental model comes into play that may be likened to the congestion in a blocked drain: if the rim is inclined toward the center, the bricks will mutually block each other from gliding down, provided, however, they form a full circle. Clearly, in the process of filling the circle, some intermediate device is needed to prevent the single bricks from falling while the mortar is still soft, such as stoppers at regular intervals along the rim. This could be realized, for example, with a particular technique of herringbone masonry, as well as with a skeleton of vertical rips across the dome.

The true challenge, however, lay in the adaptation of these basic mental models of statics to a geometrical form much more complex than a spherical dome. In fact, the approved committee model specified an octagonal, double-shell cupola whose cross-sections were defined by pointed arches. This profile made it not only possible to reach a greater height than would have been possible for a spherical dome erected over the same diameter, but it also had static advantages because a pointed cupola generates less radial thrust. Brunelleschi used a number of ingenious devices to apply and develop his knowledge of statics in order to build such a demanding structure and to secure its stability both in the process and in the completed building.

One of these devices was the system of chains built of specially designed sandstone beams that were linked by iron clamps within the internal cupola, plus an additional wooden chain. The exact design was the subject of intense consultation within the bodies of the "Opera del Duomo." Brunelleschi also implemented a rotational symmetric masonry structure within the thick internal shell. Specially formed bricks were placed along conically shaped surfaces inclined toward the center, following a herringbone pattern to enhance stability, in particular during the construction process. For the thinner outer shell, masonry rings were constructed in order to similarly impose a circular structure on its brickwork.

The construction of the dome required not only planning and constructive knowledge, but also an intimate familiarity with the materials used, together with permanent quality control and a readiness to adapt to changing availabilities and needs. The logistics involved in organizing the supply of materials and its quality control, as well as the coordination of workers operating simultaneously at the different working sites, was immense.[56] Administrative and commercial knowledge was needed to secure and deploy the funds that ensured the economic viability of the Opera del Duomo as an institution. But what kind of an institution was the Opera and how did its economy of knowledge work and develop?[57]

Initially, the project of building a new cathedral was a joint enterprise of the bishop and the commune but was quickly taken over by the commune alone. The commune in turn charged the major guilds of the city with its administration and execution. From 1331, the powerful and rich Wool Guild (Arte della Lana), took on sole responsibility, probably because of its influence in the city and its financial and administrative expertise.

The administrative structure initially comprised four *Operai*, officials or wardens who were the highest supervising authorities chosen by sortition from a list of qualified guild members for a short period of initially four and later six months. At first, they were assisted only by a treasurer, also a guildsman, in charge of controlling the acquisition and disposition of funds, and a notary scribe. The funding was secured by the commune on a long-term basis by direct and indirect taxes, as well as by providing the Opera with its own sources of revenue in form of forests that could fulfill its need for wood but also serve as a business asset.[58]

Gradually, the administration became more complex, and new institutional personae emerged, combining creative and artistic capabilities with technical expertise, but also with planning and logistic competence.[59] While this development inevitably created new dynamics within the traditional guild structure, it prepared the way for innovative forms of economic and technical practice that much later became characteristic of entrepreneurship during the Industrial Revolution. A *capomaestro* (a foreman or construction manager) was responsible for the organization of the building operations; from the 1350s, a *proveditore* (supervisor or purveyor) served as an interface between the governance structure of the Opera and its workforce.

When it became clear in 1419 that the construction of the cupola constituted a new and very grand challenge in its own right, the original governance structure was reinforced by a longer term advisory board of four hand-picked guild members, the Cupola officials, specifically dedicated to the project.[60] In April 1420, three *capomaestri* were appointed, and were soon renamed as supervisors of the cupola project, or rather *provisores operis Cupole construendi*. One of them, Battista d'Antonio, was an experienced foreman in the traditional sense. The two others were Filippo Brunelleschi and Lorenzo Ghiberti, both goldsmiths by trade, who had been the leading contestants in the competition for proposals for vaulting the cupola.[61] This corresponded to the tradition of involving artists in the design aspects of cathedral construction.

Brunelleschi soon became the central figure, and his presence on the building site was essential for the project's success, not only because of his mastery of the construction challenges and the mechanical devices he invented to confront them, but also because of his close interactions with the various components of the Opera. These interactions became necessary because many of his innovations required, time and again, the approval of the Opera's supervising authorities, but—equally important—also the creative cooperation of its workforce. Indeed, the historical archives of administrative documents pertaining to the construction of the cupola, prepared by Margaret Haines and her team, which have recently been made accessible in an online edition,[62] make plain the extent to which the entire Opera became a laboratory for probing new technologies, a laboratory that also offered new opportunities for social advancement.

In a sense, Brunelleschi's career, moving from goldsmith to become the supervisor of the cupola project, is thus paralleled on a minor scale by the many cases in which other artisans employed by the Opera contributed creatively to the project's success and exploited the career possibilities it offered. Jacopo di Sandro, for example, an ordinary stonemason working for twenty years on the cupola eventually became a supervisor of several important subprojects.[63] For instance, he took responsibility for burning the specially designed bricks, an enterprise with a strongly experimental character. Over time, he also developed initiatives as an independent entrepreneur, working on his own account when providing the Opera with wood. The Opera thus also served as a matrix for the generation of new technical and economic practices emerging from the network of activities it triggered within the city and its surroundings.[64] This development was stimulated by the mutual enablement of individuals and their institutional positions within the Opera. Stimulation by mutual enablement is obviously an alternative to the entrapment by mutual limitation discussed in the previous chapter.

One of the characteristics of the Opera's economy of shared knowledge was its reliance on public competitions for the major phases of the project. Through these competitions, the Opera summoned the knowledge of the available experts and, at the same time, generated a broad consensus around its daring endeavor. Models played a crucial role in this process as mediators between the contracting authority, experts, and the public, as tools for the design process, as experimental demonstrations of the feasibility of a project, as a guide to the construction process integrating the knowledge of the architect and that of the practitioners, and as means to secure decisions by becoming part of contractual agreements. For the Florentine cathedral alone, we know of the existence of numerous models.[65]

In 1357, for example, the Opera had to choose among three models for the nave pillars. It first convened an expert commission of five master builders and finally presented the preferred model to the public with the announcement that any person wanting to point out any deficiency should do so within the next eight days. The crucial decision of 1367 in favor of the masons' and painters' proposal was also first discussed in various commissions and subsequently approved by a public vote involving several hundred Florentine citizens.[66] In August 1418, the Opera announced a public competition for models for the vaulting of the cupola, including scaffolding and construction tools, with a guarantee of reimbursement of expenses to all contestants: "Whoever desires to make any model or design for the vault of the main dome of said cathedral . . . whether for armature, scaffolding or any other thing, or for any machine useful in the construction and perfection of said cupola or vault,

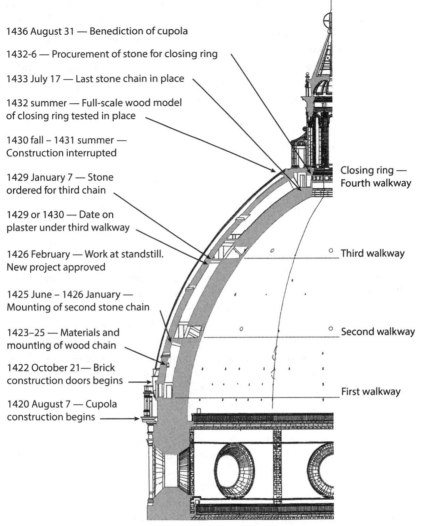

1436 August 31 — Benediction of cupola

1432-6 — Procurement of stone for closing ring

1433 July 17 — Last stone chain in place

1432 summer — Full-scale wood model of closing ring tested in place

1430 fall – 1431 summer — Construction interrupted

1429 January 7 — Stone ordered for third chain

1429 or 1430 — Date on plaster under third walkway

1426 February — Work at standstill. New project approved

1425 June – 1426 January — Mounting of second stone chain

1423–25 — Materials and mounting of wood chain

1422 October 21— Brick construction doors begins

1420 August 7 — Cupola construction begins

Closing ring — Fourth walkway

Third walkway

Second walkway

First walkway

9.8. Essential documented chronology of the cupola of S. Maria del Fiore. From Haines (2011 – 12, 99).

shall do so by the end of September and state whether he wishes to say anything to aforementioned wardens, and he will be well and graciously heard."[67] A rich prize of two hundred gold florins was offered to the winner or whoever came closest to what would be followed in the subsequent work at the discretion of the *Operai*, and all would be reimbursed for expenses.

For the occasion Brunelleschi constructed a brick and wood model of the dome in a courtyard of the Opera. He was assisted by two well-known sculptors, Donatello and Nanni di Banco, as well as by master masons. In addition, the Opera appointed its own workmen to observe the building process, whether it could be accomplished as claimed, that is, with brick masonry rising to the summit without armature. The model was the size of a small building and remained in place until 1431 when, with the actual dome nearly completed, the supervisors ordered its destruction. In 1418, however, the Opera was slow to take a definitive decision among the various proposals it received. Renouncing a wooden centering had the advantage of saving on enormous amounts of timber and avoiding the risk that a wooden support of the dimensions needed for building the cupola could warp during the building process. But the erection of a dome that would remain self-supporting even during its construction, as Brunelleschi claimed, just seemed too bold. By the end of 1418, the Opera had paid off most other models and focused on two remaining proposals: one by Brunelleschi and another by his longtime competitor, Lorenzo Ghiberti.

Only after much discussion and the presentation of new models by other prestigious consultants did the guild consuls and Operai appoint the two artists and Battista d'Antonio *provveditori* responsible for the cupola. The timing of this act coincides with the realization of a large painted wood model according to the specifications of the four Cupola Officials, which has been identified by Howard Saalman as a mockup of the existing tambour supporting the cupola in order to test the illumination of Brunelleschi's windowless dome model with light coming only from eight oculi below and the aperture at its summit.[68] The officials must have been satisfied and critics silenced on this issue for the time being.

The decision to proceed according to Brunelleschi's daring plans is only evident from a written, twelve-point memorandum dating from 1420, which specified that both shells of the cupola were to be constructed without a scaffold-supported centering.[69] But even after this fundamental decision was finally officially registered on the eve of the beginning of construction in August 1420, all further important steps in the building process remained subject to public competitions and deliberations within the decision-making bodies of the Opera. For instance, when the cupola had already reached a height of thirty *braccia*, after months of consultations in February 1426, the consuls and *Operai* would meet again to decide on the question whether the building would proceed without centering.[70] During the entire construction process, the basic technical questions would remain subject to intense discussions based on the experiences so far acquired. The realization of the cupola project was a constant learning experience for everybody involved.

Large engineering ventures with their sophisticated economies of knowledge, such as the construction of the Florentine cathedral, also became an important driving force for the development of scientific knowledge. They not only confronted traditional systems of knowledge with new challenging objects but also promoted social structures within which the integration of practical and theoretical knowledge could take place, for instance, by public competitions, by forming expert commissions, or by directly reaching out to scholars.

One example of great historical significance is the interaction between the Venetian Arsenal and Galileo Galilei around 1600.[71] In chapter 1, I cited the famous introduction to Galileo's *Discorsi* in which he refers to his frequent visits to the famous Arsenal, which, he claimed, opened up a large field of investigation. On closer inspection, however, Galileo's relation to the Arsenal did not simply involve casual visits but was the outcome of a characteristic economy of knowledge that established links between technical challenges and intellectual resources in a manner similar to that employed by the Florentine Opera almost two centuries earlier.

The Arsenal of the Venetian Republic was one of the largest military, technical, and industrial centers of the time, employing between one thousand and two thousand workers. In a single year, more than one hundred galley ships were finished or built. Like the Opera, the Arsenal was subject to a complex governance and management structure, ensuring supervision and accounting, but also the capability of transforming military and political needs into technical realities. Toward the end of the sixteenth century, such needs emerged from the confrontation of the Christian naval powers, in particular Spain and Venice, with the fleet of the Ottoman Empire in the Mediterranean. An important turning point was the famous battle of Lepanto in 1571, when the Ottoman fleet was defeated in the Gulf of Patras. The battle was a turning point also for military technology, since the increased significance of artillery made it necessary to build larger battleships, which posed, however, problems of speed, maneuverability, and stability.

At the Arsenal, these challenges led, in the early 1590s, to a systematic enquiry at the request of the military committee of the Republic. The enquiry was directed to engineers, craftsmen, and veterans, but also to external experts such as Galileo. He received a letter from Giacomo Contarini, one of the commissioners of the Arsenal, to which he responded on March 22, 1593. Contarini's letter had evidently raised questions concerning the propulsion and maneuverability of a galley and the optimal position of the oars. Galileo responded on the basis of Aristotelian mechanics, but soon turned the questions raised by Contarini into a stimulus for developing a new science dealing with the strength of materials.[72] It is one of the two new sciences presented in his conclusive work on mechanics, the *Discorsi*, the other being the science of motion discussed in chapters 4 and 6.

Galileo showed, in particular, that geometrically similar beams are not equally strong and that they become weaker with increasing linear dimensions. Clearly, his new understanding of the strength of materials was of the utmost importance not only for shipbuilding but also for architecture, since it pointed to the limits of reasoning in terms of proportions and of using scale models to predict stability properties, as Galileo himself realized: "You now see how, from the things demonstrated thus far, there clearly follows the impossibility (not only for art, but for nature herself) of increasing machines to immense size. Thus it is impossible to build enormous ships, palaces, or temples, for which oars, masts, beamwork, iron chains, and in sum all parts shall hold together. . . ."[73] When writing these lines, Galileo may have been aware of the first cracks forming in the cupola, probably resulting from the earthquakes that affected the cathedral. Today's confidence in its stability relies on both the ingenious construction techniques invented by Brunelleschi and on methods of monitoring, restoration, and engineering rooted in the new science of materials founded by Galileo. But how must building practices and urbanization be revised in a post-fossil age, when energy from fossil resources will no longer offer seemingly limitless horizons for construction? Does it make sense to return to some of the experiences of the preindustrial age? Addressing this challenge, in any case, will again require the concerted efforts of architects, scientists, and civil society, including a stimulation by mutual enablement.[74]

Chapter 10

Knowledge Economies in History

In science, for example, an "individual" can accomplish matters of general concern, and it is always individuals who do accomplish them. But these matters become truly general only when they are the affair no longer of the individual but of society. This changes not merely the form but also the content.
—KARL MARX, "Contribution to the Critique of Hegel's Philosophy of Law"

It is nonsense to think that the history of cognition has as little to do with science as, for example, the history of the telephone with telephone conversations. At least three-quarters if not the entire content of science is conditioned by the history of ideas, psychology, and the sociology of ideas and is thus explicable in these terms.
—LUDWIK FLECK, *Genesis and Development of a Scientific Fact*

Innovations are, according to the economist Joseph Schumpeter, the most striking feature of the economic history of capitalist society.[1] Innovations are not necessarily new inventions but may also consist of novel combinations of existing knowledge.[2] It is, however, a false prejudice that innovations are an exclusive feature of modern capitalism whereas ancient societies were conservative and hostile to anything new. As we have seen in the previous two chapters, the capacity of a society to generate, foster, and maintain innovations depends on its knowledge economy, that is, on the way the production, circulation, and implementation of knowledge is socially organized, including the available means of its representation and communication.[3]

In the following, I review some characteristic knowledge economies in human history (without claiming to be comprehensive). Key concepts are the carriers of knowledge and the social organization of knowledge, including the relation between knowledge and material economies, knowledge representation technologies and external representations, the social accessibility of knowledge, domains and systems of knowledge, images of knowledge, forms of knowledge integration, and reflective organization.

The typology presented in the following pages does not suggest an automatism of progress, driving the evolution of knowledge from one type of knowledge economy to the next. My main purposes here are rather, first, to illustrate how a historically specific knowledge economy can be effectively described with the help of the previously introduced theoretical framework, and second, to provide further hints of the coevolutionary character of the development of knowledge and that of societal structures mediated by knowledge economies.

Against this background, my sketch is not to be misunderstood as an abstract sociological system of knowledge economies either. All elements of the framework have a history without which they are inconceivable. Furthermore, the presentation clearly depends on and embodies a particular perspective shaped by the historical case studies on which it is mainly based. These case studies range from knowledge in nonliterate societies via the early modern knowledge societies to the growing entanglement of science and society in the twentieth century. In the next chapter, I attempt to embed this perspective within a larger, global outlook on the history of knowledge.

Knowledge Economy in Nonliterate Societies

Three Dimensions of Knowledge in Anthropology

The anthropologist Fredrik Barth has shown how attention to the mental, social, and material dimensions of knowledge may help us understand traditions of knowledge documented by ethnographic materials. In a paper from 2002, Barth argues: "I see three faces or aspects of knowledge that can be analytically distinguished. First, any tradition of knowledge contains a corpus of substantive assertions and ideas about aspects of the world. Secondly, it must be instantiated and communicated in one or several media as a series of partial representations in the form of words, concrete symbols, pointing gestures, actions. And thirdly, it will be distributed, communicated, employed, and transmitted within a series of instituted social relations. These three faces of knowledge are interconnected. Being interconnected, do they mutually determine each other? That is my claim. . . ."[4]

Barth applied this framework, for example, to a knowledge tradition he observed in his 1968 fieldwork in the Ok region of New Guinea.[5] His observations focused on the Baktaman group, who cultivate taro as their main subsistence strategy in addition to hunting, foraging, and pig raising. An important knowledge tradition in the lives of the Baktaman is a set of rituals and myths centered on growth and fertility, which guides their understanding of key aspects of their world. The mental dimension of this knowledge system consists in a network of relations established between different models of growth: "leaves on the trees, human hair on the head, the fur of marsupials, the pandanus-leaf thatching of the temple, the subcutaneous fat of pigs—linking all of them as images of the effects of an invisible force, somewhat like heat, that makes taro plants and subterranean taro corms grow."[6]

The knowledge economy in which this system is reproduced and transmitted is crucially informed by the available media—essentially objects or elements of the natural world used as symbols; nonverbal images; and acts such as prayers, songs, the oral transmission of myths, and the participation in rituals. This knowledge economy is, however, also shaped by an image of knowledge according to which all legitimate knowledge is derived from what the ancestors passed on before death. These two elements create a certain tension: the image of knowledge prohibits any change in the transmitted knowledge, but the available media make such changes unavoidable, as the performative quality of initiation acts requires some degree of spontaneous creativity. As a result, one observes slow variation in the system of knowledge through time and space, despite the claim of its invariable and direct descent from ancestral knowledge.

It is difficult to assess the role and nature of knowledge in prehistoric nonliterate societies given the limited evidence that archeology and prehistory are able to provide. Complementing this evidence with insights from studies on recent nonliterate societies may therefore be legitimate, given the apparent similarity of their material culture with that of Stone Age societies, even if inferences from contemporary observations to the distant past are highly problematic.

Nonliterate societies possess capabilities and technologies relevant to their survival that incorporate knowledge with traditional techniques for mastering primary needs such as food production, medicine, housing, and mobility. Their rich knowledge about the natural world may be illustrated by the elaborate and astonishingly adequate taxonomy and nomenclature for animals and plants available to them. Even today, some indigenous populations retain an immensely sophisticated knowledge of their local environment—about local plants used as food and medicine, as raw materials for buildings, as weapons, for making clothes, as musical instruments and various kinds of tools, or even in the context of ritual practices. Another area of highly developed knowledge is that of social behavioral patterns—as reported by some of the earliest ethnographic studies—externally represented by differentiated terminology for family and kinship relations.[7]

There is essentially no separate institution of knowledge economy and no professionalization in such societies. Knowledge is acquired by interactive communication, by participation in work processes and rituals, and by imitation during shared activities. Individuals may combine their shared knowledge in collective actions, in agriculture for instance, or when a group activity is required to erect a major building or monument. Among recent societies, this collective dimension may be exemplified by the *mutirão* regime of mutual help among the Guarani in Brazil or by the construction of men's houses among the Eipo people in New Guinea.[8] These examples also illustrate the double function of such shared knowledge for practical and cultural purposes.

External representations here comprise the means used to master the challenges of survival, the elements of the environment, and context-dependent actions, as well as the tools and objects employed in such actions: spoken language, ritualized forms of communication involving special forms of language—technical terms, poetry and songs, other forms of ritualized social behavior such as artistic or religious performances and productions—and a limited symbolic culture comprising masks, artifacts, adornments, monuments, and so forth. The cave paintings of the Altamira cave going back more than fifteen thousand years illustrate the level that such a symbolic culture could reach.

Together with orally transmitted mythologies, symbols define social identities and offer reflective resources for assessing reality. However, specialized epistemic domains such as counting or spatial orientation, in which number signs or maps may play a specific, practical role, may be exceptions to this more or less holistic use of symbols.

Knowledge is typically accessible to all members of a nonliterate society, but, again, there may be exceptions. There may be religious specialists guarding secret knowledge. A division of labor or more general gender divisions may enforce a differentiation of knowledge distribution. Some societies especially value the knowledge of old people. Among the Guarani of Brazil, for instance, knowledge is concentrated in the hands of the *Pajé*, the tribe's priest, enabling him to serve as a guide, healer, educator, seer, and sorcerer.[9] In other societies, special challenges (seafaring, for example) lead to the formation and transmission of specialized expert knowledge—a form of knowledge prefiguring developments in literate societies. For instance, traditional Polynesian and Micronesian navigational techniques have enabled specially trained navigators to undertake long-distance trips between islands that take several days out of the sight of land.[10] There are, in short, many different kinds of knowledge distribution in nonliterate societies as well, but the bulk of this knowledge tends to be broadly shared and without specialization.

The knowledge underlying traditional navigational techniques comprises elements that from a Western perspective fall into entirely different domains, such as astronomical, geographical, oceanographic, meteorological, and ornithological knowledge. Stars provide the bearing while waves and winds give an indication of the speed of the voyage; birds, currents, and the color of the water help to identify the vicinity of land. Yet, such local knowledge consists of more than a collection of isolated pieces of information compiled according to specific needs—otherwise it could hardly meet the challenges and vicissitudes of long-distance seafaring. It turns out that traditional Micronesian knowledge about navigation has a sophisticated cognitive architecture that can be described in terms of mental models that allow the "calculation" of the course of a boat on a long-distance trip.

Such situated knowledge may have the same efficacy as the knowledge of modern navigation technologies, but it is bound to a specific local context and context-specific social relations. Furthermore, it is typically embedded in comprehensive views of the world at large ("cosmologies"). In the case of Micronesian navigation by the stars, it depends, for instance, on the fact that this navigation takes place close to the equator, where stars (and planets) rise and set along paths more or less perpendicular to the horizon. Thus, in the course of the night, one star can simply take over the role of another just "underneath" it to indicate a given bearing.

In nonliterate societies, knowledge is essentially a functional aspect of action. The organization and integration of knowledge is domain and context specific because

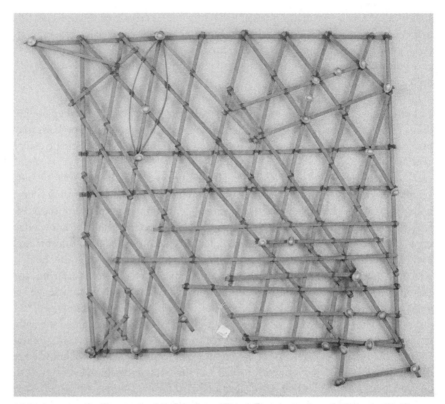

10.1. A navigation chart made of wood splints tied with thread, pandanus (palm) splints (indicating wave types), and cowrie shells (indicating islands). From the Majuro Atoll, Ratak Chain, Marshall Islands. Courtesy of the Hearst Museum of Anthropology at the University of California–Berkeley.

of the close ties between knowledge and concrete objects, actions, and circumstances. In the case of the Eipo culture, for instance, space serves as a primary principle for ordering and classifying elements of the natural environment such as materials, plants, and animals. But this space is not organized by metrical concepts of distance and direction. The comparably small geographical area in which they live and move constitutes a spatial entity composed of several thousand individual entities with their own names: mountains, woods, meadows, waters, valleys, mountain tops, caves, trees, gardens, lakes, confluences of rivers, and so on.[11]

Materials, animals, and plants are named after the locations where they can be found or to which they have a specific affinity. Needless to say, such a conception of space is neither generalizable in the Newtonian mold (space as an empty and neutral metrical container for matter) nor to other orders of magnitude such as the immediate spaces of human activities. Local space is rather structured by deictic notions with anthropomorphic origins such as *there, up,* and *down,* supplemented by communicative gestures.[12] The context-dependent geographic and local deictic notions of spatial relations make it in principle possible to communicate locations and movements in everyday life without any integration of individual perspectives into an overarching concept of space, as is characteristic of any abstract conceptual system of spatial knowledge.[13] I return to the historical evolution of the concept of space in chapter 14.

The number of mechanical tools used in the Eipo culture is so limited that these tools can easily be classified under a few categories: tools for the agricultural cultivation of the soil, various kinds of strings and ropes, various kinds of containers for the transport and storage of products and other kinds of property, tools for work on materials and the production of tools, an instrument for lighting a fire, and weapons for hunting and warfare. Mechanical knowledge is an implicit precondition for the building of houses, the bridging of rivers, and the construction of traps. The knowledge involved in the use of tools is simple and widely tool specific, but its scope covers many kinds of mechanical phenomena.

While each of these mechanical techniques may be limited to the specific knowledge involved in a certain craft, taken together they represent an impressive body of knowledge about various forms of mechanical forces, about the mechanical properties of materials, and about mechanical operations and their effects. The tools and techniques described above make implicitly extensive use of knowledge about the mechanical stability of different kinds of materials and constructions, about the effects of frictions, about mechanical tension and elasticity, and in particular about the force-saving potential of the lever. There exists a rich terminology related to the specific forces, tools, materials, and operations, but, as noted in chapter 4 when we first discussed this example, a general terminology for a generalized mechanical knowledge is evidently missing.

In general, knowledge in nonliterate societies is not subsumed under abstract categories that result from processes of iterative abstraction. Even mythological knowledge in such societies may result primarily from the projection of motives that dominate daily practice such as hunting and agriculture. (See also the box at the beginning of this chapter.) Such knowledge and its practical implementation also serves to stabilize interventions into the environment in another sense: protecting individuals and the society as a whole from dangers that might result from disturbances and violations of a larger order outside their purview and control.

Under the conditions of a nonliterate knowledge economy, the potential for building up and transmitting complex systems of knowledge is limited by the context-dependency of the available forms of representation, as well as by their transient nature. The example of Polynesian and Micronesian navigation is an exception that illustrates, at the same time, the capability of nonliterate societies to generate a differentiated knowledge economy that includes the formation of experts under challenging conditions.

Knowledge Economy in Early Literate Societies

As discussed in chapter 5, writing arose around the end of the fourth millennium BCE (ca. 3300) in southern Mesopotamia. The earliest written documents were clay tablets impressed with numerical notations and seals that indicated institutional issues and contexts. Although these documents led eventually to the development of cuneiform writing used for the representation of texts in Sumerian, Akkadian, and other languages, the earliest writing constituted a symbol system that was (probably) largely independent of spoken language and, in any case, used as an instrument of administration for the construction and control of centralized economic systems.[14] As we have also seen, on a parallel track, early writing led to calculating techniques and mathematical concepts. Early documents were very closely tied to their particular administrative contexts and do not represent background knowledge shared by the social actors in this context; in this respect, early writing exhibits much of the context dependence of face-to-face communication.

At the same time, the technologies of writing and arithmetic, in presenting systems of manipulable symbols, allowed for the emergence of new kinds of reflexivity, in particular the formation of arithmetic concepts and the formation of metalinguistic awareness, as I have discussed in chapter 5. Once writing came to represent language, it caused reflection on language itself—leading, for instance, to the creation of lexical lists—and this reflection in turn altered patterns of use, thus restructuring language. Even now, written language is typically different from spoken language, while both affect each other. Internalization of the technology of writing created a mental model that could be applied to diverse contexts, and in particular to the interpretation of all kinds of signs. Thus, the Babylonians saw "heavenly writing" in the skies and priests "read" organs in extispicy, a divination technique based on the "reading" of animal entrails.[15]

Writing constituted the first external representation of knowledge that was governed by formal semiotic rules. In principle, writing was highly suited to travel, since it was portable, durable, and reproducible. The extreme context dependence of the earliest writing, however, made it difficult for writing to move beyond the particular institutional context in which it was embedded. The potential of writing as a tool for permanently documenting spoken language was discovered only slowly and with increasing usage. When glottographic writing (i.e., writing that represents spoken language) first emerged in the Fara period (ca. 2600 BCE), it served as a mnemonic aid to recording oral genres (proverbs, incantations, hymns, and so forth).[16] Early Egyptian writing was closely associated with representational aesthetic functions, for instance, in monumental inscriptions legitimizing the authority of priests and rulers. Here as well, writing gradually assumed an ever-greater range of functions, like correspondence, historiography, and literature.

Glottography led to an increased awareness of language. Subsequently, written and spoken language developed as partly independent, partly interpenetrating systems. Glottographic writing eventually spread widely and diverged greatly in form, in response to differences of language typology, social usage, and physical media. As writing came to represent structures of spoken language and became increasingly phonetic, its context dependence decreased, and it began to spread widely. Over time, writing came to be employed in an increasing number of domains, some maintaining a parallel in spoken language, some made possible only by the technology of writing. Writing media varied, with the clay tablet predominating in Mesopotamia and papyrus important in Egypt and Greece. These media had important implications for the durability of the knowledge represented, as well as for its accumulation.

But writing was evidently not a necessary condition for such an accumulation. In early India, a purely oral culture, reflection on the sacred Vedas, facilitated by elaborate mnemonic techniques, allowed for the generation of extensive second-order knowledge, best illustrated by the fifth-century grammar of Paṇini, which consists of an elaborate system of approximately eight thousand rules (expressed in highly abbreviated sutra form) that allow for the generation of virtually all word forms in the Sanskrit language.[17] Mnemotechnics involves primarily internal representations. Yet these internal representations are structured in the context of a shared symbol-based technology that is learned, and they involve loci that are characteristically dependent on external representations.

Early literate societies such as those of Mesopotamia, Egypt, ancient China, and South and Meso-America were characterized by an expansion of knowledge systematically transmitted over generations, including administrative knowledge; complex metrological systems of measures and weights; legal knowledge; historical record keeping; calendric, astronomical, and geographical knowledge; healing and divination knowledge; and knowledge about language. Here I focus on the Mesopotamian developments.

The Example of Mesopotamia

By the middle of the third millennium BCE, writing had spread from southern Mesopotamia to the Levant; it had also undergone significant changes in the meantime and was now phonetically representing the structure of spoken language.[18] Long-term record keeping is also attested for the first time in this period. The Old Akkadian centralized state (ca. 2350–2200 BCE), incorporating various traits of its predecessors, attests to the emergence of newly ordered institutions (kingship, standing armies, palace administrative apparatus) and significant processes of standardization in writing, metrology, and other areas. During the subsequent Ur III period (ca. 2100–2000 BCE), known for its enormous administration, we find the first traces of new forms of written literature and historiography, which built to a large extent on older traditions and established a framework for the cultural identities of ensuing societies.[19]

The organization of society underwent tremendous changes in the following periods. In addition to temples, we find a largely independent state administration, as well as a tendency toward increased individualization and privatization, including private property and individual economic ventures. As far back as the Old Assyrian (ca. 1950–1750 BCE) and Old Babylonian (ca. 1850–1600 BCE) periods, we already observe a reduced number of cuneiform signs in use, which facilitated everyday communication, as attested by letters and administrative documents.[20] This process, which can be thought of as a "democratization of writing," is paralleled by the slightly later invention of alphabetic scripts in the Levant.[21] Alphabetic writing emerged as a consequence of a diffusion of knowledge and of a series of adaptations of writing to new cultural and linguistic conditions in the second and first millennia BCE, a process to which I return more systematically in the next chapter.

New forms of written knowledge that appear in this period include not only divination, grammatical, and historiographical texts, as well as various forms of lists, but also the first Akkadian literary corpus, private legal documents, and astronomical and protomathematical texts. In this period, we also find a number of multilingual lexical lists, thus enshrining the written and formalized multilingualism in the area,

10.2. A sphinx inscribed with two languages: hieroglyphs and proto-Sinaitic (the hieroglyphs on the right shoulder designate the Goddess Hathor, as do the proto-Sinaitic inscriptions). From the Middle Kingdom of Egypt, ca. 1800 BCE, Sinai, Serabit el-Khadem. British Museum. akg-images / Erich Lessing.

which is characterized by great language diversity throughout history. Some of these texts had precursors, but the level of systematization attempted, and in part achieved, during this period sets them clearly apart from earlier texts. A major part of this literature was transmitted and preserved in schools linked to the temple rather than to the palace administration (which represented the actual seat of power during this period). For the first time, we can observe a rivalry between two semiautonomous social units ("households") sharing many of the same actors, one of which enjoyed a monopoly of force. In a sense, this can be seen as the beginning of a dichotomy between state institutions and religious institutions. This opposition became crucial for the economy of creation and transmission of knowledge for the remainder of Mesopotamian history and persists to the present day.[22]

The canonization of Babylonian literature took place to a large extent during the Kassite Dynasty (ca. 1600–1300 BCE).[23] We can interpret this process as a conscious attempt to incorporate existing patterns of knowledge. This knowledge spread far beyond the borders of Mesopotamia to Anatolia, Iran, and even Egypt to some extent, influencing local knowledge traditions. As Mesopotamia became an international power from the twelfth century BCE onward, the collection of knowledge increased and became a thoroughly systematic enterprise. The emergence of this new knowledge economy led to the accumulation of vast amounts of knowledge, particularly in the areas of astronomy and meteorology.[24] In this period, Akkadian was a lingua franca and was used as a diplomatic language as well.

In spite of the canonization and systematization of accumulated knowledge, over-arching forms of knowledge integration were largely missing. The mythologies that were then being recorded in writing and systematized in the form of epics, theogonies, and so forth served as media of integration. While relating all aspects of life to each other and to society as a whole, they primarily reflected its social and political integration, rather than the underlying practices themselves. Accordingly, mythologies represented the integration of different religious traditions, typically by identifying different divinities, in terms of divine genealogies, or by imposing patriarchic hierarchies.

What was recorded was the knowledge of experts, an elite enthroned by the new division of manual and intellectual labor in the course of the urban revolution. These officials, scribes, or priests were responsible for planning activities in a redistributive economy. They organized not only the distribution of resources but also the employment of a subdued labor force, in part recruited by enslaving conquered people. Accordingly, the knowledge recorded in writing is primarily *Herrschaftswissen*, that is, hegemonic knowledge about the organization and social control of societal production, rather than knowledge about the productive processes themselves.

These extensive planning and organizational activities would not have been possible without the availability of durable and portable external representations that enabled knowledge to be communicated and stored independent of specific situations. The drawing of geometrical plans, for instance, has played a key role in architectural planning processes from the time of the ancient Orient up to the early modern period. In Egypt, work drawings with plans and cross-sections were used from the second millennium BCE onward. Approximate scale plans, partly dimensioned and with section details, have been discovered both in Egypt and in Mesopotamia.[25]

In chapter 5, I reviewed in some detail the role of accounting techniques and number systems, and the emergence of arithmetic for planning processes. I also discussed the role of schooling for the stabilization of a separate knowledge economy in which relevant knowledge was generated and transmitted. The availability of durable representations, together with the existence of such a separate knowledge economy, provided crucial conditions for the production, accumulation, and exploration of the broad variety of knowledge forms described above. Schooling thus became a starting point for what may be considered the earliest "scientific" practices in the sense of an exploration of the symbolic means of planning labor independent from the actual planning itself.

This knowledge primarily dealt with the handling of symbols such as writing and reading, and only indirectly (through those symbols' mediation) to the primary objects or actions themselves. These symbolic practices in turn represented the planning activities (such as accounting and controlling) at the center of the new knowledge. Accordingly, this knowledge was standardized and codified with the help of technical terminologies, such as the administrative terminology of Mesopotamian societies or the calendar terminology of the Mayas.[26] These symbolic practices of information processing and storage largely constituted, in form and content, the expert knowledge of early literate societies.

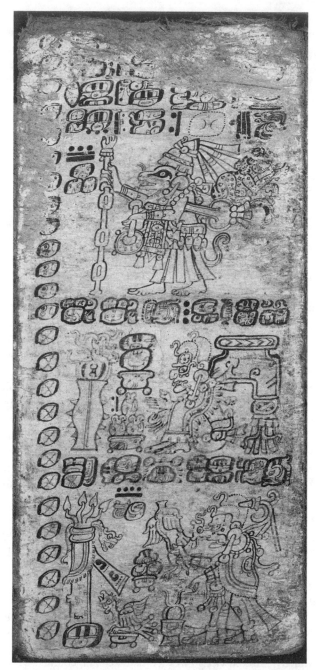

10.3. A scene describing the Yucatec New Year ceremonies of the 365-day Maya calendar. Page 28 of the *Dresden Codex*, ca. 1200–1250 CE, the oldest extant book from the Americas featuring local history and astronomical tables. See Kirkhusmo Pharo (2013, 171). Courtesy of Sächsische Landesbibliothek—Staats- und Universitätsbibliothek Dresden.

Remarkably, these practices were largely employed in an immediate way, without being themselves reflected on or giving rise to a reflective reorganization of the underlying knowledge. This may, at first glance, suggest an interpretation as mere "recipe knowledge," invariably close to immediate practice. As we have seen, however, this knowledge was actually decoupled from practice and primarily referred to cultural techniques such as writing and calculating. The fact that these practices were in turn pursued in a seemingly immediate way, lacking, for example, explicit proofs for the validity of mathematical procedures, may have been due to their institutional embedding. This embedding provided the historical actors with a self-evident background knowledge that did not have to be articulated and hence did not itself become the object of reflection. This only changed with challenges to the institutional framework, as came, for instance, with the diffusion of some of this knowledge into new cultural contexts, triggering a reflection on symbolic actions no longer ensconced in the system.

Knowledge Production in Religious Frameworks

An Imaginary Global Past: The "Axial Age"

In 1949, the existentialist philosopher Karl Jaspers suggested that in the first millennium BCE (or, more precisely, in the time between 800 and 200 BCE) profound spiritual transformations of society took place, more or less simultaneously, in Europe, India, China, and the Middle East. These transformations constituted, according to Jaspers, a novel awareness of humanity's position in the world. This novel awareness was fundamentally characterized by "transcendence," understood as the realization of a chasm between the immediately given world and a world beyond appearances—a world that could, however, be accessed through reason and ethics. Jaspers called the period of global history in which this upheaval occurred the "Axial Age" (Achsenzeit).[27] The transition was carried out by new intellectual elites gaining important societal positions outside of traditional hierarchies. This group included the prophets in Israel, the followers of Hinduism or Buddhism in India, and the philosophers in Greece and China. The protagonists were, among others, Confucius, Mozi, Buddha, Zarathustra, Elias and Jeremiah, Parmenides, Heraclites, and Plato.

Jaspers claimed that the Axial Age not only changed the cultures directly involved in it but amounted to a transformation of global culture, because the new understanding of the world eventually shaped all societies on the historical stage. For Jaspers, the Axial Age thus created a cultural communality that could serve as a global reference point for intercultural dialogue. Jaspers's idea fulfilled the widespread desire in the immediate postwar period to reconstruct a world in which such a dialogue was indeed possible.[28] The concept of an Axial Age was broadly discussed, and in the 1980s (in the

hands of sociologist Shmuel Eisenstadt), it became the core of an interdisciplinary research program.[29] In the context of this program, Eisenstadt and his
colleagues focused on comparable social processes by which, in multiple
societies, intellectual elites emerged who introduced "second-order thinking"
(to use an expression of Yehuda Elkana, who participated in the effort).[30]

Both the philosophical take on history by Jaspers and the sociological approach by Eisenstadt emphasized the parallelism of more or less simultaneous developments rather than insisting on mutual influences. While this may
seem reasonable in the absence of evidence for contemporary contact (e.g.,
between Greek and Chinese philosophers), the globalization of knowledge
could have nonetheless played a role in undergirding such parallels. Comparable developments in Greek and Chinese science (as we shall examine
them in chapter 12) may have indeed been triggered by similar social and
cultural constellations; but (as we shall see in chapter 11) they were also
conditioned by prior processes of globalization reaching farther back into
the past—processes such as the spread of food production, technology, and
literacy, which left sediment accounting for some important common conditions. The role of literacy and its institutionalization as a trigger of second-
order thinking played little role in Jaspers's original account, but was
stressed in commentaries on his thesis, such as those by the Egyptologist Jan
Assmann. Assmann pointed to comparable developments in Egypt, though
these occurred much earlier than the Axial Age.[31]

The alleged parallel developments manifested themselves in unique forms
in the various cultures of the Axial Age—as monotheistic religion in Judaism,
as philosophical inquiries into the nature of the world in Greek philosophy,
as religious transcendence in India, and as a universal ethical system in
Chinese Confucianism. This diversity raises the question of what is actually
common to these developments. Jaspers and his followers point to similar
experiences of transcendence, implying a radical break with a preexisting
continuum between the human and mythological worlds and the abandonment of a homology between those worlds. They postulated a pre-Axial past
in which humans supposedly lived in such a mythological continuum—an assumption that, on closer inspection, turns out be a historical fiction. From a
sociological perspective, Eisenstadt and his colleagues rather emphasized
the emergence of intellectuals and the transformation of elites through the
tensions raised by their awareness of this break. Contributors to this project,
such as Elkana, clearly saw the decisive role of cultural encounters in the
emergence of second-order thinking.[32]

In order to cover the entire range of manifestations of the alleged Axial Age,
Jaspers had to conceive the notion of transcendence in rather vague terms,
glossing over important differences and neglecting older developments, such
as those of Egyptian culture.[33] The normative claim that the Axial Age could
serve as a common frame for intercultural dialogue casts "transcendence" as
an intellectual awakening, leading (according to Jaspers) to the revelation of

what later came to be called "reason" and "personality," and to the possibility of renewing this experience in a renaissance. Jaspers writes about the exceptional individual responsible for this awakening: "In *speculative thought* he lifts himself up towards Being itself, which is apprehended without duality in the disappearance of subject and object, in the coincidence of opposites."[34] The European conception of a transcendental being, rising above the concreteness of experience, is thus projected into a global past.

The complex urban societies of the first millennium indeed saw the emergence of intellectual elites with special roles in their respective knowledge economies, as well as a flourishing of second-order thinking—thinking that, however, took radically different forms.[35] But the *topos* of the Axial Age concept introduces an exclusivity that obliterates the fundamental diversity that this thinking could take; it denies the possibility that such thinking could have arisen earlier or later in other cultures; and it essentially freezes history at one crucial turning point. As for its usefulness as a reference frame for intercultural dialogue, the vision of the wise men of the Axial Age turns out to be little more than an origin myth of world society.[36] It suggests that a future global convergence would be a matter of intellectual elites agreeing on some highly elusive second-order framework, rather than of global efforts that make productive use of the inexhaustible variety of human thinking on all levels.

Building on the knowledge accumulated by and spreading from early literate societies, new knowledge economies emerged in the first millennium BCE that were characterized by a further emancipation of knowledge production from the immediate purposes of the reproduction of these societies, as well as from their economic and political institutions. Religion provided an important institutional framework for this emancipation, one that could also act subversively by emancipation from the dominance of political and economic powers. In early literate societies, religious frameworks and their normative knowledge systems reflected their rising social complexity, a development that was accompanied by an increased social differentiation of political and religious elites. The potential range of such systems in space and time was considerably enhanced when the medium of writing was employed for their representation. In the third millennium BCE, religious beliefs and practices had been among the first subjects of written documentation outside the original use of writing for administrative purposes. In this way, belief systems, as well as the corresponding literary works, practices, or religious ideas (e.g., the idea of an original flood, documented in the Babylonian *Epic of Gilgamesh*) could travel well beyond the spatial and temporal confines of the societal setting in which they originally emerged.[37]

With the rise of Buddhism in India and of Judaism, Christianity, and Islam in the West, religion became decoupled from the state to a previously unparalleled degree, emerging as a source of authority separate from and potentially in conflict with that of the state, thereby developing a capacity for the global spread of world religions. Thus, religion challenged the authority of the state and ultimately transcended the limits of the state—or became a point of departure for new states. Religions embodied the authority of the state and the mechanisms necessary for knowledge production and dissemination; in effect, religions constituted virtual empires built on existing social orders. They offered a new social order greater than that of the state, but modeled on the state; hence, for instance, the concept of the Umma in Islam and the City of God in Christianity.

At some point in their development, many of these were or became state religions, such as Buddhism in northern India under King Aśoka, Christianity in the late Roman Empire under Theodosius I, and Islam under Mohammed and his followers in the Arabic world. Religions could become attractive belief systems to be adopted by states and empires seeking to regulate their social order. Under these conditions they incorporated (or helped to create, as in the case of Islam) many of the institutional and representational structures of an empire state, such as differentiated social hierarchies, comprehensive worldviews, and institutional mechanisms for preservation and transmission. The self-contained and self-organizing quality of religions, together with their reference to a transcendent authority, rendered them capable of challenging the authority of the political powers and of far outlasting their host states.

While authority was primarily asserted by the state (and grounded in physical force), religions had to place greater emphasis on justifying their authority. They developed sophisticated schemes of justification and produced extensive bodies of beliefs and knowledge through complex processes of dialectics. This new development set the stage for the accumulation and transmission of knowledge—which, while always extrinsically motivated, would be neither confined to local networks nor inseparable from immediate contexts of application. This knowledge was thus free to be repurposed or translated to new contexts. As a means for generating self-awareness and hence for reflectively constituting individual and social identity, religions imposed images of knowledge and control structures guiding the selection, appropriation, agglomeration, or exclusion of new knowledge by determining its value for the individual and collective self.

In accordance with shared epistemological frameworks, different types of knowledge were produced, transmitted, and associated with each other, often according to an explicit classification given by a respective framework. Religious traditions typically distinguish between religious and secular knowledge traditions, conceiving the latter as being ancillary to the former. Text-based religions furthered the accumulation of knowledge related to the exegesis of texts, in particular the fields of grammar, logic, and rhetoric.

Religions developed many of the features that were later characteristic of science: educational institutions such as monasteries and madrasas (schools of higher learning), recursive traditions of interpretation, commentary, confrontation, and integration with other belief systems, and a systematization of knowledge that included control structures for its legitimacy. As media for collective and individual self-reflection, religions also harbored, from the very beginning, a potential for criticizing religion itself, as may be illustrated by the discussions of anthropomorphism, that is, the interpretation of the divine in terms of human characteristics, common to various religious and philosophical traditions.[38] Take, for instance, the case of the sixth-century Greek philosopher Xenophanes, who turned against the polytheism and anthropomorphism of the Olympic gods. He ridiculed the Olympic gods for their human weaknesses such as theft, adultery, and mutual deception and suggested that gods are simply human projections and that humans always make gods in their own image:

> The Ethiopians <say that their gods> are snub-nosed and dark-skinned,
> And the Thracians that they have blue eyes and red hair.
> But if oxen, <horses> or lions had hands
> Or could draw with their hands and create works like men,
> Then horses would draw the shapes of gods like horses,
> and oxen like oxen,
> And they would make the same kinds of bodies
> As each one possessed its own bodily frame.

He instead argued for a unique god with no human qualities, anticipating Aristotle's later conception of an unmoved prime mover:

> One god, among both gods and humans the greatest,
> Neither in bodily frame similar to mortals nor in thought.
> As a whole he sees, as a whole he thinks, and as a whole he hears.
> But without any toil, by the organ of his mind [*noou phrêni*] he makes all
>   things tremble.
> He always stays in the same place, not moving at all,
> And it is not fitting that he travel now to one place, now to another.[39]

From Religious to Philosophical Knowledge Production

Greek philosophy emerged as a tradition parallel to religion, sharing its claim to resources for mastering the challenges of life, but now concentrated around elite intellectuals who developed their ideas at a critical distance from official or popular religious practices.[40] This intellectual elite was, for the most part, not directly involved in ruling society. The knowledge systems they created were decoupled from the immediate needs of societal production and organization. They were encoded in writing and span a wide range of subjects, from mathematics and astronomy through medical knowledge and physical geography to law and literature. Their hallmark was the degree of reflexivity that this knowledge assumed. The representation of knowledge by written language had become itself an object of reflection.

As a consequence, new means and forms of the reflective organization of this knowledge emerged, such as the integration of abstract concepts into elaborate systems of knowledge, the explicit *definition* of the meaning of concepts, and the organization of systems of knowledge into syllogistic and deductive structures. A reflective use of operations with symbols became the hallmark of *theories* as a characteristic new form of knowledge organization and representation. *Sophisticated argumentation and proofs* were developed that established relations within a knowledge system with the help of formal operations with linguistic forms. *Logic* emerged as a generalized theory of the operational possibilities opened up by the symbol systems of written language. Arithmetic became a theory based on reflections of operations conducted with the tools of practical calculation, and geometry likewise developed into a theory based on the reflection of operations conducted with ruler and compass.

10.4. A papyrus fragment found in a rubbish pile in Oxyrhynchus in an excavation from 1896–97 (P.Oxy. I 29). It shows one of the oldest and most complete diagrams from Euclid's *Elements of Geometry*. University of Pennsylvania Museum of Archeology and Anthropology.

Archimedes created a deductive theory of mechanics based on his reflections on operations conducted with weights on a balance.[41] The representation of the characteristics of these operations in writing became an essential prerequisite for this kind of theoretization.

Theories like arithmetic, geometry, and mechanics involved abstract categories rooted in the reflection of actions with concrete material means such as calculation tools, ruler, compass, or balance. More central to the organization of knowledge in philosophical systems such as that of Aristotle were categories founded on the reflection of symbolic actions, in particular on operations with language. Aristotle's philosophy is characterized by its encompassing accumulation, reflection, and systematization of knowledge encoded in language.[42] But key to its systematization is the reflection of symbolic and not of material practices, that is, the reflection on operations through language rather than through material arrangements, as when one performs an experiment. Therefore, the abstract categories of philosophy remain, in many cases, superficial with regard to the materials treated. In later elaborations of Aristotelian philosophy in translations, commentaries, and medieval scholastics, the meaning of Aristotle's reflectively constructed concepts often no longer referred to the primary meanings of the reflected concepts, creating a merely formal structuration of knowledge. The impressive longevity of Aristotelianism, on the other hand, was not only due to its abstract logical structure but to its rich knowledge content, based on reflections of mental models of intuitive and practical knowledge as discussed in earlier chapters. This longevity was supported by the emergence of a new literary genre, the commentary on canonic texts, which allowed newly acquired knowledge to be read into them.[43]

In Greece, traditions of natural philosophy and science initially emerged within a polycentric urban context with limited institutionalization before the Hellenistic period. The growth of this knowledge was largely sporadic, determined by the interests of a small number of individuals, despite all attempts at systematization, such as those by Aristotle and his Peripatetic successors. Both the carriers of this knowledge and the knowledge itself had a precarious status, surviving only in societal niches such as academies, courts, or the later monasteries. The fate of Greek and Hellenistic philosophical and scientific traditions was closely associated with the changing political support for their institutions, such as the Academy in Athens or the Museion in Alexandria, which were repeatedly endangered by ideological and military threats.[44]

The preservation of this knowledge can be attributed to its codification in writing and the collections of these writings by institutions such as the library of Alexandria, whose destruction illustrates the fragility of this transmission. I return to that library in chapter 13, when I discuss its role as a hub in an epistemic network of ancient scientific knowledge. In fact, much ancient knowledge was lost. What was transmitted to later eras and therefore became influential in modern times were often less the treasures of knowledge accumulated in antiquity than higher-order knowledge, that is, the reflective structures of knowledge organization characteristic of this state of a knowledge economy. Examples include the models of syllogistic or deductive organization of knowledge exemplified by the extant writings of Aristotle and Euclid's *Elements*, the idea (associated with Eudoxos) of basing a geometric model of the universe on the motion of spheres, concepts going back to Archimedes such as that of the center of gravity of bodies, or the notion of analyzing complicated mechanisms in terms of simple machines as found in the works of Heron of Alexandria and Pappus.[45]

The creation and transmission of higher-order knowledge with long-term consequences was neither the privilege of Greek philosophy nor an exclusive characteristic of science and philosophy. In chapter 11, I examine the role of "cultural refractions" in the emergence and transformation of the Greek tradition of philosophical and scientific thinking from other knowledge traditions. In chapter 12, I discuss striking parallels to the development of abstract systems of knowledge in the Chinese tradition.

Abstract concepts have shaped the traditions of religious, political, and normative thinking as well. In Judaism, for instance, one can recognize a reconceptualization of monotheism along with a tendency by the biblical prophets of the mid- to late eighth century BCE to divorce divine power from political and military might as a reaction to the aggressive ideology of the Neo-Assyrian Empire. Similarly, the emergence of Buddhism in India in the fifth century BCE occurred in the context of a reaction to Brahmanical religion and led to a highly reflective textual tradition.[46]

Rather than following the subsequent historical development of knowledge economies in Europe in detail (including the rise and fall of the Roman Empire and the long-lasting effects of ancient educational systems on the culture of the Middle Ages and their transformation under the influence of Christianity and Islam), I now jump almost two millennia in order to turn to the knowledge economy in which modern science was created. In the next chapter, I will try to compensate for this omission, at least to some degree, by pointing to the major role that globalization processes of knowledge have played in post-antiquity for bringing about this novel situation.

The Production of Experiential Knowledge

In Europe, the knowledge economy of which early modern science became an essential part was based on self-reinforcing mechanisms fostering the systematic production and circulation of experiential knowledge. From the Middle Ages onward, scientific and technological knowledge in some European societies started to become more central to economic production, social organization, and political regulations, as well as to the self-image of the leading classes, a process that eventually led to a societal transformation involving scientific knowledge to an ever-greater extent.[47] The growing importance of large-scale engineering projects in early modern Europe, which is my prime example, was related to the contemporary territorialization process and to the growing economic and political power of cities. After having recovered from the devastating effects of the fourteenth-century plagues, a vast number of local centers continuously competed for economic, political, and cultural hegemony. A more coherent territorial administration and the growth of town populations fostered infrastructural activities like the maintenance of roads and canals, and the construction of fortifications for protection against hostile occupation. Starting in the late Middle Ages, large-scale projects in civil and military architecture, in hydraulic engineering, and in mining thus became nuclei of advanced technical expertise.

The mechanization of certain work processes had been continuously growing from the High Middle Ages onward.[48] For territorial powers and other investors, economic factors may have been important reasons to provide the considerable overhead required to put large mechanical devices into operation. Mechanical devices could save labor costs; revenues could also be increased by constructing industrial devices like paper mills and then renting them out. On the other hand, the interest in mechanical technology among early modern sovereigns and the rising merchant and bourgeois classes was also fostered by cultural factors. The technical and intellectual mastery of nature according to its own principles provided a paradigm for claims to master society according to rationality. Competition for cultural prestige promoted the advancement of refined mechanical technology.

The need to organize the construction of large siege engines in the Middle Ages had provided a first stimulus for sovereigns to hire technical experts for the duration of war campaigns. Later on, technical experts were temporarily employed outside the guild system by towns or at court for civil engineering and architecture projects. Beginning in the fifteenth century, in addition to the temporary employment of engineering experts, more stable administrative infrastructures emerged that facilitated the successful realization of large-scale projects. In the industrial and political centers of Europe, positions were created in land and water infrastructure, fortification, and mechanical engineering. The incorporation of experts in these fields into the administrative infrastructure of a territory led to an increased recognition of technical problems among the political elite. Administrative frameworks for engineering tasks had first emerged in response to the challenges of the great medieval building projects discussed in the previous chapter (e.g., the large cathedrals), which were already characterized by attempts at prefabrication, standardization, and rationalization.

10.5. The lowering of the 327-ton Vatican obelisk (by Domenico Fontana) in 1586, in order to relocate it to its current location on Saint Peter's Square. This engineering feat involved nine hundred men, seventy-five horses, and countless pulleys and meters of rope. From Zabaglia and Fontana (1743, 143). Courtesy of Bibliotheca Hertziana.

Comparable structures also emerged in centers of naval architecture and are particularly evident in the case of the Venetian Arsenal.

Scientific-Technological Experts

As a response to the manifold requirements of early modern engineering projects, a new category of professionals emerged who were typically charged with the planning activities these projects implied. In that context, they accomplished technical as well as logistic and organizational tasks. These early modern artisans, engineers, engineer-scientists, and other scientific-technological experts had heterogeneous educational backgrounds.[49] In the absence of any formalized engineering instruction, most individuals with engineering expertise had been brought up within the framework of traditional craftsmanship. Their professional knowledge was predominantly transmitted orally and by immediate participation in the activities from which it was gleaned and to which it was applied.

The social status of early modern engineers and technical experts remained uncertain throughout their lives. Those employed at court remained dependent on the imponderables of patronage, with its complex system of power and prestige. Both the uncertain social status of these protagonists and the heterogeneity of their cultural heritage contributed to making this period a time of exaggerated personal ambitions and intense intellectual struggles in which opposition to the prevailing Aristotelian tradition became a hallmark of originality, despite its role as the common intellectual ground of all intellectual endeavor.

The emergence of this new institutional persona or category of professionalism was part of a self-reinforcing mechanism. Because of the limited labor force and other resource issues, these technical experts and engineer-scientists were also continuously confronted with technical as well as logistical challenges. In reaction to these challenges, they were forced to maximize the potential of traditional technical knowledge in order to create new technical means—for example, as we have seen in the previous chapter, the set of machines developed by Filippo Brunelleschi in order to build the Florentine cupola without more expensive and unaffordable scaffolding. The technical experts and engineer-scientists of early modern times were thus not only carriers of a traditional canon of knowledge (as was largely the case with the ancient administrators of large-scale projects) but were also involved in a cumulative process of innovation.

These practitioners and intellectuals contributed, at the same time, to a culture of knowledge that included a keen appreciation of inventiveness in all kinds of literary, artistic, and artisanal activities in early modern court culture. They were not necessarily (and, in any case, not completely) involved in technical practice but often specialized in reflecting on the new type of knowledge produced by this practice and, of course, in the attempt to make that reflection useful again for practical purposes. From the fifteenth century onward, innovative designs for mechanical devices began to be protected in a number of European territories by a juridical framework, marking the beginning of the modern patent system.

The Printing Revolution

Since antiquity, there had been a tradition of technological literature dealing with the knowledge of practitioners and engineers. Far into the sixteenth century, engineering treatises of all kinds circulated in manuscripts. The availability of paper and the development of printing technologies were important prerequisites for the early modern knowledge economy. Paper was made from rags by the second century BCE in China. Paper technology spread westward, reaching the Arab world by the eighth century CE and Europe in the tenth; the European manufacture of paper began in the twelfth century. I expand on the globalization of this technology in the next chapter. Paper was a necessary enabling technology for printing (and thus a key advance in the portability and reproducibility of knowledge). Print technology began in China in the second half of the first millenium.[50] Toward the end of the sixteenth century, the development of print culture in Europe intensified certain general tendencies of early modern technological literature: to write in the vernacular rather than in Latin, to address not only learned audiences and potential patrons but also literate artisans themselves, to take up ever more specialized subjects, and to enhance the social status of practitioners and engineers.

But printing also changed the early modern knowledge economy in more fundamental ways: it transformed the market into a key infrastructure for the circulation of knowledge.[51] Books quickly became a mass product that could be traded both regionally and over long distances. Publishing grew into an important economic activity with a potentially global reach.[52] It has been estimated that "in the year 1550 alone, for example, some 3 million books were produced in Western Europe, more than the total number of manuscripts produced during the fourteenth century as a whole."[53]

Because the market for academic publications was strongly shaped by the existing educational institutions, in particular the universities, alliances between these institutions and publishers emerged that in turn affected the knowledge economy, for instance, by the canonization of certain texts as standard textbooks for both teaching and learning.[54] In chapter 13, I discuss a particularly striking example of this kind of canonization and its effects: Sacrobosco's *Sphere*.

A further effect of the spread of printing was the creation of a legal and institutional framework, first protecting the privileges of publishers, and then also those of authors, a newly emerging institutional persona. The explosive growth of affordable books enabled ordinary individuals—artisans, engineers, or scientists—to build up significant libraries that had previously remained the privilege of institutions or of wealthy patrons.[55] In the fifteenth century, Leonardo da Vinci, for instance, a man with no higher education, collected almost two hundred books from all fields of knowledge.[56] The ready availability of a wide range of classical and modern works constituted an intellectual empowerment of individual perspectives, opening up novel, potentially heterodox ways of integrating the available knowledge into individual syntheses.

The codification of empirically based knowledge, first in the form of drawings, treatises, machine books, and textbooks, and later in the form of academy acts and

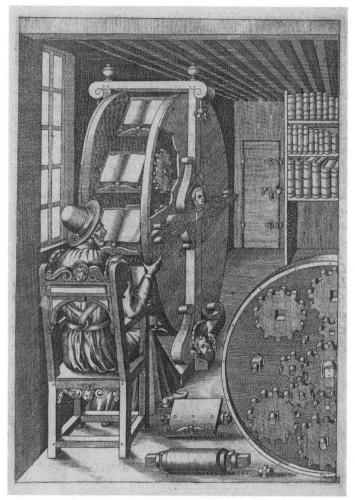

10.6. A book wheel depicted in Agostino Ramelli's *Le diverse et artificiose machine* . . . . Ramelli describes this device as a "beautiful and ingenious machine, very useful and convenient for anyone who
takes pleasure in study, especially those who are indisposed and tormented by gout." From Ramelli
(1588, 317). Courtesy of the Library of the Max Planck Institute for the History of Science.

scientific journals (plus privileges and patents) became a particularly important vehicle for knowledge integration, accumulation, transmission, and reflection.[57]
Against the background of the emerging print culture, such codification also created
a market in which not only the material applications and products of knowledge but
also knowledge itself and its representations acquired a cultural and economic value
of its own. As may be illustrated by the alliances between publishers and academic
institutions mentioned above, this codification of knowledge took place in close
association with the increasing institutionalization of its production and circulation. The institutionalization of the publishing culture was also a reaction to the

10.7. Archimedes and Hero of Alexandria as role models for the early modern engineer. Salomon de Caus, *Les raisons des forces mouvantes*, 1615. Courtesy of the Library of the Max Planck Institute for the History of Science.

intensification of the production of empirical knowledge and its increased role for numerous societal processes, be they political, cultural, or economic. In turn, the codification and commercialization of knowledge contributed to this intensification, thus forming part of the self-reinforcing mechanisms characteristic of the early modern knowledge economy.[58]

The Integration of Knowledge Resources

The problems facing early modern engineers and other technical experts necessitated the exploitation of existing knowledge resources and often brought different strains of this knowledge into contact with one another for the first time. This integration of knowledge resources of disparate origins was fostered by the representational means developed and used by early modern engineers. They also included the transformation and development of new symbolic means—for example, graphic representation, calculational operations, and representations of functional dependencies and processes of change as discussed in chapter 5, eventually giving rise to new branches of mathematics such as analytic geometry and calculus.

As discussed earlier in the context of preclassical mechanics, early modern engineers could draw on a number of shared knowledge resources, which formed a heterogeneous collection of bodies of knowledge in partial tension with one another. These resources comprised practical geometry, experience acquired in the realization of large-scale projects, and theoretical frameworks such as Aristotelian natural philosophy, Euclidean geometry, Archimedes's theory of the center of gravity, and Hero of Alexandria's theory of simple machines. Knowledge of geometry was of crucial importance to the various branches of engineering. Hero's theory became an important bridge between the theoretical and practical knowledge of mechanics, allowing complex machinery to be analyzed in terms of simple building blocks.[59] Experience in logistic and organizational skills acquired during the realization of large-scale projects proved indispensable for the engineer's additional role as a planning authority. Theoretical frameworks were made available to engineers by way of a growing corpus of ancient and medieval scientific literature. Owing to the achievements of the humanists, early modern scientist-engineers had a growing number of ancient scientific treatises at their disposal.

The impact of the resulting integration of heterogeneous strands of knowledge on the advancement of empirically based mechanical knowledge was so powerful that it eventually led not only to the revision of traditional conceptual systems (as described in chapter 6) but also to the creation of new social structures for the production and dissemination of scientific knowledge. Hence, the role of the "social context" of scientific development would be underestimated if it were considered solely as an external framing condition for a specific subculture of society instead of as an essential part of the societal dynamics itself, namely, as a self-reflection of an increasingly knowledge-based society. Some of the roots of this development go back to the foundation of the medieval universities.

The Early Modern Knowledge Society

By the late eleventh century, universities had been founded in Europe. By the thirteenth century, they had been consolidated with their own privileges, curricula, and legal regulations, assuring them a certain institutional autonomy. The faculty of arts, the common foundation of studies at a medieval university, comprised the study of the seven liberal arts, the *trivium* of grammar, rhetoric, and dialectics, and the *quadrivium* of arithmetic, geometry, astronomy, and music. It shaped an understanding of science that was also applied to the higher studies of juridical, medical, and theological matters. Given the relative autonomy of the university as an institution, as well as the heterogeneous yet systematized contents of the learning it promoted, it became a focal point of political, religious, and intellectual tensions and conflicts in late medieval and early modern Europe.[60]

Basic competency in reading, writing, and calculating were also indispensable to a rising merchant economy, as well as in managing the administrative procedures of late medieval and early modern engineering projects. In an economically thriving city like fourteenth-century Florence, about a tenth of the adolescent population learned reading and writing.[61] A select group of boys was also taught calculational techniques in preparation for becoming merchants; an even smaller group learned grammar, logic, Latin, and philosophy. The growing institutionalization of elementary education was accompanied by the spread of treatises and textbooks, codifying practical knowledge and enhancing its dissemination across traditional social divides.

In Italy, new institutions were created for training practitioners, such as the Accademia del Disegno in Florence, in which theoretical knowledge was transmitted alongside practical knowledge.[62] Some of the earliest academies (such as the Accademia Telesiana) had dealt mainly with philosophy, literature, and the sciences in the ancient model of the Platonic academy.[63] Initially, academies were weakly institutionalized and represented informal circles, often depending on patronage. The further development of academies was shaped by existing models of institutionalization such as the guilds, religious orders, and the system of privileges, as well as by the demand for production in natural knowledge and technological achievement. Academies also became attractive institutions for the early modern knowledge economy because knowledge production in other bodies (such as the universities and religious orders) was more confined because of the restrictions imposed by their curricula or by political and religious power structures. Academies contributed to the development of the public sphere and civil society. They fostered the emergence of scientific communities and a society open to their insights. Their regularly published acts launched the first modern scientific journals, such as *Le Journal des Sçavants* and *Philosophical Transactions of the Royal Society*.

Toward the end of the sixteenth century, the problems investigated by craftsmen and engineers had created numerous points of contact between practical and theoretical knowledge, in particular in fields like geometry, mechanics, astronomy, and optics. Such contact was fostered by the fact that engineering projects did not take place as isolated technical ventures. They rather formed part of a social fabric comprising an ever more complex division of labor ranging from artisans to administrators to engineer-scientists primarily concerned with the intellectual side of such endeavors. It was in fact the lack of rigid class distinctions within this social fabric that allowed for the integration of disparate components of knowledge—characteristic of technology and science in the early modern period. One possible framework for an exchange between practical and theoretical knowledge was, as mentioned above, provided by the integration of engineering responsibilities into communal or territorial administrations.[64]

Among the points of contact between practical and theoretical knowledge were challenging objects, such as the pendulum or the trajectory of projectile motion, but also newly developed scientific instruments, such as the telescope, the microscope, or the thermometer, all products of modern technology. It is indeed a defining feature of the early modern knowledge economy that the means of production were now being purposefully used as means for the production of knowledge. The motivation for the new knowledge was the specific possibility of technically realizing natural processes offered by new technologies of production. The introduction of experiments was hence not the presupposition of an empirically based science but the result of a reflective abstraction from this practice, which complemented existing procedures.[65] For a long time, practitioners and engineers had, for instance, created devices such as models of machines with the aim of better understanding how they worked. With the increasing emancipation of a knowledge economy directed at the creation of empirically founded knowledge, such practices eventually turned into a method for testing claims and assertions generated by systems of knowledge that were now being explored in their own right.[66]

Given their provenance as a theoretization of the experiences of engineers, technicians, metallurgists, chemists, and other practitioners, these systems of knowledge could in turn anchor the creation of new forms of empirically based theories—which then became the hallmark of the knowledge systems of modern science. By means of experiments and methodologically controlled observations, their assertions acquired an unprecedented intersubjective reliability. These new systems of knowledge comprised reflectively constructed symbol systems (like calculus) and cognitive operations structured by reflected experience (like chemical formulas). Eventually, the self-reinforcing mechanisms connecting the production of scientific knowledge with socioeconomic growth were stabilized and substantially expanded in the context of the Industrial Revolution.

The new scientific knowledge was, however, not just the result of a reflection of practical knowledge; it had deeper historical roots. It was, in particular, the result of an integration of practical and theoretical knowledge with a long prehistory. Higher-order knowledge about reflective forms of knowledge organization inherited from antiquity was indeed crucial for the integration of experiential knowledge into new knowledge systems. The ancient heritage of knowledge comprised knowledge systems with high systematicity—such as Aristotelian philosophy, Archimedean mechanics, Ptolemaic astronomy, and Euclidian geometry—which served as points of departure and models for the transformative endeavors of knowledge integration in the early modern period.[67] Since the foundation of the medieval universities, the transmission, elaboration, and integration of ancient knowledge systems with new knowledge had received unprecedented institutional support. The result was the knowledge system of scholasticism with its universal claims of validity—which became, for the same reason, increasingly sensitive to epistemic challenges.

The Western Christian scholastic system of knowledge was thus exposed to a constant process of transformation, but because of the primacy of religion in the control structures of the knowledge economy, it was, at the same time, subject to externally imposed limitations. This constellation helps explain why the sixteenth century reform of astronomy by Copernicus—placing the Sun rather than Earth at the center of the universe—could have had such far-reaching ideological consequences: it occurred within the context of a dominant knowledge economy imposing the universal and exclusive validity of certain forms of knowledge. Eventually, the growing mass and complexity of the rapidly accumulating knowledge challenged this knowledge economy and the worldview it supported. This knowledge "explosion" counteracted all attempts to confine the expansion of knowledge and eventually helped to foster the creation of the new institutions of knowledge production mentioned above.

The religious context acted, on the other hand, as a background for the generation of shared images of knowledge about the feasibility, viability, and legitimacy of the new modes of knowledge production. In a sense, the claim of religion to counterbalance the authority of the state in terms of its own, internal logic encouraged science to challenge the authority of religion. Early modern philosophers, scientists, and artists could, for instance, appeal to the possibility of penetrating into the secrets of nature and gaining certain knowledge about it, while claiming that both nature and humans operated on similar terms and used the same language, owing their existence to the same creator. The Christological dogma that God had become human could thus be effectively turned around by claiming godlike faculties among humans to attain certain knowledge. On the First Day of his "Dialogue concerning the Two Chief World Systems," Galileo Galilei is quite explicit on this point:

> SALV[IATI]: You put the point very sharply, and to answer the objection it is best to have recourse to a philosophical distinction and to say that the human understanding can be taken in two modes, the *intensive* or the *extensive*. *Extensively*, that is, with regard to the multitude of intelligibles, which are infinite, the human understanding is as nothing even if it understands a thousand propositions; for a thousand in relation to infinity is zero. But taking man's understanding *intensively*, in so far as this term denotes understanding some proposition perfectly, I say that the human intellect does understand some of them perfectly, and thus in these it has as much absolute certainty as Nature itself has. Of such are the mathematical sciences alone; that is, geometry and arithmetic, in which the Divine intellect indeed knows infinitely more propositions, since it knows all. But with regard to those few which the human intellect does understand, I believe that its knowledge equals the Divine in objective certainty, for here it succeeds in understanding necessity, beyond which there can be no greater sureness.[68]

The new vision of knowledge emphasized an explanation of nature based on its immanent principles. In response to the dominant religious worldview, this knowledge assumed the character of an all-embracing interpretation of the world. Among the earliest examples of such constructions of self-contained world systems are Bruno's infinite universe of innumerable worlds, Descartes's system of turbulent world matter in eternal motion, and Galileo's simplified Copernican cosmos.[69]

Initially, the experience of successfully intervening in nature had been effectively limited to a few particular fields of knowledge, such as mechanics. In this sense, the newly constructed systems of knowledge displayed a mismatch between the methodically guided establishment of specific domains of knowledge and the attempts of natural philosophy involving atomism, mechanism, or other programs to offer a common conceptual framework and to integrate them into an overarching scientific worldview. Systems of natural philosophy such as Kepler's *harmonia mundi*, Newton's active principles, Leibniz's doctrine of monades, Kant's metaphysics of forces, or Lamarck's theory of biological transformations soon turned out to be incapable of integrating the rapidly growing body of empirical knowledge.[70] Nevertheless, the "mechanization" of the worldview, in the sense of taking mechanics as a basis or at least as a model of scientific explanations, continued through the seventeenth and eighteenth centuries. It was not until the end of the nineteenth century that the mechanical worldview completely lost its position as the unquestioned basis for scientific endeavors.[71]

Sovereignty, Representation, and the Emergence of the Modern State

The emergence of the state in early modern Europe changed the conditions for the production and dissemination of knowledge. The state demanded that extant knowledge economies refocus to meet territorial needs and efficiently exploit the available resources on a new scale, as emphasized at the time by Giovanni Botero.[72] This stimulated the creation (or transformation) of institutional structures such as academies and universities. In this context, the knowledge economy assumed a new function: it now became an important mechanism supporting the new political order. This development also affected systems of knowledge subject to the new knowledge economy by privileging useful knowledge and organizing it in forms that conformed both to the challenges of new tasks and to an education anchored in traditional knowledge economies.

This may be illustrated by the example of preclassical mechanics (as discussed in chapter 6), a conglomerate emerging from encounters between new technological challenges and ancient systems of knowledge such as Aristotelian natural philosophy and Archimedean mechanics.[73] In the beginning, the internal organization of the new knowledge often directly reflected the political order (e.g., when themes were arranged according to the interests and power structures of the ruling court or courtly literary traditions).[74] Eventually, with the growing autonomy of emerging scientific communities, such forms of representation were overcome in favor of field-specific canons and more or less standardized infrastructures that came to be part of a toolbox for the knowledge economy of a sovereign state.

The sovereign state not only demanded and fostered a new knowledge economy; it was itself the result of knowledge development.[75] In sixteenth-century Europe, the concepts of sovereignty and the state acquired a new meaning, resulting from changing power relations and a theoretical reflection of these political changes. They yielded new forms of governance (a "matrix" in our terminology) in connection with the confessional wars that required an authority above and beyond those of the traditional order, which was itself considered to have been bestowed by God. Since each faction claimed to fight for God's will, a new order that could bring peace required the development of a political sphere in its own right, separate from and above the traditional authorities, worldly and ecclesiastical.

This new order was conceptualized by the jurist and trailblazer of French absolutism Jean Bodin, who picked up preexisting terms of sovereignty and the state and gave them a new meaning while weaving a theoretical framework around them.[76] This framework was then widely used in turn to encourage and justify the new political order. The basic idea of sovereignty was the transfer of power to a highest authority that served the peace of society and was hence endowed with exclusive privileges, in particular that of jurisdiction.

Thereafter, sovereignty became a key concept in describing the new form of political rule that concentrated power in the hands of a single ruler while limiting it to a specific territorial reach. Represented by the sovereign ruler (l'état c'est moi), the political sphere itself became intelligible in a new way—not as the organic composition of a society by its different members, each having its own particular function—but one that instead was legitimized by reference to the godlike powers of the ruler.[77] The autonomy and formability of this new form of dominion were the hallmarks of sovereignty as a new political concept. While sovereignty was initially legitimized by its alleged compliance with divine order, its actual function as a representation of the political sphere of society eventually became the center of political discourse.

According to Thomas Hobbes, the absolute power of the sovereign ruler was conferred to him by the people for whose protection he was responsible. In his Leviathan, or The Matter, Forme, and Power of a Common-Wealth, Ecclesiasticall and Civill from 1651, written in the shadow of the English Civil War, Hobbes described the absolute power of the sovereign ruler as resulting from a social contract.[78] In his Two Treatises of Government of 1689, John Locke justified the state by referring to its task to secure natural rights (such as individual freedom), and admitted that a government that does not fulfill this function may be overthrown.[79] The understanding of the political sphere as the place where the relations between individuals and society are negotiated suggested a convergence of the sovereignty of the state and the autonomy of the individual, as is found in the work Jean-Jacques Rousseau. Rousseau indeed concluded that, if one wants to preserve individual autonomy in the creation of a state, "sovereignty cannot be represented."[80]

This, however, raises the question of the relation between sovereignty and representation, and the source of power legitimacy. More generally, negotiating the relation between individuals and society is inconceivable without mediation involving different forms of representation. As discussed in chapter 8, such representations may not only serve as an external representation of knowledge, but also as an external representation of an institution. Both cannot be reduced exclusively to their functional roles as cognitive or social technologies; they also possess symbolic or symptomatic aspects indicating unresolved tensions—for instance, when the representation of an alleged "collective will" is not really susceptible to individual interventions. The baroque forms of representation of the absolutist state, for example, not only served to underpin the alleged divine powers of the sovereign but also gave this pretension a visibility that placed a question mark behind it, revealing its fragility.[81]

10.8. The state as a body composed of bodies. Frontispiece by Abraham Bosse of Thomas Hobbes's *Leviathan, or The Matter, Forme, and Power of a Common-Wealth Ecclesiasticall and Civill* (1651). See Bredekamp (2006). Wikimedia Commons.

Questioning the legitimacy of the ways in which sovereignty is implemented and legitimized eventually helped to lay the groundwork for political actions such as the American and French Revolutions, which insisted that a representation of sovereignty should not just be postulated but should be directly subjected to a popular will. Remarkably, in both cases a written constitution, offering a framework for the exertion of legislative power, played a decisive role as an external representation of this popular will.[82]

Knowledge Production in the Age of Industrialization

Since the early modern period, the range of science has expanded dramatically, not so much because an alleged scientific method was applied to new domains of experience but because ever-new objects came into contact with the developing network of scientific knowledge and because of the cognitive dynamics of this network. Thus, newly discovered specimens from exploratory voyages, new technological devices, or social and behavioral phenomena that acquired practical relevance, such as population statistics, became challenging objects for the existing scientific framework that eventually acquired global relevance.[83] This extension of the domain of science eventually included new disciplines of the social and behavioral sciences as well, such as psychology, educational research, or sociology, which emerged from the framework of philosophy and adopted (following the model of the natural sciences) their own methods of empirical research. In the course of this process, all dimensions of science underwent massive expansion alongside an increasing specialization, fragmentation, and commodification of knowledge.

In the core disciplines of the natural sciences as they emerged between the early modern period and the late nineteenth century, a handful of abstract concepts yielded a vast array of scientific knowledge. The concepts of space, time, force, motion, matter, and a few others played this role for classical physics; the concept of chemical compounds assumed a similarly foundational role for chemistry; and the concepts of species, gene, selection, variation, and adaptation structured classical evolutionary biology. As I have argued in chapter 7, such core groups of concepts usually achieved their privileged position in the organization of knowledge only after a long process of knowledge integration and reorganization. In the eighteenth and nineteenth centuries, the assimilation of new insights to these disciplinary knowledge systems was no longer primarily the achievement of individual researchers but rather the result of intradisciplinary communication processes regulated by a knowledge economy with its own epistemic institutions, articulated control structures, and images of knowledge.

With the establishment of specialized disciplines, overarching philosophical integrations of knowledge became largely obsolete.[84] They were replaced by a pluralism of scientific endeavors embodied in the institutionalization of disciplines, including the possibility of interdisciplinary ventures, the emergence of new disciplines, and other forms of differentiating between disciplinary sciences. This pluralism is part of the "classical image of disciplinary science" comprising default expectations about the complementarity and unity of the knowledge thus produced.[85] Disciplinary structures supposedly guarantee, by an implicit automatism, a competent and comprehensive distribution of labor in science. This complementarity, however, has rarely been realized in actual practice. On the other hand, it did remain an ideal for the modern research university, going back to Wilhelm von Humboldt's initiative for the foundation of the Berlin University in 1810.[86] It was supposed to combine general education according to humanistic principles with research within an ever more professionalized disciplinary framework.[87]

The development of disciplinary science took place against the background of a growing mutual dependency between knowledge and material economies, which accompanied the rise of sovereign national states (which employed scientific knowledge to stabilize their economic and political power) and the dawn of industrialization. Joel Mokyr and Margaret Jacob have argued that the so-called Industrial Revolution relied crucially on the new knowledge economy that had emerged with the Scientific Revolution of the early modern period. Mokyr writes: "Rather than focus on political or economic change that prepared the ground for the events of the Industrial Revolution, I submit that the Industrial Revolution's timing was determined by intellectual developments, and that the true key to the timing of the Industrial Revolution has to be sought in the scientific revolution of the seventeenth century and the Enlightenment movement of the eighteenth century."[88]

The relevant developments comprised several dimensions. One, as Jacob writes, was the new knowledge itself: "Knowledge of the physical universe gave entrepreneurs a singular advantage when economic decisions were being made. Their participation in the scientific culture of the day meant they could approach coal mining, or cotton and linen production, or spindle making and engine construction, armed with knowledge once assumed to be irrelevant to *homo economicus* as classically formulated. They understood in systematic ways phenomena such as air pressure, the force of a lever, the problem of friction—all but a small part of the relatively new science of applied mechanics."[89] Other dimensions of the new knowledge economy relevant to the Industrial Revolution were new institutions of learning and new media and mechanisms for access to knowledge and knowledge transmission: "The driving force behind progress was not just that more was known, but also that institutions and culture collaborated to create better and cheaper access to the knowledge base. Technology in the nineteenth century co-evolved with the new institutions of industrial capitalism."[90] According to Mokyr, "the knowledge revolution in the eighteenth century was not just the emergence of new knowledge; it was also better access to knowledge that made the difference."[91] He has characterized the new knowledge economy as an "Industrial Enlightenment," which, according to him, formed the missing link between the Scientific and the Industrial Revolutions.[92]

Even more important is his insistence on a feedback loop between different forms of knowledge, which, in turn, acted as a driver of the feedback between knowledge and material economies: "The historical question is not whether engineers and artisans 'inspired' the scientific revolution or, conversely, whether the Industrial Revolution was 'caused' by science. It is whether practical men could have access to propositional knowledge that could serve as the epistemic base for new techniques. It is the strong complementarity, the continuous feedback between the two types of knowledge, that set the new course."[93]

In the next chapters, I come back to the global, environmental, and material contexts in which these developments took place, which included the unlikely but vital circumstance of the geological formation of coal, and the superabundance of productive energy it afforded. Over the course of the nineteenth century, science began to have a major, global impact on human life. New fertilizers, new means of transforming energy (e.g., the steam engine), new means of communication (e.g., the telegraph), new measures against widespread diseases (e.g., antibiotics and vaccination), and new materials would have been inconceivable without the close association between science, technology, and social and economic development. In turn, new technologies represented challenges for the further development of science, such as the need to produce cheap steel to satisfy the growing demand for the construction of buildings, ships, railroad tracks, machines, and weapons.

Similarly, the sudden importance of fertilizer ingredients such as phosphates or nitrogen defined new fields of research, technology development, and economic productivity. Natural resources such as Peruvian guano were eventually substituted by chemical products.[94] I come back to the latter example in chapter 15, when I discuss the pathways that have led us into the Anthropocene.

The discovery of the role of microbes in the infection of wounds stimulated the development and production of disinfectants and antiseptics.[95] Chemistry also generated new materials, such as industrial rubber (made possible by the vulcanization process), the first synthetic plastic, and bakelite. The development of artificial dyestuffs provided, at the same time, a major boost in the development of the chemical industry and for that of the science of organic chemistry. Indeed, all of the abovementioned achievements constituted driving forces for their further scientific exploration, which opened up new technological and economic possibilities in turn.

In addition to the consolidation and specification of the academic disciplines, the Industrial Revolution also saw a further differentiation of modes to produce scientific knowledge, in particular between research more or less closely associated with technological and industrial applications. In connection with the rising societal and economic demand for science, new institutional formats were created during the nineteenth and early twentieth centuries.[96] At the same time, types of "useful" or "practical sciences" had been institutionalized, such as the mining sciences and the agricultural sciences, later designated as engineering sciences and technological sciences, thus realizing in a new way the coupling between the practical and theoretical knowledge achieved by the scientist-engineers of the early modern period. With the help of scientific-technological experts, the technological sciences played, for instance, an important role in the early industrialization of Prussia.[97]

The Industrial Revolution had a major impact on the technologies of knowledge representation and dissemination.[98] New technologies extended printing along several vectors. Hot metal typesetting hastened automation by replacing the process of manual composition (in which type was picked one by one from a typeface) with the keyboarding of text. The typewriter, first commercially manufactured in the United States in the 1870s, eliminated the centralized ownership of the means of mechanical production of texts and allowed mechanical technology to be used for the creation of even ephemeral documents. In 1848, the first rotary press was built in the United States. The invention of the linotype machine between 1886 and 1890 made it possible to cast and set whole lines at a time with the help of a keyboard. The mimeograph, largely replaced in the second half of the twentieth century by photocopy technologies, lowered the barriers of cost, skill, and time associated with the reproduction of printed documents and thus allowed for the flourishing of popular self-published literature. Teletype machines, which originated around 1907, allowed for the remote transmission and printing of text. Jacquard's punch card controlled loom (1804) and Hollerith's tabulating machines, developed to deal with the massive data that needed to be processed for the 1890 US census, first exemplified modern techniques of information processing.

The new strategic function of scientific knowledge for economic and political processes also imposed new demands on education. Thus, the modern institutions of general education and professional training emerged with their own integrating knowledge systems. School attendance became obligatory, and school knowledge was canonized at all levels from elementary school to advanced education, with an attendant appreciation for technical and scientific learning. The traditional universities underwent reforms. New engineering colleges, technical universities, and technical schools were founded, while hospitals combined clinical medicine with the development of medical techniques. In addition, extra-university research institutions were supported both by state and private funding, while the role of industrial laboratories expanded beginning in the middle of the nineteenth century. One example of a state-funded institution is the German Physikalisch-Technische Reichsanstalt, which established science-based technological standards. The institutes of the Kaiser-Wilhelm Society, many of which transformed practical challenges into research problems, also relied on private funding; they also established a new model of institutionally supported basic research.[99]

Around the same time, institutionalized science policy emerged as a new form of reflection on the rapid development and increasing significance of science organization. General societal aims and specific economic, political, or military interests directed research strategies, educational policies, and the management of large-scale research projects with technological aims. In particular cases only, an integration of knowledge from different disciplines took place in the implementation of scientific knowledge, as in the form of expert judgments. Expert knowledge became even more aloof because the process of specialization went along with a divergence between general knowledge expressed in natural language and scientific knowledge encoded in technical language and complex symbol systems specific to a particular area of knowledge. Its dissemination therefore turned into a problem that could only be solved pragmatically under specific circumstances.[100]

The Second Industrial Revolution

Since the middle of the nineteenth century, science-based industries, such as the chemical and the electrical industries, had come into being, turning both industry and the market, alongside the military, into a major driving force of innovation in science. As a consequence, the economy of resources and the economy of scientific knowledge became ever more closely intertwined, thus preparing the ground (after the middle of the nineteenth century) for what has been called the Second Industrial Revolution, which is associated first with this rise of science-based industries and later with the global spread of electronic appliances.[101]

The Second Industrial Revolution fundamentally changed the relationship between technological change and science and significantly enhanced the dependence of developed societies on the production of scientific knowledge. It led to a veritable "institutionalization of innovation."[102] When it commenced, the knowledge economy was still porous enough to allow for substantial contributions from tinkerers and individual inventors, yet institutionalized enough to take up inventions as scientific and technological challenges to be systematically improved. Eventually, the role of independent inventors diminished, being ever more dominated by an integration of scientific innovation in the industrial process.[103] Science, in turn, became part of an accelerating process in which science-based technical and commodified applications—such as photography, telecommunication, and later computer storage—were systematically employed as tools for further exploration. The conditions for scaling up and globalizing the development of science were accordingly a consequence of science and its technological implementation in the context of industrial capitalism, including the emergence of modern transportation and communication technologies.

The use of steam power for transportation, which had driven the development of the railroad system, received a further boost after the invention of the steam turbine in 1884, leading to a steep decline in marine transport costs. An entirely new development began with the invention of internal combustion engines, stimulated by an improved scientific understanding of thermodynamic machines, in particular of the so-called Carnot cycle. The idea of using heated gas for generating motive power was not new, but it became viable with the availability of an appropriate fuel—gasoline, obtained by crude oil refining as developed in the 1860s. A four-stroke gas engine was invented in 1876 and, less than ten years later, transformed into a gasoline-burning engine, thus laying the groundwork for the modern automobile.

The spread of the car as a consumer device also affected the broader economy of shared technical knowledge, since the knowledge needed to build, improve, or repair cars also spread widely, leading to cumulative, evolutionary improvements and developments of basic principles. The history of transportation and traffic in the Second Industrial Revolution, from the steam engine to the gasoline-powered internal combustion engine, illustrates the crucial role of technological systems in such transformations.[104] Just as the steam engine relied on coal mining and the processing of iron, the success of car traffic based on the internal combustion engine depended on the emerging oil industry, which it in turn reinforced.

Similar features can be identified in the further technological exploitation of electricity, both for power transmission and for communication.[105] Some of the principles behind the technical application of electricity, such as its use for lighting, electric motors, dynamos, and telegraphy, go back to the early nineteenth century and were further developed in parallel with James Clerk Maxwell's formulation of a full-fledged theory of electromagnetism in 1865.[106] The practicability and

10.9. Gottlieb Daimler (*rear right*) next to Wilhelm Maybach at the handlebar (*front right*), and Adolf Groß (*front left*), director of the machine factory in Esslingen, in a Daimler belt-driven car, which still resembled a horse carriage, 1897. Courtesy of the Daimler Public Archive.

implementation of the basic ideas depended on numerous technical improvements and inventions but also on the realization of a systemic infrastructure to which individual appliances and devices could be adapted. The use of electricity as a power source required technology for generating electrical energy from other sources; devices to transform electric energy into kinetic energy, light, or heat; and an infrastructure to transport electricity over long distances. The creation of such an infrastructure required decisions about fundamental alternatives, which induced both technological momentum in the sense of a growing power of technology over society but also the social determination of the path of the further technological development. The choice between alternating and direct current is a good example.[107]

The use of electricity for communication quickly took on an industrial dimension. In 1851, the Western Union Telegraph Company was founded. It soon realized the first transcontinental telegraph line across North America. In the same year, the first submarine cable was laid between Dover and Calais. Alexander Graham Bell invented the telephone in 1876, as prepared by investigations on the reproduction of sound by scientists such as Hermann von Helmholtz. The development of wireless telegraphy followed Hertz's 1888 experiments confirming Maxwell's predictions about electromagnetic waves.[108] The news reached the twenty-year-old Guglielmo Marconi through an electrical journal, stimulating him to build equipment for radio communication and ultimately to commercially exploit his inventions by founding the British Marconi Company in 1897.[109]

10.10. Italian electrical engineer Guglielmo Marconi (1874–1937) at work in the wireless room of his yacht *Electra* in 1920. Photo by Topical Press Agency/Getty Images.

Inventors such as Marconi, Alexander Graham Bell, and Thomas Alva Edison exploited the opportunities presented by scientific developments and their mass availability through scientific and popular publications, which in turn used some of the advancements in knowledge reproduction technology described above.[110] The inventor of the Second Industrial Revolution was a new institutional persona, knowledgeable about science but also aware of the long journey between scientific ideas and technological and economic practice. Inventors sought protection of the rights to the commercial benefits of their inventions by filing patents but often also became entrepreneurs themselves. Another option was to create their own laboratories and to act as contractors to large enterprises, as did Edison with his laboratory at Menlo Park, New Jersey, contracted by Western Union, which became a kind of predecessor of modern consulting institutions.

But the pace of innovation, the growing complexity of technological systems, and the long path from initial ideas to working products eventually brought the knowledge economy of the independent inventor to its limits through lack of capital, lack of qualified personnel, and lack of the capacity to cope with the systemic character of technology. Large companies, on the other hand, strove for greater control over the innovation process, facing, for instance, the danger that crucial components of a technological system would be patented by a competitor. As a result, larger industrial laboratories emerged in which scientists more and more replaced individual inventors, in which innovation became an integrated component of the industrial process, and for which the realization that devices typically belong to larger technological systems was fundamental.

In the course of the Second Industrial Revolution, science raised and spread growing expectations concerning general societal progress. Systems of higher education gradually opened up to include women. In many parts of the world, women fought to become part of academia, and in some countries, were finally able to enter colleges and universities, as well as to forge careers in technical professions, traditionally reserved for males. The struggle to attain equal opportunities and hence to reach a gender-balanced knowledge economy continues to this day.

In the second half of the nineteenth century, we see the emergence of new institutional structures in reaction to the challenge of accelerating innovations and their profound impact on societies, including the very environment for innovations. These profound transformations necessitated a new level of regulation concerning the development of technological systems, from the traditional water supply and sewage systems to the new systems of railroads, telegraphy, electrical power, and telephony.[111] Governments and regulation authorities established the rules of the road for these systems, from electricity voltages to the standardization of typewriter keyboards.

These regulations affected both the material economy and the economy of knowledge by fostering new institutions for knowledge production such as the Physikalisch-Technische Reichsanstalt mentioned above, which dealt, among other things, with the standardization of light sources.[112] As seen in chapter 7, the related measurements of black body radiation eventually triggered the creation of quantum physics. The exploration of this challenging problem demonstrated, as has been discussed in detail, the limits of classical physics.

The exploration of technological systems at the turn of the twentieth century pushed the limits of societal practices and institutional regulations in several other regards. The system of manufacturing, for instance, became itself an object of scientific analysis and technological improvement under the label of "scientific management."[113] Manufacturing became more efficient through the mass production of individual parts from which complex products could be assembled, thus exploiting an economy of scale. Another innovation was the introduction of continuous-flow production, in which tasks moved to the worker, thus enhancing control over workflow and minimizing the loss of time between operations.

Another regulatory system that was pushed to its limits was that of patent regulations, which protected the creative achievements of individual inventors but increasingly also acted as a roadblock impeding the integration of large technological systems. The further development of wireless communication, after the establishment of spark telegraphy, is a case in point.[114]

By the beginning of the new century, it had become clear (in particular to the large companies involved) that the next step was the wireless transmission of sound, and in particular of the human voice. While the technical feasibility was demonstrated by 1906, the technological integration and economic implementation of an entire working system was hindered in the United States by patents on individual components held by competing players. This stalemate was only overcome by an intervention of the US government on the eve of World War I, when radio transmission became a military asset for the navy.[115]

War as Catalyst of the Knowledge Economy

Science in the twentieth century was characterized by new knowledge economies that emerged under the pressure of economic, political, and military demands.[116] These demands imposed external constraints on the organization of science and technology, acting as selective forces on the further development of knowledge. The wars of the twentieth century saw an increasing mobilization of science and technology in the service of military advantage. They enabled technological innovations that would have been more difficult to realize if driven by market forces alone and thus generally enhanced the dependence of the societies involved on those science-based technologies, which, once unleashed, could hardly be contained.

I have mentioned the critical role of the military in the development of radio transmission. Another example from World War I is the development of artificial fertilizers based on the Haber-Bosch process, a case to which I return in chapter 15. The process served the production of both fertilizers and explosives; without the war, it would not have been scaled up to industrial dimensions so quickly. The war also triggered the development and spread of poison gas, modern artillery, airplanes, machine guns, tanks, and submarines. Yet even with all its destructions and the creation of novel destructive technologies, the war did not efface the achievements of the Second Industrial Revolution with regard to the economy of knowledge.

The understanding of infectious diseases, for instance, was not just an isolated scientific achievement.[117] It was successful in reducing communicable illnesses because it profoundly changed the economy of knowledge through the spread of household knowledge about how to prevent infections—by vaccination and by fighting carriers of diseases but also by boiling and filtering water, and by food preparation, preservation, and proper cooking. These changes in the economy of common knowledge comprehensively resulted in a steep decline in mortality rates in developed countries. In the next chapter, I turn more extensively to the global aspects of these and other scientific and technological developments. In chapter 16, I specifically deal with their role in colonial violence and domination.

Here just a few remarks may suffice to illustrate the general scaling-up effect that World War I had on developed countries, as did World War II and the Cold War thereafter. One global consequence of World War I was the rapid spread of the model of in-house research laboratories, which were initially characteristic of just a few large companies. Another, more important consequence was the increased integration of science into national economies, an integration that took, however, very different forms in different countries. In the United States, the war generally strengthened the importance of science for industry. In countries like France and Italy, research was less integrated into the firms, while a larger role was played by consultants and state-supported research organizations, such as the Centre National de la Recherche Scientifique in France.[118] In the Soviet Union, science came ever more under the centralized control of the ruling party and the state.[119] A dominant position was assigned to the Academy of Sciences, which in the 1930s was transformed from an honorary society into a large body of research facilities, centrally controlled and linked with the political and economic planning authorities of the country.

10.11. Between the military and science during World War I. Fritz Haber (inventor of the Haber-Bosch process for synthesizing ammonia and "father of chemical warfare") pointing to canisters containing chlorine gas with Oberst Leopold Goslich of the German Gas Pioneer Regiment, and Officer Fuchs at the left (1914–18). Courtesy of the Archive of the Max Planck Society, Berlin-Dahlem.

All countries needed to tap into the resources generated by science as an international endeavor, albeit on the varying timescales in which research typically produces tangible results.[120] Accordingly, they developed highly differentiated economies of knowledge (sometimes also discussed under the label "national innovation systems") with different contributions from and encounters between firms, the military, private and public institutions of research and education, and political and regulatory authorities. Under authoritarian regimes, as in Nazi Germany, science was massively taken into the service of military, economic, and political operations, including in horrific crimes against humanity like the Holocaust. This did not happen exclusively, and not even predominantly, by way of the top-down mobilization of resources by states and enterprises but rather through a self-mobilization of science in response to new funding and career opportunities.[121]

Big Science

If Big Science is the pursuit of science on an industrial scale, with massive investments in equipment and personnel, with an elaborate distribution of labor, and governed by management processes, then its origins can be traced back to the science-based industries of the Second Industrial Revolution and their scaling-up under the aegis of governments and the military in World War I. But by the early twentieth century, even basic science had gone "big," as scientists pushed the development of ever larger instrumentation in order to answer fundamental questions.

Early examples are the large astronomical telescopes conceived by George E. Hale around the turn of the century and the cyclotron invented by Ernest O. Lawrence in 1929 to investigate the structure of matter.[122]

Nuclear physics, launched by the discovery of nuclear fission in 1938,[123] played a decisive role in driving Big Science and the amalgamation of military, industrial, and scientific interests. The Manhattan Project, led by the Unites States and started in 1939, produced the first nuclear weapons by bringing together research, engineering, production, and military facilities on an unprecedented scale, ultimately employing more than 130,000 people. It played a key role in establishing what departing President Dwight D. Eisenhower called, in his 1961 farewell address, the "military-industrial complex," warning against the potential "domination of the nation's scholars by Federal employment, project allocations, and the power of money."[124]

The military-industrial complex, or rather the "military-industrial-scientific complex," turned out to be a persistent legacy of World War II and was reinforced, rather than loosened, during the Korean War of the early 1950s, after the Sputnik shock of 1957, and in the course of the long Cold War. This particular form of knowledge economy left profound marks on the further development of science and technology, and even on the social sciences, which underwent a scientistic turn, shaped by the recruitment of humanists and social scientists for military-relevant problems.

After World War II, large-scale instrumentation also affected the life sciences, for instance, when ultracentrifuges, electron microscopy, electrophoresis, X-ray crystallography, UV-spectroscopy, and other experimental techniques helped scientists to understand biological phenomena on a macromolecular level.[125] A decisive role was played by a few hubs in the initially rather thinly spread network of scientific cooperation. These hubs were formed by laboratories with unique pieces of equipment or with a unique combination of skills. They served as catalysts for integrating knowledge from a diverse array of disciplines. This was the golden era of molecular biology. Imagined or real opportunities for engineering and commercial applications emerged only later. This new perspective encouraged large-scale organized cooperation (such as the human genome project[126]) but also gave rise to a new fragmentation of knowledge production as a result of commercial and cultural boundaries.

The military-industrial-scientific complex constituted a model for the integration of science into large-scale societal structures so that we may speak more generally of "socioepistemic complexes" with a general tendency to persist and enforce the dependence of societies on scientific knowledge. Another example is the biomedical complex comprising pharmaceutical industries, hospitals, universities, professional organizations, and government agencies.[127] It was rooted in the upturn of biomedical research during World War II and became a major driving force of the life sciences after the war. Clinical practice and laboratory research became closely intertwined, with statistical methods increasingly dominating over practical clinical knowledge. Public investment and support for research and professional training played an important role in the expansion and further development of this socioepistemic complex.

The Transformative Power of Electronics

Another outcome of the military-industrial-scientific complex was the boost of electronics, first through the invention of the transistor and then through the development of digital computers, both having in turn major impacts on the transformation of the knowledge economy.[128] I have mentioned above that World War I played an important role in the development of the voice-carrying radio, which soon became a mass phenomenon. The next technological challenge, tackled in the 1930s, was television. The mass media of radio and television allowed for the extremely quick dissemination of knowledge to unprecedented numbers of people—but the ease with which they could be controlled and their low interactivity also made them ideal tools of propaganda.

Although the potential of the electron tube, on which these technologies were based, was not yet exhausted, as the development of radar during World War II illustrates, bottlenecks were already evident in the 1930s. The power demand and heat generation of electron tubes limited the further development of large telephone networks. To address this challenge, Bell Laboratories, which became one of the largest corporate laboratories in the United States, with over five thousand people working there in the 1940s, initiated a long-term research program focusing on semiconductors as a hitherto neglected alternative to the electron tube.

While the goal of replacing the electron tube was defined by its importance in existing communication networks, its replacement by a semiconductor device turned out be a long way off. The development was delayed because the war set other priorities, such as radar research. The task was taken up again immediately after the war, and its realization was not a matter of turning pure into applied science. It required fundamentally new insights that could not have been anticipated. Bell Laboratories offered a circumscribed space for an economy of knowledge in which specific technical problems of economic interest could be addressed in exploratory basic research. The process gave rise to another transformation of the scientist-engineer: he or she became an institutional persona characterized by the need to master the theoretical and experimental techniques of cutting-edge quantum physics.

Yet when the invention of the transistor was announced in 1948, its technological and commercial success could not be taken for granted, in spite of the decision to entice early adopters with a liberal license policy. Owing to high cost and unreliability, the early adoption of the transistor was limited to niches in which these impediments were less of a problem, such as in hearing aids for those who could afford them. The breakthrough came with the adoption of the new technology by the military, for which the small size and low energy consumption of the transistor were important advantages, in particular in the context of the missile program. It also helped that cost was no object. Eventually, the transistor not only found its way into consumer products but also opened up entirely new applications beyond those of the electron tube it was originally intended to replace. It created, to use my terminology from earlier chapters, a new horizon of possibilities by breaking the oligopolistic structure of an electronics industry focused on the radio.

The research laboratory that had generated the new technology became a model emulated by other large firms, but subsequent major innovations, such as the silicon transistor or integrated circuits, came from outsiders.[129] But even in a world of small and start-up firms (epitomized by what later came to be called Silicon Valley), large research laboratories such as Bell Laboratories, major universities such as Stanford, and the constant thirst of the military and the space program for new technology provided crucial resources feeding the innovation process with shared knowledge, qualified personnel, and sufficient capital. Rather than confined breeding spaces for innovations, the big centers now acted more like hubs in a larger network, allowing for more flexible exchanges of these resources.

The key innovation enabled by the miniaturization of semiconductor devices and integrated circuits was the electronic digital computer. This innovation, rooted in a long tradition of conceiving and building calculation machines, was also driven by military demand, then by advanced professional uses (including for scientific purposes), and finally by an expanding consumer market for personal computers and a growing number of appliances incorporating digital technology. The further improvement of semiconductor technology was closely coupled with that of computers, leading to the kind of accelerated development described by the prediction now known as Moore's law: The number of transistors in an integrated circuit doubles in a period between one and two years. This rule of thumb has been more or less confirmed since its formulation in 1965 but is now beginning to break down owing to physical and economic reasons.[130]

The first digital computers greatly augmented human capabilities in managing knowledge in military and engineering contexts, as well as in administration, economics, and the natural sciences. Computers led first to advances in the culture of calculation. Their application to text and language processing followed—at first only slowly—but eventually led to a major transformation in which the computer came to augment human mnemonic and linguistic capacity. Attempts to realize "artificial intelligence" using new information technologies have extended the possibilities of representing and processing knowledge in far-reaching ways since the concept was introduced in the mid-1950s.

Computer networks developed in parallel with the transformation of computers from calculation machines to knowledge-encoding and processing devices. Network technology can also be traced back to the military, with pioneering efforts funded by ARPA—the Advanced Research Projects Agency of the US Department of Defense— in the mid-1960s. The ARPANET was one of the first computer networks employing what came to be the basic Internet technologies of packet switching and TCP/IP protocol, critical to the transmission of data in a network.[131]

The development of both the Internet and later the World Wide Web was driven by the quest to overcome the limits of localized computing power and data storage and to facilitate information sharing, first for military and scientific purposes and then as a general-purpose information network.[132] Without powerful computer networks, large-scale scientific projects, ranging from particle physics to the Human Genome Project, would be impossible. But their impact on knowledge potentially reaches much farther: by transferring the operative functions of human thinking into

10.12. US Army employees holding parts of the first four army computers. *Left to right:* Patsy Simmers (mathematician/programmer) holding ENIAC board; Gail Taylor holding EDVAC board; Milly Beck holding ORDVAC board; Norma Stec (mathematician/programmer) holding BRLESC-I board. 1962. US Army Photo No. 163-12-62.

an electronic medium with the opportunity (offered by the World Wide Web) to create a global representation of human knowledge, digital network technology is changing the knowledge economy dramatically. In chapter 16, we shall discuss the ambivalent character of these changes in detail.

In the wake of the new information technologies, new fields of research emerged not only in the technical domain (e.g., computer science) but also in humanities like cognitive science, being motivated by studies of artificial intelligence.[133] Here traditional questions of rationality and science are addressed in a new way, reconstructing a logic of thinking that does not presume its universality and that no longer abstracts from its contents. Such approaches have made it evident that formal logic and a cognitive psychology abstracting from the contents of knowledge are inadequate to express the operative functions of human thinking. In this book, I have made extensive use of many of these insights.

The Problem of Measuring Scientific Productivity

In the course of the twentieth century, the knowledge economy of modern science was challenged by the growing size of the scientific enterprise. A widespread response to this challenge was to strengthen the values underlying the disciplinary system of science with the help of an increasingly extended institutional scaffolding. The scaffolding reinforced these values by imposing externally controllable formal and, if possible, quantifiable criteria. This institutionalization of scientific standards (exemplified by such phenomena as scientometrics and peer review in scientific publications) began to play a prominent role in the second half of the twentieth century and has ensured a high level of professionalism and quality control in scientific production, in spite of its enormous growth and worldwide spread.[134]

At the same time, the new forms of assessing scientific productivity weakened the underlying values by curtailing the significance of intellectual exchange in favor of mechanisms of social control, replacing personal judgments with formal evaluations, reading with counting, quality with quantity. The standardization of publication formats in academia has fostered a culture that takes the quantity of publications or their "impact" factor (how often and where one is cited) as measures of achievement, although these are, at best, weak proxies of intellectual merit, and at worst constitute an economy that rewards a high output of low-quality work. The growth of science has enhanced tendencies to standardize, institutionalize, and even to automatize many of the control procedures that, at the beginning of modern science, were merely an expression of more or less informal reflections about the quality of scientific achievements within a community of peers.

The Rise of New Disciplines

The knowledge economy of modern science is, however, not just subject to unidirectional tendencies of growth and specialization. With the expansion of science, new opportunities for knowledge integration and unification emerged, as well as new perspectives from which knowledge could be judged and evaluated. Borderline problems (as I have described them in chapter 4) continued to act as triggers for the transformation of the classical system of scientific disciplines. An example is the transformation of biology along the fault lines with other disciplines, one after another taking center stage in the great wars of the twentieth century. Thus, biology first amalgamated with chemistry after World War I, giving rise to biochemistry; then with physics after World War II, triggering the birth of molecular biology; and finally with cybernetics, computer science, and automatization, fostering the conceptualization of biology in terms of information theory; and giving rise later, in the aftermath of the Cold War, to the development of biotechnology.[135]

The rise of the earth sciences in the twentieth century constitutes a major conceptual and institutional transformation emerging from reorganizing previously fragmented pursuits into a unified discipline.[136] The history of the earth sciences thus illustrates the protracted character of such transformations, as emphasized in chapter 7. Just as it was the case for the so-called Darwinian revolution in biology or in the establishment of general relativity, a comprehensive and generally accepted new framework was established only decades after the emergence of the original idea that appears in hindsight to mark the crucial, alleged "paradigm shift."

In the case of the earth sciences, this original idea goes back to the work of the German geologist Alfred Wegener in the first third of the twentieth century. Wegener

10.13. The last photo of Alfred Wegener (1880–1930), creator of the continental drift theory, taken during an expedition to Greenland in 1930. © Archive of the Alfred-Wegener-Institut, Bremerhaven.

proposed that the continents of today had moved over geologic time to their current positions, breaking away from an original supercontinent.[137] But the theory of continental drift only became generally convincing after a complex and extended process of knowledge transformation. This was not just a matter of confirming the original hypothesis but of establishing, in the 1960s, the modern theory of plate tectonics, now confirmed by geophysical evidence of a novel character matching, in particular, the requirements of the US postwar scientific community.[138] This is in striking parallel to the epistemic transformation of the original theory of general relativity over a similarly extended historical period, an example to which I return in chapter 13. The protracted character of the transformation of geology in the twentieth century was not simply due to the multitude and heterogeneous character of empirical knowledge to be gained and integrated but, as we shall see, also to the peculiarities of the economy of knowledge guiding this integration.

Geology was not a new science and had long dealt with such questions as the formation of mountains, the distribution of continents and oceans, the constitution of the interior of the earth, and the age of the earth. Among its challenging problems were the presence of marine deposits on dry land or the large slabs of rock found in the Alps that had been transported over enormous distances.

By the end of the nineteenth century, many of the key problems of geology had assumed the character of borderline problems emerging along the fault lines of the various approaches to understanding the nature and history of the earth. There were, on one hand, approaches that tried to understand its history mainly in terms of physics and astronomy, such as those of Georges-Louis Leclerc de Buffon and William Thomson, the later Lord Kelvin. On the other hand, a well-established tradition of geologists, such as Abraham Werner, Georges Cuvier, or Charles Lyell, went out into the field and focused on the evidence from the rock record. In addition, paleontology, with its interpretation of the fossil record in terms of Darwin's theory of evolution, also had strong implications for the understanding of the history of the earth.

A prominent example for a borderline problem of geology was the debate on the earth's age.[139] In the 1860s, on the basis of thermodynamic calculations, Kelvin argued that the earth could not be much older than a few tens of millions of years. This was in stark contrast to the hundreds of millions of years and more that biologists thought was needed for the evolution of life on Earth. This was also hard to reconcile with the interpretation of the rock record in terms of "uniformitarianism," that is, the principle that during the history of the earth, the same forces were active that could be observed today. Eventually, it turned out that this borderline problem could only be resolved on the basis of new physics, that is, the nuclear physics required to understand the energy sources of the sun.

Wegener's model of continental drift, first proposed in 1912 and then elaborated in his 1915 book *The Origin of Continents and Oceans*,[140] addressed the distribution of continents and oceans as a borderline problem, taking into account both geological facts, such as the jigsaw-puzzle fit of the continents and supporting stratigraphic evidence, *and* the fossil record pointing to an extensive interchange of species between continents during earlier geological periods. Wegener's theory also presented a new solution to the problem of mountain formation, avoiding weaknesses of earlier proposals, such as the assumption of a cooling earth.

The mental model of a cooling, solid earth, as defended by Kelvin, had been used earlier by geologists to explain the characteristics of the earth's surface. According to the Austrian geologist Eduard Suess, the cooling, shrinking earth caused portions of its crust to collapse, which then formed the ocean floor.[141] He speculated about the existence of a giant supercontinent that he called "Gondwanaland," which eventually broke apart. This assumption could also account for the existence of similar fossil records in the now-separated continents, as these had been formerly connected by "land bridges." For his American contemporary James Dwight Dana, the basic features of the earth's topography were, in contrast, permanent, primordial features that had emerged as a result of the unequal melting temperatures of different minerals when the surface first became solid.[142]

Another mental model, the model of "isostasy," characterized the relationship between continents and oceans according to Archimedes's principle of flotation. The features of the earth's surface, such as mountains or continental blocks, were supposed to be in hydrostatic equilibrium with some underground medium so that elevated areas were compensated by a mass deficit below them. This idea had come up in the middle of the nineteenth century, in the context of a geodetic survey of British colonial holdings in India.[143] Precise measurements had to take into account the gravitational attraction of large mountain massifs such as the Himalayas, but the effect turned out to be less than expected, pointing to a mass deficit underground. The mental model of isostasy allowed for two different specifications: either the mountains behaved like icebergs and were supported by "roots" underground (Airy model) or hydrostatic equilibrium was achieved by a variation of the density in proportion to the variation in elevation (Pratt model). The Airy model required a fluid medium for the continents and mountains to float upon, as icebergs do.

Wegener's model of continental drift combined aspects of Suess's model with the Airy model of isostasy. It incorporated Suess's insight into the contiguity of continents, as well as the suggestion from isostasy that continents, supposed to be composed of less dense materials than the ocean basins, could float while maintaining hydrostatic equilibrium with a highly viscous medium underground. In Wegener's conception, Gondwanaland was not a sunken continent, a hypothesis that would have been difficult to reconcile with isostasy, but was broken up in pieces. His assumption of horizontal motions of the continents allowed, at the same time, for an explanation of the emergence of mountain chains, rift valleys and island arcs. He could even explain challenging problems such as the evidence for large lateral movements in the Alps.

These and other mental models served as cores around which, in the first half of the twentieth century, geological theories were elaborated and new evidence accumulated. But which theories were elaborated and how new knowledge was evaluated crucially depended on the relevant knowledge economies and their constraints, among them images of knowledge and path-dependencies resulting from prior theoretical or practical commitments. These knowledge economies varied with national contexts, and what varied with them were the default assumptions underlying these selective processes.

The crucial role of the knowledge economy is evident from the fact that accumulation of novel empirical knowledge was driven not only by research questions but also by the institutional frameworks supporting the pursuits of geologists, in particular those responsible for large-scale geodetic surveys. In the United States, for instance, such surveys played an important role for the investigation of geophysical questions, beginning with the US Coast and Geodetic Survey, originally founded under Thomas Jefferson in 1807. In the twentieth century, the Carnegie Institution of Washington became an important institutional patron of American earth science. Its geophysical laboratory fostered, in particular, the physical and chemical aspects of earth science. During World War II and afterward, substantial geophysical and oceanographic work was also supported by the US Navy. These institutional frameworks decisively shaped the default assumptions mentioned above.

An important source of new empirical knowledge was the methodological advancement made possible by the development of novel instrumentation, as well as progress in other fields, in particular physics and chemistry. The biogeographical and paleontological patterns supporting Wegener's hypothesis were essentially known since the late nineteenth century. The discovery of radioactivity at the beginning of the twentieth century opened up a new perspective, undermining Kelvin's argument of a cooling earth and pointing to the geological role of radiogenic heat.

That the earth contained a mobile layer underneath the crust became a consensus among geologists in the mid-1920s. The results of a US submarine expedition to the Caribbean in the late 1920s suggested that ocean basins were geologically active regions where major regional stresses of the earth's crust could be identified. The presence of such stresses and their role in crustal down-warping was confirmed in the 1930s, also by seismic evidence. Meanwhile, the British geologist Arthur Holmes had pointed to convection currents in the substrate as a possible explanation of such down-warping.[144] Magnetometers developed during the war at the Carnegie Institution made it possible to detect weak magnetic fields. Studies of paleomagnetism showed in the 1960s that the earth's magnetic field has several times reversed its polarity, a finding that could be used to confirm the hypothesis of seafloor spreading by geophysical data.

The accumulation of new empirical knowledge did, however, not by itself entail a decision among competing theories. Otherwise, Wegener's theory could have become the consensus among geologists much earlier, by the latest in the 1940s, when overwhelming, albeit mostly qualitative evidence was available in its favor, while other geological models had turned out to be problematic. At that time, Wegener's theory had been systematically expounded, together with recent evidence and a plausible explanation for the mechanism of drift, in the comprehensive books by Alexander du Toit (*Our Wandering Continents*, 1937) and Arthur Holmes (*Principles of Physical Geology*, 1945). But, as we have seen in the case of Einstein, the evaluation of a borderline problem strongly depends on perspective. As Naomi Oreskes has convincingly demonstrated for the case at hand, the evaluation of Wegener's thesis of moving continents was profoundly shaped in the United States by the peculiarities of the local economy of knowledge that effectively delayed its acceptance until the 1960s.

The reasons for this delay, she argues, were connected with (in the terminology of this book) the images of knowledge, default assumptions, and path-dependencies prevailing in the economy of knowledge underlying the pursuits of American geologists.[145] Among the relevant images of knowledge was an adherence to a theoretical pluralism that entailed skepticism with regard to such a grand unifying theory as that of Wegener; the commitment to an inductive, experimental methodology stressing the primacy of quantitative evidence and precision; and an inclination to emulate physics and chemistry—that is, tendencies that can also be observed in contemporary developments of the life sciences.

The default assumptions of American geologists and the path-dependencies of their theoretical commitments were dominated by a geophysical perspective reinforced by the institutional contexts of American geology. They were also shaped by the privileged role of the Pratt model of isostasy in the theoretical efforts of American geologists and by the paradigmatic function of local mountain formations such as the Appalachians for their reasoning. In contrast, the new empirical knowledge becoming available in the 1960s was essentially free of such prior theoretical commitments, while complying with the demand for quantitative geophysical data. It could thus be considered as evidence in favor of a fresh approach to the wandering of continents, now dubbed "plate tectonics." As a matter of fact, the emergence of such a stable, unifying theory relied, in the end, on an integration of all the available

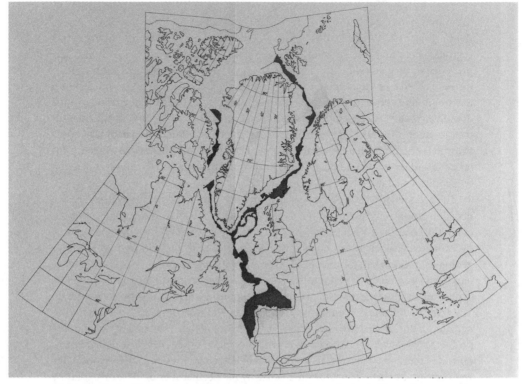

10.14. The "Bullard fit." In 1965, Edward Bullard published the result of calculations determining the best fit of the coastlines of North America, South America, Africa, and Europe. This result was at the time considered as important evidence in favor of plate tectonics. From Bullard, Everett, Smith, et al. (1965, fig.7). Courtesy of The Royal Society.

knowledge, finally resolving the tensions between historical geology, paleontology, and physics that had initially triggered this development.

The further development of the earth sciences in the second half of the twentieth century was driven by strong economic and military interests—in particular, by the search for natural resources and geostrategic advantages in the Cold War. A turning point was the International Geophysical Year 1957–58. Although its official support by the United States was motivated to a significant extent by military interest in gaining global data of strategic relevance, it nevertheless encouraged international cooperation within a global scientific community, inaugurating an era of international observation and regulation of the Earth system.[146] In chapter 13, I examine more closely such community-forming processes with the help of network theory.

In the 1960s, the rise of the earth sciences fostered a global perspective on threats to the environment that was articulated in political agreements such as the restriction of mining and military activities in Antarctica, limited nuclear test bans, and the capability of the international community to react to such challenges as ozone depletion, an example to which I return in chapter 15. This global perspective was also articulated in scientific and popular literature, as in Rachel Carson's famous book *Silent Spring* from 1962.[147] *Silent Spring* denounced the devastating effects of pesticides, which became one of the touchstones of the international environmental movement.

Challenging the Authority of Science

By the 1970s, it had become widely realized that the very concepts of progress and innovation, being images of knowledge, involve societal interests and views about what science means and where it is leading. It had become clear, in particular, that political, economic, or military decisions—as well as the market and public opinion—could affect the path of scientific development with long-term and sometimes devastating consequences.

The great pitfalls of science in the twentieth century had made it evident that scientific innovation is not just the business of science and its irresistible progress. Furthermore, the continued existence of traditional societies still living in a preindustrial age suggested that the development of human societies is not necessarily linked to an inevitable acceleration of technology, and not at all with the emergence and cultivation of science. Obviously, science was only one of many possible forms of expressing human culture, a realization that, in the humanities and social sciences, suggested approaches to science studies that no longer accepted unidirectional concepts of modernity and even cast doubt on the rarified status of science as a privileged form of knowledge.

Meanwhile, however, much of the inner workings of modern societies, their economies, their political systems, their cultural traditions and mind-sets, and even their mechanisms of biological reproduction had become the object of science and science-based interventions, sometimes with immediate self-regulatory consequences, for instance, by the use of statistics in politics, economics, medicine, and education.[148] On the other hand, the knowledge most crucial to a society's future has often not been generated by its academic institutions, in particular regarding fundamental technological, infrastructural, social, or economic decisions that impact the environment; if it had been available, it may not have been implemented because of the incapability of the society as a whole to absorb or implement this knowledge.

Until the 1960s, for instance, national programs for the civil use of nuclear energy were hardly complemented by nuclear waste policies.[149] This only changed with the imperative to prevent the proliferation of nuclear arms, the need for industrial policies concerned with energy provision, and the rise of antinuclear movements in Europe and the United States. As a result, new forms of entanglement between science and society emerged. Governments invoked science as an authority to justify political actions, often with the aim to depoliticize controversial issues. Scientists became involved in environmental politics both as experts (often dependent on vested interests) and political activists.

It is evident that no simplistic, rationalistic-technocratic model of policymaking of the "speaking truth to power" type adequately describes this situation. It is generally not the case that science first identifies a problem, then offers a solution that politics finally has to implement. That simply does not reflect the societal dynamics of dealing with knowledge. "Truth" is not produced in an area free of interests, values, and uncertainties, and "power" is more than simply the adoption and implementation of expert knowledge regardless of normative and broader societal considerations.[150]

In democratic societies, public funding of science requires justification; social expectations may affect the direction of the development or even impose severe limitations on it. Matching the very different timescales and perspectives of science with the short-term expectations of other societal domains (including demands for the credibility, public accessibility, transparency, and active participation of science in public dialogue) has become a major challenge for societies that rely on scientific advancement.

Since the 1960s, social movements have challenged the authority of science in numerous fields, ranging from nuclear physics to pesticide chemistry to psychopharmacology. Scientists associated with large corporations or conservative political groups, on the other hand, manipulated the conversation to keep controversies open to debate, even after a broad scientific consensus had formed on such issues as tobacco smoking, acid rain, DDT, and climate change.[151] They thus contributed to preventing scientific knowledge from becoming integrated into a broader societal economy of knowledge. In chapter 16, I come back to this issue, discussing such suppressed scientific knowledge under the label "dark knowledge."

Academic Capitalism

The period after the 1970s was characterized by a "liberalization" in the flow of capital and goods, the emergence of new forms of transnational and domestic governance (the "New Public Management" movement),[152] the emergence of new information technologies, the dissolution of the Communist bloc, and the emergence of Asian economies, but also by an increased awareness of the limitations of global natural resources and by the above-mentioned skepticism with regard to scientific progress as a self-evident component of modern industrial societies.

In the knowledge economy of modern science, public images of science play an important role, acting as mediatory instances modulating the interaction between science and other societal subsystems. They often reflect the interests and views of specific groups—at least as much as the actual processes by which scientific knowledge is generated. Since the 1970s, the economization of all societal domains has promoted an image of science as an activity to be organized and conceptualized in economic terms, thus fostering what has been aptly called "academic capitalism."[153]

With the Bayh-Dole Act of 1980, for instance, a uniform patent policy was introduced in the United States that enables universities and nonprofit organizations to register patents for inventions made in federally funded research projects. The intention was to encourage universities to engage in technology transfer and to increase commercialization, thus strengthening US economic competitiveness.[154] As a consequence, numerous technology transfer offices were created in the United States, Europe, and Asia. While this policy has led to the creation of thousands of spin-off companies contributing billions of dollars to the American economy, it has also tended to limit perspectives to short-term profits rather than minding the benefits of society at large.

This tendency has been reinforced in the United States by diminishing public funds for higher education, forcing universities to raise the tuition for their students and forcing faculty to seek external funding, which often comes with expectations related to profitable outcomes. As Robbert Dijkgraaf has observed, "These days, the added burden of industrial research is crowding out basic research at many universities. Meanwhile, governments are increasingly directing research funding to tackle important societal challenges, such as the transition to clean sustainable energy, battling climate change, and preventing worldwide epidemics, all within flat or decreasing budgets. As a consequence of the priorities and politics of the time, basic research is too blithely given short shrift, its budget often ending up as the remainder of a growing series of subtractions."[155]

However, the reduction of the economy of knowledge to a single-minded market model hardly promises long-term success in either a strictly economic or a broader societal perspective. This is at least what the many examples that I have discussed suggest, which point to the protracted character of innovation and its dependence on public investment (be it military or civil), as well as to the crucial role of the freedom of thought for any form of innovation. Commenting on Abraham Flexner's essay "The Usefulness of Useless Knowledge," Robbert Dijkgraaf wonders, "Indeed, in today's metric- and goal-fixated culture, how can we meaningfully convey the 'usefulness of useless knowledge'? How many points must one count along research's lengthy, circuitous, and surprising path, often with many dead ends and hairpin turns that lead to further unexpected vistas? How does one articulate a potential outcome of an idea without at the same time boxing it in?"[156] In his own remarkable essay on the usefulness of useless knowledge, Nuccio Ordine cites the French mathematician and philosopher Henri Poincaré: "Without a doubt you have often been asked the use of mathematics and whether these delicate constructions, entirely the fruit of our mind, are not artificial and spring merely from our fancy. I ought to make a distinction between those who ask this question: practical people ask us solely how to earn money from it. This does not merit a reply. It would be more to the point to ask them the use of amassing great riches and whether, in order to do so, it is worth neglecting the arts and sciences, for only they enable us to take pleasure from them, *et propter vitam vivendi perdere causas* [and for the love of life to lose the very reason for life itself]."[157] Future discussions, also about the reliability and trustworthiness of scientific claims, could benefit from taking notice of the more general concept of the knowledge economy I have developed here.

PART 4

HOW KNOWLEDGE
SPREADS

Chapter 11

The Globalization of Knowledge in History

The works of Hippocrates, Galen, and Avicenna occupied whole shelves of every good fifteenth-century library from Baghdad to Oxford to Timbuktu, but these three giants of medicine had not a word to say about syphilis.
—ALFRED W. CROSBY, *The Columbian Exchange*

As a general principle, I believe that the longer the time which has elapsed between the first successful achievement of an art or invention in one place and its appearance in another, the more difficult it is to entertain the idea of a purely independent invention.
—JOSEPH NEEDHAM, "Chinese Priorities in Cast Iron Metallurgy"

Columbus accidentally ran into America but thought he had discovered part of India. I actually found India and thought many of the people I met there were Americans. Some had actually taken American names, and others were doing great imitations of American accents at call centers and American business techniques at software labs. Columbus reported to his king and queen that the world was round, and he went down in history as the man who first made this discovery. I returned home and shared my discovery only with my wife, and only in a whisper. "Honey," I confided, "I think the world is flat."
—THOMAS L. FRIEDMAN, *The World Is Flat*

The Need for a Global Perspective

In the previous chapter, I examined knowledge economies from a diachronic perspective that emphasized societal conditions for knowledge production, sharing, and implementation. This may have given the impression that the history of knowledge follows an almost inevitable series of stages of continually rising levels of integration and reflection. In terms of historical reality, however, the transition from one knowledge economy to another is anything but inevitable.[1]

While it is true that the possibility of higher levels of reflection and integration of knowledge depend on the tools and experiences accumulated over the course of history, the process of accumulation itself, like the exploitation of the options it offers, depends on additional factors that cannot be arranged into an unambiguous logic of progress. In particular, the knowledge economies I have considered so far never existed in isolation from a larger world of knowledge transmission and exchange. Ultimately, the evolution of knowledge can only be understood as a global process.

In recent years, investigations into the migration of knowledge and comparative historical studies have become active fields of research in the history of science.[2] With few exceptions, however, the emphasis is placed mostly on local histories: detailed studies of political and cultural contexts, social practices, and the construction of knowledge. The resulting rich but rather fragmented picture may lead us to underestimate the extent to which the world has been connected, for a very long time, by knowledge. In fact, the development of languages, food production, resource and energy use, architecture, writing, and calculating have been part of long-term and indeed global processes since very early times and can only be properly understood from a more comprehensive perspective.

Recent archeological findings (based on novel technologies such as ancient DNA analysis) have, for instance, shown that the spread of agriculture from Anatolia to southeastern and middle Europe in the late seventh and early sixth millennia BCE was a result of the migration of farmers, who marginalized indigenous hunter-gather populations.[3] Wheel and chariot spread in the fourth millennium, appearing at about the same time at the North Sea and in Mesopotamia. Evidently, innovations—such as the knowledge about the construction of wheels and vehicles, about how to yoke a team of oxen to a chariot, about how to mine for copper ore or how to cast axes or bracelets—could travel via interconnected regional networks across Eurasia with extraordinary speed.

I therefore propose to speak of globalization as a process reaching far back into prehistory. I come back to this often-neglected phenomenon of ancient globalization below. Globalization is here understood as referring to the global spread of material and symbolic culture, ranging from goods and technologies to language, knowledge, and beliefs, as well as political and economic institutions, even if that spread is part of a long-term development. Crucial aspects of globalization are the resulting accumulation of means of survival and social cohesion and the intensification of interdependencies between distant regions.

But what is the importance of a history of knowledge for understanding modern globalization processes? I argue that the succession of waves in which innovations spread across the continents left a layered structure of cultural achievements, and in particular of knowledge, similar to the strata of geological sedimentation. This layering does not simply amount to a linear accumulation but may display twisted and fragmented structures, as does the geological record. Still, earlier processes of globalization, such as the spread of farming, of basic technologies such as pottery or metallurgy, or of writing, religion, or science condition later globalization processes by their varying regional impact or failure to take hold. They also help to explain, as we shall see in a comparison of ancient European and Chinese science in the next chapter, striking similarities that are more plausibly explained by similar presuppositions created by earlier globalization processes rather than by direct transfer.

The long-term dynamics of the globalization of knowledge are shaped by an interaction of intrinsic and extrinsic processes that potentially enhance its social dominance, its range of application, and its degree of reflexivity or, alternatively, destroy its autonomy and reduce its complexity. The intrinsic dynamics of knowledge is, as we have seen in part 2, characterized by the interaction between knowledge forms and representation structures, triggering developmental processes within a given economy of knowledge, such as the transformation of Aristotelian natural philosophy into the system of knowledge known as preclassical mechanics. Extrinsic processes refer, in contrast, to migration, trade, conquest, colonialization, missionary work, or other expansive activities by which systems of knowledge spread, including their more or less violent imposition on new populations.

Intrinsic and extrinsic developments are closely intertwined, and their interplay is a decisive mechanism of the global economy of knowledge. The intrinsic development of a knowledge system, such as its exploration in a given social and cultural context and its subsequent restructuration, may become an extrinsic cause of knowledge globalization. Intrinsic developmental processes may enhance the technological potential of a society and thus its capacity to engage in extrinsic globalization processes, or they may lead to knowledge that is more suitable to travel because of its greater independence from context. The development of geographic, astronomical, and navigational knowledge in early modern Europe, facilitating long-distance maritime seafaring and being universally applicable independent of geographic location, is an example of both cases. The opportunity to colonize, discussed in greater detail below, also depended on achievements of intrinsic knowledge development, such as progress in astronomy, navigation techniques, or other technologies bolstering military or economic superiority. In chapter 13, I review how some of these achievements accumulated in the epistemic networks of late medieval and early modern Europe.

Extrinsic processes, such as the transfer of a given system of knowledge to another natural and cultural setting, may in turn foster intrinsic processes through new experiences, which act as triggers of change, as when the European colonialization of other continents confronted both European and non-European knowledge systems with novel challenging objects. An example of such an extrinsic process of globalization triggering an intrinsic development is the opening of Japan to the West in the middle of the nineteenth century. Japan had closed itself to the world in the seventeenth century in order to defend its internal political stability against the threats represented by foreign missionary and commercial intrusions. In the 1850s, warships of the US Navy arrived in Japan with the mission to open up the country after more than two centuries of self-imposed closure to commercial and diplomatic relations with the West, if necessary by military force. This extrinsic pressure triggered a process of profound change. The so-called Meiji Restoration after 1868 launched a process of modernization that included the introduction of Western science and technology, but it did not simply imitate Western models. This development rather constituted an intrinsic process of globalization in which both the social and political order as well as the knowledge economy were transformed in a way that turned Japan into a global player in politics, economics, science, and technology in the twentieth century. I come back to this example in chapter 16.

11.1. The picture shows the arrival of American steamships under the command of Matthew Perry in Japan on July 8, 1853. The expedition had the aim to break Japan's century-long isolation and to establish contact and diplomatic relations with the Japanese government. It became a historical turning point for Japan's relations with Western powers. From Tsukioka Yoshitoshi, *Kōkoku isshin kenbunshi* [Chronicle of the imperial restoration], 1839–92, Meiji period (1868–1912). The Metropolitan Museum of Art.

In summary, the global history of knowledge is neither simply the triumphant ascent of ever more rational, effective, or otherwise superior knowledge, nor can it be explained as a long-term consequence of the initial ecological conditions of the process.[4] It is rather a result of the complex interdependence of intrinsic and extrinsic processes and the unfolding of this interaction over time.

Toward the end of this chapter, I return to a well-studied example, the many renaissances that Greek science has experienced over time, stressing a few points that are crucial to a global history of knowledge. One of these points is the process of what I call "cultural refraction." It here refers to the transformation of Mesopotamian and Egyptian knowledge imported into the Greek context and the conditions for this transformation. One of these conditions is the articulation and reflection of structures of knowledge previously embedded in institutional structures that did not alone provide motivation for such an explication.

I further argue, referring to the example of the medieval science of weights, that the appropriation of Greek scientific knowledge into the Arabic world was not like passing the baton in a relay race. It rather led to genuinely novel concepts shaped by a focus on particular material objects, a specific selection of Greek source texts, and a new form of discourse. The resulting reconceptualization of Greek science left a profound mark on early modern science in the Latin world. It serves as another example of the transformation of contingent circumstances into an internal engine of knowledge development. In this context, I also discuss characteristics of translation processes as a method of knowledge transmission, emphasizing their dependence on the knowledge economies of the source and target societies, their network character, the role of multilingualism, and, in particular, the autocatalytic dynamics they typically unleash.[5]

Finally, I comment on the period of Iberian globalization as a critical moment in our journey into the Anthropocene because of the sphere of global interaction to which it gave rise, but also because of its devastating consequences for the indigenous population of the Americas and for the Africans forced to work as slaves on American plantations, as well as the transformative changes of the global environment it brought about.

Recent Globalization of Science

Since the nineteenth century, educational and scientific institutions following European models were adopted worldwide in the footsteps of European imperialism, commerce, and colonialism.[6] The German research university became a model that was adopted in England, France, and—toward the end of the nineteenth century—the United States and Japan. The context of this development was competition among rival nations but also global industrialization and colonialism—major forces driving the planet into the Anthropocene, as I discuss in chapter 15. The capitalist world system proceeding from Europe, Europe's colonial dominance, and the system of knowledge developing through Europe's experience abroad can hardly be divorced from one another, as we shall see in more detail below.

Technological and scientific standards spread in the course of industrialization. International meetings contributed to creating worldwide networks of scientists. Throughout the world, science developed into a strongly integrated subsystem of society, increasingly becoming the foundation on which political power and decision making were based. States recognized that educational policy was instrumental in asserting their own claims to power in the socialization of their populations, in their political formation, and in the storage and propagation of knowledge. According to Jürgen Osterhammel, an emerging global consciousness provided a framework for societies to engage in self-diagnostics with regard to their current situation.

Much of today's knowledge, whether scientific, technological, or cultural, is shared globally. Science in the twenty-first century benefits from the creation and exploitation of new social and technological structures, which enable the global flow of knowledge and expertise. The Web has become a global knowledge infrastructure for scientific knowledge as well. It was invented in 1989 at the huge European laboratory CERN (Conseil Européen pour la Recherche Nucléaire) in Geneva—a prominent example of a globally relevant scientific hub. CERN was set up after World War II with the explicit purpose of facilitating the maximum possible international cooperation in high-energy physics.

High-energy physics is essentially the science of subatomic particles, an expensive venture with little immediate economic or social impact. Higher and higher energies are required to penetrate deeper into the structure of matter, and hence, larger and

11.2. The installation at CERN of the vessel of the Big European Bubble Chamber at the beginning of the 1970s. By 1984, the end of its operative period, it had delivered 6.3 million photographs to twenty-two experiments dedicated to neutrino or hadron physics. Around six hundred scientists from about fifty laboratories worldwide had jointly analyzed the three thousand kilometers of film it had produced. © 1971–2018 CERN.

larger facilities become necessary. CERN demonstrates the possibility of large-scale international cooperation on knowledge production under very special boundary conditions, that is, in the absence of immediate political, military, or economic implications. Not least for this reason, it became a test ground for new global knowledge infrastructures. CERN was not only the venue for the invention of the Web but also for the development of grid computing and the initiation of the open access movement.[7]

One important result of the interaction between intrinsic and extrinsic processes in the globalization of knowledge over the long twentieth century is the emergence of global objects of science, in particular global human challenges such as climate change, scarcity of water, global food provision, global health issues, reliable energy supply, sustainable demographic development, and nuclear proliferation. The mitigation and handling of such global challenges to humanity are inherently connected both to the development of policies and to the production of scientific knowledge on a global scale. The Intergovernmental Panel on Climate Change (IPCC), for instance, was set up with the intention to assess knowledge about climate change and provide advice for policymakers.[8] But no comparable organization as yet exists for dealing with many other global challenges, such as that of energy supply.

The Intergovernmental Panel on Climate Change (IPCC)

The IPCC was founded in 1988 by the World Meteorological Organization (WMO) and the United Nations Environment Programme (UNEP). Its task is to take into account all available scientific information in order to assess climate change and its consequences, as well as to formulate possible counterstrategies. The *Principles Governing IPCC Work* state, more precisely, that its mission is ". . . to assess on a comprehensive, objective, open and transparent basis the scientific, technical and socio-economic information relevant to understanding the scientific basis of risk of human-induced climate change, its potential impacts and options for adaptation and mitigation. IPCC reports should be neutral with respect to policy, although they may need to deal objectively with scientific, technical and socio-economic factors relevant to the application of particular policies." In 2007, the IPCC was awarded the Nobel Peace Prize.

The work of the IPCC informed the United Nations Framework Convention on Climate Change, which was signed at the Earth Summit in Rio de Janeiro in 1992. Its declared objective was the "stabilization of greenhouse gas concentrations in the atmosphere at a level that would prevent dangerous anthropogenic interference with the climate system. Such a level should be achieved within a time-frame sufficient to allow ecosystems to adapt naturally to climate change, to ensure that food production is not threatened and to enable economic development to proceed in a sustainable manner."[9]

Today, globalized scientific knowledge crucially shapes policies and politics. Policies, on the other hand, shape the organizational form of science and determine research priorities. Any flow of scientific knowledge that comes to be associated with the international policy of individual states or with multinational actors unavoidably takes on a global character. But it is as yet unclear which international arrangements are most effective to meet collective international problems, which arrangements actually bring about scientific advancement, and whether political coordination is under all circumstances favorable to the advancement of science, and what happens to science when it becomes a resource that plays a role in structuring international regimes, shaping global images of knowledge, or enabling new forms of global governance.[10] Globalized science in fact often serves as an ideological tool for legitimizing collective political actions. According to Yaron Ezrahi, invoking seemingly "pure" science is a way of "depersonalizing" or "depoliticizing" the exercise of political power.[11]

As I discussed at the end of the previous chapter, the rising societal demand to gain revenues from publicly funded science in an increasingly competitive world is currently leading to a transformation of the academic world in accordance with global economic models. Around the turn of the millennium, the so-called Bologna Process was established with the aim to standardize higher education in Europe. Later, the Lisbon Strategy aimed to turn Europe into "the most competitive and dynamic knowledge-based economy in the world."[12] Asian universities have a strong orientation toward international (American-biased) rankings, such as the Shanghai Ranking and the Times Higher Education World University Rankings, and adjust their policies accordingly.

Political discussions of present globalization processes mainly refer to the economic processes globalizing the markets for goods, capital, and labor that take place in short historical time spans; conversely, the long-term global diffusion of technical innovations and bodies of knowledge is often considered as a mere presupposition or consequence of economic, political, and cultural processes. However, as I argue in the following, globalization involves knowledge in more significant ways, in particular by way of shaping the perspectives of its actors and their potential interventions.

Contradictory Aspects of Globalization

Discussions about current globalization processes often emphasize two apparently contradictory characteristics of such processes: homogenization and universalization on the one hand, and their contribution to an ever more complex and uncontrollable world on the other. Indeed, the economic power of globally organized transnational corporations increasingly translates into a standardization of mass culture and universal tendencies of the wasteful consumption of natural resources. Contrastingly, owing to the unequal distribution of wealth (among other factors), the same pressures of homogenization provoke an increasingly diverse spectrum of strategies to cope with these pressures, which leads to an increasingly complex patchwork of social relations. National and regional institutions and traditions in fact play an often neglected, mediatory role in filtering and transforming the effects of globalization.

Such observations point to the possibility that the alternative between an increasingly homogenized "flat world"[13] and an increasingly complex network of social relations may be insufficient to capture the dynamics of globalization processes. Globalization in fact results from a variety of processes, all characterized by the tension between universalization and growing complexity. Economic globalization, for instance, extends the dominance of the world market over local patterns of production and distribution and, at the same time, provokes counterstrategies for developing diverse local patterns of economic subsistence under the new conditions. Globalization homogenizes culture and destroys local customs, but it also stimulates morally grounded antiglobalist or nationalist countercultures, as well as religious and political fundamentalism. In the field of political decision structures, globalization leads to a growing number of international institutions whose task it is to deal with problems transcending the influence of political institutions of national states. While

11.3. Indian woman sitting on a bench at a McDonald's fast food restaurant in Varanasi, Uttar Pradesh, India. Roberto Fumagalli / Alamy Stock Photo.

globalization thus challenges national autonomy through global pressures, national integrity is, at the same time, also under pressure from a search for new (or old) regional interests and identities.

The contrast between the tendency toward an ever "flatter" and an increasingly "fractal" world (exhibiting similar complex patterns at different scales) suggests that comprehensive globalization processes result from a superposition of various processes, such as the migration of populations, the spread of technologies, economic and political structures, the dissemination of cultural or religious ideas, traditions of multilingualism, or the emergence of lingua francas. While these processes each have their own dynamics and history, it is their mutual impact on each other, and in particular on knowledge, that marks globalization as we observe it in the present.

Goods, tools, technical skills, and ingenious solutions circulate among human groups at different rates of diffusion, and typically faster than languages, values, rituals, systems of knowledge, religious frameworks, or administrative and political institutions. These variations of rate account for the characteristic retardation of globalization processes after the realization of their initial incentives. Goods and the technologies that produce them often spread independently of each other, each being associated with discrete systems of knowledge that make them relevant and accessible to a given culture. The transfer of the knowledge necessary for producing and inventing tools often requires, in particular, linguistic capabilities and frameworks of ideas that can only be built up once globalization processes of other types have taken place.

Furthermore, the various stages that occur over the course of comprehensive globalization processes do not occur in mechanical succession, otherwise one could be certain that the globalization of markets, for example, implied a globalization of a certain political system, which is clearly not the case. Rather, interactions among the various processes may lead to very different types of globalization. All these complex interactions are also indicative of the crucial role of knowledge in these processes.

This is evident, for instance, in the development of economic policies after World War II.[14] In 1944, the Bretton Woods Conference, under US and British leadership, established rules for international economic policies, reversing the prewar emphasis on protectionism and fostering the expansion of international trade. State interventionism according to the Keynesian model was an essential part of the Bretton Woods system, dominating international economic policies until the beginning of the neoliberal era in the early 1970s. The International Monetary Fund, the World Bank, and the World Trade Organization were all founded in its wake. On closer inspection, however, it did not amount to spreading a universal paradigm that was then identically reproduced; rather, it left room for experimentation, adaptation, and cooperative learning among different countries. As illustrated by the rapid spread of the central banking system (a key element of the Keynesian model of state interventionism), local experiences and their exchange—in particular among developing countries—could play a key role in the transformation of this model.

The history of knowledge influences other globalization processes—including the formation of markets—by shaping the perspectives of its actors as well as its critics. In the following, I first explore the role of knowledge in globalization by considering further examples. I then return to the various components of globalization processes and in particular to the sediments or layers left by the succession of globalization processes in the course of history.

The Role of Knowledge in Globalization

It is generally accepted that knowledge is involved in globalization processes, since it constitutes a specific condition for every form of their realization. On the political level, the spread and improvement of education is therefore considered to be critical for mastering the challenges of globalization. Similarly, the spread of knowledge is clearly a consequence of globalization processes, just as the exchange of goods or the diffusion of a language also transports knowledge. Education is thus considered both a precondition of globalization processes as well as a consequence of their realization. But the transmission of knowledge through education is only one—and not necessarily the decisive—type of social interaction that determines the development and diffusion of knowledge in globalization processes. More generally speaking, knowledge does not simply constitute one more aspect of globalization as its precondition or consequence, but represents a critical factor with its own dynamics that shape the interaction of other components of globalization.

One example of the critical role of knowledge, and also of the need for a broader understanding of knowledge, is the challenge of matching new technologies and traditional behavioral patterns. This challenge can hardly be addressed by focusing only on traditional school education. Globalization processes such as the spread of technology or migrations of people obviously presuppose the diffusion of different forms of knowledge—in this case, the knowledge of how to deal with the technology and the knowledge of how to establish a life under new circumstances.[15]

In international developmental aid projects, one can observe how a culturally specific, Western understanding of accountability and moral responsibility is often transformed into a globalized standard and then imposed by the more powerful Western partners in what I have called an extrinsic process of globalization. The root of such norms is ultimately local knowledge emergent from specific experiences in the economy of Western urban societies, which is then extrapolated and imposed—in an extrinsic, politically steered process—onto developing countries.[16]

The assumption that developmental aid, directed, for instance, at improving the water supply in a developing country, should establish waterworks as independent, economically viable units and keep them functioning as essentially self-sufficient enterprises has become part of such a globalized standard. This standard is, however, rarely complemented by an intrinsic development of knowledge about water management in the diverse social settings in which it is imposed.

According to anthropological studies of such an infrastructure transfer in the 1990s, the technological improvement of waterworks in Tanzanian cities through developmental aid was indeed not accompanied by a successful transfer of the organizational structure required to run these waterworks according to globalized standards. But the failure ultimately led to follow-up projects that dealt with this issue in a way that was no longer fully directed by these standards. Instead, the follow-up projects opened up new opportunities for the emergence of local knowledge in response to the challenges of globalization.

Similarly, the knowledge of how to deal with the HIV/AIDS infection was spread by international campaigns in developing countries beginning in the mid-1980s. These campaigns often include supposedly globalized standards on how to forge appropriate and "healthy" ways of dealing with the disease—standards that are actually modeled on specific Western ideas of autonomous, "empowered" individuals.[17] This is, on the one hand, a spread of globalized medical, scientific, and technical knowledge. But it also constitutes a globalization of local knowledge that is anchored in specific, non-universal, Western conceptions of the relation between the individual and society. Now these culture-specific conceptions are imposed on diverse social environments by the extrinsic dynamics of developmental politics, backed by economic and political globalization.

But since such knowledge transfer is conditioned by an interaction between different, encompassing knowledge economies, it does not necessarily lead to changes in social behavior that correspond to the expectations originally associated with it. Even when knowledge about HIV/AIDS becomes widely available in a country through governmental and nongovernmental campaigns, the behavior of local people toward the disease frequently remains largely governed by a traditional knowledge economy shaped, for example, by the dynamics and conflicts of family and clan relationships—and even by mentally assimilating the disease into witchcraft and disorders supposedly caused by the non-observance of ritual prescriptions.

The interaction between knowledge economies ultimately shapes the results of an encounter between globalized and local knowledge, but it may also result in new, locally emerging forms of social behavior responding to challenges (such as those of a disease), and of both the knowledge and technologies for coping with it. The partial disintegration and remodeling of globalized epistemic frameworks in the course of such encounters—between local and global knowledge—can act as a source of innovation in comprehensive globalization processes, including their political and economic dimensions. In chapter 16, I come back to the productive role that local knowledge emerging from globalization (sometimes also called "glocal" knowledge) can have in responding to the challenges of the Anthropocene.

But there is also another, more general aspect of the role of local perspectives that is worth discussing in this context. Resisting the dominance of global epistemic frameworks and of global political and economic regulations by using practices and perspectives that strengthen local or regional forms of sustainability may possibly contribute to stabilizing the entire human-Earth system, that is, the complex system that has resulted from global society having become dynamically coupled to the Earth system. One may ask, for instance, whether islands of local stability in which democratically legitimated institutions regain control over global markets can damp out some of the negative effects of capitalist globalization.[18] They can perhaps help to prevent resonance effects, which result from such globally effective feedback mechanisms as financial or energy crises, or attacks on cybersecurity causing breakdowns that have devastating consequences on a global scale. But while local initiatives alone can hardly address the irreducibly global challenges of the Anthropocene, they may still contribute to connecting these global challenges to local experiences of how to change the world for the better. In chapter 15, I discuss human-Earth interactions more extensively as aspects of a complex dynamical system.

Ancient Globalizations and Their Sediment

As discussed in chapter 8, actions leave sediment that changes the conditions for subsequent actions. This also holds, of course, for the succession of globalization processes in the course of history, resulting in a layered structure of such sediment that shapes the conditions for further globalization processes. It is particularly so for the globalization of knowledge. Globalization is more complex and more ancient than meets the eye when focusing only on the economic processes of recent history. A one-way transfer of knowledge from Europe or other developed regions of the world to other parts of the world—by means of goods, institutions, technologies or knowledge—never took place.[19] Instead, the history of knowledge transfer has always been multidirectional and multilayered. Technology transfer is often more appropriately conceived as a "technological dialogue," as Arnold Pacey has argued.[20] Such dialogues began much earlier than European expansion and remained multidirectional, even during European colonialism and imperialism.

Our situation today results from processes reaching far back into history. In order to illustrate my point, I will therefore discuss a few examples from the ancient world, the spread of knowledge through migrations, the role of multilingualism and writing, the introduction of new means of transport, and some globalization processes that eventually contributed to the success of early modern science.

Long-distance, indeed intercontinental, connections with an attendant spread of knowledge are even older than *Homo sapiens* itself. There is considerable evidence that humans and their close hominid kin moved out of Africa in several waves. More than 1.8 million years ago *Homo erectus* spread from Africa to Eurasia. About three hundred thousand to two hundred thousand years ago *Homo sapiens* emerged in Africa.[21] From there, anatomically modern populations began to migrate, permanently colonizing other continents since at least fifty thousand to sixty thousand years ago, by the beginning of the so-called Upper Paleolithic. Their migrations also depended on the vicissitudes of the unstable climate of this period.[22] Knowledge transfer of intercontinental, pan-Eurasian proportions can be readily documented in prehistoric times. The spread of early humans was concomitant with a spread of knowledge embodied, for instance, in the stone tool technology that led to the creation of a wide range of Upper Paleolithic tool traditions.[23]

Spoken language has always constituted one of the chief means of transmitting knowledge. Of special note are two types of linguistic situations that were as frequent in the ancient world as in the modern: multilingualism and lingua francas. Multilingualism and language contact give rise to phenomena such as linguistic borrowing (where a word from a foreign language becomes integral to the transmission of a foreign concept) and translation (where a text is transferred from one language to another and is inescapably altered—in both form and meaning—in the process). Lingua francas constitute a strategic solution to the problem of linguistic pluralism, in which parties come to adopt a single language (e.g., Sumerian, Akkadian, Aramaic, Greek, Latin, English) as common currency; this language may be the mother tongue of only some of the parties. Typically, lingua francas have emerged in response to the exigencies of trade, but they also play a key role in knowledge (languages of learning), law (diplomatic languages), and religion (sacred languages, *linguae sacrae*). But in becoming a lingua franca, not only does a language change its value (in a social sense), but its terms frequently change their value (in a linguistic sense) as well.[24]

Knowledge also spread with the later expansion of agricultural technologies relating to the domestication of cereals and animals. Intensive gathering of wheat and barley in the Fertile Crescent eventually led to agricultural practices that resulted in the genetic modification of cereals (domestication) about ten thousand years ago. Evidence for the domestication of small livestock (sheep, goats, pigs, cattle) dates this practice to at least 8000 BCE.

Within a few millennia, these agricultural advances together with the domesticated cultivars spread through demic migration (i.e., population diffusion), probably in waves, to southeastern Europe, arriving in Greece by the middle of the seventh millennium BCE, and spreading from there northwestward, arriving around 5600 BCE in the Upper Rhine lowland plain, from where the next wave of settlers then proceeded to the Lower Rhine area.[25] Rice cultivation followed a protracted domestication process completed about 6,500 to 6,000 years ago in China and about two millennia later in India; the related diffusion processes are still under discussion.[26] Summarizing recent research on the spread of Neolithization with regard to the different modalities of diffusion, the archeologist Mehmet Özdoğan remarks, "Analytic or synthetic explicative models such as migration, colonisation, segregated infiltration, the transfer of commodities and of know-how, acculturation, assimilation, and maritime expansion that are seemingly mutually contradictory actually took place simultaneously as distinct modalities."[27] I return to the so-called Neolithic Revolution in chapter 14 as a major step in cultural evolution.

The diffusion of farming knowledge across the Eurasian landmass proceeded in waves of overland expansion by small groups of migrants and the ensuing exposure of hunter-gatherer groups to their technologies. Since the fourth millennium BCE, the domestication of equids (*Equus asinus* and *Equus caballus*) and camelids (*Camelus bactrianus* and *Camelus dromedarius*) increased the ability of disparate groups to communicate with each other over great distances. These transport animals, also used for riding, constituted a new, faster means for the spread of people, goods, as well as knowledge, enabling new waves of migration.[28]

Boats had been in use probably since the times of *Homo erectus*. Maritime travel also enlarged the reach of early farmers. Evidence points to the use of early watercraft by the first Neolithic settlers who arrived in Crete carrying plants and animals around 7000 BCE. Early seafaring was not necessarily limited to coastal sailing. The discovery of banana phytoliths at the site of Munsa (Uganda) in the interior of Africa in contexts some five thousand to six thousand years old—together with the absence of banana at any intervening sites in Southeast Asia, India, or the Arabian Peninsula—suggests that the banana was transported by sea from its origin in Papua New Guinea. Trans–Indian Ocean sailing was evidently a reality at least six thousand years ago. Some 1,500 years later, long-distance sailing between India, southeastern Arabia, and Mesopotamia was becoming routine.

By the end of the fourth millennium, Eurasia was well connected by trade routes running along east-west and north-south axes. These routes allowed for economic, technological, and epistemic interchange.[29] Similar processes took place in the Americas, such as the domestication of plants and animals, sedentariness, and the development of technologies like ceramics and metallurgy, and ultimately even urbanism and writing—but the extent to which these developments were exchanged was more limited. In the Americas, greater geographical obstacles created fundamental limits, impeding long trade routes. The climatic diversity along the north-south axis of the continents limited population contact as well as the transfer of agricultural achievements, as Jared Diamond has argued.[30]

From the third millennium BCE at the latest, the existence of trade routes connecting centers of early urbanization—for instance, Egypt, Mesopotamia, and the Indus Valley—is well documented. Technical innovations such as the development of bronze technology enhanced the need for raw materials like copper and tin, which had to be procured through an extended network of trade routes. Thus, local technical, economic, and political developments and the growth of networks reinforced each other. As discussed in chapters 5 and 10, the technology of writing began to spread, probably from Mesopotamia and Egypt; it may have spread almost immediately to Iran and Syria, then possibly a thousand years later to the Indus civilization, and it may even have influenced, another thousand years later, the development of writing in China. We see an adaptation of the idea of writing to local settings in the spread of cuneiform to Elam or Anatolia (to write Hittite), or the spread of the

11.4. Assyrian scribe writing Akkadian in cuneiform script on a clay tablet next to an Assyrian scribe writing Aramaic in alphabetic script on a piece of papyrus or leather (pergament). Reconstruction of a wall painting from Til Barsip, eighth century BCE. See Cancik-Kirschbaum (2012, 134). Louvre. Photo by Florentina Badalanova Geller.

Phoenician alphabet to Greece. This spread led to enormous possibilities for transmitting knowledge, discussed in the previous chapter.[31]

By the second millennium BCE, large empires had emerged in western and eastern Asia, from the Egyptian Empire in neighboring Africa (via the empires of Mesopotamia and the Hittite Empire in Anatolia) to Shang and later Zhou in China. In the middle of the first millennium, the Persian Empire extended from the Nile to the Indus and constituted an important conduit between western Asian and Indian cultures. The Achaemenid Persian Empire, in fact, encompassed both Mesopotamia and parts of West India and Pakistan, where the easternmost Achaemenid-controlled satrapies were located. This political "umbrella" created the conditions in which knowledge and technology transfer could occur within the boundaries of a single empire.[32]

By the outset of the first millennium BCE, achievements such as agriculture, pottery, architecture with stone and brick, metal working, water management, urbanization and statehood (including warfare), sophisticated administration, writing, literature, art, and the beginnings of science had become the common property of many cultures in the ancient Near East. In spite of extensive warfare and periods of destruction and decline—such as the so-called Dark Ages of ancient Greece from 1200 to 800 BCE—the knowledge behind these achievements was spreading in a way that eventually contributed to their long-term stabilization and further development.[33]

Later, the Hellenistic empires of Alexander the Great and his followers, the Roman Empire, Islamic rule, and the Mongol Empire established extended interaction spheres between different cultures. The impact of such empires on social and cultural connectivity, their territorial expanse, their reliance on extended commercial exchanges, and their continuous struggles with neighbors and nomadic populations supported the spread of knowledge with more or less practical relevance to their functioning. This happened in spite of the fact that empires also made attempts to keep certain knowledge secret, such as smelting technologies among the Hittites, or Greek fire (an incendiary liquid) in the Byzantine Empire.

In summary, over long periods of human history, knowledge was disseminated in connection with migration and trade, as well as with power and belief structures. Knowledge spread across long distances or over vast areas as a by-product of other diffusion processes, for instance, the expansion of empires or the spread of religions. Some of these processes were of a transregional and cross-cultural character, but some of them were more like corridors connecting distant regions by a thin, often indirect and fragile chain of transmission, such as the "Silk Road" network connecting Europe with China.

The Dynamics of Transmission Processes

As knowledge was often but a "fellow traveler" with other globalization processes, participating in their dynamics without governing them, the results of transmission were sometimes only transitory, but a long-lasting sedimentation of at least some achievements nevertheless occurred. This accumulation reached beyond the eras of single empires, since their succession usually involved, even in instances of major destruction, the adoption of preexisting infrastructure, the inclusion of at least some members of the intellectual elite, and a continuity of local technological or cultural achievements. As a result, the historical succession of large empires from the Mesopotamian, via the Persian and the Roman empires, to the Islamic and Mongol empires comprised significant processes of knowledge accumulation. For example, Roman regulations for agriculture found their way into the corpus of Islamic law; similarly, Persian economic achievements served as a model employed by the Arabs in their conquests.[34]

As long as knowledge is but a fellow traveler in the spread of power and belief structures or of migratory or commercial activities, its transmission is governed by extrinsic dynamics. Yet the transmission of knowledge typically also involves intrinsic dynamics, strengthening the significance of knowledge as it proceeds—for instance, by stimulating the creation of new media and institutions for its transmission. The transmission of knowledge via writing, for example, required the transmission of literacy, typically by instruction. As writing spread to different cultures, which spoke different languages or had different writing media available, the technology was adapted to local conditions. The transmission of knowledge in the course of translation movements (such as those into Arabic and Latin discussed below) fostered the creation of new institutions of learning or the strengthening of existing ones. Examples include the foundation of libraries and centers of translation in the early Islamic world and the expansion of the curriculum of medieval universities as they absorbed the knowledge transmitted through the translation of Arabic texts.

Transmission processes in which knowledge spreads as a fellow traveler often increase the significance of knowledge and may actually turn into intentional and directed processes of knowledge transmission. One explanation is the systemic quality of at least part of the transmitted knowledge, with the consequence that one piece of knowledge points to others and hence tends to reconstitute a system. For example, Gerard of Cremona, the famous translator from Arabic, active in twelfth-century Toledo, followed a more or less systematic program shaped by al-Farabi's tenth-century classification of the sciences.[35]

Another reason for the autocatalytic quality of the transmission of knowledge is the empowerment that the gain of knowledge typically signifies for the receiver—or is expected to bring with it. Thus, for example, Jesuit scientists transmitting European astronomical and technological knowledge to China between the sixteenth and the eighteenth centuries attained high positions at the Imperial Court because their knowledge was expected to strengthen the ruling dynasty.

I come back to this particular case in the next chapter, in which I show that knowledge may also become weaker in transmission processes, in particular when larger bodies of knowledge are transmitted between profoundly different knowledge economies. Knowledge may become atomized when spreading primarily as a fellow traveler—in the case of missionary activities, for instance—so that only isolated chunks are transmitted, while being recontextualized within the knowledge economy at the "receiving" or appropriating end.

The conditions that initially favor the transmission process impose constraints that may eventually hinder the transmission of knowledge, in particular when other transmission processes do not follow suit and when the production of new knowledge remains more or less unilateral. The Jesuit mission to China, for instance, was motivated by the quest to spread Christianity using European cultural achievements (scientific and otherwise) to get a foothold at the Chinese court. But, as discussed in the next chapter, the transmitted knowledge changed its character when assimilated to a different knowledge economy; it failed to trigger a major scientific upheaval, let alone a transformation of Chinese society. The initial religious motives driving the transmission process on the European side eventually contributed to its demise, when, in the mid-eighteenth century, the Catholic Church no longer accepted the accommodation policy of Jesuits with regard to Chinese traditional rituals.[36]

Successful transmission processes typically presuppose that originator and receiver already share common aspects of material culture and basic knowledge. Even more important, their sustainability presupposes some degree of reciprocity between originator and receiver. When knowledge economies differ between a receiver and transmitter of different cultures, the transmission of knowledge unavoidably amounts to a transformation of knowledge, since motives and epistemic structures are different in the target culture. As a consequence, the transmitted knowledge is reconstituted in novel ways and governed by the dominant knowledge economy of the target culture, often with repercussions for the originator. Under certain circumstances, the transmission of knowledge may also induce a change in the knowledge economy of the target culture, in particular when its knowledge economy is unstable or in crisis. I return to these points in the next chapter with the example of the transmission of European scientific knowledge to China.

In spite of the transient and even ephemeral quality of knowledge transfer processes in history, they kept large parts of the world connected over long periods of time by the spread of economic, cultural, and religious traditions, through the exchange of technologies, practices, and ideas, or through knowledge encapsulated in writing. This connectivity, however, did not lead to uniformity among the cultures in which knowledge was being produced, disseminated, and appropriated. This is also because of the layered character of globalization processes mentioned above.

Knowledge transmission processes are also layered in the sense that the introduction of a new process (such as the exchange of knowledge through written texts) does not lead to the eclipse of earlier processes (such as the exchange of knowledge by the diffusion of material culture or through interpersonal contacts). In the ancient civilizations, writing remained an elite phenomenon dealing only with certain subjects, while artisanal knowledge continued to be transmitted by interpersonal contact. We are thus invited to consider the history of globalization processes from a perspective similar to that of geology, wherein one must also consider, in principle, any local environment as being shaped by a layered global history. And just as in geology, we should not presume that these strata are always neatly displayed one on top of the other. Rather they may be distorted, squeezed, or even inverted, and may reach the surface in surprising ways.

Traditionally, historians of technology have focused on key innovations and their spread, neglecting the fact that cutting-edge technologies are always just one stratum on top of a complex layered structure that manifests itself at different places in different forms. When considering, for instance, the rise of new forms of mobility such as trains, cars, and airplanes in the nineteenth and twentieth centuries, one should not forget, as David Edgerton has pointed out, the continued and even growing role of horses as a means of transport during industrialization. In Britain, for example, more horses were used for transport in the early twentieth century than ever before or afterward. Horses were also massively involved in military actions in both World Wars. Horses, camels, wooden plows, or handlooms should therefore not be considered simply as technologies of previous historical times, Edgerton argues, but rather as historical substrata of a layered global presence. Investigations into the global history of knowledge should hence take the maintenance and reproduction of these underlying strata over time just as seriously as the more obvious innovations.[37]

As a result of the multilayered structure of the transmission of knowledge, including the sedimentation processes that extend over many generations, even today the global availability of technologies and technological products does not necessarily imply that globalized knowledge systems are uniform. Both global and local conditions may shape them, and knowledge spread can even be blocked, as when Japan isolated itself from foreign influences for more than two centuries.

11.5. Horse-drawn oil tanker wagon, Vancouver, 1911. City of Vancouver Archives.

In general, however, transmission processes cannot be prevented, though they often lead to unexpected or unintended consequences. When, in the middle of the sixteenth century, Bartolomé de las Casas, the first resident bishop of Chiapas, was ordered to prepare a census of New Spain, he gathered the knowledge on which he eventually based his famous report for Charles I of Spain, "A Short Account of the Destruction of the Indies," demonstrating the catastrophic impact that the Spanish colonialization of Mexico had on the indigenous population.[38] I return to the devastating consequence of this particular globalization process at the end of this chapter.

As we can learn from the above-mentioned examples (the transmission of water technology and knowledge about HIV disease to developing countries) the spread of a globalized technological infrastructure or of medical treatments is not necessarily paralleled by the same kind of changes in social structure and the corresponding social knowledge that accompanied the original establishment of these achievements in more technologically advanced societies.

This is so because an epistemic vacuum does not exist: a scientific theory, a mental model, or a conception of the world is, even under adverse conditions, typically not discarded before a new one takes its place. Similarly, when actions must be performed under the conditions of an externally imposed or otherwise heterogeneous knowledge economy—for instance, under colonial occupation or in developmental projects—they are still accompanied by reflections lending them meaning. Transferred knowledge is always matched with local knowledge. This encounter typically triggers transformation processes with novel consequences, occasionally at the level of first-order knowledge and virtually always at the level of second-order knowledge about the meaning of the knowledge transferred, as well as about the cultural and social identities of the carriers.

Globalization as a Condition for
the Emergence of Early Modern Science

The emergence of early modern science in Europe crucially depended on the globalization of knowledge in history, from the global spread of basic cultural techniques such as writing, the diffusion of technologies such as gunpowder and firearms, enhanced geographical knowledge, the worldwide circulation of biological specimens, to the transmission, enrichment, and transformation of Greek science (discussed below).[39] But the crucial contribution of globalization processes to establishing early modern science as a broadly shared societal practice with its own globalization potential may perhaps best be illustrated with two striking and well-known examples: paper and Indo-Arabic numerals.

I briefly mentioned the history of the invention and dissemination of paper as a cheap writing material in the preceding chapter. Paper took on a role that had been played in antiquity by clay and papyrus. It helped to spread literacy, in contrast to the limited literacy in medieval Europe partially caused by the reliance on parchment and skins for writing materials. Papermaking was known by the second century CE in the Han dynasty in China and then traveled to eastern and western Asia, following the Silk Road. By the end of the seventh century, papermaking had reached the Indian subcontinent, and by the middle of the eighth century Samarkand, transmitted by Chinese prisoners to the Abbasid conquerors. Subsequently it spread to the rest of the Islamic world.

The diffusion of paper illustrates the layered structure of globalization processes and also the associated retardation effects mentioned earlier. Paper began to spread from China only after it became widely used. It was first introduced to other areas as a commodity and only reproduced by local technology much later. Paper was known to the Arabs by the seventh century, whereas the technology for manufacturing it arrived more than a century later. By the tenth century, paper had entirely replaced the use of papyrus. Similarly, it was known in Europe no later than the tenth century, whereas paper mills were constructed only about two centuries after its initial introduction.

It took another century before Europeans realized that the Chinese also used paper and much longer before they became aware that it was actually a Chinese invention. The extent to which the connectivity provided by the Islamic Empire accelerated the spread of technological inventions by the Chinese is remarkable when compared with their spread before the rise of Islam. This is also true of another key example of the accumulative potential for the globalization of science: the invention of the Indo-Arabic numeral system, now in universal use.[40]

The oldest numbers written in the place-value system with base ten are found on the Gujarat copperplate inscriptions from about 595 CE. There is, however, textual evidence that the place-value system originated much earlier. At least since the mid-third century, a distinct number system was used in India that associated numbers with physical or religious objects and arranged them in a place-value system. Remarkably, one of the earliest texts testifying to the use of this system is an astrological treatise based on a Greco-Babylonian astrological tradition. The Indian place-value system may thus have roots in the Babylonian tradition, but it may also go back to the use of counting boards with an intrinsic decimal place-value structure as they were used in China, from which they may have been brought by Buddhist pilgrims.

While the autonomous development of the system in India cannot be ruled out, its emergence from transmission and transformation processes of older knowledge brought into new contexts and represented by new media is not entirely unlikely. Among the new media were Indian literary texts that made use of the above-mentioned concrete number system for reasons of style and required synonyms for ordinary number words to preserve the scansion of a verse. Such literary contexts may have hence preceded the use of the new number system for calculation. By the seventh century, the decimal place-value system had reached Syria to the west and Cambodia, Sumatra, and Java to the east. By the late eighth century, the Indo-Arabic numerals were known in the Islamic Empire.

In 773 CE, a group of ambassadors from India visited Baghdad, including a scholar with astronomical and mathematical expertise. One of the first authors of mathematical treatises in Arabic was Al-Khwarizmi, from Chorasmia, who worked in the first half of the ninth century under Calif Al-Mamun in Baghdad at the House of Wisdom. His arithmetical treatise is the first known Arabic work using the Indian decimal-place system. While Indian numerals became known in the West as early as the late tenth century, it was the twelfth-century Latin translation of this treatise that caused them to be widely adopted in western Europe. In the first half of the thirteenth century, easily accessible introductions to calculating with the new number system were written and became adopted as textbooks in the newly founded universities and grammar schools. They also became the basis for widely spread mathematical training in vernacular languages. Without the globalization of paper and Indo-Arabic numerals, but also without other technologies acquired in the course of globalization processes, such as gunpowder and the compass, some of the breakthroughs of early modern science would have been hardly conceivable.

Greek Science as the Result of Cultural Refractions

A striking and important example for the transformation of knowledge by intercultural transfer processes is Greek science. It has often been hailed as part of the "Greek wonder," as if it were a creation out of nothing, but it was actually rooted in a transfer of knowledge from the ancient Near East to the Greek context.

During the first millennium BCE, the globalization of knowledge in the ancient Near East had been an intermittent cumulative exchange and diffusion of local developments within a much wider sphere of circulation, connecting, for instance, the literary world of India with the cuneiform traditions of Mesopotamia or the Anatolian cult of the Great Goddess with Greek religious traditions. Evidently, these exchange processes also involved an active appropriation of the transferred cultural achievements on the part of the "receivers," characterized by those achievements' transformation and recontextualization. This process not only gave rise to cultural hybrids (as has often been emphasized) but also to profoundly new forms of knowledge, such as Greek science.[41]

The genesis of Greek science is to be found in Asia Minor, not far from the cultural centers of Mesopotamia. As a consequence of the transfer of Mesopotamian medical, astronomical, and mathematical knowledge to a different cultural area, that knowledge itself took on another form. While the justification of Mesopotamian or Egyptian scientific knowledge was largely inherent in the institutional and representational structures in which it was generated, it became the subject of explicit normative reasoning in the Greek context, geared to a public discussion of political decisions and their justification.[42] One could characterize this transition as a shift from the practical to the theoretical justification of scientific claims.

In Babylonia and Egypt, mathematical knowledge was, as we have seen in chapters 5 and 10, deeply embedded in the practical knowledge of accounting, planning, and measuring. Through schooling it eventually became somewhat detached from its immediate purposes so that mathematics emerged as an activity in its own right.[43] Since knowledge was embedded in the age-old institutional and practical contexts of Mesopotamian or Egyptian culture, there was, however, simply no motivation to make the reasoning behind certain claims explicit in a way that we recognize today as an argument, "proof," or scientific justification.

A new stage was reached after the transmission of some parts of this knowledge to the new cultural context of sixth-century Greece. Here it was brought into contact with a new cultural technique, that of rhetorical contests in which new methods based on rules and definitions were made to consciously operate within the meaning of statements.[44] As discussed in chapter 5, when these methods were applied to mathematical knowledge, an abstract concept of number emerged and was used in inferences based on these new methods. This way of operating with numbers was no longer closely associated with practical knowledge. On the contrary, in the hand of thinkers like Pythagoras it became an esoteric art completely detached from applications. Such knowledge for its own sake became a pillar of elitist education in the Greek polis, a circumstance that explains why it did not remain a mere episode but became part of a long tradition.

In summary, Greek philosophy and science are inconceivable without the globalization of knowledge processes during the second half of the first millennium BCE. It was not least also a matter of the transmission and appropriation of first-order practical and technical knowledge, of instruments and data, which provided the foundation of new second-order knowledge and thus for the breakthroughs of theoretical reflection in Greek science, still frequently considered to be the true birth of the exact sciences.

The Renaissances of Greek Science

It is telling that Greek science had to be rediscovered so many times, that there were so many "renaissances." Each of them exposed science to a new level of globalization, integrating it with knowledge traditions of other origins. In fact, a rediscovery always constitutes a spoliation, a placing of older knowledge into a completely new context. Greek science was first transformed into Hellenistic science and then, in the eighth and ninth centuries, appropriated and developed in the Arabic-speaking world under the rule of the Umayyad and Abbasid caliphs. When Greek science (or rather its transformed legacy,) was eventually appropriated in early modern Europe, so much had changed in the meantime—notably, the technology of writing had diffused, diversified, and been altered by the new technology of printing—that instead of Greek science being reborn, Greek science contributed to a process that resulted in a new type of science: modern science.[45]

How did the globalization of Greek science happen? In the period following the disintegration of the Western Roman Empire, the wider Mediterranean world remained highly connected, with more distant territories eventually becoming strongly related to it, such as the Arabian Peninsula, the Indian subcontinent, and Central Asia, mainly following the expansion of the early Islamic caliphates. In the centuries often characterized as "post-antiquity," people, material objects, ideas, and knowledge continued to migrate across vast geographical spaces.

As the work of Sonja Brentjes has shown, following up on groundbreaking research by Dimitri Gutas, during the eighth and early ninth centuries, a likely limited number of Middle Persian, Syriac, and Greek texts were translated into Arabic, the content of which seems to have come very close to the spectrum of texts translated in previous centuries into Syriac, Middle Persian, or Armenian. Hence, it seems that the first wave of translations into Arabic exhibited impressive similarities with the earlier and in part parallel translations into these other languages. Only in the later eighth century did things slowly begin to change. Caliph al-Mahdi (ruled 158–68 CE / AH 775–85) opened the scholarly world of his courtly advisers to the Christian communities in Iraq, thus integrating Syriac-speaking scholars into the intellectual, administrative, and political world of the caliphate.

The rise of the Barmakid family from Bactria as viziers and leading administrators of the dynasty brought with it patronage for translations of Indian as well as Greek texts on medicine, astronomy, and mathematics, and the foundation of the first hospital in Baghdad. Translations of Greek philosophical texts of greater number and thematic breadth seem to have been undertaken only in the ninth century. These have traditionally received the most attention, while scholars have neglected the role of previous translations of Middle Persian texts and the activities of Iranian and Syriac scholars at the Abbasid courts of the eighth century. The emphasis on philosophical and scientific texts and pursuits has also tended to obliterate the key role played by astrology and its cultural function as an explanatory theory, as a professional context for scholars, and as a political instrument.

Before continuing to discuss the Arabic appropriation of Greek science, some remarks on translation processes may be useful in order to highlight important general features: Translation processes have always been a prominent medium for the transmission of knowledge, in particular when knowledge is represented in writing. They are always faced with a double challenge: a mapping between two languages, and a mapping between conceptual systems that are not usually coextensive or even compatible with each other. Translation processes involve, in any case, a combination of both linguistic and technical competencies that often require intercultural cooperation. Addressing the second challenge not only requires mastery of the contents, but often also requires new linguistic resources created in the course of the process.

11.6. This stone relief, originating in the Peshawar area of present-day Pakistan, probably shows a scene related to the Greek story of the Trojan horse, possibly adapted to a Buddhist context. It has been dated between the second and third century CE, that is, the time of the Kushan Empire. It thus illustrates a "cultural refraction" typical of the so-called Greco-Buddhism in the Gandhara tradition. See Foucher (1950). Courtesy of the British Museum.

These may range from the creation of a lexicon of technical terms to that of elaborate grammars and other forms of reflection on language.

An example of the creation of new linguistic resources is the spread of Buddhism, which has been studied under the perspective of knowledge transfer processes by the historian of religion Jens Braarvig.[46] This case shows many characteristic features of translation processes and may hence warrant a short digression: On the whole, Buddhism spread rather randomly, even after it had become a state religion in the third century BCE under Aśoka, the ruler of Magadha, who also fostered missionary activities. Some of its conceptual assets, such as the story of the life of Buddha or the notion of hell, may have even spread by way of a highly mediated and indirect transmission to Christianity, detaching their origin from a completely different system of thinking.

Yet when Buddhism first took hold in a new context, for instance, in China, it typically triggered activity toward the acquisition of new texts and new knowledge. In Tibet, the decision of the king to adopt Buddhism in the seventh century CE, followed by the adoption of an appropriate writing system and comprehensively organized translation activities, resulted in a complete transformation of Tibetan culture. As the history of Buddhism illustrates in an impressive way, the transmission of knowledge could become the source of new knowledge about language.

The way such knowledge was created very much depended, however, on the relevant knowledge economies, in particular that of the target area. While in the Tibetan case of translating Buddhist scriptures, systematic aids were created that embodied linguistic meta-knowledge, such knowledge remained largely implicit in the case of translations into Chinese, being represented instead by paradigmatic examples. The very possibility of translation depends on the existence of a partial cultural overlap, that is, of shared elements of material culture and knowledge, of compatible motives, and of the existence or the possibility of creating or exploiting multilingual environments. This brings us back to the transmission and transformation of Greek science, which strongly depended on such communalities.

The Arabic translators of Greek philosophical or scientific texts were not simply dealing with a dead culture but with a living tradition involving active scholars. Knowledge, practices, and perhaps even memories of institutions with roots in the Sassanian Empire, if not the institutions themselves, may well have survived among Christian, Zoroastrian, and other religious communities in Iraq, western Iran, and possibly elsewhere in the East of the Abbasid Caliphate. Multilingualism, which is attested among Arabic, Middle Persian, as well as Greek native speakers, was an important skill for the early translators into Arabic. But in addition, some of the early translators also had at their disposal technical knowledge in astronomy or astrology, arithmetic, logic, philosophy, and alchemy.

Translation processes involve a negotiation between different and often mutually exclusive goals. These goals range from a faithful rendering of the original linguistic structure to the reconstruction of the original content in a new medium. As a consequence, either basic knowledge about the content is transformed and partly lost in translation, or its original linguistic representation is distorted, interfering with the possibility of rendering certain higher-order connotations of the original meaning. These connotations are in fact often represented by semantic links within the wider field of language and hence reside in the context rather than in the text. As a consequence of these considerations, one might claim that a perfect translation is only really possible when it is no longer necessary.

When new terminology is created to faithfully render the "technical" contents of an original source in the target language, this terminology has little chance of resonating in the same way within the semantic context of the new language as the original terminology could within the source language. Conversely, the creation of such new terminology may create new semantic fields, hence effectively changing the target language in a way that is shaped by the transmitted contents.

The transmission of Greek science through Arabic appropriations and translations amounted to a profound transformation of knowledge. Thabit Ibn Qrra, for instance, a scholar active in Baghdad in the second half of the ninth century, reworked, together with his dialogue partners and students, Greek texts on mechanics—among many other subjects in which he was interested.[47] This involved philological and terminological work with textual fragments from Greek and Hellenistic antiquity, in all likelihood the search for further material of Greek provenance, and an exchange with practitioners experienced in building balances and performing practical calculations. They pursued the vision of something like a science of weights, making contact with the widespread practice of constructing and using steelyards (an unequal-arm balance, discussed in chapter 14), while placing their work within an Aristotelian philosophical framework. In the context of extended, seminarlike discussions, they tried to build conceptual bridges between disparate worlds of knowledge.

What Thabit eventually created, as a broker between these different worlds of knowledge, was a work without direct precedent in Greek mechanics. This original

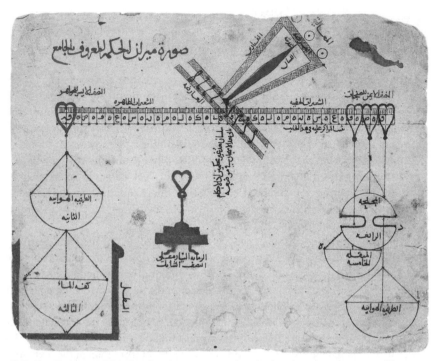

11.7. The "Balance of Wisdom," with its parts as described and explained in a twelfth-century manuscript of a book by 'Abd al-Raḥmān Khāzinī, ca. 1121. MS LJJ 386 (unpaginated), Lawrence J. Schoenberg Collection of Manuscripts, Kislak Center for Special Collections, Rare Books and Manuscripts, University of Pennsylvania.

contribution was shaped by focusing on the steelyard as a challenging object of mechanics, by a compilation of particular elements from ancient texts on the balance, and by reorganizing this ancient knowledge within a seminar context.

The Arabic work on a science of weight was eventually taken up in the Latin West by scholars such as Jordanus de Nemore, one of the most important mathematicians of the Middle Ages, who worked before 1260.[48] He represents for the mathematical sciences a parallel figure to his contemporary Albertus Magnus, who established Aristotelianism as a frame of reference for theological and philosophical discourse, also benefiting from the Arabic-Latin translation movement of the preceding century. Jordanus flourished in a period in which Latin Europe had established its own institutional and intellectual structures capable of absorbing the rich knowledge inherited from the Arabic world. By bringing subjects such as the science of weights into a more rigorous, Euclidean form, he elevated them to the scientific standards of the emerging scholastics, a transformation that did not take place without leaving traces on the contents with which it was concerned. One such trace that later shaped early modern European science was the concept of *gravitas secundum situm* ("positional heaviness"), which Jordanus introduced following his Arabic sources in order to distinguish between a weight and its positional effect.[49] On a lever, for instance, a weight has a larger positional effect the further it is from the fulcrum. Similarly, on an inclined plane, the steeper it is, the greater the positional effect of a weight. Such considerations played an important role in the differentiation of the original concept of weight and the emergence of new concepts such as momentum, torque, and the vector character of forces. This process extended over many centuries during which the transmitted practical and theoretical mechanical knowledge was repeatedly reorganized under different perspectives. These perspectives also carried with them the legacy of their respective cultural contexts, for instance, the disputation culture of Thabit's circle, the scholastic world of Jordanus, or the world of early modern scientist-engineers such as Galileo. The globalization of mechanical knowledge was thus a long-term, path-dependent process in which past historical circumstances left their mark on the emerging scientific concepts.[50] Let us now take a closer look at the overall dynamics of the Arabic-Latin translation movement as an important moment in this long-term development.

Networks of Translation

The Arabic-Latin translation movement amounted to another transformation of knowledge partly rooted in Greek science. In the Latin world, knowledge from the South had been gradually diffusing to the North since the tenth century, with first translations from the Arabic and with instruments such as astrolabes spreading in the Iberian Peninsula. In the eleventh century, this knowledge then moved from monasteries in northern Spain to western France, southern Germany, and Switzerland. The transmitted knowledge comprised practical knowledge about instruments but also astrological, astronomical, and arithmetical knowledge. It contributed to a particular image of knowledge that stimulated the further acquisition of knowledge. This image of knowledge was embodied in the impression of an alien culture in supposed possession of superior and perhaps secret knowledge—knowledge that was believed to potentially constitute a powerful asset for Latin culture as well. In Salerno, not far from the Byzantine settlements in Puglia, Greek texts on medicine also began to be translated into Latin from the eleventh century onward. These earlier translations became an important factor in the knowledge economy of western and central Europe, for instance, in book markets. The earliest translations thus unfolded autocatalytic dynamics, since the products of the translations stimulated further translation activities, culminating in the twelfth and thirteenth centuries.[51]

The Arabic-Latin translation process was the result of a network of activities.[52] (In chapter 13, I come back to the role of networks in a more systematic way.) Hubs of this network typically emerged at the boundaries between the Roman Catholic world and the Islamic and Byzantine Empires and in areas that were multicultural and multilingual, such as Spain and southern Italy, particularly Sicily. In the twelfth century, Palermo, on the outskirts of Europe, became a meeting point for Latin and Arabic scholars who generated translations as well as new joint contributions.[53] In the twelfth and thirteenth centuries, a network of scholarly migrations began to develop between these centers at the boundaries, and now also included urban centers that sustained scholarship in the heart of Europe, such as Oxford and Paris.

Soon after the conquest in 1085, Toledo had become a hub of the translation network because it offered room for contact and cross-cultural collaboration. The existence of such boundary spaces with comparatively high political and religious tolerance was an important condition favorable to the transmission of knowledge across cultural borders. Some of the translators were self-appointed men, mostly from the lower clergy, who traveled from different parts of Europe to the emerging translation centers in an effort to gather new knowledge from their translation activities.[54] But in their work, they mostly relied on local people familiar with Arabic and the local Latin dialects, as well as with the contents of the texts to be translated. As in the case of the Greek-Arabic translation process, multilingual competence was crucial. It was offered mostly by Mozarabs, that is, Arabicized Christian inhabitants of the Iberian Peninsula, and Jews, who shared Arabic culture and education. The various translators often depended on Jewish scholars familiar with the contents of the texts. The Jewish elites of Toledo also commissioned the reproduction and distribution of scientific texts from Arabic, partly performed by prisoners of war.[55]

The transmission and translation process from Arabic to Latin, as well as to other languages such as Hebrew or the vernacular languages, was driven by a number of local and global motives. Among the more local driving forces was the culture of the Mozarabs, who not only brought their multilingual competence to the process, but also their interest in preserving their own culture. Among the more global motives was a widespread interest in magical, astrological, and divinatory knowledge in Catholic Europe. Most extant translations refer to such themes, which were apparently central to twelfth- and perhaps even thirteenth-century translation activities.[56]

The translation efforts set in motion a self-accelerating process of knowledge acquisition. Triggered by the systemic character and the images of knowledge attendant with some of the translated knowledge (e.g., philosophical systems or classification schemes of the sciences), the missing pieces were sought for. Eventually, adjustments of the receiving knowledge economy took place, including the development of the medieval universities, with the consequence that the transmitted knowledge was reproduced and extended in a new context. The rise of scholasticism from the second half of the eleventh century as a Europe-wide network of higher learning and scholarly exchange also depended on the stimulus of the knowledge appropriated from the Arabic-speaking world.

Iberian Globalization

With the European voyages to the Americas and the circumnavigation of Africa, a truly global interaction sphere emerged at the end of the fifteenth century, dominated by European, and at first primarily Spanish and Portuguese, conquest and colonialization. It was later extended by Dutch, British, and French commercial and colonial interventions that not only enhanced the global circulation of people, goods, technologies, and knowledge. Together, they also challenged existing knowledge systems all over the world.[57]

The impact on the history of knowledge of what one may call the "Columbian turn" was, however, even more profound because of the immense consequences of

11.8. Mongols besieging Baghdad in 1258. Rašīd al-Dīn Fazl-ullāh Hamadānī. *Ǧāmi' al-tavārīḫ.*
Rašīd al-Dīn Fazl-ullāh Hamadānī, ca. 1430–34. Supplément persan 1113, fols. 180–81. Bibliothèque
nationale de France, Département des manuscrits, Division orientale.

Christopher Columbus's voyage of 1492. It included the extermination of millions of
people in the Americas and the rise of the transatlantic slave trade, the ensuing breakdown of indigenous knowledge traditions, and dramatic changes of the very world
that had become the object of knowledge (even before the advent of industrialization)
by the global spread of plants, animals, and diseases through human intervention.[58]

The colonization of the Americas was itself an outcome of globalization processes.
In Eurasia, the Mongol conquests of the thirteenth century had their own devastating consequences on the preservation and transmission of knowledge from earlier
historical periods. Entire knowledge economies reaching back to the ancient world—
such as the irrigation culture of Mesopotamia—vanished because of the destruction
of the infrastructure and the loss of lives, in particular during the 1258 siege of Baghdad. The important library there was also destroyed, many of its treasures lost forever. At the same time, the Mongols created a Eurasian interaction sphere reaching
from the Far East to central Europe, creating novel opportunities for the mobility
of people, technology, goods, and knowledge, but also for diseases such as the Bu-
bonic plague, which in the fourteenth century killed an estimated fifty million people
in Asia, Africa, and Europe.[59]

The enhanced connectivity of Eurasia offered Europeans, on the other hand, new
opportunities for exploration and commerce with China, India, and other parts of
Asia, as may be illustrated by the reports of European travelers, such as William of
Rubruck or Marco Polo, who benefited from the Pax Mongolica of the second half
of the thirteenth century.[60] Owing to the rise of the Ottoman Empire in the fourteenth

and fifteenth centuries, its conquest of Constantinople in 1453, and its control of land routes to southern and eastern Asia, which had been vital to the spice trade of the Italian city-states, the search for a sea route to India became a growing concern and an opportunity for the Atlantic powers, Spain and Portugal. The European capability of sending exploratory and military expeditions overseas was, as I have stressed, built on scientific and technological advances that had resulted, in turn, from globalization processes of knowledge, including the spread of geographical knowledge and mapmaking, of navigational techniques such as the use of the compass, and of military technologies such as the use of gunpowder for cannons and guns.[61]

In the thirteenth century, knowledge of gunpowder, which had been invented in China during the ninth century, spread across Eurasia, being employed, for example, by Islamic armies and the Mongols in the form of gunpowder bombs thrown from catapults.[62] The first reports about gunpowder arrived in Europe in the middle of the thirteenth century. The development of the canon also goes back to China, where precursors are found as early as the twelfth century. In Europe and the Islamic empires, canon technology began to spread from the beginning of the fourteenth century.

The Iberian globalization was special in that it affected not only human affairs, but, in a sense, also the planet at large—at least its biosphere, on which it had an unforeseeable impact.[63] It connected biota that had been separated for millions of years and dramatically changed their ecosystems. Animals, plants, and pathogens circulated between the Old and the New Worlds with devastating consequences, creating the need to establish new ecological balances. At the time of Columbus's arrival, the humans inhabiting the New World counted in several tens of millions, of which the large majority perished as a result of contact with the Old World. Old World animals such as horses and cattle thrived in the New World, as did the new settlers depending on them and the invaders who led the way.

The diverse biological heritage of the two worlds favored the new arrivals who had coevolved with the domesticated animals and pathogens they brought along and had developed resistances to the diseases that rapidly killed a large fraction of the indigenous population.[64] In the struggle for survival and domination that unfolded in the New World, this biological legacy was one of the greatest assets of the new arrivals. It is at least questionable whether their technological and military superiority alone would have sufficed to shift the balance of power so drastically to their advantage in an environment alien to them and given their initially small numbers.[65] Rather quickly, however, the European conquerors and settlers came to dominate the entire American double continent, in spite of its conspicuous population and some highly sophisticated civilizations, extracting its riches and turning them into fuel for what came to be a world market. The superiority allowing Europeans to dominate the emerging world system of the sixteenth and seventeenth centuries was therefore not so much a precondition of the conquest of the Americas but, to some extent, also its consequence.

The colonization of the New World eventually turned an accidental biological circumstance into a global political and economic advantage for Europeans, boosting their extractive economies as well as the hegemony of their knowledge economies to a global scale. A significant contribution to this development was the merging of the

11.9. From Theodor de Bry's *Das vierdte Buch von der Neuwen Welt* (1613), which gives an account of Spanish intrusions into several South American regions. This drawing (plate 18) depicts a Spanish military excursion into the hinterland of Cartagena de las Indias (now Colombia). De Bry's account makes use of several sources, such as those of Girolamo Benzoni, and uses them to portray to a broader public this dark account of how the Amerindian population was abused by the Spanish. Internet Archive.

agricultural systems of the Old and New Worlds, with white and sweet potatoes, maize, and cassava eventually becoming staple foods in Europe, Asia, and Africa, while wheat and animal protein from cattle, sheep, and pigs became important foodstuffs in the Americas. Rice had, by 1700, become a basic provision for the slaves carried from West Africa to the Americas.[66] The exchange of crops and domesticated animals amounted to a globalization of the underlying systems of agricultural knowledge and became an important factor in the long-term growth of the world population between 1750 and 1930, again with a clear bias in favor of people with European roots.

The Iberian globalization also concerned other forms of knowledge, including scientific knowledge, which was being transformed in the process. In the sixteenth century, religious orders founded colleges, some of which grew into universities, in numerous places, including Santo Domingo, Lima, Mexico City, Bogotá, and Quito.[67] They produced knowledge of a global character from the start, as knowledge circulated between different localities across the colonial empires.[68] This worldwide flux of knowledge was governed by a complex knowledge economy comprising colonial and ecclesial institutions, as well as numerous local infrastructures still relying on indigenous traditions. One of its characteristics was that considerable amounts of the knowledge accumulated in this circulation remained unpublished, censored, kept secret, and in any case inaccessible to the larger public, being stored in state or church archives, libraries, or personal collections.

Within this colonial knowledge economy, knowledge was, in the beginning, primarily systematized by European religious, philosophical, and scientific frameworks and traditions, such as those of Aristotelian scholasticism, natural history, or Galenic medicine; but it became increasingly diversified by the vast amount of new experiences and challenges encountered in the process of globalization, for instance, hitherto unknown plants and animals or indigenous medical practices. It affected the subsequent transformation of European knowledge economies and systems, from scholasticism to the modern system of scientific disciplines, which, in turn, left its mark on the organization of knowledge in the colonial world. The later transformation of natural history into the modern earth and life sciences is, in any case, hardly conceivable without the knowledge made available by the Iberian globalization. It enabled, for instance, the exploratory voyages of Louis Antoine Bougainville, James Cook, and Georg Forster, as well as the syntheses of Carl Linnaeus, Alexander von Humboldt, and Charles Darwin.[69]

In the late seventeenth and eighteenth centuries, the European Enlightenment touched the colonized American and Asian territories, stimulating the emergence of new institutions of scientific knowledge as well as new images of science related to societal and economic progress. The spread of education, and in particular of Enlightenment ideas and scientific knowledge, was, however, restricted and often repressed by the colonial authorities. Even in the late eighteenth century, for instance, American and Spanish residents of Cuba were prohibited by law from studying or writing on subjects related to the colonies.[70] Meanwhile, however, Spanish colonial domination in the Americas had been shattered by wars and revolutions in Europe and North America. Following of the Declaration of Independence of the British colonies in 1776, the French Revolution of 1789, the Haitian revolution of 1791, and the Napoleonic wars in Spain, the early nineteenth century saw Latin American countries liberating themselves from the colonial yoke. They gained independence as nation-states, adopting and transforming European models.

In the beginning of this chapter, I referred to intrinsic and extrinsic processes in the globalization of knowledge and the crucial role of their interplay. The global sphere of scientific and technological interaction emerging in the wake of the colonization of the Americas was clearly a result of both. It had been fostered by the accumulation of globalized knowledge by the European powers, which then spread as a fellow traveler through conquest and colonization. In the next chapter, I discuss the fellow traveling of scientific and technological knowledge and its effects through the example of the Jesuit mission to China.

The global circulation of knowledge enabled and facilitated subsequent globalization processes. It catalyzed the establishment of a world market by acting as a backdrop and amplifier for different forms of protoindustrialization and industrialization—primarily but not only in the European centers. It included the worldwide circulation of technologies employed in the extraction of resources and

11.10. Oil painting by Eduard Ender (1822–1883) depicting Alexander von Humboldt (seated) and Aimé Bonpland by the Orinoco River (during the expedition in 1799–1800 to Venezuela). © akg-images.

the exploitation of human labor. An example is the invention and spread of the amalgamation process that uses mercury for the extraction of silver from ore, which turned out to be crucial for the rapid expansion of silver production in Latin America after the mid-sixteenth century. The expansion of mining activities and other components of the colonial economy (such as sugar plantations and other monocultures) brought with it environmental degradations such as the spread of toxic mercury deposits and deforestation, marking clear steps into the Anthropocene.

The colonial empires further hastened globalization processes by laying the groundwork for a knowledge economy that enabled the collection of global data, both on natural phenomena—meteorology or ocean currents, for example—and on human populations. But this knowledge economy also opened up perspectives on the global change process itself, including the role of humans as major drivers of environmental transformation. In the late eighteenth and early nineteenth centuries, these novel, global perspectives were articulated by a number of protagonists, among them the mining expert Alexander von Humboldt—who had the privilege and the capacity of seeing more of this world than most of his contemporaries.[71] In 1865, Charles Darwin still remembered how Humboldt's travel reports had influenced him: "I have always thought that Journals of this nature do considerable good by advancing the taste for Natural history; I know in my own case that nothing ever stimulated my zeal so much as reading Humboldt's Personal Narrative."[72]

Chapter 12

The Multiple Origins of the Natural Sciences

In ancient times Yi originated the bow, Chu armour, Hsi Chung the carriage, Ch'iao Ch'ui the boat. Does it follow that the armourers and the wheelwrights of today are all gentlemen, and the four inventors were all vulgar men? Moreover anything that you follow must have been originated by someone, so that on your own showing you follow in everything the way of vulgar men.

> —MOZI, in response to the Confucian precept: "The gentleman follows and does not originate." (Quoted in Angus Graham, *Later Mohist Logic, Ethics and Science*)

From Galileo I desire so much, as I have also written on other occasions, the calculation of the eclipses, in particular of the solar ones, which can be known on the basis of his many observations. All of this is in fact extremely useful for us for the correction of the calendar. And if there is any writing on which we can rely in order to avoid being expelled from all parts of the Empire, then it is only this one.[1]

> —JOHANNES SCHRECK. (Quoted in Isaia Iannaccone, *Johann Schreck Terrentius: Le scienze rinascimentali e lo spirito dell'Accademia dei Lincei nella Cina dei Ming*)

The Need for an Evolutionary Perspective

How does a global history of knowledge affect our view of science? In the preceding chapter, I emphasized the role of globalization processes for the emergence of modern science but still focused on the well-known example of the transformations in Greek science as an important source for much of what followed. But how many times have the natural sciences actually emerged in history? Did it only occur once in ancient Greece, as has been assumed in the grand narrative of European science? Or do its roots lie in various cultural and historical contexts? But one may also wonder whether such questions make sense at all, as it has proven difficult to formulate universal demarcation criteria to draw a distinction between science and other approaches to studying the natural world. What kind of insights could one expect from answers to the question, "How many times have the natural sciences emerged?" For instance, what was their contribution to the "Great Divergence,"[2] that is, to the emergence of the Western world as the most powerful civilization in the nineteenth century?[3]

In the following, I address this question on two levels: the level of scientific practice in the sense of a particular knowledge economy characteristic of science, related to the exploration of material culture for the sake of knowledge acquisition, and the level of the interaction between such a particular knowledge economy and the economy of a society at large. Regarding the first level, I argue that there have indeed been several origins of the natural sciences in the course of human history, and I specifically address the cases of ancient Greece and China as examples.

Concerning the second level, I argue, on the background of the preceding two chapters, that the economy of scientific knowledge became entangled with the knowledge and material economies of societies in a protracted global process. In this sense, the emergence of the natural sciences as major components and drivers of other societal and, in particular, economic developments is a virtually unique world historical event. This is due also to the powerful global impact that these developments had within a relatively short historical time span, diminishing or even eradicating opportunities for the growth of alternatives.

In chapter 10, we saw the long slope of increasing entanglement between science and society with a focus on European developments. In chapter 11, I emphasized the global interdependencies and consequences of this process, specifically in rapidly fostering Western domination in the age of colonization. That this process culminated in particular places at a particular time, shifting the balance of power in the way that it did, is of course as much an outcome of such global dynamics as it is the result of the numerous contingencies to which I have alluded, including the clash of ecologies as a consequence of the European conquest of the Americas.

This is what makes the question, "How many times have the natural sciences emerged?" so attractive, yet so intractable. The peculiar character of this question may become clearer if one asks a structurally similar question, but one that constitutes a much more prominent topic of discussion than the question concerning the plural origins of the natural sciences: "How many times has the human species emerged?"

"What is man?" was already a topic of discussion in the time of the ancient Greek philosophers. Diogenes Laertius describes Diogenes of Sinope's mocking reaction to Plato's answer to this question: Plato defined man thus: " 'Man is a two-footed, featherless animal.' He was much praised for the definition. So Diogenes plucked a cock and brought it into Plato's school, and said: 'This is Plato's man.' On this account, an addition was made to the definition: 'With broad nails.' "[4]

As with answers to the question of the emergence of the natural sciences, answers to the question concerning the emergence of the human species were informed for ages by concepts of origin—on the one hand, of course, by religious concepts of origin (sometimes echoed in the origin myths of the natural sciences and their heroes) but, on the other hand, by purportedly scientific claims regarding the allegedly progressive character of biological evolution. Darwin, however, saw the relation between the human species and other forms of life as one of continuity, a continuity without which an answer to the question of the emergence of the human species can

12.1. Plato's man: a featherless cock. Drawing by Laurent Taudin.

indeed only be found in origin myths.[5] By embedding the history of science within a broader history of knowledge, we create the possibility of establishing a similar continuity for our purposes—a continuity that might release us from the necessity of a clear-cut origin.

Against the backdrop of evolutionary concepts, it becomes particularly apparent why there can no more be a unique criterion for drawing a distinction between scientific and nonscientific knowledge than there can be a single characteristic that distinguishes human beings from animals. It is certainly not the "broad nails" from Diogenes's anecdote; we cannot draw the distinction by means of a single trait, nor our ability to learn, our use of tools, our language, our empathy, or our proclivity for violence. Many of these features also exist in the animal kingdom in one incipient form or another.

Recent studies on human evolution have shown that not only was there a lengthy transition period in the transformation of animal to human but that the genus *homo* itself had many more branches than suggested by the traditional idea of evolutionary "progress" toward modernity. For long periods of time, there were various hominids either competing with each other or living in isolation from one another. Concerning their individual representatives, it is sometimes not so easy to say what may have distinguished their cognitive capacities from those of modern humans. We also know that there were crosslinks among human ancestors, who interbred and thus exchanged genes, as has recently been demonstrated. They possibly also learned from

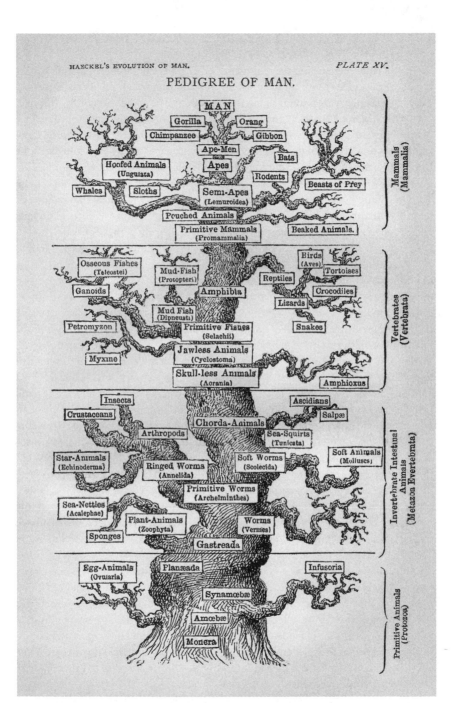

PEDIGREE OF MAN.

12.2. Ernst Haeckel's "tree of life," Darwin's metaphorical description of the pattern of universal common descent. From Haeckel's *The Evolution of Man* (1879). Wikimedia Commons.

encounters with one another, and they indirectly influenced one another by occupying certain ecological niches and through migration.[6]

Furthermore, even after the emergence of anatomically modern humans, there is no striking evidence for at least one hundred thousand years of any characteristically modern behavior distinguishing these humans from other hominids. This time lag between biological and cultural evolution, which has recently even been extended by the finding of *sapiens*-like fossils in Morocco dating from over three hundred thousand years ago, has been described as the "sapient paradox" by archeologist Colin Renfrew.[7] Consequently, the outcome of this evolutionary process cannot be adequately described by claiming that one species of humankind (namely, ours) managed to triumph over others, owing to whatever superior biological characteristics, and that the others rightfully perished because of their inferiority. After all, recent genetic analysis has shown that some of the genetic legacy of these perished breeds (in particular of Neanderthals and Denisovans) is still present in modern humans.

What ultimately prevailed in the course of this long and multibranched evolution of hominids was not necessarily a superior type of human but a new type of process, which molded the evolutionary process of natural selection and to some extent pushed it to the background.[8] This new process was that of cultural evolution, essentially based on the transmission of a material culture and, at the same time, constituting an evolution of the knowledge retained and embodied therein. Can we imagine this evolution moving one step further, to a point at which we may speak of scientific knowledge and find an answer to the question of the plurality of its origins?

Considered from such an evolutionary perspective, different "origins" of the natural sciences are to be expected. Yet these should not be conceived as isolated events competing with one another but rather as belonging to trajectories, all of which contributed to the emergence of the natural sciences as addressed by the question posed at the beginning of this chapter. If we fail to consider the global history of knowledge, the question of the emergence of the natural sciences will be distorted by origin myths from the very outset. Without examining the processes of reflection by which, in various historical situations, knowledge has been transformed or not transformed into new forms designated as "scientific knowledge"—including considerations of missed opportunities, acts of repression, or potential alternatives—the historical memory (or rather, anamnesis) can only generate origin myths.

A comparison of European and Chinese science will help to address and clarify two points relevant to the question of the emergence of the natural sciences. First, over the course of history, the natural sciences (i.e., societal practice directed at producing knowledge by reflecting in a sustained and systematic way on human material interventions in nature and art) emerged independently more than once. Reflective thought processes may take numerous forms and yield distinct and even singular outcomes that cannot be fit into a logic of linear progress terminating with the modern Western sciences.

Second, the integration of the natural sciences within the larger epistemic and material economies was the result of a long-term process that, in the early modern period, began to have a major and ultimately global impact on societal developments, becoming constitutive for this development on a global scale, at the latest with the Second

Industrial Revolution. The way this globalization unfolded in different places at different times very much depended, however, on the local economies of knowledge—which, of course, also changed in the process. In the final part of this book, I come back to the constitutive role of scientific knowledge for global society, marking the onset of what I call "epistemic evolution" as a new stage of cultural evolution.

Early Texts on Mechanics in Greece and China

The oldest scientific work on mechanics in the European tradition is a text titled *Mechanical Problems*,[9] which can be approximately dated to the period between 330 and 270 BCE. It has been traditionally attributed to Aristotle and originated, in any case, in his school, probably during his lifetime.[10] The bulk of Aristotle's writings had become known to the Latin West, partly through transmission by Arabic translations, by the beginning of the thirteenth century. Aristotelian mechanics was also transmitted to the Arabic-speaking world, but reached the Latin West through manuscripts originating in tenth-century Byzantium. The *Mechanical Problems* became widely known only after its first printing as part of the *Corpus Aristotelicum* by Aldus Manutius in Venice in 1497. Because of its concern with technical devices, it raised great interest among early modern scientist-engineers and became one of the most influential works about ancient mechanics of the period.

The text consists of an introduction and thirty-five sections called "problems," which are often merely one paragraph in length and almost always begin with the phrase, "Why is it that . . . ?" The treatise is centered on the question: "Why is it that small forces can move great weights by means of a lever?"[11] The literary form of the *Mechanical Problems* reflects the Greek *problemata* tradition, which probably emerged from a real dialogical situation. Obviously, the author addresses the topic of mechanical instruments primarily because it constitutes a provocation to the Aristotelian system of natural philosophy. In fact, the text addresses the challenges to Aristotelian physics represented by technical devices that produce effects that seem contrary to nature. Mechanical devices appeared to produce such effects because of their ability to overcome a seemingly basic principle of physics: the equivalence of cause and effect. How can it be that, with the help of a mechanical device such as a lever, a small force suffices to lift a large weight?

In the context of the project of systematizing various areas of knowledge, Aristotle—or a member of his Peripatetic school—analyzed this challenging problem. Particular attention was paid to lifting practices, with their techniques for *outwitting* nature so as to raise greater loads with smaller forces, and to weighing practices, with their techniques of establishing *equality* of weights. Was it conceivable that, because balances always imply equivalence, such balances held the key to understanding how the mysterious capability of the lever could be reconciled with the equivalence of cause and effect? The reasoning of the text is essentially based on this intuition.

The argument is shaped by a particular mental model: the balance-lever model. This mental model of practical knowledge resulted from a combination of two familiar mental models first encountered in chapter 6: the equilibrium model associated with the ordinary balance with equal arms, and the lever model, which is a specific version of the *mechanae* model. The two models could be integrated because there happened to be a device to which both were applicable: the unequal-arms balance, relatively novel at the time. In chapter 14, I say more about its invention and the history of weighing practice as an example of technological evolution. For our current purposes, it is sufficient to realize that it was possible to combine two mental models of practical knowledge because of this novel device, which simultaneously functioned as a lever and as a balance. In the case of an equal-arms balance, weight differences are balanced by weights; in the case of an unequal-arms balance, they are balanced by changing the position of the counterweight along the scale or, as described in the Aristotelian text, by fixing the counterweight at the end of the beam and changing the position of the suspension point.

This led to an insight into the equilibrium model, namely, that weight can be compensated not only by weight but also by distances. By equating the lever with the beam of a balance, this device made it possible to understand the mysterious force-saving effect of a lever as an equivalence of cause and effect, achieved by a compensation of differences in force by differences in length. It was this practical knowledge related to the unequal-arms balance that provided the empirical basis for the formulation of the law of the lever.

While the *Mechanical Problems* thus reflects the basic knowledge of practitioners, the text is clearly not motivated by practical concerns. The development of a theoretical science of mechanics would not have been possible without a preexisting theoretical tradition concerned with the explanation of natural phenomena. The emergence of mechanics, the realization of the theoretical possibilities of the lever-turned-balance, took place in a particular cultural setting in which theoretical reflection was already pursued in a number of other venues of experience and in which there existed a strong culture of disputation and persuasion by justification in the form of arguments. The political and juridical customs of the Greeks (at least in Athens) made the success of projects depend on the effective persuasion of others by rhetoric and argument.

The pioneering role of the Aristotelian text on mechanics has been compared, in joint work with Matthias Schemmel and William Boltz, with the earliest text on mechanics from the Chinese tradition, a couple of sections from the so-called *Mohist Canon*, dating from about 300 BCE.[12] The Mohists may be described as a philosophical school of the fifth to third centuries BCE, a time of competing philosophical schools in a fragmented political landscape, but they were actually more than that: they were a community held together by a commitment to rational debate, quasi-religious beliefs, and a particular form of political engagement. While we know little about Mozi, the legendary founder of the school, some key texts of his school have survived, albeit in mutilated form. The Mohists criticized traditional doctrines (such as those of Confucius) on the basis of logic and attempted to set up a system supporting rational argumentation. They adhered to a principle of universal love and rejected military aggression—though they were also experts in military defense technologies, to which part of the extant writings are dedicated.

The text of interest to us has an intricate internal structure, resembling a hypertext with its rich internal references.[13] It deals with logic, ethics, and science, including geometry, optics, and mechanics. The text suffered a most dramatic transmission history after the disappearance of the Mohist school (most probably related to the reunification of China under the Qin dynasty in 221 BCE), the subsequent suppression of philosophical schools, and the establishment of a stable political order under the Han dynasty after 206 BCE, supported by its adoption of a Confucian ideology. One problem with the text's transmission was the fact that it had apparently required a teacher to explain its sophisticated order and peculiar terminology. After the interruption of oral transmission, the text remained an enigmatic remnant of the past that could only be restored to its original meaning through a sustained philological effort.

The preparation of a text for the authoritative Han Imperial Library in the last century BCE exacerbated the situation. This redaction seriously mixed up the text's internal structure, rendering it essentially unintelligible. During the Sui dynasty (i.e., in the period between the late sixth and early seventh centuries CE) a very reduced version of the Mohist corpus began to circulate and all but replaced the full version. The full text survived, however, in the Imperial Library at least until the time of the Sung dynasty, which reigned from the tenth to the thirteenth century. It was printed in this period as part of the more than one thousand titles of the so-called Taoist Patrology, which also includes the works of other unorthodox thinkers who had gained some respect among the Taoists. While the full text was basically ignored for almost a thousand years, it began to circulate again in the mid-sixteenth century, just before European missionaries established contacts with Chinese officials and scholars. In the nineteenth century, Chinese scholars realized the striking parallels between the *Mohist Canon* and Western scientific thinking, reviving the interest in a lost tradition of knowledge.

The intricate structure of the *Mohist Canon* has been reconstructed by the Sinologist Angus Graham, who based his work on earlier achievements of mainly Chinese philologists. It contains "sections," which are constituted by a "canon" and an "explanation" referring to it. In the canon, certain basic terms are defined; in the explanation, complex problems are treated. The sections cover four branches of knowledge. The first branch may be called logic, though it is not a logic of syllogisms but rather a reflection on language offering procedures for consistent description in order to avoid paradoxes. The second branch is on ethics. The third branch, which interests us here, is concerned with science. It comprises the sections on mechanics, while the fourth branch deals with the art of disputation.[14]

The text represents a striking parallel to Aristotelian physics in the West. It states, for instance:

> The beam: If you add a **weight** to its [i.e., the beam's] one side then [this side] will necessarily hang down. This is due to the **effectiveness [of the weight]** and the **weight** matching each other.

> Level [both sides] up with each other, then the base is short and the tip is long. Add equal **weights** to both sides, then the tip will necessarily go down. This is due to the tip having *gained* **effectiveness [of the weight]**.[15]

The central question addressed in these sections is: How can it be that one and the same heavy body has, under certain circumstances, a different effect from the one it normally has? It is answered by introducing a pair of abstract terms, "weight" and "effectiveness," that differentiate the term "weight" in order to account for its occasionally inconsistent behavior; in other words, the natural behavior of an object is confronted with the modified behavior it displays under certain artificial circumstances. We encountered a similar issue in the previous chapter in connection with the transformation of Greek science through its appropriation by scholars from the Arabic and Latin Middle Ages. In this process, the concept of positional heaviness emerged, allowing observers to distinguish between a weight and its positional effect. The ancient Chinese text evidently addresses a similar problem.

On the basis of the intuitive mental model according to which a greater force has a greater effect, the structure of this reasoning may be understood as follows: The model implies that equal weights have an equal effect. But practical experience gained in handling mechanical devices may violate this model, for example, when it turns out that equal weights laid on a beam have different effects because of their different positions. The Mohists' theoretical reflection now aims at resolving this apparent conflict between practical knowledge about mechanical arrangements and the intuitive expectation about what would occur naturally (i.e., without a mechanical device). The basic idea is that, by differentiating the notion of cause with regard to its "effectiveness," which is the new concept introduced here, the original mental model can be restored and enriched.

The *Mohist Canon* describes several mechanical arrangements that produce circumstances in seeming conflict with expectations of what would naturally occur. These arrangements cannot always be reconstructed, and sometimes remain obscure. But their significance always lies in their interference with the anticipated course of things. In short, they typically yield a puzzling outcome. We thus encounter, for instance, a beam that does not bend although it is burdened with a weight, or something that leans and cannot be set upright. The natural tendency of a weight would be to move down vertically by itself, but interference via an artifice, such as pulling a weight up or to the side or supporting it from below, prevents this from happening in the contrived circumstances introduced in the text.

This is not yet a quantitative explanation of the type we encounter later in Archimedes's works, but it resembles the scheme used in the *Mechanical Problems* attributed to Aristotle. Both texts, the *Mohist Canon* and *Mechanical Problems*, depict mechanical effects as surprising and in need of explanation, as though they were paradoxical. Subsequently, this apparent paradox is resolved by introducing consistent technical terminology. The *Mohist Canon* also addresses optical problems according to this scheme.[16]

As was the case for the Aristotelian *Mechanical Problems*, the arguments by which these puzzles are addressed and resolved in the *Mohist Canon* can be understood as resulting from a reflection on shared practical knowledge in the context of a culture of disputation. The artifices that pop up in the mechanical sections represent mechanical devices that evidently played a role in contemporary technology—for example, in the techniques of military engineering, in which the Mohists are believed to have excelled.

While the knowledge documented by these texts thus has a practical background, the issues they raise are clearly not practical problems but a matter of theoretical reflection, drawing on the means offered by contemporary philosophical discussion. The Mohists apparently opposed the derivation of ethics from natural tendencies, as was advocated by the Confucian tradition. They rather strove to demonstrate that ethics could consistently be grounded in the sphere of human intentions and actions. In a similar way, the Mohists' occupation with mechanical problems appears to have been motivated by the philosophical concern to show that, while mechanical processes induced by humans may not occur as intuitively expected, they still remain rationally comprehensible.

In China as well as in Europe, neither the existence of a culture of disputation nor its specific concerns had, in the first place, anything to do with practical mechanical knowledge. In this sense, the emergence of theoretical mechanical knowledge was a contingent historical event that was dependent on specific cultural circumstances. As it turned out, however, the similar discursive practices in both cultures shaped reflection on practical mechanical knowledge in a similar, albeit nonidentical way. It resulted, in fact, in different abstract conceptions, such as the concept of the effectiveness of a weight in the Chinese context and the law of the lever or the concept of center of gravity in the European tradition. As I have mentioned above, in the Western tradition, a concept corresponding to the Mohist notion of "effectiveness" played an important role only later in the form of the concept of positional weight.

Both in Europe and in China, the specific character of theoretical mechanics was originally as transient as the historical context that brought it about. What remained as the substance of scientific mechanics in the long run was not a specific literary form shaped by this context but rather, first of all, the mental models of intuitive and practical knowledge that remained stable because of the continuity of craftsmanship and engineering, and second, the mental models of theoretical knowledge and the abstract concepts associated with them—at least as long as they were handed down in a theoretical tradition dependent on the transmission of written texts.

Does this constitute an independent second emergence of the natural sciences? Would the Chinese and the European approaches converge in the further course of history? If we conceive of the natural sciences as a sustained and systematic endeavor to gain knowledge about the natural world by exploring the available material means, then this example suggests that the first question must be answered affirmatively. As far as the second question is concerned, I refer to my introductory remarks about the history of the *Mohist Canon*. The intellectual tradition for which it stands seems to have been suppressed with the emergence of centralized control in China under the Qin dynasty from 221 BCE onward; as a result, the Mohist tradition practically vanished from the later intellectual history of premodern China. A practical tradition of mechanical devices continued to exist in China and became an important common reference point in the encounter between Chinese and European scholars in the early modern period. But the tradition of theoretical mechanical thinking was broken off at an early stage, making it difficult to answer the second question about a possible convergence of Chinese and European approaches to science.

Still, what we know about the beginnings of Chinese science allows us to analyze some peculiar communalities and differences with regard to the European case, suggesting that science can indeed take distinct pathways in its development, depending on its cultural contexts. Turning first to the similarities, it is striking that both the Greek and the Chinese traditions were preoccupied with puzzles and paradoxes. This does not seem to be a coincidence, as both traditions emerged within a culture of disputation.

In the Greece of the second half of the fifth century BCE, we encounter, for instance, the so-called Sophists, a group of itinerant intellectuals preoccupied with the establishment of paradoxes who debated and lectured publicly. Roaming through Athens and other city-states, they offered educational services to wealthy youth.[17] A culture of disputation had also emerged in the China of the fourth century BCE, during the so-called Warring States period, when China was splintered into many small states and different schools challenged the doctrines of Confucius.[18] The Mohists were among them, offering their advice and services to rulers and courts, competing with other thinkers in the hope of achieving political influence. They were a community also composed of merchants and craftsmen and had an affinity for practical knowledge, which must have influenced their choice of topics.[19] Later in the fourth century, the intellectual debate intensified, with new competitors entering the scene. The ambition of the Mohists to cover all spheres of human knowledge and to sharpen their intellectual weapons by developing a method of disputation can be seen against this background. It seems that their aim was to demonstrate that consistent thought was possible.

Another parallel is the kind of knowledge that became the object of reflection in the Greek and Chinese traditions, namely (though not exclusively), practical knowledge related to the material culture of the time, such as mechanical arrangements and devices. Reflecting on the linguistic representation of such knowledge generates theoretical knowledge; its objective is not necessarily that of solving practical problems but primarily aims, as in this case, at organizing and structuring the knowledge itself.

These commonalities in the knowledge that became the subject of reflection are, on the one hand, due to universal material conditions that thus transcend culture. They are, on the other hand, traceable to a comparable level of technological development, such as the availability of the specific mechanical arrangements discussed in the two texts. The material culture may in part also be the result of a previous transfer of knowledge—when, for instance, instruments and equipment (and consequently the knowledge inherent therein) were exchanged. This possibility illustrates what I described in the preceding chapter as the layered structure of knowledge transfer.

However, in all likelihood the reflective thinking in each of these societies occurred independently, as there is no evidence of contact or of the exchange of relevant theoretical texts. This also suggests why these two processes of reflection produced such different results. For instance, the Mohists used the term "effectiveness" of a weight to explain the lever effect, whereas the Greek *Mechanical Problems* ascribes this effect to differences in the velocity of circular movements described by the radius of the circle.

Differences of this type could also indicate more fundamental disparities in the social conditions surrounding the emergence of theoretically reflected knowledge. The *Mechanical Problems* attempts to explain mechanical paradoxes using the framework of Aristotelian natural philosophy. The Mohists apparently had no such preexisting natural philosophy framework as a foil against which to develop their arguments. The schools with which they competed had not yet developed such a system. The context of their analysis rather appears to have been the theme of correct action and the correct use of language. Alongside the scientific reflections, ethical and logical considerations essentially make up the *Mohist Canon*.

At the beginning of their activities as a school in the fourth century BCE, the Mohists had developed a basically utilitarian tradition. Later, the so-called metaphysical crisis of the middle of the century (raised by questions about the role of the individual) had forced them and other schools to rethink their principles and to justify them on a new level of reflection. In the *Mohist Canon*, they attempted to "[lay] the foundations of an ethical system independent of the authority of Heaven, built on the actual benefit and harm, desires and dislikes of individuals."[20] The necessity of a consistent explanation of the world apparently arose from this quest to explain human action without referring to authorities or metaphysical entities.

In China, during the Warring States period, universal worldviews and cosmologies based on natural philosophy were only just beginning to emerge, whereas in Greece, the creation of such worldviews preceded the occurrence of theoretical reflection on language.[21] The nexus between mathematical traditions and deductive reasoning, which is documented in rudimentary form in the *Mohist Canon*, also differs significantly between the cultures of ancient Greece and ancient China.[22]

In summary, both cultures gave rise to forms of theoretical reflection about practical knowledge that amounted to a systematic exploration of regularities in the material environment—in short, to an incipient form of scientific practice. The different backgrounds and contexts under which these practices evolved affected their outcomes on a conceptual level. The fact that one tradition, the Greek, continued (albeit in an intermittent way and through the various cultural refractions discussed in the previous chapter) while the other tradition was discontinued was grounded in external circumstances, not in the intrinsic properties of the traditions. Evidently the conditions for the genesis of a scientific tradition are different from those for its long-term survival.

The Transmission of European Science to China

The case we have considered constitutes an alternative evolutionary pathway, or rather, at least the beginning of such a pathway. Could this have eventually led to a "scientific revolution" in the subsequent course of history? Could it, in other words, have resulted in a similarly dynamic rate of knowledge development as was observed in early modern Europe under different conditions? The question as to why China did not experience a scientific revolution is commonly referred to as the Needham Problem, named after the English pioneer of Chinese history of science Joseph Needham. The Needham Problem was one of the guiding principles behind his comprehensive work, *Science and Civilization in China*.[23]

There have been many attempts to answer this question but also some strong objections to the practice of counterfactual history in the first place. Here I do not attempt to solve Needham's problem but rather discuss a related question, namely, why the massive transfer of European scientific knowledge by Jesuit missionaries to China in the seventeenth century did not trigger a major transformation of Chinese systems of knowledge. This allows me to address the question of how many times the natural sciences have been invented on the second level introduced at the beginning of this chapter: the level of knowledge economies. The occurrence or not of such a major transformation cannot, in any case, be linked to a single factor (such as the presence or absence of an ancient scientific tradition to be built on) but is, as I shall argue, a matter of the larger knowledge economy of a society.

The Jesuit order was founded in 1540 and took on a significant role in the European knowledge economy in the age of the Counter-Reformation.[24] Jesuit colleges rapidly spread all over Europe, offering free education focusing on a humanistic canon and, in the higher grades, including philosophy and theology in the scholastic tradition. While this offering was directed at those ready and capable of engaging, it also meant an opening of access to knowledge for lower social classes, from the sons of officials, artisans, and urban citizens to the occasional talent discovered by a country priest. By 1600, hundreds of colleges had been founded in Europe, as well as in the Spanish and Portuguese colonies.

By the mid-seventeenth century, the order played a dominant role in the educational institutions of Catholic Europe. This included its role in the production of scientific knowledge, from the order's leading position in the arts faculties of Catholic universities to the engagement of its members in the new sciences. The *ratio studiorum*, outlining the basic rules to be followed in Jesuit education as prepared at the Collegio Romano at the end of the sixteenth century, gave, under the influence of Christopher Clavius, a prominent role to mathematics.[25] Jesuits contributed to all aspects of contemporary science, producing some twenty thousand books on nontheological subjects by the end of the eighteenth century. Jesuit colleges and universities became an important institutional backbone for the expansion of scholarly and scientific activities, pursued not least with the intention to gain and defend cultural hegemony in contest with the Reformation. Thus, as mentioned in chapter 10, Jesuit intellectual activities were constrained by a dogmatic theological framework and the corresponding implementation of control measures, such as censorship. Yet Jesuit scientists were able to make significant contributions to early modern science, owing in some cases to their unique access to their own global intellectual network.

Clavius participated in the papal commission responsible for the 1583 revision of the calendar, the so-called Gregorian calendar reform established by Pope Gregory XIII. This calendar reform had become necessary because of discrepancies between the Julian calendar, dating back to antiquity, and the apparent motion of the sun, by then amounting to a difference of ten days. This was relevant for determining religious festivities, because the vernal equinox was used to set the date for Easter. In the next chapter, I return to this issue as an example for the role of networks in the transformation of knowledge systems. Here the calendar is of interest because of its key role in the development of relations between European and Chinese intellectuals.

In China, the calendar was, from ancient times, an expression of the mandate of heaven and therefore of crucial significance to the emperor and his court.[26] For an agrarian society, a functioning calendar also had, of course, a practical, economic meaning as the dominating influence on agrarian cycles. The practical and symbolic meaning of the calendar had coevolved since antiquity, with the symbolic meaning taking on an ever-larger significance for the ruling of the Chinese empire. A reliable calendar allowed the emperor, the son of heaven, to articulate and confirm his mandate in accordance with celestial events such as solar and lunar eclipses. Unpredicted events, on the other hand, could be interpreted as signs of lacking virtue or even of an imminent decline in the emperor's reign.

For this reason, sophisticated methods of calendar computation had been developed, as well as institutions where such computations were performed, in particular the Astronomical Bureau, belonging, in the period here under consideration, to the Ministry of Rites. In the fourteenth century, the Ming dynasty took over the institution from the preceding Yuan dynasty established by the Mongols. In 1271, the Yuan had also set up a separate astronomical bureau, performing calendar calculations in the Islamic tradition, using methods going back to Greek astronomy. The Ming did not maintain a separate office but kept the Islamic calendar practice alive within a single Astronomical Bureau, without, however, integrating this tradition on a deeper intellectual level with the Chinese tradition.

Beginning in the fifteenth century, officials of the Astronomical Bureau noticed discrepancies and repeatedly failed to predict eclipses accurately, but they nevertheless were unable to convince the authorities that a calendar reform was necessary. As Benjamin Elman has carefully traced, this situation began to change gradually. In the course of the sixteenth century, however, prediction errors and their consequences for ritual ceremonies became ever more noticeable.[27] This realization of a crisis coincided with the arrival of the first Jesuit missionaries, who came to China and established contact with Chinese scholar-officials.

The Jesuits arrived in Asia and Latin America as fellow travelers with and protagonists in the Iberian globalization discussed in the preceding chapter. Goa, India, a center of Portuguese colonial activities in Asia, also became a central base for Jesuit missionary activities. They arrived in Goa from Lisbon, then went on to Japan or China. In 1583, the first permanent Jesuit residence was established in China, close to Macao. A key figure of the mission was Matteo Ricci, who had studied theology, philosophy, mathematics, astronomy, and cosmography at the Collegio Romano with, among other teachers, Clavius. Not long after the arrival of the Jesuits, Chinese scholars discussed the possibility of securing them as experts in discussions about the reform of the calendar; the earlier adoption of Islamic methods of calendar computation had set a precedent that now helped to legitimize the appropriation of such foreign knowledge.

The Jesuits quickly realized the opportunity that the calendar problem constituted for their missionary activities. Ricci requested additional support and expertise to be sent to China. In 1612, his fellow Jesuit Nicholas Trigault returned to Europe to report about the state of the activities and to organize a scientific mission to take back to China specialized works, scientists, and instruments. This enterprise met with great resonance and the support of the pope and the European courts. The project was also acclaimed by the European scientific community, which admired the prominent role of scholar-officials in the Chinese Empire. Many European scholars volunteered to join the mission. Eventually, thousands of books were shipped to China, astronomic instruments were built, cannons were cast, and Jesuit calendar and map production experts were appointed.

Among the experts returning with Nicholas Trigault to China in 1618 was the Jesuit scientist Johannes Schreck, who had studied with some of the most prominent contemporary European scientists, among them Galileo. When, in 1621, he finally arrived in Guangzhou (Canton) after an adventurous voyage, the Chinese calendar problem had become even more aggravated. This helped to create a window of opportunity for the Jesuits, who had meanwhile encountered considerable resistance to their missionary activities. Schreck began to work at the Astronomical Bureau in 1629, but he died the following year. His work on calendar reform was continued by other Jesuits who filled prominent positions at the imperial office, even after the Ming dynasty collapsed in 1644 and was replaced by the Qing dynasty, that is, the new Manchu rulers of China. Schreck's fellow Jesuit Adam Schall von Bell then became the first European to head the Astronomical Bureau.

For both the Jesuits and their Chinese contacts, the interest in the transfer and appropriation of European scientific and technical knowledge was, however, not limited to the calendar issue. Since the early days of Matteo Ricci in China, Jesuit and Chinese scholars had cooperated on numerous translation and compilation projects to make the foundations of European science and its latest findings accessible to readers in China, from the edition of the first six books of Euclid's *Elements* and a translation of Clavius's commentary on Sacrobosco's *Sphere* (an elementary introduction to astronomy and cosmography to which I return in the next chapter), to books on surveying, hydraulics, and agrarian management, to Galileo's telescopic discoveries. Even fields such as theoretical mechanics, which had fallen into oblivion since the Mohists and were no longer prevalent in the Chinese tradition of knowledge, became part of this campaign.[28]

Between 1626 and 1627, Schreck, who had himself mastered many of these subjects, was waiting in a church in Beijing for the emperor's orders concerning his astronomical projects. During this period, he met Wang Zheng, a scholar-engineer with similar interests who also waited in Beijing, expecting to take the imperial examinations to become an official. While they waited, they worked on a common book project, the *Yuanxi qiqi tushuo* (Diagrams and explanations of the wonderful machines of the Far West), which appeared in 1627. It constitutes a unique compendium of mechanical knowledge, beginning literally with Adam and Eve and progressing to Aristotelian cosmology and simple machines from the Greek tradition of mechanics and then to the latest theorems of Simon Stevin and Guidobaldo del Monte. The book

12.3. Tapestry: The Astronomers, from L'histoire de l'empereur de la Chine series, wool and silk, Beau-vais manufacture, 1722–32. The templates were designed between 1686 and 1690 under the direc-tion of Philippe Behagle (1641–1705). In the center is the Chinese emperor Shunzi (1638–61), and the bearded figure to the right is the Jesuit court astronomer Adam Schall von Bell (1592–1666). The teacher and the child are probably Schall von Bell's and Johannes Schreck's (1576–1630) succes-sor, the Jesuit astronomer Ferdinand Verbiest (1623–88), with the son of the Emperor Kangxi (1654–1722). Inv. Nr. BSVQ0140. Bamberg, Neue Residenz mit Rosengarten, Room 10 c. © Bayerische Schlösserverwaltung, Rainer Herrmann / Maria Scherf / Andrea Gruber, München.

also contains the first explicit formulation of the law of the lever known from Chi-nese history.[29] No contemporary European book offers a similarly comprehensive and compact account of both theoretical and practical mechanics. Remarkably, Schreck and Wang succeeded in reorganizing this knowledge and adapting it to Chi-nese collections of mathematical problems, rather than following a theoretical ex-position structured on European terms. This reorganization nevertheless succeeded in essentially preserving the substance of the knowledge it transferred.

In light of all this, it may seem plausible to expect that the massive transfer of the most advanced scientific and technical knowledge from Europe to China, through personal communication, books, and instruments, could have triggered something like a scientific revolution, at least if one assumes that knowledge by itself or the meth-ods of its generation are the critical factors in such a transformation process. There can hardly be a question that the knowledge transfer was successful on an intellec-tual level—the Chinese books that emerged thereafter were much more than mere translations and fragmented imitations of European sources. Rather, the books co-authored by European missionaries and Chinese scholar-officials constituted an ef-ficient vehicle for the appropriation of this knowledge by a Chinese audience. As illustrated by *Qiqi tushuo*, the above-mentioned book on the "wonderful machines of the Far West," the scientific and technical knowledge appropriated by Chinese intellectuals received a new form suitable to the Chinese context when it was brought in touch with familiar literary forms, intellectual traditions, and local technologies.

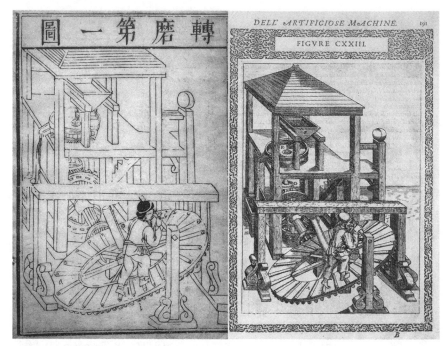

12.4. Transmission of engineering knowledge: a treadmill in the *Qiqi tushuo* from 1627 (*left*) and in Agostino Ramelli's *Le diverse et artificiose machine*, 1588 (*right*). Courtesy of the Library of the Max Planck Institute for the History of Science.

The depth and scope of European scientific and technical knowledge were not diluted in a shadowy "trading zone," confined to local rules of exchange in the presence of global disagreements about the very meaning of the knowledge exchanged. Rather, the resulting, genuinely novel representations of knowledge were meaningful beyond the locally and temporally limited exchange process. Because of their connections to broader resources of shared knowledge, they could in fact be more generally used (as their later reception illustrates) to pick up and further develop both the theoretical and practical traditions incorporated in these works.

As a matter of fact, however, a scientific revolution similar to Europe's did not take place in China during the seventeenth and eighteenth centuries. One explanation, as Matthias Schemmel has pointed out, was the very different nature of the two knowledge economies, which served both as driving forces and selective filters for the transmission process.[30] Because of these differences, the Jesuit program of using science and technology as vehicles for missionary purposes was doomed to fail, though the strategy of gaining influence by convincing members of the political elite was a tidy match for the hierarchical structure of Chinese society. The problem was, however, that the underlying systems of knowledge and their embedding in distinct knowledge economies were characterized by very different constellations of science, religion, and politics, as well as by different dynamics.

Despite the fact that the transmission of scientific knowledge was a secondary concern for the missionaries, they were, in a way, more successful in transmitting science than in transmitting their faith. At least among the learned Chinese, their science aroused much more interest and reception than did their religion. The missionary effort in China thus led to the encounter of two intellectual elites, each representing a culture with highly sophisticated knowledge systems and advanced technologies. But while the Jesuits succeeded in attracting the attention of learned Chinese and in interesting them in the achievements of European civilization, Western science was not integrated, on a larger scale, within the Chinese knowledge economy of the early modern period. While the Jesuits attempted to convey mathematics, the sciences, Aristotelian philosophy, and Christianity as inseparable parts of an integrated worldview, this worldview was dissected within the Chinese economy of knowledge, with the various parts being received and appropriated very differently. European knowledge was largely perceived as a complement to Chinese traditions; Europeans were seen as good calculators but not as proponents of an all-encompassing new system of knowledge.

Accordingly, the transmission was, as we have seen, most effective in the domains of mathematics and calendar astronomy—areas in which a deficit could be perceived from within the Chinese knowledge economy. In addition, as mentioned above, the earlier adoption of Islamic methods of calendar calculation had set a precedent for dealing with foreign knowledge in such cases. The limited institutional scope of the Astronomical Bureau offered a clearly circumscribed space for coping with this problem without the risk that it would proliferate into other domains of the knowledge economy. But for the other parts of the transferred knowledge, however, no similarly powerful structures or mechanisms existed, in spite of the attempts of individual Chinese intellectuals to engage in a broader transmission process. A famous example is Xu Guangqi, an early convert and influential official, who was in charge of the calendar reform until his death in 1633. Xu Guangqi had been keenly interested in using European knowledge for broader institutional and technological changes in China, for instance, in the domain of agriculture, and participated in several book projects conveying the pertinent European knowledge.[31]

But in spite of the massive influx of new knowledge, the Chinese knowledge economy remained largely unchanged. Its remarkable stability, at least when compared with the contemporary European situation, may also be illustrated by its resilience in the transition from the Ming to the Qing dynasty in 1644—essentially leading to foreign rule by the Manchu conquerors—which also saw the continued, and even strengthened role of the Jesuits as scientific advisers on astronomical matters. Yet the new knowledge did not trigger a major institutional change. Rather, it was the Chinese knowledge institutions, such as the Astronomical Bureau at the imperial court and the state examination system, that shaped the fate of the new knowledge. European scientific and practical knowledge, in particular that of calendar making and the prediction of astronomical events, was selected by and assimilated to Chinese systems of knowledge in a way governed by the requirements of the Chinese knowledge economy, without the sweeping cultural dynamics of the European case. The elements of Western knowledge that could be assimilated into the Chinese system were accepted, while other elements, such as theoretical mechanics, remained marginal.

As I remarked in chapter 10, an important role was played by the specific architecture of the knowledge systems involved. In Europe, practical knowledge, scientific knowledge, and religious views were closely interwoven within overarching worldviews. This was not just a matter of lofty intellectual construction. It was deeply ingrained in the shared knowledge produced and reproduced in the wider knowledge economy—in the numerous Jesuit colleges, universities, and academies, for example. The resulting connectivity of knowledge made challenges resonate throughout the knowledge economy. Consider the repercussions provoked by the Copernican heliocentric system, not only on astronomy but also on natural philosophy and religion. In Europe, the question of the true world system, whether heliocentric or geocentric, had far-reaching religious and political implications. That a similar controversy failed to materialize in China was not a matter of lacking knowledge, but rather of different knowledge economies and architectures.

With the exception of the question of the predictive power of the calendar, it was much more unlikely for such knowledge to come into conflict with traditional worldviews in China because it belonged to a different sphere, a sphere that was largely separate from the moral and political values of state orthodoxy. To some extent, this even held for religious beliefs. From the Chinese perspective, the Jesuits' religious mission was legitimate as long as it could be assimilated into the familiar pattern of Buddhist sects, seeking self-cultivation, without impinging on the neo-Confucian state orthodoxy. That this was not self-evident may be illustrated by the so-called Nanjing incident of 1616, when Jesuit missionaries were put on trial for infringing on Confucianism and essentially had to renounce public activities between 1616 and 1622.[32]

In summary, the transmission of European scientific knowledge to China took place between two fundamentally different knowledge economies. The embedding of scientific knowledge in broader systems of knowledge and, in turn, the embedding of these systems in the knowledge economy of a society shaped the outcome of this transmission and helps to explain why Chinese society did not undergo a larger transformation triggered by the knowledge brought in from Europe. While this does not constitute an answer to the Needham question of why China did not experience a scientific revolution, it has become clear that such a question, if it makes sense at all, cannot sensibly be treated by focusing on a particular epoch, but only within the context of the long-term evolution of knowledge and knowledge economies.

Eventually, the same global constellation that had engendered the massive transmission of European scientific and technological knowledge to China imposed restrictions that hindered its continuation. The Jesuits had gained a prominent position among Chinese intellectuals and at the imperial court in part because they had accommodated Chinese customs and tolerated Confucian rituals. This accommodation allowed them to become part of the Chinese knowledge economy in the first place, bringing their scientific expertise to bear on the treatment of one of its intrinsic challenges. In this way, however, they eventually reached the limits of their own framework for action. In 1704, after a long controversy on the question of whether Chinese ritual practices were compatible with Catholic belief, the church condemned Confucian rituals, with the consequence that Christian missions were eventually banned from China.

Transmission in the Age of Colonialization

A second phase of the introduction of Western science to China began in the middle of the nineteenth century.[33] While the first phase saw the Jesuits as fellow travelers of Iberian colonialism, the second phase brought a much larger group of missionaries and experts to China as fellow travelers of British and French imperialism, alongside American, Russian, German, and Japanese interlopers. Before the Opium Wars of 1840–42 and 1856–60, the Qing Empire had maintained its seclusion politics and forbidden any missionary activities. Its defeat in the First Opium War forced China to accept unequal treaties with Western powers, obliging it to open five port cities to international trade. After the Second Opium War, China had to allow foreign missionaries to freely proselytize in China.

With the advent of an evangelical revival in the English-speaking world, thousands of Protestant missionaries entered China in the second-half of the nineteenth century. Just as their Catholic predecessors in the seventeenth century, they engaged in the translation of scientific and technical literature, which they also saw as conducive to accepting a Christian worldview. Remarkably, the renewed translation activity took up exactly where the Jesuits left off. One of the first translations, appearing in 1857, was dedicated to the remaining books of Euclid's *Elements*, thus completing the work left unfinished by Matteo Ricci and Xu Guangqi 250 years before. Jesuit efforts at conveying astronomical and mechanical knowledge found their sequel, for instance, in the translation of William Whewell's introduction to mechanics, which exposed China to an analytical formulation of classical mechanics.[34]

The initial institutional bases for these translation activities were missionary activities in combination with Western publishing houses, in particular the London Mission Press, which was active in Shanghai—one of the five newly opened treaty ports. Later Chinese institutions were created in response to the challenges of dealing with the militarily superior Western powers. For instance, in 1862, a new governmental school was established in Beijing. It was originally planned as a language school for interpreters dealing with the foreign powers, but it eventually added mathematics and science to its curriculum, spurring further translation activities pursued jointly by Westerners and Chinese. In the same decade, a translation bureau was set up as part of the Jiangnan Arsenal in Shanghai. This shipyard had been created in the context of the so-called Self-Strengthening Movement as part of a modernization campaign aimed at restoring national power by adopting Western military technology and infrastructure.

Continued external pressure—such as China's defeat in the war with Japan in 1894 and its repression by foreign powers following what has become known in the West as the Boxer Rebellion at the turn of the century—triggered further reform measures during the late Qing dynasty. In the aftermath of these events, the Chinese economy of knowledge eventually attained a new stable structure, albeit after a long and strenuous process of reorganization. In this new structure, the modern natural sciences were closely connected with other societal and economic developments, and ultimately contributed to the creation of an industrialized society.

If this connection to the larger economy of knowledge is included in the question of how many times the natural sciences emerged, then it happened only once, as the result of an extended historical process in which the global economy of knowledge was eventually transformed through its increasing dependence on scientific and technological knowledge. That this global development surfaced in different regions at different times and in varying forms is hardly surprising and should not be misinterpreted or simplified in causal terms.

The history of the transmission of Western science to China is just one thread in this global development, a thread linear only in appearance. The impression of some early-twentieth-century Chinese scholars that Mohist doctrines constituted the beginning of an alternative Chinese pathway to the natural sciences in this more encompassing sense may therefore not have been entirely mistaken.[35] The unlucky fate of this doctrine just shows how vulnerable such pathways were at the outset— just as the transformation of the traditional Chinese knowledge economy under the pressure of Western military and technological superiority illustrates how powerful the new knowledge economy was once it had become mature.

Chapter 13

Epistemic Networks

Almost all of us have had the experience of encountering someone far from home, who, to our surprise, turns out to share a mutual acquaintance with us. This kind of experience occurs with sufficient frequency so that our language even provides a cliché to be uttered at the appropriate moment of recognizing mutual acquaintances. We say, "My it's a small world."

—STANLEY MILGRAM, "The Small-World Problem"

How does individual behavior aggregate to collective behavior? As simply as it can be asked, this is one of the most fundamental and pervasive questions in all of science.

—DUNCAN WATTS, *Six Degrees: The Science of a Connected Age*

Social Network Analysis

What we have learned from the preceding discussion is that the emergence of the natural sciences as powerful components of industrialized societies is a world historical event that can only be understood against the background of global history. This development did not, however, lead to a uniform landscape, which rather continued to be influenced by local traditions and contexts. Nor can the emergence of the natural sciences be associated with a single origin, even when we focus on the European situation alone. Indeed, here as well the development of new economies of knowledge was a protracted process and, in particular, a matter of the confluence of heterogeneous and partly independent traditions of knowledge. As discussed in previous chapters, these traditions dealt with phenomena and processes as diverse as astronomical regularities, mechanical technology, or chemical transformations, which became the objects of interrelated, but highly diversified societal efforts directed at knowledge production and accumulation.

The development of novel endeavors of knowledge production involves changes in societal as well as in epistemic structures, just as the emergence of the natural sciences at large, in the sense discussed in the previous chapter, was simultaneously an epistemic and a societal transformation on a global scale. In the following, I therefore examine this cotransformation or coevolution of social and epistemic structures in greater detail.[1] For this purpose, I make use of a specific tool that I must adapt to this context: social network analysis. Its reach is still limited because of the complex mathematical and conceptual problems of modeling the interaction of social networks with the more stable economic, political, and societal underpinnings of these networks, which change on different timescales. New approaches in multilevel network analysis aim at taking these complex interactions into account, but many of their aspects are still under investigation.[2] Thus, the use of network analysis for a historical theory of knowledge evolution is breaking new ground; hence, some of the material presented in the following is still in a preliminary state.

But as some of the case studies to be discussed in the following illustrate, the language of social networks is able to disclose important new insights into the transformation of knowledge structures. It opens up a relational perspective on societal and mental processes, making it possible to investigate knowledge evolution in depth by bringing individual interactions within a knowledge economy and overall changes of knowledge structures into the same theoretical framework and, what is more, into a framework that allows for quantitative analysis.[3] With the help of network analysis, one can capture the self-organizing dynamics of epistemic communities, conceived as communities organized around shared systems of knowledge. Investigating these dynamics is also of great interest to the question of what knowledge, which knowledge economy, and which epistemic communities may be capable of confronting the challenges of the Anthropocene. I return to this question in chapter 16.

The presence of network concepts in sociological explanations is not new, and their usefulness for analyzing knowledge transfer processes is rather evident. Classical papers such as Mark Granovetter's "The Strength of Weak Ties" from 1973 or Duncan Watts's book *Six Degrees* from 2003 have gained an increased relevance in a world in which emerging, non-obvious, and flexible connections are becoming ever more relevant.[4] Network analysis is also central to the way in which a vast amount of information is ordered and encountered online. Sociological investigations have suggested remarkable insights into the role of network structures for the emergence and diffusion of novelties.[5] Historical investigations with the help of networks concepts, such as the study of the rise of the Medici family by John Padgett and Christopher Ansell, have been equally productive in terms of new insights.[6] But the community of historians has, in general, taken up the use of network tools more hesitantly. Quantitative network analyses of historical processes, going beyond the metaphorical use of network concepts, have only recently been pursued more intensely, combining network analysis with historical interpretation.[7]

So far, network analysis has been mostly limited to specific historical networks, such as the social network of Medici family, as well as maritime networks, correspondence networks, information networks studied with the help of citation and co-citation analysis, networks of the diffusion of innovations, and coauthorship networks as a proxy for scientific collaborations.[8] In our context, however, we must envisage a more ambitious aim: to conceptualize, in terms of network analysis, the connections between the different dimensions relevant to a historical theory of knowledge evolution, that is, the social, material, and epistemic dimensions discussed in previous chapters. In the following, I attempt to sketch, on the basis of joint work with Roberto Lalli, Manfred Laubichler, Matteo Valleriani, and Dirk Wintergrün, basic concepts of a theory of epistemic networks encompassing all of these dimensions. Epistemic networks are here understood as social networks involving

A Centralized   **B** Decentral   **C** Distributed

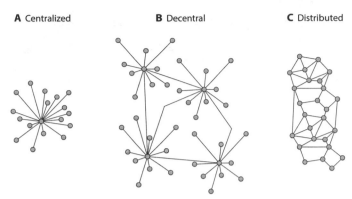

13.1. Centralized, decentralized, and distributed networks. See Baran (1964), which was published as part of a RAND Corporation study to create a robust and nonlinear military communication network.

the transmission and transformation of knowledge. They constitute an essential mechanism of any knowledge economy.

In addition to their promise for a historical theory of knowledge evolution, network concepts (but also concepts of nonlinear dynamics, chaos, complexity, and evolutionary theories) harbor the potential for recognizing connections or parallelisms between hitherto unrelated fields, thus constructing novel bridges between mathematics, the sciences, and the humanities. This prospect, combined with the capacities of the information technologies, also includes new possibilities of dealing with the challenges of quantitative analyses and large data sets in the social sciences and the humanities. Network analysis has become an important research instrument in this context, but it has to be treated with some care because of the risk of blurring conceptual distinctions or losing the intellectual depth of other traditions and approaches in the humanities.

Let me introduce a few basic notions, terminologies, and studies in social network analysis that are particularly helpful to outline a historical theory of knowledge evolution: Networks comprise nodes and edges, which may or may not be directed. The degree of a node is the number of connections it has to other nodes. Nodes with a higher degree than the average are also called hubs. The probability distribution of degree over the entire network is the network's characteristic degree distribution. An important special case is a so-called scale-free network, a term introduced by Albert-László Barabási and Réka Albert in 1999 to describe networks whose degree distribution follows a power law, that is, a law according to which the change in one variable by a factor produces a change by a constant factor of another variable.[9] Power laws are therefore scale invariant, because a change of scale simply generates a power law proportional to the original. In scale-invariant networks, hubs are common. Networks are said to be self-similar when they display self-repeating patterns on scales of all lengths.

As early as 1965, the physicist, historian of science, and information scientist Derek J. de Solla Price analyzed the networks of citations between scientific papers,

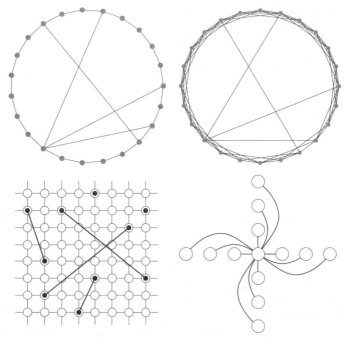

13.2. Real networks are often characterized by two factors: they are large in size, but the average length of the path between two nodes is relatively small. A "small world" model to describe the creation of these networks was introduced by Watts and Strogatz (1998). Examples from a wide field of applications are discussed by Albert and Barabási (2002) (e.g., the World Wide Web, Science Networks, or Ecological Networks). The figures show examples of "small worlds," which emerge by introducing only a few shortcuts, based either on a circular network (*top diagrams*) or on underlying patterns (*bottom examples*). Adapted from Newman and Watts (1999, fig. 1).

finding that it had a distribution we would today describe as that of a scale-free network.[10] He later attempted to explain this property as being attributable to a "cumulative advantage," or preferential attachment of edges to hubs that already have many edges, an effect that is known in sociology as the Matthew effect, according to which the rich get richer and the poor poorer.[11]

From the mathematical theory of graphs, one may derive insights regarding the stability, redundancy, and connectivity of a network. Basic topological structures of networks are relevant for understanding the stability and redundancy of an epistemic network. Scale-invariant networks, for instance, are rather robust, whereas highly centralized networks tend to be more fault-prone. The failure of the central hub can lead to a complete collapse of the network.

As a further description of the inner structures of a social network, Mark Granovetter proposed the distinction between "strong ties" and "weak ties."[12] Strong ties are relationships between the nodes in a network that are designed to hold for the long term and display a high degree of interaction. As a rule, these relationships have to be ensured by high investments of economic and social resources. A typical example is a long-term business relationship. Weak ties, by contrast, are occasional meetings that do not lead to direct business relationships, such as an individual's relationships to an extended circle of friends or meetings at trade fairs or exhibitions.

Duncan J. Watts adopted the idea of a "small world" (introduced by Stanley Milgram in the 1960s) and used it as the basis for a network theory primarily conceived for analyzing the connectivity generated by electronic media, and especially by e-mails.[13] Building on Granovetter's ideas, Watts's model is characterized by the observation that in a network in which the actors are only tied to their immediate neighbors, one can achieve a considerably higher connectivity—in the sense of being reachable via "average short paths"—by introducing just a few additional weak ties. In social and historical investigations, this is regarded as a characteristic phase in the temporal development of a network, a phase that is, following Watts's definition, also called "small world." The state of maximal connectivity is achieved here when an actor is directly tied to all the other actors.

Three Dimensions of Epistemic Networks

If we are to combine the instruments of social network theory with an analysis of the dimensions and dynamics of knowledge evolution, we must pursue a more fine-grained description of these networks. Epistemic networks have (just like knowledge) social, material, and mental dimensions, which are closely intertwined. These dimensions are represented by distinct networks, which I call social, semiotic, and semantic networks in the following. Alternatively, I also refer to the social, semiotic, and semantic dimensions of a multidimensional epistemic network. Let me now describe the basic features of each of these networks or dimensions.

First are social networks of action, whose nodes are represented by individual or collective actors while the edges represent their interactions or other relations between them. In the context of epistemology, I focus on actors possessing knowledge and involved in producing, exchanging, disseminating, or appropriating knowledge; in such cases, I speak of knowledge-related or simply epistemic actions. Among actors, processes of social and communicative exchange take place. Such exchange processes are, as discussed in chapter 8, subject to the general conditions for the reproduction of a society and typically involve a division of labor within the society and its component institutions. These conditions become manifest, in particular, in the form of traditions, rules, conventions, and norms, but also in terms of constraints and power structures that strengthen, weaken, facilitate, or impede ties within social networks.

Second are networks whose nodes are the real objects, material instruments, artifacts, or external representations (such as systems of symbols) and whose edges are the actions, physical transformations, or other relations by which a connection between them is established. For simplicity's sake, I call them "semiotic networks," even though they are not limited to signs but rather include the entire material context of action.

Concrete examples of semiotic networks include technological artifacts generated on the basis of the technological knowledge of a producer, the books collected in a library, the articles in a scientific journal, the signs of an alphabet, or the formal linguistic systems of mathematics, chemistry, or physics. In these cases, we can recognize specific regulatory mechanisms at work, for instance, the production technology of a company, the editorship of a journal, or the rules for the adequate use of a symbolic system such as the formulas of mathematics and chemistry. For historical scholarship, semiotic networks are typically the starting point for the reconstruction of the other aspects of epistemic networks. Thus, for example, museums and archives collect objects in order to allow for the reconstruction and illustration of their original meaning and use in their historical contexts.

Third, mental or cognitive structures of knowledge may also be analyzed in terms of network structures, which I call, following a long tradition, "semantic networks," although I may use this term more generally than is customary.[14] Nodes represent building blocks like concepts, mental models, or the elements thereof that obtain meaning from their role in the interpretation of experiences as well as from their relation to other nodes. The edges of a semantic network consist of the thought processes or cognitive links by which the cognitive building blocks relate to one another.

In historical contexts, semantic networks are not directly accessible to analysis but must be reconstructed from external representations, such as written texts. Such reconstructions make use of the fact that concepts have representations in language, called "lexicalizations." Typically, however, these lexicalizations have no one-to-one relation to either concepts or other cognitive building blocks; one and the same concept may correspond to different terms in language, while the same term may represent different concepts. The contiguity of words in a written text and, in particular, its sentence structure may serve as a proxy for the edges in a semantic network. While texts are linear representations of knowledge, in the sense that they do not make explicit complex semantic network structures, they still allow a reader to reconstruct at least parts of the semantic network carrying their meaning, with missing parts typically supplied by preexisting knowledge. In the case of historical texts, such preexisting knowledge may no longer be available, in which case it must be reconstructed from other sources documenting the shared knowledge of a particular historical situation.

Here I treat networks as a more general social structure from which institutions, as they have been considered in chapter 8, emerge as a more specific social form by adding regulative controls—controls that determine, for instance, which relations are legitimate or not; which relations are required, facilitated, or hindered; which relations entail which other relations; and so forth. From this perspective, institutions lend a systemic nature to parts of networks because their regulative properties allow them to maintain and reproduce relations among the pertinent elements of the network. An epistemic community, for instance, may emerge as a loose network of social interactions. At some stage, rules for these interactions may be articulated, turning the network into a more organized and permanent form of cooperative practice.

Regulative controls may operate not only in the social domain but in all three dimensions of an epistemic network: we can imagine a group of social actors agreeing on regular meetings, the creation of an organization, or some other form of governance, or we can think of semiotic networks, in which systemic structures are implemented by certain regulatory mechanisms, be they natural laws, technical relations, or conventional rules for the usage of signs, such as the rules for correctly using a writing system. Also with regard to semantic networks, we may consider ordering principles, such as a deductive structure leading to those networks' "institutionalization" (i.e., systematization). In this way, a system of knowledge is formed in which previously unrelated or only loosely related chunks of knowledge come together in a highly organized system of knowledge. An example is the bringing together of ancient geometrical knowledge in Euclid's deductively organized *Elements*.

As some of the following examples illustrate, the emergence and development of epistemic communities are shaped by interactions between the three types of networks. Thus, we observe how an institutionalization (or systematization) of one network can support the creation of regulative structures in the other two. An institutionalization of social networks chunks knowledge into knowledge systems. Highly organized systems of knowledge or the canonization of knowledge by the conventions of its external representation make it easier for a network of actors to organize themselves around a body of shared knowledge.

On a qualitative level, these interdependencies may seem to be intuitively plausible (or not), but with the help of quantitative network analysis it now becomes possible to substantiate such broad heuristic claims concerning the transformation of systems of knowledge and epistemic communities on the basis of large empirical data. In principle, quantitative network analysis can therefore go beyond the still prevailing local case study approach to such questions, being applicable to major, and indeed global change processes as well.

Network analysis thus offers, at least in principle, the possibility of integrating the vertical and horizontal axes of our investigation, that is, the long-term, diachronic perspective of part 3 with the global, transversal perspective of part 4. Networks have indeed played a key role in the transmission of knowledge over long time periods and across cultural boundaries, often alongside commercial and political connections, being favored or hindered by geographical and ecological conditions. The globalization of knowledge as described in chapter 11 may be conveniently rephrased in network terminology. I therefore return once more to the example of the transmission of Greek science in order to illustrate the concepts introduced above with the help of a familiar example. I then turn to more specific historical case studies employing network analysis.

Greek Science as a Network Phenomenon

While mathematicians and philosophers were scattered throughout the Greek world, the epistemic network of Greek thinkers was characterized by the central role of certain hubs, such as (in chronological order) Miletus, Athens, and Alexandria.[15] The importance of such centers was conditioned by geographic, political, and economic factors. Hence, the occurrence of cosmological thought in Milesian thinkers such as Thales, Anaximander, and Anaximenes is related to the position of Miletus at the heart of Asia Minor, a cultural crossroads to which the cosmological knowledge of the

Babylonians would most likely have found its way. Similarly, the wealth accumulated by the maritime empire of Athens, together with the trade and political connections that were established there, provided the socioeconomic conditions that led to the flourishing of the arts and sciences in the Age of Pericles.

While networks favored the exchange of knowledge, as well as the inclusion and spread of innovations, the long-term accumulation of knowledge remained fragile as long as there were no institutions dedicated to its systematization and preservation. In Greece, traditions of natural philosophy and science initially emerged within a polycentric urban context with limited institutionalization before the Hellenistic period. The growth of scientific knowledge was largely sporadic, determined by the interests of a small number of individuals and institutions, such as the Platonic Academy or Aristotle's Lyceum.

A more substantial institutionalization of science and attempts at a systematic accumulation of knowledge began in the Hellenistic age, but these were limited by the dependence on a few large hubs that were not part of a robust network and constituted critical points of failure. Nonetheless, as discussed in chapter 11, Hellenistic science was able to make significant advances in areas such as astronomy, partially as a consequence of the fact that the Hellenistic world now included Babylonia; Greek thinkers therefore had direct access to Babylonian texts and the knowledge of Babylonian practitioners.

The interaction between the three different networks (social, semiotic, and semantic) may be illustrated by a familiar example: the famous Museion of Alexandria, founded by Ptolemy the First in the third century BCE.[16] The Museion quickly became a prestigious center of learning that would strengthen the central role of Alexandria for Hellenistic culture and forge a new cultural identity by collecting texts from all parts of the known world. The project may have been conceived on the tail of Alexander's plans to build a huge library in Nineveh in keeping with his imperial ambitions in the domain of culture and learning.

The value of the collection was enhanced by systematic efforts to exploit and augment the accumulated knowledge. The texts may hence be considered as the nodes of a semiotic network whose edges are the actions by which connections among them had been established, that is, all actions pertaining to the acquisition, storage, cataloging, and systematization of these items. For the historian, the very limited surviving records (extant writings, catalogues, narrative accounts, traces of editing or translating efforts) may serve as proxies for the elements of this semiotic network.

The relevant social network may be defined as being constituted by authors, editors, translators, librarians, and scribes, as well as by their patrons and the owners or carriers of scrolls. Their knowledge-related actions essentially coincide with the actions just described. But from the perspective of the social network, the important role of political initiatives, institutions, and regulations for the knowledge economy becomes particularly visible. According to Galen, Ptolemy III sent letters to sovereigns all over the world, asking them to send him their book collections. Traders were asked to bring back books from their travels. When a ship entered the harbor of Alexandria and happened to carry scrolls, these were confiscated and delivered to the Museion, where they were copied—the original was kept and the copy restored to the owner. Messengers were sent to courts all around the Mediterranean and beyond to acquire texts and bring them to Alexandria.[17]

On the basis of the knowledge thus brought together and in the context of scholarly work pursued at the Museion, partially in association with technological ventures, experiments, and expeditions, new insights were generated that had a major impact on subsequent developments. In what were to become key domains of the Renaissance and the early modern Scientific Revolution—philology, mathematics, mechanics, medicine, optics, geography, and astronomy—the Hellenistic knowledge economy anticipated later achievements.[18]

The semantic networks incorporating this knowledge as they can be reconstructed from surviving texts such as Ptolemy's works on astronomy and geography, Galen's works on medicine, or Hero's works on mechanics may be considered as resulting, at least in part, from an internalization—in the sense of an intellectual reflection, systematization, and synthesis—of the knowledge brought together by the semiotic and social networks just described.[19] The new systems of knowledge represented by these influential works were indeed based on an integration and reflective reorganization of the distributed knowledge resources of the ancient world that had been assembled by the unique epistemic network of the Museion.

This example therefore also illustrates how, in one specific case, different networks interacted. The highly organized knowledge economy of the Museion left its traces on the knowledge systems and external representations it produced. Its institutionalization of social networks accumulating the scattered knowledge of the ancient world contributed to the emergence and transformation of knowledge systems documented by writings that later became in turn the nuclei for new epistemic communities.

Medieval Knowledge Networks

Another example showing how epistemic networks foster the integration of dispersed knowledge into new systems of knowledge is the accumulation and transformation of cosmological and geographical knowledge in the Middle Ages, as recently studied by Matteo Valleriani.[20] From the eleventh century onward, traveling and mobility in Europe became ever more widely spread. They were stimulated by the emergence of wide-ranging trade relations dealing with products such as spices, silk, and wool; mediated by crusades and other military campaigns, long-range pilgrimage voyages, and the rising significance of learning; and fueled by the foundation of universities, the spread of major building activities, and contact and competition of Latin Europe with the Byzantine and Islamic worlds. The growing network of social relations in which urban and religious centers acted as important hubs was, at the same time, an epistemic network favoring the exchange and accumulation of knowledge.[21]

Knowledge exchange in this network was regulated by powerful institutions such as the Catholic Church and stabilized by teaching institutions such as monastery and cathedral schools and the emerging universities. These regulative instances led to the canonization and spread of particular bodies of knowledge, for example, cosmological views based on ancient geocentric traditions, or specific ways of preparing calendars that conformed to religious dogmas and liturgical rules. Such knowledge was, however, not just distributed from a single center; it had to be locally reproduced at different sites within the network.

This is particularly clear for calendar making, which, in the thirteenth century, had to comply with rigorous demands imposed by the Roman Catholic Church but also had to account for locally unique geographical conditions. What spread in this case was therefore not a uniform calendar but centralized knowledge about the making of calendars. This knowledge was interwoven with other domains of knowledge, in particular with elementary astronomical and geographical knowledge. This knowledge was canonized and spread by central institutions such as the University of Paris—a hub closely associated with the Catholic Church. Yet, knowledge was not only locally reproduced but also enriched and extended by activities at different sites of the network. Beginning in the thirteenth century, treatises on cosmology, calendar making, and other subjects were copied all over Europe and augmented with commentaries and new materials. These novelties spread over the network in turn so that an accelerating Europe-wide accumulation of knowledge was engendered.

While the overall structure of knowledge remained confined over centuries by the canonization and restrictions imposed on it by central institutions, the choices and priorities of knowledge accumulation and its long-term consequences can only be understood as a cumulative network effect, rather than as the results of singular political, religious, or economic interventions. Thus, the enhanced mobility of merchants, soldiers, scholars, and explorers mentioned above favored the spread of knowledge useful to travelers for geographic orientation, whether on land or on the sea. Mathematical geography based on the grid of coordinates introduced by Ptolemy, nautical innovations as they were made by the explorers of the Southern Hemisphere, and new knowledge about climate zones were important subjects spreading within the network and enriching it with empirical knowledge about the natural world.

The practice of calendar making had—through network effects—a similar effect, enhancing the role of empirical knowledge in societal practices.[22] The spread of locally unique calendars made it evident that they deviated from both a liturgical order under the centralized control of the Roman Church and the natural order of astronomy. The religious authorities perceived this diversity as a threat to their central control. It therefore became a powerful motive for strengthening the compatibility between the calculations involved in calendar making and actual astronomical observations. This greater emphasis on empirical knowledge was thus in part a consequence of network dynamics, that is, of the interaction between the network of the diverging local calendars, the semantic network of centralized time reckoning, and the social network of control imposed by the Church. This tension eventually led to the major calendar reform of the sixteenth century that created the Gregorian calendar.

This example suggests that it makes sense to investigate the massive diffusion of scientific knowledge in late medieval and early modern Europe with the help of network analysis. In this way, an intuitively plausible picture can be substantiated, extended, or revised on the basis of quantitative data. As discussed in chapter 10, this massive diffusion of knowledge was fostered by the spread of universities, academies, and educational institutions, framed by new canons of knowledge, and supported by the epistemic communities that were familiar with it. The emerging knowledge economy was evidently favored by economic and social conditions, but it also exhibited a self-organizing behavior that can be understood as a network effect. This is suggested by the lack of effective central control of scientific practice and by the existence of strong social and mental boundaries blocking the diffusion of knowledge, including those caused by religious conflicts. Nevertheless, scientific knowledge accumulated at an astonishing rate and traveled quickly across geographical, political, and cultural borders.

The growth and mobility of early modern scientific knowledge apparently resulted from a network in which most protagonists were in contact with only a few others. But there were also a few protagonists who were in contact with very many others, acting as network hubs. The French polymath Marin Mersenne, for instance, formed the hub of large network of correspondents in the first half of the seventeenth century.[23] Such networks possessed similar connectivity at the level of institutions that sponsored and promulgated scientific knowledge, such as courts, religious societies, the homes of wealthy patrons, universities, and the newly founded scientific societies, such as The Royal Society. Again, most institutions had direct relations with only a few others, but a small number of them were hubs with numerous direct connections.

The presence of such similar structures at the level of individual scientists and the level of institutions suggests that the relevant networks exhibit the properties of self-similarity and scale freeness discussed above. Positive network externalities (i.e., the effect that one user of an entity has on the usefulness of that entity to other users) fostered the inherent dynamics of spreading scientific knowledge so that the more people engaged in it, the more useful it became. If this picture is correct, the spread of scientific knowledge across epistemic networks contributed to social and economic conditions favoring in turn the further development of science and thus its own propagation.

Epistemic network analysis not only reveals detailed insights into the dynamics of knowledge integration but also into the formation of epistemic communities. Since, in the history of an emerging field of knowledge, we are typically not dealing with a given "scientific community" (Kuhn) or an existent "thought collective" (Fleck) the understanding of the emergence and transformation of such communities is a crucial issue for which the analysis of epistemic networks turns out to be quite helpful.[24]

The Illuminating Example of the *Sphere*

This may be illustrated by an investigation of the spread of the famous cosmological treatise titled *Tractatus de sphaera* by Johannes de Sacrobosco, as investigated by Matteo Valleriani and his collaborators. The *Sphere* was written in the thirteenth century

quos máifeſtos ut circulũ æqnoctialé & ecliptica & circulos paralellos eclipticæ
duoſ:quélibet p ſex gradus ut zodiaca zona hébat i q̃ diſtiguamus ſigna duode
cim.Ité ſignabimus tropicos & circulos articũ & antarcticũ & ſex circulos co-
luros tráſeũtes p púcta.xii.ſignog̃ zodiaci & polos eius:reliquos uero circulos
palellos eclipticæ leuiſſime deſcribemus:& ſilſ q̃ tráſeũt p polos zodiaci & p oés
gradus zodiaci:ut ſeref i ca.ſcdo huius:p quos lõgitudíes & latitudíes ſtellag̃ fſ

xag̃ ſignareuale
amus.Ná cũ hũ
erimus tabulã
uerificatã ad tp̃ſ
nr̃m i ipa ſphæ
ra ſituare ſtellas
poterimus hoc
mõ.quoniã ſiſit
ſtella lõgitudís
graduú.l.& lati
tudís. xliiii.gra-
duú:ubi in ipſa
ſphæra colurus
lõgitudís.l. gra-
dus:ſecat palel-
lũ ſeptétriõalé ſi
latitudo ſteꞁlæ é
ſeptétriõalís uel
auſtralé ſiauſtra
lis : ſignabimus
ſtella illá:quã in
forma uel colo-
re uel q̃titate di
ſtiguimus : ut i
quo ordine ſit i
telligaſ& reliq̃s
mõ cõſili. Hori
zõté uero rectũ
& obliquũ atꝗ
méidianũ &axé
ſine corpe terre
huic ſpæ ſolide
adaptabimꝛ:uti
in ſpa pcedéti q̃
ex circulis octo
eꝛat cõpoſita.

13.3. Illustration of an armillary sphere. From Johannes de Sacrobosco, *Sphaerae mundi compendium soeliciter inchoat* (Venice, 1490). Courtesy of the Library of the Max Planck Institute for the History of Science.

at the University of Paris following earlier Latin and, in particular, Arabic traditions. Over a period of more than three hundred years, derivatives of the original treatise spread an ever-expanding collection of astronomical, physical, geographical, and mathematical knowledge across Europe and well beyond the narrow scholarly community.

Often considered a conservative text, the *Sphere* significantly contributed to a transformation of shared knowledge that eventually helped to prepare the ground on which the new astronomical system of Copernicus was received and controversially discussed, not just as a solution of technical problems of interest to a few experts but as part of a new worldview comprising many domains of knowledge. The *Sphere* tradition did so both by integrating different domains of knowledge into its expanding knowledge system and by forming an epistemic community familiar with it.

The *Sphere* had been reprinted over three hundred times by the end of the seventeenth century and circulated in the hundreds of thousands of copies. Tracing its spread across Europe with the help of network analysis therefore means to capture a large quantity of data regarding its printing, revision, and reception.[25] Information about print events and other data relevant to the evolving social network involved has to be combined with semiotic and semantic information about the changing structure and contents of the treatise. The original treatise is divided into four chapters. The first describes the basic setup of the geocentric worldview, with the sphere of Earth, its axis, and its poles. The second chapter describes the circles on this sphere as well as the celestial sphere. The third chapter discusses the zodiacal signs, the diversity of days and nights, and the division into climates. The fourth chapter introduces further basic features of the geocentric view (such as the circles and motions of the planets) and also explains the causes of eclipses.

The treatise thus presents the essential features of the mental model of the cosmos underlying the geocentric tradition of Aristotle and Ptolemy but leaves aside all sophisticated technicalities and complications that had been integrated into this model over the ages by professional astronomers and philosophers. Although it was not a classical text, the *Sphere* was quickly adapted to the practice of commentary that made it suitable for university usage, making it a standard introductory text in all European universities until the seventeenth century.

Because of these features, the *Sphere* could act as a seed crystal for knowledge accumulation. Its suitability to a particular knowledge economy (the teaching context of universities and other institutions of learning) guaranteed its success. It became a widely read standard text that could easily be expanded in accordance with the literary tradition of commentary supported by this institutional context. Its focus on a basic mental model of the world and its particularly lean representation of this model offered cognitive possibilities for enriching the basic model with new knowledge that could then spread across the institutional networks.[26]

The nodes of the social network of the *Sphere* tradition were authors and institutions such as universities, and, after the invention of the printing press, early modern printers. They received and generated knowledge, and they consumed, produced, and disseminated treatises, externally representing this knowledge in revised editions of the *Sphere*. The resulting social network had a small number of central hubs, such as Venice, Paris, Wittenberg, and Antwerp, whose relative dominance changed over time. Weak ties were important for quickly disseminating new knowledge across the network.

Traveling individuals, for instance, could realize such weak ties, as when the French humanist Elié Vinet visited the famous Coimbra college in Portugal. There he evidently became acquainted with the work of the Portuguese cosmographer Pedro Nunes on a mathematical treatment of climate zones according to Ptolemy's *Geography*. Nunes had included this treatment as well as other materials from

Ptolemy in his commentary on the *Sphere* of 1537.[27] His mathematical treatment of climate zones was then translated by Vinet and included in a 1556 edition of the *Sphere*, published by Cavellat in Paris.[28] From this great hub it rapidly spread across the European network, establishing a widely shared new mathematical framework for the conception of climate zones. Because of its important role in the university curriculum as well as in a wider economy of knowledge, the *Sphere* tradition shaped an epistemic community characterized by its familiarity with such new conceptions, which also included knowledge gained from the great exploratory voyages of the time.

The Epistemic Community of General Relativity

Social networks shape knowledge systems just as much as knowledge systems may in turn trigger, via social networks, the creation of epistemic communities concerned with this knowledge. The result of such a process of self-organization, in which a new system of knowledge emerges alongside an epistemic community practicing it, may be called (abusing the notion introduced by Kuhn) a "paradigm," in the sense of a unity of practice and practitioners. As discussed in chapter 7, a new paradigm is, however, not to be understood as the product of a sudden and unstructured gestalt switch attributable to a single ingenious scientist, as Kuhn initially believed,[29] but of a protracted reorganization of knowledge that is usually a community effort. With the help of network analysis, we can trace both the epistemic and the social dimensions of such reorganization processes in greater detail.

A striking case in point is the emergence, after World War II, of a community of physicists working on general relativity in the context of the so-called renaissance of general relativity, briefly mentioned at the end of chapter 7. This renewed interest came into being as a delayed reaction to the creation of the theory half a century earlier. What caused the delay, and why did general relativity see a spectacular rise after the war, at which point it became, with quantum physics, a second foundational pillar of modern physics?

The basic ideas and mathematical equations of general relativity had emerged with Einstein's work early in the twentieth century. As discussed in chapter 7, the establishment of the new theory was itself the result of a process of knowledge reorganization in which not only Einstein but a small community of collaborators and competitors were involved. Yet, in 1915, when Einstein published the field equations of general relativity, it was not yet a paradigm in the sense defined above. Not only did an epistemic community practicing it not yet exist; the theory itself had not yet been elaborated into a comprehensive system of knowledge broadly applicable to the physical world. Its range of application was initially limited to a few astronomical phenomena, such as the perihelion anomaly of Mercury or the deflection of light in a gravitational field. Many of its conceptual implications remained unexplored, in particular those that were in conflict with the knowledge system of classical physics from which it was built. Only gradually did general relativity emerge as a universally applicable conceptual framework profoundly distinct from classical physics. This development was not an achievement of Einstein but of an emergent community. It only came to preliminary completion in the early 1960s.[30]

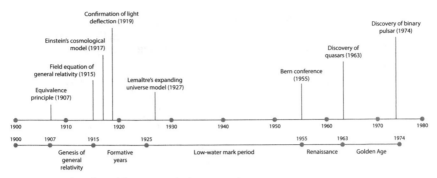

13.4. Essential timeline of the protracted relativity revolution.

Allow me to briefly recapitulate the main phases of the history of relativity (see the final section of chapter 7). The genesis of the theory of general relativity by Einstein and a few others in the period between 1907 and 1915 was followed by a short phase of excitement during its formative years between 1915 and the 1920s. In this "formative period," a gradually emerging community of physicists, mathematicians, and astronomers explored the immediate implications of the theory for physics and astronomy, such as the bending of light by a gravitational field that was confirmed in 1919.

Then, during a "low-water-mark period" period (Jean Eisenstaedt), lasting until the 1950s, this growth came to a halt, and work on the theory was, with the exception of work in cosmology, framed by mainstream concerns, including attempts to supersede it by more comprehensive frameworks, such as a unified field theory.[31] General relativity became a marginal issue for most theoretical physicists compared with other branches of research (e.g., quantum mechanics, nuclear physics, and quantum electrodynamics). As a consequence, only a few scientists, mostly working in isolation from one another, invested serious efforts into developing specific aspects of the theory. Those who did so, as did Albert Einstein himself, often regarded general relativity merely as an intermediate step on the path to a larger theoretical framework.

Many of general relativity's far-reaching implications for the conceptual foundations of physics and for the understanding of astrophysics and cosmology—such as the notion of space-time singularity or the concept of black holes—thus remained essentially unexplored until the early 1960s. The rise of professional journals and associations dedicated to general relativity, new astronomical discoveries such as that of quasars, cosmic microwave background radiation, and pulsars, as well as the public visibility of key representatives of the field such as John Wheeler, Roger Penrose, and Stephen Hawking mark what has been called the "renaissance" (Clifford Will) of general relativity after the war. Its hallmark was the emergence of a community of scientists for which general relativity was their main professional concern.[32] The renaissance was followed by the "golden age" (Kip Thorne) of general relativity, culminating in the discovery of gravitational waves.[33]

It thus seems that general relativity had, eventually, created its own epistemic community. How did this happen? Plausible explanations include the argument that the renaissance was stimulated by astronomical breakthrough discoveries made possible by new observational technologies, or that it was a trickle-down effect of the general boom of physics caused by the affluence of the physics community after World War II, when it massively benefited from military funding.

On closer inspection, however, the story—investigated in cooperation with Alexander Blum and Roberto Lalli—was more sophisticated and is best captured by conceiving it in terms of epistemic networks rather than in terms of explanations that reduce the social to the cognitive or the cognitive to the social or material dimensions of knowledge.[34] In this case, the nodes of the social network are the scientists working on aspects of general relativity or related areas but also the institutions supporting research in this field and determining its direction. The nodes of the relevant semiotic network are the external representations serving as vehicles for knowledge transmission, such as articles, books, preprints, and correspondence between the actors. They include the codification of methods and tools applied by the scientists active in this field of research and also the material means of knowledge production, such as observatories or computing machines. The semantic network includes the concepts of general relativity and related knowledge domains in physics, mathematics, and astronomy.

From the 1920s to the mid-1950s, the social network was a loosely connected group of scientists interested in specific aspects of general relativity. Only a few institutions, particularly the Institute for Advanced Study at Princeton, housed small research groups dedicated to the field. Moreover, as a result of rigid disciplinary boundaries between mathematics, physics, and astronomy and the dividing effect of national borders, there was only a limited exchange of information among these few groups and institutions. Prior to 1955, for instance, there were no significant efforts to organize meetings dealing with all aspects of general relativity.

The semiotic network was just as poorly developed. Between the 1920s and the early 1950s, only a few comprehensive monographs had been published. Most of them had been published before 1925, and even these usually only represented the personal approach of the author in question. Not one of these works was in a position to become a central hub of the semiotic network. Because scientists active in this field belonged to various disciplinary communities and different nations, their results were scattered in the publication series of national societies or in journals of various disciplinary focuses, such as mathematics, astrophysics, and physics. The theoretical methods applied by the various actors were highly dependent on their education, as well as on their research goals and contexts, and hence were quite distinct from one another. As a consequence, the density of connections in both the semiotic and the social networks was inadequate to ensure a rapid dissemination of new knowledge.

As for the semantic network, it was split along the lines of distinct research programs pursuing their own objectives. Among the main directions of research were attempts to expand general relativity in order to formulate a unified field theory; the program to quantize Einstein's field theory of gravitation in the sense of a theory of quantum gravity; and the development of relativistic cosmology, in addition to parallel mathematical work on differential geometry, which was also more or less disconnected from other research. As a consequence, it was hardly possible to identify a clearly defined core of knowledge shared by all scientists working within these diverse traditions. Thus, no fully connected epistemic network of general relativity existed before its renaissance.

In order to understand how this situation eventually changed, one has to take into account the interaction between the social network, constituted by a scattered but growing group of actors, the initially weakly connected semantic network of intellectual resources for general relativity, and the semiotic network of publications dispersed over a diverse set of venues.

This interaction took place within a rapidly changing knowledge economy shaped by the political and economic conditions of the postwar period. As a consequence of the fundamental role of physics in World War II and its significance during the global arms race during the Cold War, the number of physicists increased explosively—resulting in the exponential growth of postdoctorate programs in physics.[35] At the same time, a long and mobile postdoc phase became established as a normal stage in a career path. In the 1950s, a number of centers existed in which various research programs related to general relativity were pursued. The new tradition of a long postdoc phase permitted more rapid transmission of knowledge from one of these centers to another and—albeit to a lesser extent—from one disciplinary tradition to another. Many of these young scientists worked successively in three or four of these centers and brought their own knowledge resources with them. At the same time, they remained in contact with the perspectives and research concepts of other institutions. Below, I come back to the important role of these postdocs in shaping the epistemic network of general relativity.

The political East-West split during the Cold War naturally made connections between certain parts of the network more difficult. But East-West connectivity increased with a period of détente starting in the mid-1950s, favored also by collaborations across the Iron Curtain. The Institute of Theoretical Physics of the University of Warsaw, headed by Einstein's former collaborator Leopold Infeld, for instance, was an important hub of the international network, hosting postdocs and sending scientists abroad.[36]

The semiotic network also experienced a rapid multiplication of nodes in the postwar period: the number of articles on general relativity grew rapidly, with a steep rise beginning in the mid-1950s. Nevertheless, the network remained siloed into isolated subnetworks, since the culture of publication was still largely characterized by national differences and local traditions. Between 1948 and 1962, approximately 1,500 papers were published in more than two hundred journals and six different languages.[37]

The renaissance of general relativity involved the interaction of all the network types that I have introduced. The turning point came with the establishment of weak ties connecting the diverse subnetworks owing to the activities of a small number of actors prompted by contingent circumstances. One contingent but consequential activity was the celebration, in 1955, of the fiftieth anniversary of the special theory of relativity at its birthplace in Bern.[38] It amounted to the organization of the first conference dedicated exclusively to problems of relativity theory. On this occasion, which coincided with the death of Einstein, several participants became aware for the first time that research on general relativity had made significant progress in recent years and that they were now, as a community, responsible for its pursuit. The Bern conference led to an enhanced self-awareness of the dispersed scientists working on different aspects of general relativity and to a rapid increase of connectivity in all dimensions of the epistemic network.

Until the mid-1950s, the research centers mentioned above more or less followed a stable research agenda shaped by a few prominent leaders. They were, initially, loosely connected by the exchange of postdocs, an exchange that did not significantly affect the topology of the epistemic network at first, in particular at the level of its semantic dimension. This changed after the Bern conference. Now the postdocs linking different research centers played an ever more important role in the integration of fragments of knowledge and research perspectives into a coherent picture of general relativity. Their mobility established weak ties in all dimensions of the epistemic network. The postdocs enhanced the connectivity not only of the social network but also of the semiotic and semantic networks. Wherever they went, they brought their preprints and discussed how to relate knowledge developed in particular contexts to solve new problems. This increased connectivity contributed in turn to a collective recognition of the untapped potential of general relativity and eventually resulted in an institutionalization process nurturing its revival.

Within just a few years, the structure and size of the social network changed as the result of conscious decisions and measures taken by historical actors who recognized the theory's developmental potential. In the course of this process, the theory itself underwent a profound transformation that turned it, for the first time, into a universally applicable physical framework in which objects to which it applies could be sensibly defined. Only then could something that might merit the characterization of a "paradigm" of general relativity be born. This transformation was consolidated through an institutionalization of the relevant epistemic community by creating, in 1959, the International Committee of General Relativity and Gravitation. In light of the new understanding of general relativity, gravitational waves, for instance, could be recognized as a measurable physical phenomenon predicted by the theory. Similarly, black holes emerged as physically meaningful objects that could now be defined within the theory and explored with astrophysical methods.

When major new astronomical discoveries opened up new astrophysical applications of general relativity, they fell on fertile ground prepared by a global network of scientists capable of addressing these discoveries with the intellectual resources assembled by this network. General relativity eventually became *the* theory of space and time, as well as the recognized basis of astrophysics and observational cosmology, constituting the core of a rapidly evolving semantic network sustained by a robust institutional framework.

Historical Network Analysis

As the examples here presented merely refer to specific case studies that do not cover the ground laid out in the previous chapters, a few remarks may be in order on the future prospects of network analysis for a history of knowledge as I see them.

The examples of the coevolution of systems of knowledge and epistemic communities discussed in this chapter suggest that network analysis, suitably adapted to a history of knowledge, can be helpful in linking micro- and macro-history, as well as diachronic and transversal approaches to the development and spread of knowledge. Network analysis may, at first glance, appear to be merely a reformulation of familiar historical processes veiled in sophisticated language. But it can actually serve to combine the precision of micro-history grounded in "thick" descriptions of historical contexts with the sweeping generalizations of global history by validating such generalizations through rich historical data, their interpretations, and the mathematical analysis of their relations.

The challenge of reaching this ambitious goal lies as much in the elaboration of appropriate theoretical tools, mathematical instruments, and software technologies as in finding ways of assembling and preparing the relevant data. The data required for such an approach are perhaps not typically Big Data, in the sense of the natural sciences, but they are certainly "rich data." Rich data are essential to the computational humanities, which still fall behind in the exploitation of such data, compared, for instance, with bioinformatics, which deals with similar problems. Rich data are often quantitatively more substantial than the data relevant to a typical historical case study and qualitatively more refined and contextualized than the Big Data of many social science studies, in part because in history, the outcome of change processes is often known. Rich data are characterized by a high number of attributes and are typically very heterogeneous. Many network models, in contrast, are developed for data defined by a low number of attributes.

As mentioned at the beginning of the chapter, dealing with multilevel and multidimensional networks, such as the social, semiotic, and semantic dimensions of an epistemic network (with each dimension comprising many levels), opens up a number of theoretical and technical challenges. An intriguing feature of epistemic networks is their self-referential character, that is, the historical actors' realization that they participate in a specific network activity may shape their thinking and acting, and thus influence, in turn, the network activity itself. This realization can act as a turning point in the emergence of regulative structures operating on a network, as the examples of calendar reform and the renaissance of general relativity illustrate. It may be considered an act of collective reflexivity. The understanding of such acts typically requires a mixed approach: a combination of traditional scholarly analysis and interpretation with novel quantitative methods directed at dealing with the complexity of the underlying interactions. Since such acts will become increasingly relevant to addressing future challenges, it is worthwhile to extend the limited case studies presented here to larger strands of the global history of knowledge. In chapter 16, I come back to the role digital media may play in supporting such acts of reflection and self-organization.

PART 5

ON WHAT KNOWLEDGE OUR FUTURE DEPENDS

Chapter 14

Epistemic Evolution

Light will be thrown on the origin of man and his history.
— CHARLES DARWIN, *On the Origin of Species*

In the social production of their existence, men inevitably enter into definite relations, which are independent of their will, namely relations of production appropriate to a given stage in the development of their material forces of production.
— KARL MARX,
"A Contribution to the Critique of Political Economy: Part One"

... organisms must be analyzed as integrated wholes, with *Baupläne* [construction plans] so constrained by phyletic heritage, pathways of development and general architecture that the constraints themselves become more interesting and more important in delimiting pathways of change than the selective force that may mediate change when it occurs.
— STEPHEN GOULD AND RICHARD LEWONTIN,
"The Spandrels of San Marco and the Panglossian Paradigm"

Science Becomes Existential

We have seen various appearances of the natural sciences—for instance, in ancient Greece and in ancient China. I have emphasized the unique characters of the emerging science, arguing that the development of science is not a unidirectional or deterministic process. Modern science is evidently one outcome of a global history of knowledge that cannot be understood without taking into account the interaction of knowledge with a variety of other societal structures, in particular the rise of capitalist and later industrial economies. The coupling of science with economy has itself a long history, reaching back at least to the self-reinforcing mechanisms of the early modern knowledge economy.

In chapter 12, I raised the question of the origin of modern science and compared it to the question of the origin of the human species. The answer to the latter question involved considerations about the emergence of a new evolutionary process: cultural evolution. Cultural evolution began as a peripheral phenomenon of biological evolution—beginning with tool use and social learning—before it eventually became the dominant process of human history. Can the question about the origin of modern science, not as a marginal activity, but as a game-changing event of global history, be addressed in a similar way, as a question about evolutionary processes and, in particular, the onset of a new evolutionary process, complementing biological evolution and modulating cultural evolution?[1]

In this case, it becomes plausible that science could emerge *and* become a dominant factor of our current state without being the necessary result of some initial conditions. From such a perspective, scientific knowledge is just as unlikely to be an inevitable consequence of some given initial conditions as, according to Darwin's theory of evolution, the emergence of human beings is but a contingent result of the processes of evolution. At the same time, considering the succession of cultural and biological evolution makes it conceivable that this cascade of evolutionary processes might be continued by new forms of evolutionary processes, first emerging as marginal effects of the currently dominating evolutionary process and then taking over.

There are speculations that evolutionary processes beyond cultural evolution will involve the self-organizational capacities of technological systems, based in particular on future developments of artificial intelligence. In this case, evolution beyond cultural evolution will primarily involve not the biosphere but the technosphere, that is, the global ensemble of technical systems and infrastructures initially created by humans but that meanwhile may have begun to develop its own dynamics, pushing human actors into a marginal role. In chapter 16, I come back to the concept of the technosphere and its role for assessing our predicament in the Anthropocene.

In my view, the dystopian vision that cultural evolution is about to be replaced by some kind of cybernetic evolution of large technological systems, perhaps in combination with technically upgraded human beings, glosses over the important transformation that cultural evolution has already undergone through the involvement of science and technology. This involvement has made human actors both more powerful and more dependent on the technological niches they have created for themselves, including the use of intelligent systems. It is not least through the development of science and technology and their implication in industrial economies that humanity has been able to expand to its current size of more than seven billion people, and it will only be possible with the help of science and technology to maintain cultural evolution for a humanity of this order of magnitude. With regard to the envisaged possibility of artificial intelligence taking control over the planet, human knowledge is currently still in a role to tip the balance one way or the other. In short, cultural evolution has become crucially dependent on the global knowledge economy and, in particular, on the scientific, technological, social, political, and other kinds of knowledge it generates and distributes.

A particular role is played by the knowledge necessary for creating, maintaining, and further developing the large social and technological infrastructures on which global cultural evolution has become dependent. Some of this knowledge can only be produced in the sophisticated knowledge economies we have come to see as being characteristic of science. In this sense, humanity has entered a new phase of cultural evolution, or possibly into the next stage of a cascade of evolutionary processes, once this specific type of knowledge economy has turned from an accidental into a necessary condition for preserving, sharing, and developing the achievements of cultural

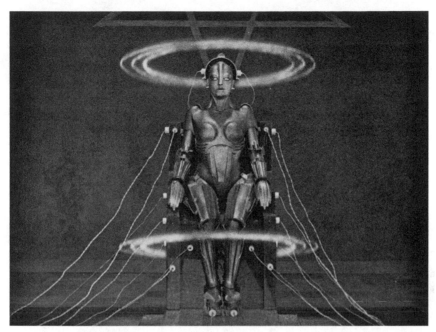

14.1. Maria, the android, in Fritz Lang's expressionist 1927 film *Metropolis*, one of the first robots featured in cinema. Walter Schulze-Mittendorff—WSM Artmetropolis.

evolution on a global scale. As a shorthand, I use the term "epistemic evolution" for this new, distinctive phase and evolutionary process.

There is much to suggest that we have already entered into this new evolutionary logic, perhaps at some point during the nineteenth century, if we consider, for instance, that without production processes made possible by the natural sciences and technology, we would no longer be able to feed the human population, an argument to which I return in the next chapter. This new evolutionary logic is characterized by couplings between the natural sciences, industrial production, and other aspects of societal development, and has since shaped the history of science and technology in a specific way. We have seen the powerful consequences of such an entanglement of epistemic and societal structures in some of the examples already discussed.

To avoid a possible misunderstanding: Science and technology may still be considered merely specific aspects of cultural evolution, just as early forms of tool use and social learning in hominid evolution did not stop biological evolution. But the existential role the new learning processes eventually assumed in human history—social learning in the case of cultural evolution and the accumulation of scientific knowledge in epistemic evolution—justifies, in my view, the introduction of a new term for this major transition. This is all the more the case as the survival of human cultures as we know them will not only depend on the continued accumulation of scientific and technological knowledge but on the ways in which this accumulation will be shaped in the future and on how the knowledge it generates will be put into practice in order to ensure this survival. That this might also involve major societal transformations, if not revolutions, is another matter.

To avoid another misunderstanding: Science was, by itself, not the major driving force catapulting humanity into epistemic evolution. As discussed in chapter 10, the pursuit of scientific practices was, for most of human history, dependent on contingent external circumstances such as patronage. Only through the capitalist mode of production did science become an important factor of production. In the course of the Industrial Revolution, science and technology became involved in the expansive dynamics of capitalist economies, with planetary consequences. But in the course of this process and because of its planetary impact, science and technology have been transformed from boundary conditions of economic development into indispensable factors of cultural evolution, on whatever form of economy or societal organization it will be based in the future.

This is comparable to the situation at the beginning of human history, when the generation of cultural knowledge turned into an existential condition for the evolution and further existence of *Homo sapiens*, thus becoming the starting point for the many pathways of cultural evolution. Just as cultural evolution was grounded in biological evolution, so this new form of evolution—epistemic evolution—is grounded in cultural evolution. With each new evolutionary process in this cascade, the preceding ones eventually become to some extent dependent on the following layers. Thus, the continued existence of our species in a biological sense became dependent on cultural evolution once the latter had reached a global level, and with the globalization of an economy dependent on science and technology, the survival of human culture as we know it becomes dependent on epistemic evolution.

In the remaining chapters, I refer to further hints pointing at the onset of a new evolutionary process beyond cultural evolution. This perspective may help us to rethink the concept of the Anthropocene with respect to the evolution of the human species. Perhaps one can even identify the onset of epistemic evolution with the much-debated beginning of the Anthropocene without falling into the trap of origin questions mentioned in chapter 12—just as the concept of the Anthropocene stimulates us to reconsider the biological (or natural) and cultural dimensions of future human development. But even if the question of the emergence of a new, distinctive evolutionary process must remain speculative, it does suggest that we rethink cultural evolution in light of the role of knowledge as it has been discussed in this book. For this purpose, I use the following to explore ways of reformulating cultural evolution as involving knowledge in essential ways. I illustrate such reformulations with a number of examples from the history of knowledge, leaving the question of the nature of epistemic evolution to future investigations.

One of my starting points is the fundamental role of material culture in shaping the way in which humans reproduce themselves within certain socioeconomic structures.[2] The transmission of this material culture from one generation to the next constitutes a basic continuity of human history. Of similar importance to what follows is the distinction between historically bestowed conditions of a society that it can reproduce by itself and other external conditions (e.g., ecological or social resources) for which this is not the case. These concepts and insights go back to Marx and have often been viewed as amounting to a deterministic account of human history. Such a misinterpretation overlooks, however, that Marx's conception merely implies that the material means of production open up and condition a historically specific horizon of possibilities for social action, not an automatism of historical development, let alone guaranteed progress.

How can one conceptualize cultural evolution in a way that takes into account our insights into the development of knowledge? On the foundation of what has been discussed in previous chapters, the above picture can be supplemented (though not completed!) by taking into account the role of knowledge. A historically inherited material culture not only confines the range of possible social structures but also the range of knowledge that can be attained at a given stage, as well as possible knowledge economies. The knowledge and the knowledge economy available in a given situation confine, in turn, the possibilities of a society to reproducing certain historically given conditions as opposed to others. From this perspective, knowledge may be recognized as an important regulative mechanism of cultural evolution, alongside other regulative mechanisms, such as institutional structures.

But however desirable an evolutionary framework for a history of knowledge may be, the important question is whether it is actually possible to recognize an evolutionary logic in the historical records—without imposing it by an exaggerated analogy with biology and without ascending to a level of abstraction where all cats become gray. I believe that the historical findings examined in the preceding chapters point in such a direction, in particular the long-term, cumulative aspects of knowledge development, its dependence on contingent societal contexts, and the profound transformations of the architecture of knowledge.

Examples are the emergence of new systems of knowledge from a reorganization of preceding systems; the sedimentation and plateau-building processes of knowledge economies; the transformation of contingent circumstances and challenges into internal conditions for the further development of knowledge systems, accounting for the path dependency and layered structure of this development; and the feedback mechanisms that may arise between knowledge economies and knowledge systems, giving rise to the emergence of new epistemic communities.

Just like the evolution of life, knowledge development has direction but is not globally uniform. It is neither deterministic nor teleological. Chance events may have long-term effects by becoming incorporated into the developmental process. Knowledge development is self-referential insofar as it contributes to shaping its own environment by processes of sedimentation and plateau formation corresponding to niche construction in biology. It is also a layered process, in the sense that later forms of knowledge do not necessarily replace earlier ones. External representations shape the long-term transmission of knowledge, ensuring its continuity, while their exploration under different circumstances opens up possibilities for variation and change.

Extended Evolution

There are many ways in which such a generic description of our findings may be articulated in terms of an evolutionary theory. But ultimately, the evolution of knowledge will simply be a particular perspective on cultural evolution. Cultural evolution displays features that have gained prominence with recent developments of evolutionary theory, such as the role of niche construction and that of complex regulatory networks. In the following, I argue that both can be integrated into a conception of "extended evolution."[3]

In particular, human societies transform their environments by means of their material culture, which forms a "niche" and decisively shapes their historical evolution. Furthermore, human societies do not vary randomly but in ways that are governed by societal structures regulating the behavior of individual and collective actors. The institutions and cognitive structures discussed in earlier chapters are examples of such regulative structures. The niches that humans have constructed in the course of history do not just affect what biologists call fitness landscapes (i.e., the conditions shaping selective processes)[4] but provide crucial regulatory effects. A concept of extended evolution should therefore integrate regulatory network and niche construction perspectives within one framework of interacting causal factors. A crucial aspect of such a framework is that niche construction depends not only on complex regulative structures but in turn also shapes them. The constructed niche may thus function as an extended regulatory system. I therefore also speak of a "regulatory niche." Let us now consider how this works for cultural evolution.

Human practices change the environment in ways that are characteristic of an ensemble of actors, such as a society and its regulatory structures, including its social organization and knowledge. The resulting transformation is shaped by both the regulatory structures and the nature of the material objects and means involved. The transformation also affects both the actors and their material environment. The material environment often reveals ways of repurposing aspects of a society's material culture through tapping into its inherent potential. Biologists call such a repurposing of existing features an "exaptation"[5]—for instance, when, according to one theory, feathers evolved as a means of temperature regulation and were eventually repurposed for flight.

I refer to the generation of the material, external results of this complex interaction between a society and its environment as "externalization." The converse process affecting its regulatory structures is called "internalization." Internalization may be realized by different mechanisms in biological and cultural evolution, and also in societal and individual developments. It is important not to misunderstand externalization as some kind of projection of internal structures onto an outside world, nor internalization as the direct reflection of this outside world—they always depend on both the transformers and the transformed.

Human material interventions do not leave their social and cognitive systems indifferent but constantly change their internal structures, resulting in what one might call an "endaptation." In the simplest case, an endaptation is limited to the assimilation of new experiences to systems of shared knowledge and the regulative structures of existing institutions. In general, however, new experiences may lead to the transformation of these structures. Just as exaptation amounts to the emergence of new external functionalities of existing features, endaptation refers to the emergence of new internal regulatory possibilities offered by an existing environment. The latter may also be illustrated by the seriation task discussed in chapter 3, showing how general cognitive structures are built up from specific experiences.

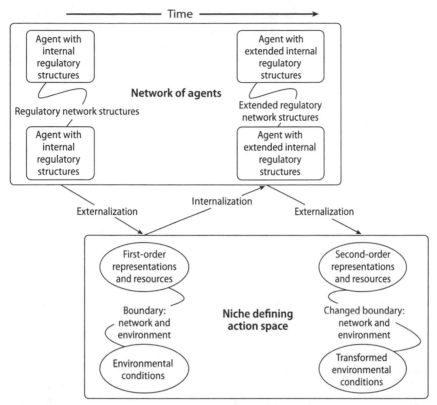

14.2. Networks of agents evolve, including their internal regulatory structures and niches. The niche itself has a network structure induced by the primary network. Its nodes are those aspects of the environment that condition, mediate, or become the target of actions—in short, the environmental resources and conditions of the internal system. The extended network, including the environment, defines an action space that shapes possible innovations, canalizes the evolutionary process, and delimits the structure of the inheritance system. See Renn and Laubichler (2017).

As discussed in the context of the emergence of abstract concepts in chapter 3, what is being internalized in this process is not the environment per se but the ways in which human actors practically relate to it. Like externalization, internalization is always mediated by both the actors and their material means and objects. Endaptation refers to the effects of this complex interaction on the regulative structures and also depends on both the transformers and the transformed.

These specifications may appear to be subtleties, but they are of fundamental importance. They should make it clear that one cannot separate the regulatory structures of human societies from their material embodiments and that the material environment always means more than an embodiment of human regulatory structures or the context of human practices. The materiality of the environment imposes, so to speak, its own logic on any interaction. The human material interventions in their environment that I have described with the help of the concepts of externalization and internalization do not therefore form coupled processes that always perfectly complement each other to constitute a predetermined developmental process. Rather, the historical processes I am describing here from an evolutionary perspective are open, not only in the sense that they leave room for contingent circumstances to become "internalized" within this evolutionary process, but also in the sense that their outcomes always depend as well on the much wider contexts of nature and society we have not been considering here.

Still, an understanding of the interplay between environmental transformations and the regulatory structures of human societies offers insights into the dynamics of cultural evolution that may be compared with attempts to understand the regulatory functions of niche construction in biological evolution when the environment becomes part of an extended biological system. What is more important here is the possibility this perspective offers to bring social and epistemic regulatory processes into the same framework centered on the material nature of human actions and practices.

The implications of this framework for cultural evolution may be succinctly summarized in terms of a famous quotation from Karl Marx: "Men make their own history, but they do not make it just as they please; they do not make it under circumstances chosen by themselves, but under circumstances directly encountered, given and transmitted from the past. The tradition of all the dead generations weighs like a nightmare on the brain of the living."[6] Our focus is human practice in the context of a given material and social culture. This culture may be considered as a "niche" resulting from prior actions, which is what Marx refers to when he says that men make their own history, yet "they do not make it under circumstances chosen by themselves, but under circumstances directly encountered, given and transmitted from the past." The cultural evolution of a human society involves its dealings with challenges and experiences, which are then internalized in its mental and social structures. In Marx's words: "The tradition of all the dead generations weighs like a nightmare on the brain of the living." The mechanism giving rise to specifically human systems of social organization and thinking, mediated by material culture, may thus be described as an iterative process of transformations of human practices and their internal and external conditions.

This scheme, involving the processes of externalization and internalization described above, generalizes the features of transformation processes that have been discussed in the previous chapters, yet remains specific enough to guide historical explanations that must, of course, always be founded on an analysis of the historical sources—as is the case for evolutionary explanations in biology. Now it would be plausible to elaborate this scheme in terms of the network structures discussed in chapter 13 and to consider a system and its regulatory niche as an extended regulatory network. Actions are then regulated by the structure of this extended regulatory network. They may be directed at the environment or at other agents and would always involve environmental resources. Note that the distinction between niches and environments is scale-dependent, that is, that the niche for an internal network at one scale may be part of the internal network at another. Regulatory evolutionary changes leading to genuine novelty can then be explained as a consequence of the creation of additional regulatory modules (called "epicycles" by Tomlinson) or by network transformations. Major evolutionary transitions may be characterized by the emergence of new platforms that allow systems to perform new kinds of

14.3. From the *Book of the Dead* (an ancient Egyptian funerary text used from ca. 1550 to ca. 50 BCE): The Papyrus of Hunefer. *Above:* Hunefer kneeling in adoration before a table of offerings in the presence of fourteen seated gods, serving as judges. *Below:* The Judgment, or weighing of the conscience; the jackal-headed Anubis examines the pointer of the balance. British Museum, London. © akg-images / Erich Lessing.

functions. But rather than going into more technical detail of the concept of extended evolution, let us now discuss some examples.[7]

The Evolution of Technology

Consider, for instance, the evolution of technology as a particular aspect of cultural evolution.[8] From our perspective, technological devices may be considered material representations of the institutional and cognitive regulatory structures of the societies that invent, produce, and use them; they shape these structures in turn by creating spaces of action that determine what people can and cannot do with them under the given historical circumstances.

The emergence and transformation of weighing technologies may serve as a relatively well-studied example. Weighing technology was originally introduced for regulating social and cognitive processes dealing with the exchange of goods. The external representations of these regulative processes comprised, among other things, a simple balance, standard weights, and a specialized technical terminology. Reflection on these external representations gave rise to an abstract and quantitative concept of weight, distinguished from other bodily characteristics such as bulk or material quality.

Weighing technology emerged when the administrative and economic developments of early urban societies began to involve standards for exchange values, as discussed in chapter 3. In Egypt and probably slightly later in Mesopotamia, the lever balance with equal arms of fixed length was introduced at around the turn from the fourth to the third millennium. In Mesopotamia, standardized weights have been preserved dating from the mid-third millennium BCE. In the context of the political and economic globalization processes of the first millennium BCE, the use of standards of value and of weight spread widely. By the middle of the first millennium, coined money was common in Lydia, Greece, and India, and somewhat later also in China.

14.4. Greco Buddhist bas-relief from the Gandhara culture (between the second and third centuries BCE), which flourished in what is today a border region between Afghanistan and Pakistan. The relief shows a scene from one of Buddha's former lives as King Sibi. On the left, flesh is cut from one of the king's legs and offered to a hawk in order to ransom the life of a pigeon. The flesh is weighed with what is obviously a bismar, a balance with a moveable fulcrum. British Museum, Inv. Nr. 1912, 1221.1. © Trustees of the British Museum.

Human thinking is deeply affected by the availability or nonavailability of simple artifacts like balances and standard weights, and this availability or lack constitutes a "niche" in the evolutionary parlance introduced above. That becomes particularly evident if we consider the diversity of basic concepts still induced today by this elementary aspect of material culture. In an intercultural study of intuitive physical thinking, the psychologist Katja Bödeker compared weight conceptions of European children with those of children growing up on the Trobriand Islands, an archipelago of coral atolls in the Solomon Sea, east of the coast of New Guinea, a classical site of ethnological research from the times of Malinowski.[9]

While certain aspects of intuitive physics are similar in Europe and on the Trobriand Islands (such as the interpretation of motion by the motion-implies-force model discussed in chapter 4), Bödeker found remarkable differences in the physical understanding of weight. In contrast to European children acquainted with balances, for children on the Trobriand Islands, unfamiliar with such measurement devices, objects such as hair or feathers have absolutely no weight, even if a lot of them are piled up.[10]

Without balances, there is neither occasion nor reason to form an extensive concept of weight. The islanders possess, on the other hand, many forms of knowledge of their own—for instance, a sophisticated knowledge about time and time measurement that is distributed in the community among experts and laymen and closely interwoven with agricultural concerns and other social activities.[11] While such local knowledge may be on the decline as a result of cultural contact, it may be of nonnegligible relevance precisely in the context of the global challenges of the Anthropocene to which I return in the next chapter.

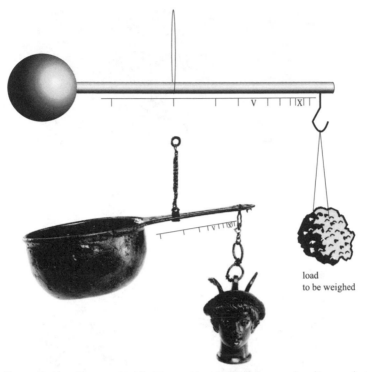

14.5. *Above*: Graphic representation of a bismar with moveable fulcrum and nonlinear scale. *Below*: An improvised bismar from Pompeii: a pan converted into a balance by cutting a slit into its handle in which a suspension chain can slide. The pot serves as counterweight, while the load is hung from the end of the handle (to the right). The head of Mercury now attached to the handle is a reconstruction error. See Damerow, Renn, Rieger, et al. (2002, 99). Museo Archeologico Nazionale, Naples, inv. 74165.

The basic principle of balances remained static for millennia: the weight of the item to be weighed on one arm of the balance was compensated (or literally "balanced") by the identical weight of one or more standardized balance weights placed on the other arm of equal length.[12] This only changed when a new type of balance emerged on the basis of a different principle: the balance with variable arm length, more commonly referred to as the unequal-arm balance. This type of balance is recorded in the late fifth century BCE in Greece and may have been in use at the same time or somewhat later in India. In balances of variable arm length, the counterweight remains unchanged. Instead, the distances at which the forces act from the fulcrum are varied to bring about equilibrium.

The earliest evidence of the introduction of balances with variable arm length comes from a play by Aristophanes, *Peace*, which was first staged in Athens in 421 BCE. In the play, a maker of war trumpets is ridiculed because he cannot figure out what to do with his surplus trumpets. Trygaeus, the central character of the play, suggests pouring lead into the bell and adding "a scale pan hung with cords, and you'll have just the thing for weighing out figs. . . ."[13] Despite being rather abridged, this description of the transformation of a trumpet into a balance makes it rather

14.6. A steelyard or Roman balance with two different fulcrums and their corresponding suspension hooks. The steelyard has a moveable counterweight (to the left), while the load is placed in the scale pan (to the right). When the other fulcrum is used, the steelyard is turned around and so presents a different weighing range. Such sophisticated balances were extremely widespread throughout the Roman Empire and used in the Mediterranean area until recently. See Damerow, Renn, Rieger, et al. (2002, 102). Museo Archeologico Nazionale, Naples, inv. 74041.

unambiguously clear that the trumpet is turned into a specific type of unequal-arm balance: the bismar.

In this kind of balance, equilibrium is produced by altering the position of the fulcrum with respect to the beam, for instance, by a moveable thread from which the beam is hung. In the more familiar Roman balance, also referred to as the steelyard, equilibrium is reached by varying the distance at which a movable counterweight acts from the fulcrum. In India, bismars are attested to as early as the end of the fourth century BCE. The archaeological records suggest that the bismar enjoyed a continued tradition. It can still be found in use today.

The allusion in Aristophanes's play to the transformation of an everyday item such as a trumpet into a bismar underscores the comparatively scant requirements posed by its construction. The bismar is indeed a rather simple instrument; its most palpably complicated feature is its nonlinear scale, where the scale intervals representing equal weight differences follow a harmonic division. In antiquity, bismar scales were, however, not theoretically established but empirically constructed by gauging. Their construction therefore poses only minimal requirements for the underlying mechanical knowledge.

Yet the establishment of a balance with unequal arms as a socially acceptable technology was contingent on specific preconditions that formed the niche in which this innovation became possible. This preexisting level of knowledge and of societal regulations involved the abstract concept of weight, its representation by a series of standard weights as part of the material culture, and a widespread familiarity with weighing practices. These practices provided the cognitive and social regulations needed to empirically gauge this new kind of balance and to make this gauging acceptable to its users without requiring any further sophisticated knowledge. Their firm establishment created the specific regulatory niche (or platform) in which such a counterintuitive new type of balance could be invented and established.

A Brief History of Space

Our evolutionary framework makes it possible to identify and succinctly characterize stages in long-term developments, such as that of the human mastery of space investigated by Matthias Schemmel and his research group.[14] According to their results, this long-term history may be roughly divided into five stages that match the stages of knowledge economies already discussed, only now we recognize more clearly the evolutionary mechanism leading from one stage to the other. All stages may be characterized in terms of mental and social regulative structures together with their external representations, including language.[15] They evidently build on each other, but in such a way that higher-order or later structures of spatial thinking and practice do not necessarily or completely substitute for prior ones. As was emphasized in chapter 10, the transitions from one stage to the next are not predetermined but contingent on external circumstances and thus highly path-dependent. They involve processes of reflection that depend, in particular, on the availability of external representations forming a regulatory niche. Here are the five stages:

1) *Naturally conditioned space* in the sense of schemata of action based on the similar biological constitution of all humans and the fundamental similarities in their physical environments. These schemata of action are rooted in sensorimotor intelligence that allows for spatial inferences to be drawn in the context of practice and perception but are otherwise inaccessible to the actors.

2) *Culturally shared space*, which is still shaped by elementary mental models controlling the action and perception of "naturally conditioned space," such as the so-called permanent object and landmark models familiar from studies in developmental psychology. The first allows for the handling of bodies in our vicinity, while the second underlies cognitive mapping skills, allowing for navigation through various environments. But now these mental structures are endowed with cultural meaning. Thus, culturally shared large-scale space is spanned not only by landmarks, places, regions,

and their relations, but also by the meanings attached to these entities. Aspects of the natural and cultural environments, culturally conditioned practices, and language serve as external representations.

Knowledge about the environment is explored and accumulated by societies over the course of many generations. The resulting immense cultural diversity is due to the fact that different knowledge systems and institutions may represent responses to the challenges of very different ecologies, but it is also due to the different evolutionary trajectories along which this exploration takes place.

3) *Administratively controlled space* as it was conceived and practiced, for instance, in the ancient civilizations of Mesopotamia, Egypt, China, or India. The social control of space involved mental models of practical knowledge related to building activities, urban planning, surveying, and field measurement. They might be externally represented by buildings, models, instruments, measuring tools, graphical representations, or symbol systems, for example. In ancient societies, new forms of spatial knowledge emerged from exploring these tools, as well as by exploring the external representations employed in administrative practices.

In this way, for the first time, units of length, area, and volume were integrated into metric systems spanning spaces of different scales, whereas no previous relation had existed, for example, between spatial dimensions in the bodily realm and the spatial dimensions of a journey. However, such integrations of different domains of spatial knowledge may have taken different forms in different cultures. They took place, in any case, within a niche shaped by the preexistence of first-order representations supporting historically specific regulatory structures of spatial thinking and allowing for their extension by exploring the potential inherent in these representations.

4) *Higher-order concepts of space* as they are externally represented by written texts, possibly comprising diagrams, formalized language, and other symbol systems. Beginning in Mesopotamia and Egypt, and then more markedly in Greece and China, the division of societal labor generated a knowledge economy with new groups of actors and structures of social interaction such as schooling and disputation. Here, existing concepts of space and their representations could be further explored. Thinking about space was regulated, to begin with, by the first-order concepts of space emerging at the preceding stage and by their external representations. Reflecting on practices involving these external representations in turn gave rise to higher-order concepts of space as they are represented, for instance, in Euclid's *Elements*.[16]

Owing to the representation of this higher-order knowledge of space by written texts, a tradition could be picked up even centuries after it had last been actively pursued. Different evolutionary lines may be distinguished: Mathematical higher-order knowledge, in particular, resulted from reflection on practical spatial knowledge and the use of instruments.

Philosophical higher-order knowledge resulted from reflection on the linguistic representations of intuitive spatial knowledge. Further explorations of second-order concepts of space and reflections on their results led
to a proliferation of theories of space both in philosophy and geometry.

5) And finally, *empirically controlled spatial concepts and practices* as they
emerged in the multiplication of spaces of experience by political expansion, trade, exploration, and engineering. This expansion changed (as an
"externalization") both the natural and social environments, as well as the
world of symbolic and instrumental representations—a transformation
that, in turn, triggered an "internalization" in the form of new regulative
structures, both in the social and the cognitive realms.

In early modern Europe, for instance, the accumulation of experiential knowledge took place in part within institutions specifically designed
for the purpose of knowledge acquisition, such as academies or universities. Intellectually, the accumulating empirical knowledge was organized
in integrative structures based on symbolic and formalistic languages:
numerical coordinates, analytic geometry, calculus, and differential equations. These structures stabilized or brought about empirically controlled
spatial concepts and practices that are highly counterintuitive, for instance,
the knowledge that Earth has a spherical shape and land masses and water
distributed all over it.[17] The theoretical knowledge resulting from the
expansion of experiential spaces also had repercussions on other layers of
knowledge, as well as on regulatory structures of societal practice. An
example is the impact of global, geographical coordinates on navigation
techniques, in particular on deep-sea navigation. This last stage also comprises the emergence of new space and time concepts in the context of
Einstein's theories of relativity, discussed in chapter 7.

An Even Shorter History of Time

The similar biological configuration of all humans and the fundamental similarities of their physical environments suggest that there are structures of temporal cognition that do not vary significantly across epochs and cultures, and
therefore constitute the basis of all temporal thought. On the other hand, the
cultural development of human society from simply structured social groups
all the way to city-states and empires led to new and diversified forms of
social experience and to the symbolically mediated control of time—for instance, in the form of rituals, calendars, written records, and clocks.

The concept of time expanded significantly as the horizon of human experience was extended through explorations into history, geology, and
biology. As a whole, research on history and natural history did for the concept of time what the discovery of new continents had done for the concept
of space. In the development of modern physics, the horizon of experience
on which the concept of time is based was extended once again (and much

further) into both smaller and larger units of time, with occasionally revolutionary consequences.

Developmental psychology has identified some elementary conditions of temporal cognition: the ability to form sequences of actions and events in one's mind, the ability to comprehend movement, and the ability to coordinate sequences and movements with one another.[18] All three abilities are common to animal behaviors in varying complexity. They involve memory, causality recognition, and the differentiation of distinct movements, and they develop in human ontogenesis over the course of the progressive coordination of patterns of action.

The development of human culture subjects these three dimensions (memory, causality recognition, and movement differentiation) to essential changes: memory changes through the development of external representations, which make even isolated actions and events comparable; causality evolves according to ever-new possibilities of intervening in one's environment; and movement differentiation is altered by the scope of knowledge that expands as a consequence of human culture.

The character of the human understanding of time thus consists not primarily in some specifically human, biologically rooted "time organ" but rather in the human capacity for social cognition. An elementary form of sharing temporal knowledge is joint action. But it is language in particular that allows individuals to exchange information about temporal experiences. Most languages code temporal sequences at the most basic syntactic level. In general, the linguistic representation of the elementary structures of temporal knowledge does not, however, cover any context-independent abstract terms for time and its properties.

It is hardly self-evident that the experience of temporality in music, the experience of temporality during a journey, or the experience of temporality in a generational shift should be covered by an overarching concept of time. Instead, the development of such a concept depends on two essential prerequisites: the occurrence of real processes that mediate between different domains of temporal experience, and the possibility of representing such experiences in a medium that allows reflection thereon.

An essential stage in the development of the concept of time was therefore the emergence of regular activities determined by animal migrations, harvesting opportunities, and later by crop cultivation and animal husbandry, and the requisite grasp of the way these relate to seasonal and astronomical cycles. Certain human and natural processes thus entered into a real correlation that found its expression in constantly recurring sequences of acts and their coordination with natural events. An important external representation of this coordination was the calendar in all of its forms, including rituals and ceremonies (see the discussion of the calendar in chapter 13). The mental correlation to this coordination is a concept of time that covers both the earthly and the heavenly spheres, as was (and is) reflected in the religious and astrological ideas of many cultures.

Depending on the specific representations and the availability of tools such as arithmetic, the calendar allowed humans to relate the periodicity of the seasons, the changing phases of the moon, and the alternation of day and night to one another, and to identify the cycles of their coincidence. In this way, a concept of time was created that was characterized by the cyclical repetition of events. While the Mesopotamian, Egyptian, Chinese, Meso-American, and Greco-Roman cultures all began with basically cyclical models, they nevertheless came to articulate very different concepts of time. This diversity reflected the ways in which different domains of temporal experience were brought together and integrated in overarching schemes.

The further development of the concept of time was decisively shaped by the invention of instruments for determining time—such as the gnomon, sundial, and water or sand clocks—as well as by the media available for representing and integrating various temporal experiences—such as rituals, mythological narratives, iconic representations, or philosophical worldviews. Sundials in Greece, for example, were not only widely spread but also came in very different forms, even spherical. This form had a particular advantage in that the spatialization of time became easily discernable, matching the uniform celestial motions in an apparently spherical sky.[19]

14.7. A spherically cut sundial with central gnomon point from Pompeii. This form of sundial had the advantage that the spatialization of time accomplished by such instruments took on a particularly simple form that may have supported the idea that celestial motions are uniform and take place in an apparently spherical sky. Berlin Sundial Collaboration, Image of Dialface ID 25, Pompei Inventory Nr. 34218.

In the ancient world, the systematic observation of sequences of astronomical events and their integration into models of celestial motion (such as those of Greek natural philosophy) eventually led to a new understanding of time that associated the passage of astronomical time with uniform and continuous circular movement.[20] Along with the geocentric worldview, the idea of uniform time passage underlying all celestial motions was thus established, offering a foundation for numerous scientific developments.

In early modern science, the conception of a universal time eventually assumed a fundamental role similar to what Euclidean geometry did for the science of space.[21] For a long period, however, astronomical time remained a far-from-universal reference for the social time of practical life. Indeed, well into the age of modern industrialization, uniform astronomical time was only one of many possible times, each of which claimed predominance in its given domain.[22] Further transformations of the physical concept of time are discussed in chapter 7; time as a cultural abstraction is mentioned in chapter 8.

The Emergence of Human Language

The framework of extended evolution, developed jointly with Manfred Laubichler, helps us to understand what may be called "XXL transitions," that is, major transitions that fundamentally affect the position and role of humans in their environment, as in the cascade from biological to cultural to epistemic evolution. The emergence of human language and the development of food production by agriculture and domestication are examples of such XXL transitions that I shall now briefly discuss before finally turning to the XXL transition represented by our entry into the Anthropocene.[23]

The evolutionary mechanism giving rise to specifically human ways of thinking is often described in terms of distinct thresholds.[24] Instead of postulating distinct evolutionary steps leading from biological to cultural evolution, our evolutionary scheme rather suggests a continuously working feedback mechanism. In this feedback mechanism, ecological circumstances acting as evolutionary driving forces are themselves partly created by the regulative structures of human evolution through niche construction. This is because the material aspect of human practices turns out to be crucial for the transmission and transformation of the evolving regulative structures. They could function as an external memory, as catalysts for the emergence of different perspectives, and as triggers for reflection; as such, they affect all the dimensions of thinking processes that have been discussed as important components of characteristically human thinking.

In *A Natural History of Human Thinking*, Michael Tomasello, who analyzed the characteristics of human thinking on the basis of extensive experimentation with nonhuman primates and human children, investigates the origin of the cooperative nature of human thinking.[25] He postulates two key evolutionary steps. In the first step, a novel type of small-scale collaboration in human foraging led to socially shared joint goals and "joint attention," creating a possibility for individual roles and

14.8. Tai chimpanzees using sticks as extractive foraging tools to gain access to otherwise unobtainable food sources such as honey (*left*) or termites (*right*). The tools are chosen for properties such as flexibility or diameter and are adjusted by the chimpanzees for suitable length. Photos courtesy of Liran Samuni, Tai Chimpanzee Project.

perspectives within ad hoc situations. In the second step, which is characterized by growing human populations competing with each other, humans developed "collective intentionality," enabling them to construct a common cultural ground by means of shared cultural conventions, norms, and institutions. The evolutionary mechanism is thus described in terms of ecological circumstances driving humans into more cooperative ways of life and fostering adaptations for dealing with problems of social coordination.

Against the background of extensive empirical studies involving comparisons between children and nonhuman primates, Tomasello identifies the specific cognitive abilities emerging in these two evolutionary steps, which are designated as joint and collective intentionality, respectively. Joint intentionality is characterized by the fact that humans can conceptualize the same situation from different perspectives, that they can make recursive inferences about each other's intentional states, and that they can evaluate their own thinking with respect to the normative perspectives of others. Collective intentionality extends joint intentionality to include a conventional dimension of these cognitive capabilities, which are now broadly shared within a culture and no longer a matter of ad hoc situations.

Tomasello refers to the famous treatment of major transitions in evolution by John Maynard Smith and Eörs Szathmáry,[26] pointing out that "humans have created genuine evolutionary novelties via new forms of cooperation, supported and extended by new forms of communication. . . . And humans have done this twice, the second step building on the first."[27] For Maynard Smith and Szathmáry, the major transitions involve changes in the way information is stored and processed. Their own evolutionary account of language, for example, involves the explanation of the genesis of grammatical structures by genetic assimilation, essentially turning cultural into biological inheritance. Such an argument may fit well with human cognitive evolution being characterized by major evolutionary transitions, as Tomasello assumes, claiming that language "plays its role only fairly late in the process. . . . Language is the capstone of uniquely human cognition and thinking, not its foundation."[28]

This view is, however, not undisputed. The language scholar Stephen Levinson and his collaborators argue, in contrast, "that recognizably modern language is likely an ancient feature of our genus pre-dating at least the common ancestor of modern humans and Neandertals about half a million years ago."[29] Be that as it may, they stress that human evolution is a protracted and reticulated process, involving both vertical and horizontal processes of gene-cultural coevolution and leading to the multilayered regulatory structure of the human communication system observed today. Indeed, the human communication system comprises gestures, facial expressions, pointing, pantomiming, and vocalizations, all as part of one integrated multimodal system of communication.[30] Levinson and collaborators have suggested that this multimodal system is the result of a superposition of evolutionary layers involving what they call "an infrastructure of communicational abilities."[31]

But what kind of evolution could produce such a layered structure? The difference between an assumption of distinctive evolutionary steps and our evolutionary scheme becomes clear when we analyze the emergence of language in the context of collaborative and communicative structures. Generally speaking, communicative structures must have arisen in collaborative situations, depending on the specific structure, size, and ensuing challenges of the relevant communities. Even before the first protolinguistic communication systems came into being, there must have existed some regulative patterns of cooperation, such as situative action coordination mediated by visual and other material clues.[32]

At least 1.8 million years ago, at the time of *Homo habilis*, such regulative structures had already been shaped by a shared material culture of tool use and transmission. Communication systems, including gestures, pointing, facial expressions, pantomiming, and (much later) vocalizations, would have initially only marginally supported such regulative structures without representing the full range of cooperative possibilities. Rather, they may have begun as sporadic, domain-specific, and highly context-dependent communicative interactions, which complemented other regulative structures and inherited their "meaning" therefrom.[33]

While communication systems presuppose certain cognitive capabilities on the side of the actors (such as joint attention), they also affect the development of these capabilities by opening up an explorative space in the sense of Vygotsky's "zone of proximal development."[34] This exploratory space exists precisely because communication systems constitute, like the underlying material culture, external representations of regulative structures that typically have a larger horizon of applicability than that given by their initial purpose or circumstances of application. The zone of

14.9. Kanzi, a male bonobo, was born in 1980 and featured in several studies on great ape communication abilities. In this 2006 photo, he converses with primatologist Sue Savage-Rumbaugh using a portable "keyboard" with symbols. For the story of Kanzi, see Savage-Rumbaugh and Lewin (1994). Wikimedia Commons.

proximal development refers to the difference between what an actor can do spontaneously without help and what the actor can do with support from a favorable environment. This difference is not only observable in children but also in acculturated nonhuman primates brought into a human environment.[35] In the evolution of populations, systems of external representation may offer the conditions for an elaboration of implications that functionally correspond to Vygotsky's zone of proximal development in individual learning. The exploration of the inherent potential of an incipient communication system may therefore give rise to a "bootstrapping" of developmental possibilities.

This bootstrapping process opened up the possibility for an iterative process of language evolution. It was based on an interplay between regulative structures of social and cognitive interactions and their external representations. The bootstrapping process must have been born from some contingent ecological context that constituted an external scaffolding for human social interactions, like conditions favoring collective foraging, hunting, and tool use. These initially fragile social interactions probably involved context-dependent signaling, by means of gestures mimicking actions, for instance.

The next step would have been a gradual exploration and extension of this situative action coordination, including a discovery of new possibilities, such as the ritualization and conventionalization of gestures and vocalizations, opening up the possibility of an internalization of the experiences with these external representations of thinking and social interaction. This would have also led to an amplification of the employment and systematization of signs, including not only their decoupling from immediate contexts of usage but also from genetic control, owing to the buffering effect of cultural evolution with regard to biological evolution.

The point is that the exploration of these new possibilities effectively changed the environment in which social interactions took place, creating a new niche with feedback on the action coordination itself—in particular on the possibilities for articulating the goals of actions and hence the separation of their planning and execution.

As a consequence, actions could be performed in a new context and must have been accompanied by ever more extended communicative practices, which in turn acquired systemic features in terms of an increasing internal organization of sign systems. Ritualization may have played a special role in promoting this systematization, as Gary Tomlinson has argued in his account of the origin of human language: "In nascent ritual, the iteration of activities opposed and yet akin to the daily making of the Paleolithic taskscape introduced to it something new, an additional fold in its formation that arose from the complexities of emerging cultural systems."[36] There was no sharp borderline between ritual and habitual practices:

> We can picture the pedagogy needed to convey a complex technological sequence from one Neandertal generation to the next as a one-on-one sharing of attention . . . involving a skilled toolmaker and observant novice. If we extend this only a little, we can picture a group of novices watching and learning from the toolmaker; there is nothing we know about Neandertal society that would render implausible such a scene. Already in the group scene, however, we have shifted to a higher degree of social organization, with perhaps a clearer distinction of it from other activities around it and a clearer marking of the special status of the teacher. A further shift, slight but significant, might see a glimmering awareness of the special import of the group pedagogy, marked by distinct social practices—the decorating of the novices with body paint, for example, or the burning of meat or leaves. Now a metafunction of the cultural system is forming, and with it the special arena of ritual comes into view.[37]

Once a cultural system was established, it changed the meaning of its constituent parts: "If the bead-making system found in many early human cultures was attendant on its constituent parts (shells, teeth, etc., and ideas and practices of social difference), it also enforced a new way of experiencing those elements after it had formed. Shells were no longer merely shells but potential markers, and social difference was now linked to a particular regimenting of material substance."[38] This extension and systematization of communicative and ritual practices in turn enriched the possibilities of action coordination. What may initially have been sporadic, situation-dependent signals within a marginally relevant communication process were eventually transformed into elements of a more and more self-sustaining system of communication, comprising conventionalized gestures and vocalizations that could be used outside of their native contexts. These elements hence received their meaning not only from the original contexts of action in which they were applied, but also from their role in the emerging communicative system.

One immediate effect of such an internalization of external contexts is the stabilization of fragile social scaffoldings, which originally may have been highly dependent on contingent external conditions, for instance, specific ecologies. Indeed, stability seems to have been a hallmark of other Paleolithic cultural practices as well, for instance, of the characteristic Levallois-type stone knapping, which showed a remarkable uniformity from the Lower to the Upper Paleolithic periods for over more than two hundred thousand years.[39] At the same time, however, the long-term development of technological practices displayed an increasing degree of internal organization and systematization of operational sequences. This may have been due to an iterative process of technological evolution as described above.

Such an iterative evolutionary process may indeed account for the emergence of the multimodal system of human communication of which modern language forms the capstone. It is, in any case, a characteristic feature of our evolutionary scheme that new layers do not replace earlier ones but are rather integrated with the preexisting layers in an ever more extended regulative architecture. Witness the fact, for instance, that our vocal language continues to be accompanied by body language. The emergence of language fundamentally changed the position of humans with regard to other species by dramatically enhancing their capacity for cooperation, for the planning of actions, for the division of labor, for the accumulation of knowledge, and hence for massive interventions in their environment.

The Emergence of Language in Recent Times

The evolution of language is difficult to reconstruct. Our hominid ancestors and siblings are long extinct and have left us with no record of earlier forms of language. The emergence of new languages in recent history is therefore of particular interest to researchers attempting to reconstruct the origins of human language, especially in addressing the question of the relationship between the biological prerequisites of language and the role of cultural developments for its emergence. Two prominent examples of new languages that emerged not from the transformation of existing languages but from more elementary building blocks are creole (from pidgin languages representing limited communication systems) and fully developed sign languages in communities of deaf people (from more basic forms of gestural communication).[40]

It is problematic, however, to draw parallels between the phylogenetic evolution of language, on the one hand, and historical and ontogenetic developments, on the other. The latter developments have all taken place in environments shaped by established human languages, while the relevant learning processes involved human brains that evolved much earlier to understand and produce such language. Nevertheless, historical cases of language emergence from more elementary building blocks may illustrate the interplay of preexisting cognitive dispositions and the social construction of shared meaning and language structure in processes of extended (cultural) evolution.

The emergence of Nicaraguan Sign Language in the 1980s has been discussed as a particularly striking case of the ex-novo construction of a language by deaf children who were essentially left to themselves. One of the first linguists to draw attention to this exceptional case was Judy Shepard-Kegl who was invited in 1986 by the Ministry of Education to visit Nicaragua as a deaf-education consultant.[41] The case was even interpreted as providing clear-cut evidence for the innate character of key structures of language. At that time, literacy in Nicaragua was low, and deaf children in particular had few opportunities to receive special instruction. Unlike in other American countries, American Sign Language (ASL) was not widespread in Nicaragua. Despite these circumstances, a group of children and young adults were apparently able to create a sign language entirely on their own. How was this possible?

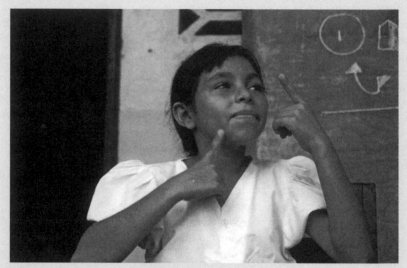

14.10. A deaf girl using Nicaraguan Sign Language at the Esquelitas de Bluefields, Managua, Nicaragua, 1999. Photo Susan Meisalas / Magnum Photos / Agentur Focus.

After the Somoza dictatorship was overturned during the Nicaraguan Revolution in 1979, novel opportunities for teaching the deaf opened up. But official education focused on an "oralist" method, which was successful only in exceptional cases (i.e., teaching lip-reading of a spoken language with the intention to integrate deaf children into the Spanish-speaking majority). What turned out to be more beneficial to the deaf community were, as Laura Polich argued, newly created possibilities to participate in normal working life, as well as opportunities for the self-organization of this minority group.[42]

A novel sign language in Nicaragua is thus difficult to consider as a creation from nothing, or the mere unfolding of a preexisting biological program for language production. In reality, it emerged in a context where not only new experiential and cultural resources became available to the deaf community, but where for the first time the formation of such a community became possible. During this process, the social status of deaf children transformed: no longer objects of shame, they became socially conscious members of a community.

But the cognitive processes related to the emergence of a new language were not simply the product of external circumstances, or of biologically determined automatism. Instead, they were the outcome of an interplay between the externalization of cognitive and social structures and the internalization of their external representations, in the sense of a development of regulative cognitive structures for sign language production and comprehension. This interplay took place in subsequent cohorts of speakers.

The first cohort consisted of students brought together in a school in Managua between 1979 and 1983 who had been using only gestures and "homesigns" (i.e., systems of communication developed by deaf children to communicate with their hearing parents by means of gestures such as pointing to objects or the mimicking of actions). Owing to the lack of dynamic interaction within a peer group, their repertoire was rather limited and mainly adapted to their communicative function within an isolated household. The second cohort, who joined the first after 1983, were exposed to a not-yet fully developed language that began, however, to represent a community achievement. According to Ann Senghas and Marie Coppola, two pioneers who investigated the birth of Nicaraguan Sign Language, "These initial resources were evidently insufficient for the first-cohort children to stabilize a fully developed language before entering adulthood. Nevertheless, over their first several years together, the first cohort, as children, systematized these resources in certain ways, converting raw gestures and homesigns into a partially systematized system. This early work evidently provided adequate raw materials for the second-cohort to continue to build the grammar."[43]

A critical factor for the creation of a full-fledged language capable of expressing arbitrary content was evidently the emergence of an interacting community of speakers. Two crucial aspects of the maturation of Nicaraguan Sign Language were the transformation of preexisting "holistic" signs into building blocks of a comprehensive communication system, and a "grammaticization" (Morford) of the language in the sense of the emergence of system-internal grammatical properties related to an "automatization" of language processing.[44]

This automatization (a process in which younger learners were particularly proficient) may be considered as an internalization of the available language resources, enriched by accumulated experiences of the language learners. It may well have included a special role for the adolescent language learners, whose experiences were evidently broader than those of the first homesigners because of their exposure to working life under the new conditions, which were created in particular by teachers at a vocational center for adolescents.[45] The maturation of the novel sign language in turn affected the cognitive capabilities of its users, influencing, for instance, their spatial cognition, as has been demonstrated by psycholinguistic studies.[46]

The emergence of Nicaraguan Sign Language may thus be seen as the result of a coevolutionary process, comprising the self-organization of both a linguistic system and a community of its users. This process did not take place in a vacuum but was supported by what Vygotsky called a zone of proximal development, acting as a niche with regulative functions for linguistic, cognitive, and social developments.[47] This zone of proximal development was brought about by engaged teachers such as Gloria Minero,[48] who encouraged the children and young adults to create a self-help group, as well as by cultural resources shaped by the existence of developed languages, including sign languages, in their closer and wider environments.

The Neolithic Revolution

The second example of an XXL transition deals with the so-called Neolithic Revolution.[49] The domestication of plants and animals has often been considered a major leap in the advancement of human culture, enabling sedentariness, urbanization, and state formation—in that order. But even apart from the fact that this linear sequence of "progress" does not hold up to the archeological evidence (which indicates, for instance, that sedentariness long preceded food production from domesticated plants and animals), there is another, deeper problem with it: on closer inspection, the transition from an existence as hunter-gatherers (or pastoralists) to that of sedentary farmers does not look like progress at all! Hunter-gatherers lived on a much greater variety of food resources, and they had a mobile life, not being confined to crowded settlements stricken by infectious diseases and not forced to spend their days in hard labor, surviving by the sweat of their brows. They were also, as the archeological record suggests, usually taller and probably healthier than their farming contemporaries.[50] If that was the "progress" of civilization, one may wonder why anybody ever voluntarily went along with it. The varied life of hunter-gatherers has often been contrasted with life under the conditions of a "more advanced" society and even been praised as a utopia by Marx—with suitable amendments and additions, of course:

> For as soon as the division of labour comes into being, each man has a particular, exclusive sphere of activity, which is forced upon him and from which he cannot escape. He is a hunter, a fisherman, a shepherd, or a critical critic, and must remain so if he does not want to lose his means of livelihood; whereas in communist society, where nobody has one exclusive sphere of activity but each can become accomplished in any branch he wishes, society regulates the general production and thus makes it possible for me to do one thing today and another tomorrow, to hunt in the morning, fish in the afternoon, rear cattle in the evening, criticise after dinner, just as I have a mind, without ever becoming hunter, fisherman, shepherd or critic.[51]

The mobile lifestyle of hunter-gatherers has, as a matter of fact, been characteristic of the largest part of our species's history. It only changed late and gradually. Even today what might be the last representatives of this life are still with us, while their ecological niches rapidly disappear. On a global scale, sedentary, food-producing societies probably became the dominant mode of existence for most people only during the early modern period. But why did people abandon hunting-gathering in the first place? Why did they ever take on the hardship of becoming first farmers and then workers in industrialized societies? Was this an inevitable development, or did we simply take a wrong turn at some point in our history? An analysis of this XXL transition from the point of view of extended evolution may help us to answer these questions.

First of all, the fact that humans have profoundly changed their environment is not fundamentally traceable to the beginnings of the domestication of plants and animals, fixed-field agriculture, and urbanization. Humans had already been involved in niche construction much, much earlier, even prior to the emergence of *Homo sapiens*. It is indeed reasonable to claim that niche construction itself was crucial for their emergence. Widespread use of fire is attested at least as early as four hundred thousand years before the present, with its large-scale ecological effects and profound feedback on human genetics. The use of fire may, however, go back in time much farther, up to a million years before the present. Human control of fire was the result of another protracted transformation process. It shaped landscapes by deforestation, making room for desirable mushrooms and plants such as grasses and shrubs. Fire could also be instrumental in hunting. And through cooking, fire allowed humans to externalize part of their digestive process and thereby manage with a much smaller gut than their primate relatives.[52] True, at the time of the onset of the Neolithic Revolution (around 10,000 BCE) humans numbered only about four million. Nevertheless, their large-scale impact on the earth was already beginning to be discernible. While William Ruddiman speaks of an "Early Anthropocene" induced by early farming,[53] James Scott introduces the notion of a "Thin Anthropocene" beginning long before with the extensive landscape management made possible by fire; only later is it followed by a "Thick Anthropocene."[54] In the next chapter, I return to the question of the beginning of the Anthropocene.

A curious aspect of the transition to agriculture was that the multiple subsistence strategies of hunter-gatherers were, under some conditions, abandoned in favor of a reliance on a few crops and domesticated animals. How did this happen? Clearly, it was not the case that an earlier stage necessarily implied a later one. From the perspective of extended evolution rather than a teleological development, we see a highly path-dependent process in which one stage offers a horizon of possibilities, while the next stage may turn accidental (e.g., environmental) conditions and triggers of change into new internal regulatory functionalities and necessary components of its own preservation and further development. This is the essence of what I have called "endaptation" in the "Extended Evolution" section above.

Just as there were likely many pathways leading to early communication systems, there were also many routes that led to food production in different parts of the world—as well as many others that did not. Here I concentrate on the emergence

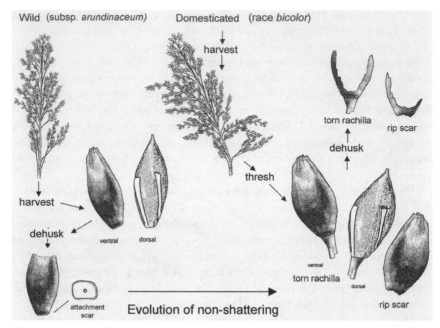

Wild (subsp. *arundinaceum*)   Domesticated   (race *bicolor*)

harvest

torn rachilla   rip scar

dehusk

thresh

harvest

dehusk

ventral   dorsal

ventral

torn rachilla

dorsal

attachment
scar   **Evolution of non-shattering**   rip scar

14.11. Evolution of wild into domesticated sorghum. A key property of domesticated crops is reduced shattering (the dispersal of a crop's seeds upon ripening), which makes harvesting more effective. See Winchell, Stevens, Murphy, et al. (2017). Image by D. Q. Fuller.

of food production in the Fertile Crescent around 10,000 BCE, as well as on the rise of the first cities in southern Mesopotamia. Developed agriculture is a comprehensive subsistence strategy involving intensive human labor. It represents an economic system by which certain human societies foster and regulate the production of a large part of their food and other conveniences (such as clothing) from domesticated plants and animals. Domesticated plants such as cereals are adapted to human nutritional needs and even rely on human intervention for their reproduction.

But long before humans began to sow harvested seeds, they cultivated wild cereals and pulses such as lentils, peas, and chickpeas.[55] I have mentioned above early forms of communication such as protolanguages as forming a kind of plateau for the emergence of real language. In chapter 5, I discussed the role of protowriting. I now refer to this predomestication cultivation in a similar way.[56]

Unlike fully developed agriculture, predomestication cultivation in the sense of the manipulation of wild plants and animals did not in itself constitute a complete subsistence strategy but only one component of a much richer spectrum of subsistence strategies common among hunter-gatherers. Though it had evidently existed in human history for a very long time, it played a more or less marginal role for food production in the same way that early communication systems must have initially played a rather marginal role in human cooperation. This component of the subsistence strategy was certainly not motivated by the later outcomes of domestication—which could not have been foreseen—but must have constituted a practice with its own rationale and dynamics.

Predomestication cultivation was an intermediate stage that did not itself entail a transition to full reliance on food production. It does, however, offer another example of the principle that the horizon of applications of a given means is larger than the end for which it was originally put to use. This also applies to some of the instruments and techniques employed in the early cultivation of plants, such as sickles, threshing baskets, winnowing trays, mortars, grinding stones, and even irrigation techniques. In any case, the period between 8000 and 6000 BCE not only saw the planting of the so-called founder crops but also the appearance of the first domesticated goats, sheep, pigs, and cattle.[57]

In the Fertile Crescent, predomestication cultivation eventually developed into full domestication. One reason may have been that sedentariness—which at certain locations long preceded domestication and may have been encouraged by the surplus of food resources at these locations—favored cultivation practices bound to local environments.[58] Given the investment of labor into these practices at the expense of other subsistence strategies, such local predomestication cultivation practices may have in turn stabilized sedentariness, thus creating what the archeologist Dorian Fuller has called "traps" along the protracted path to domestication.[59]

This mutual reinforcement between sedentariness and cultivation would thus be similar to the stabilization of prelinguistic communicative systems resulting from resonance effects between external conditions and the internal structure of the evolving system. In any case, there was initially no guarantee that predomestication cultivation would necessarily lead to domestication proper. Only at some points along some trajectories may "tipping points" have been reached that then drove the further development in a particular direction, whereas other trajectories may have been aborted or remained in intermediate stages. Accidental external circumstances were thus occasionally transformed into conditions for the internal stability and further development of social formations in which the domestication process took place.[60] For the evolution of domesticated plants and animals, human labor practices constituted a new ecological niche to which they adapted.

Since cultivation was part of a network activity taking place in an extended geographical area (and not just in a small core region, as traditionally assumed), migration and exchange among different sedentary communities eventually contributed to a diversification and enrichment of cultivars across many locations.[61] The resulting recontextualization of cultivation may have helped to separate the wild from the cultivated, thus contributing to a process by which human-defined plant or animal populations were ultimately transformed into biologically defined populations. Eventually, domesticated crops were no longer bound to the local contexts in which their ancestors were originally found but spread into other areas and ultimately across the world.

Early Neolithic settlements must have been precarious institutions in the sense that their sustainability depended not only on a set of environmental factors, but also on labor investment that would not yield immediate benefits. The benefits would be evident only after some time, and only if conditions remained stable. New symbolic practices may have provided scaffolding to support these communities, to cope with the awareness of uncertainty, and to strengthen social cohesion in such a way as to keep larger communities together.[62] It may have been due to these symbolic practices that large-scale building projects could be accomplished—projects far beyond the capacities of individuals or spontaneous collective activity. In the case of these building projects, the capability to conceive and sustain labor chains also created the conditions for such innovative developments as the invention of bricks, an example discussed in chapter 9.

Nevertheless, for more than four thousand years, the emergence of domesticated plants and animals did not lead to the rise of cities and states—a remarkable delay to which the political scientist James Scott has recently drawn attention.[63] Rather, we still see a scattered landscape of people pursuing multiple subsistence strategies, but now among them "stable and highly sustainable subsistence economies based on a mix of free-living, managed, and fully domesticated resources," as one of the leading theorists of domestication, Melinda Zeder has formulated it.[64] These sedentary villages eventually constituted the raw material from which, on the fertile alluvial soils of southern Mesopotamia, the later cities and states could be established.

But again, this transition was no necessary implication of an intermediate stage that was itself doomed, from the outset, not to last. Instead, temporary ecological circumstances, in particular precipitous declines in the water levels of the Euphrates River in the period between 3500 to 2500 BCE, may have acted as triggers of change, creating the conditions under which the matrix for crafting a city-state emerged.[65] These circumstances may have led to a concentration of the population in a smaller area, prefiguring what would later become an "urban environment" surrounded by a limited amount of arable land that now needed to be exploited more intensely, in part through the use of labor-intensive irrigation techniques. By 3200 BCE, the state of Uruk with its stratified society and redistributive economy was firmly established. It set the stage for the development of an elaborate administration based on the division of physical and intellectual labor, and gave rise to the sophisticated knowledge economy discussed in earlier chapters.

James Scott has suggested that it was no accident that archaic states relied on specific staple foods such as grain or rice. He argues that these foods were particularly well suited not only for storage and distribution but also, because of the ease of determining their maturity and harvesting time, for state control by taxation and accounting.[66] Indeed, regulatory structures are always closely tied to the material means for creating and implementing them. In any case, the very possibility of imposing such regulatory structures, shaping social and biological rhythms as well as coupling them, only emerged as an unintended result of the preceding domestication phase.

But the implementation of such control structures in the context of the archaic state now definitively transformed the domestication process from an accidental byproduct of the exploration of environmental niches (as merely one aspect of a multifaceted subsistence strategy) into a necessary constituent to the further development of the state. Reliance on a few staple crops and a small number of domesticated animal species became the dominant subsistence strategy under these conditions—with huge consequences for the further coevolution of human societies and their ecologies. The very self-preservation of the early states relied on the maintenance of their labor force by coercion, which precluded the mobility associated with the broad subsistence strategies of hunter-gatherers or the "multilateralists" of the pre-ceding phase.

The people of the early city-states suffered not only from the exhausting drudg-ery of their work and the ensuing wear and tear of their bodies. Urban societies achieved new levels of population density and thus facilitated the evolution of a num-ber of infectious diseases, which in turn had a tremendous effect not only on the regulative structures of these societies but eventually on the global distribution of power and wealth, as discussed in chapter 11. In addition, these societies were fragile constructs, always liable to collapse because of infectious disease, loss of population, resource depletion, warfare, migratory movements, or climate change. For millen-nia to come, states remained islands in a sea of other societal formations. Why did these fragile and coercive societies nevertheless come to prevail? Was it primarily a matter of demographic growth (which was initially rather slow, probably owing to the high mortality rate in urban environments, but then gradually took off)? James Scott comments, "The world's population in 10,000 BCE, according to one careful estimate, was roughly 4 million. A full five thousand years later, in 5,000 BCE, it had risen only to 5 million. This hardly represents a population explosion, despite the civilizational achievements of the Neolithic revolution: sedentism and agriculture. Over the subsequent five thousand years, by contrast, world population would grow twentyfold, to more than 100 million."[67]

Demographic growth certainly played an important role in the long run. More generally, however, I see the globalization processes discussed in chapter 11 as a driv-ing force behind this development. The early cities and states may have been islands, but they tended to be connected by trade, travelers, migrations, or war, even if these only constituted weak ties within an interregional network. A slow but relentless growth of interdependencies and means of social cohesion was achieved by demo-graphic expansion, migration, trade networks, and conquest, by the spread of the sediment of material culture such as tools or weapons, the spread of social technolo-gies such as slavery or the employment of mercenaries, and—not to be underestimated—the spread of cultural abstractions such as economic value, repre-sented by money or religious and political concepts (god, empire).

Whatever material or mental products emerged from the initially few and insular cities or states—with their sophisticated economies of knowledge, enabled by the division between physical and intellectual labor—these products triggered a feed-back loop of intrinsic and extrinsic globalization. As discussed in chapter 11, the in-trinsic globalization of knowledge is the result of processes of iterative abstraction that took place on an experiential basis that substantially depended also on processes of extrinsic globalization. It led to forms of knowledge unbound from specific local contexts, such as when geography came to be based on the model of a spherical earth, which allowed observations of astronomical events at any two places to be consid-ered in relation to each other. The emergence of this model depended on the extrinsic globalization processes that had made it possible to acquire experiential knowledge of a sufficiently large portion of the earth in the first place. And it enabled, in turn, further extrinsic globalization processes by facilitating navigation, and hence trade and conquest. The global applicability of this geographic knowledge stands in stark contrast to more locally situated knowledge, for instance, the navigational knowledge of Micronesian sailors, which was confined, as discussed in chapter 10, to regions close to the equator.

The expansive quality of this feedback loop between intrinsic and extrinsic globalization processes is ultimately due to the exploration of horizons of possibility emerging in different times and at different places. Because we are living in a large but finite world, many niches were available for diverse local developments for most of human history, but these developments could not stay isolated forever. Instead, they gave rise to this feedback loop, which eventually led from urbanization and state formation to the creation of large empires, to colonialism and global capitalism.

This expansive quality should, however, not be confounded with progress, not only because the process was shaped by contingencies and remained highly path dependent but because there is no guarantee that the feedback loop of globalization will actually enable us to cope with a finite world. Ultimately, therefore, we have to confront the boundary conditions under which these dynamics have unfolded since the very beginnings of our species. This is the key challenge of living in the Anthropocene, to which I now turn.

The example of the emergence of food production illustrates the importance of bringing the evolution of knowledge and of other regulatory structures of human societies into an encompassing framework of cultural evolution centered on human material practices and accounting for what I have called XXL transitions, which fundamentally change the position of humans in their environment. From this vantage point, in the next chapter I investigate the way scientific knowledge affected agriculture as a system of human subsistence at the beginning of the twentieth century as one among several pathways into the Anthropocene and perhaps beyond cultural evolution.

Chapter 15

Exodus from the Holocene

I find it very important in disembarrassing ourselves of our vanity, short-sightedness, biases, and ignorance in general, in respect to universal evolution, to think in the following manner. I've often heard people say, "I wonder what it would be like to be on board a spaceship," and the answer is very simple. What does it *feel* like? That's all we have ever experienced. We are all astronauts. I know you are paying attention, but I'm sure you don't immediately agree and say, "Yes, that's right, I am an astronaut." I'm sure that you don't really sense yourself to be aboard a fantastically real spaceship—our spherical Spaceship Earth."

—RICHARD BUCKMINSTER FULLER, *Operating Manual for Spaceship Earth*

The concept of progress must be grounded in the idea of catastrophe. That things are "status quo" *is* the catastrophe.

—WALTER BENJAMIN, *The Arcades Project*, in *Selected Writings*, vol. 4

The Holocene Bubble

Science fiction authors have imagined alien worlds and wondered about lives or even cultures in such worlds. Think of a world full of water with hardly any rigid boundaries, or a world without visible light, or worlds in which life is based on different chemistry. Naturally, in such fantasies we always take our own world as the standard. But what we have learned in recent years is that such alien worlds may not just be out there but that some of them actually existed in our own past—for instance, when oxygen was toxic for life or when the entire earth was covered with ice.[1] And what we are gradually realizing is that even stranger worlds may be ahead in the future of our own planet.[2] We just have to look over the fence at our neighbors in the habitable zone of the solar system: Mars may have once been warm and moist but is now a barren desert, while Venus, our sister planet, has turned into hell as a result of a runaway greenhouse effect. The image of Earth from outer space—the blue marble that became so iconic during the space age— also conveyed the awareness of our planet as a small, fragile sphere, limited in its resources. We all became aware that the world we inhabit is the "spaceship earth," an island indeed, yet not only in space but also in time.[3] The environment in which human civilization has thrived for around twelve thousand years, the Holocene, will hardly last forever.[4]

We slowly begin to realize how much of our mastery and understanding of the world, and of ourselves, may depend on the contingent and transient circumstances of the Holocene.[5] It has provided us, for instance, with a more or less moderate climate that did not make it too difficult to find shelter and food; in short, under the favorable and relatively stable conditions of the Holocene we have been able to consolidate our expansion across the globe by inventions such as agriculture and the domestication of animals. These resources, as well as stable environmental conditions, have favored the long-term, cumulative growth and global diffusion of technologies and infrastructures, such as the processing of metals or the use of wood as a renewable building material and energy resource. Their spread was supported by long-distance mobility across land and sea, which was in turn based on these resources. Later, the conditions of the Holocene facilitated access to fossil resources such as coal, gas, and oil, as well as their exploitation and transport.

Similarly, our conceptions of the world, as diverse as they may be across ages and cultures, have been deeply shaped by the conditions of the Holocene and the rather small window through which we have observed our world so far. Indeed, on a geological scale the existence of humanity is a recent phenomenon. Even more recent is the time in which we have been pursuing scientific observations of the earth. Meanwhile we have learned—for instance, from the eight hundred thousand–year record from the EPICA Dome C ice core—about the dramatic changes of the Earth system. For the most part of human history, however, the larger natural environment has appeared to us as essentially stable and as a resource humans can make use of, with different views, of course, as to its limitations or to what we owe to the world in compensation for its appropriation, be it sacrifice or allegiance. We have also become used to the expansion of our own species, unaware that we ever were and might again become an endangered species. We take it for granted that we can shape our immediate environment according to our needs, while our larger environment remains out there, perhaps untouched and friendly, perhaps alien and hostile, unknown or simply indifferent to us. Or, as historian of the Anthropocene Jeremy Davies succinctly puts it, "The Holocene matters because it is the only geological epoch so far in which there have been symphony orchestras and hypodermic needles, moon landings and gender equality laws, patisseries, microbreweries, and universal suffrage—or, to put it plainly, the agricultural civilizations that eventually made all of those things possible."[6]

Living, as Davies formulates it, at the interstices between two different worlds (the Holocene and the Anthropocene) raises new questions. Under the aegis of Western capitalism, the character of nature as a cheap resource has been foregrounded, while cultures and traditions that have instead emphasized the need for striking a balance between humans and their environment have been marginalized. Can a particular social organization and exploitation of nature be held responsible for destroying the stable conditions of the Holocene, and can a different social organization stop

15.1. Photo taken during a German expedition by Alfred Wegener, Johannes Georgi, Fritz Loewe, and Ernst Sorge to Greenland in 1930–31, showing Georgi holding an anemometer. Nachlass Dr. Johannes Georgi, Archiv für deutsche Polarforschung.

this destruction? Do certain views and values associated with human progress, such as individual freedom, justice, and democracy, depend on the way we are dealing with and conceive of nature and ourselves? Do they depend, in particular, on the abundance of energy resulting from the exploitation of fossil resources, and does sustainable development require different ideals, values, and self-regulatory processes—in particular, the renunciation of privileges? How many of us can this planet actually carry and under what living conditions? As radical as such questions may sound, they still presuppose that we as humans have choices and could ultimately become (if we "get our act together") a prime mover or at least the peaceful tenants of the Earth system, which is our habitat.

Perhaps not. What we have largely ignored so far is that our habitat is a highly coupled nonlinear system whose past has seen dramatically different stages with no less dramatic transitions between them.[7] Such systems do not behave regularly but may rather exhibit chaotic dynamics under the influence of external force, such as human interventions. The variability of the Earth system extends over diverse timescales and includes major transitions even in periods shorter than the well-known glacial/interglacial cycles. Consider, for instance, the fact that the Sahara became a desert less than ten thousand years ago.

The Earth system is characterized by numerous couplings and feedback loops that remained essentially unknown to us until rather recently.[8] The terrestrial and marine biospheres are coupled not only by the water and carbon cycles, for instance, but also by the generation, transport, and deposition of dust. When the climate becomes cooler and drier, less vegetation grows, enhancing the spread of iron-containing dust from barren soils. This dust, transported by wind to the oceans, there acts as a fertilizer to the phytoplankton, leading to an increased absorption of carbon dioxide from the atmosphere. This absorption in turn drives the climate to become even drier and cooler, resulting in a positive feedback chain.

The water cycle, the carbon cycle, and all the other couplings happened to keep our environment and climate relatively stable during the Holocene, but they are now massively affected by human interventions. Against the background of the larger and much more dramatic history of the Earth system as we have gradually come to uncover it, it now becomes clear that we cannot take for granted the comfort zone of the "Holocene bubble" in which we have been living. It may well turn out to be a less secure basis of our existence than we had assumed.

When Did the Anthropocene Begin?

Have we already entered into the Anthropocene? When did or will it begin?[9] This is a typical "origin question." Since ancient times such questions have been associated with questions of guilt. From this perspective, the transition from the Holocene to the Anthropocene amounts to an expulsion from paradise. But who is responsible? Among the potential culprits are clearly an extractive economy, an exploitative conception of nature, and capitalism, but also modern science and technologies and—why not?—perhaps even cultural evolution as such.

It is important to remember, however, that the Anthropocene, with all of its provocative impact, is first of all a descriptive concept, a controversial geological *terminus technicus*.[10] The question of when exactly the Anthropocene began as a stage in earth history is a specifically geological question, related to the characteristic tools, criteria, and standards used by this discipline to establish its temporal classification scheme. The concept therefore lacks explanatory power and does not tell us what the driving forces behind the current exodus from the Holocene are, nor how these forces operate and function. How humanity became a geological force is a question than cannot be answered by the earth sciences alone.

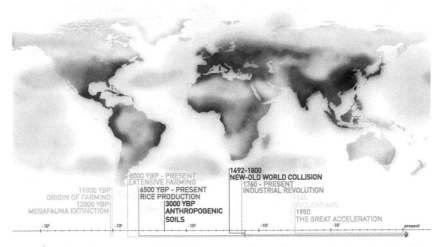

15.2. A timeline of some proposals for the beginning of the Anthropocene. Image created by Chris Reznich.

The Onset of the Anthropocene: Meshing Historical and Geohistorical Timescapes

From a geological standpoint, a precise dating of the beginning of the Anthropocene acquired through "rock solid" stratigraphic evidence is of crucial importance. Currently, the most promising proposal focuses on the mid-twentieth century, combining the plutonium signal at the start of the Atomic Age with the "Great Acceleration" taking off around 1950.[11] This, however, does not imply that earlier historical events—or even slower, accumulative processes—are any less important for understanding the transition into this new epoch.

There is much that points to a diachronous onset of the Anthropocene. For instance, the alteration of the pedosphere (the outermost layer of Earth, composed of soil) is a key indicator of human impact on the global environment. As a solid-phase stratigraphic boundary, it marks the division between humanly modified ground and natural geological deposits, yet in an often diffuse, gradational, or mixed form.[12]

Second, there is more than one perspective in which historical and geological times are intertwined. One example is the rise of fossil fuel use during the Industrial Revolution, which led to the rapid consumption of energy resources that accumulated over large geological time spans. Another is the so-called sixth extinction, which refers to the mass extinction of biological species currently under way. The human-induced erosion of biodiversity has been ongoing since at least the late Pleistocene with the eradication of large mammals but has continued throughout the entire Holocene period. Some scientists argue that the current rate of species extinction is by orders of magnitude higher than the normal background rate throughout geological time, leading to comparisons of the ongoing annihilation with the other five major extinction events in earth history.[13]

The assessment of the Anthropocene as a technical term of the earth sciences depends on a number of criteria, preferably on a "synchronous base," specifying a time or, more correctly, a chronostratigraphic layer that must be the same everywhere around the globe; a position in the sedimentary record that defines this synchronous base—also known as a "golden spike" because a golden metal plate is often placed into the relevant rock layer (Global Boundary Stratotype Section and Point or GSSP); and a specified rank in the stratigraphic hierarchy (stage, epoch, period, era).[14] More specifically, the golden spike refers to a material reference point defining the lower boundary of a stage on the geologic timescale.

As the geologist Jan Zalasiewicz and his colleagues have suggested, the entry into the nuclear age (signaled by the detonation of the Trinity A bomb in New Mexico on July 16, 1945) could be used to mark the beginning of the Anthropocene because of the stratigraphic evidence from artificial radionuclides but also from the sediment of global industrialization such as fly ash, concrete, and plastic. This was a unique event, comparable to the meteorite impact causing the extinction of the dinosaurs and marking the end of the Mesozoic era.[15]

Several proposals have been made for the onset of the Anthropocene. Some of them refer to the earliest human interventions with large-scale consequences for the planetary environment. One such intervention is the mastery of fire by hominids. Another proposal refers to the mass extinction of the megafauna in the late Pleistocene, some thirteen thousand years before the present. Around this time most continents saw unprecedented extinctions of megafauna, possibly attributable to a combination of climate change, human hunting of large mammals, or other more indirect effects of human presence.[16]

Another suggestion, also claiming an early beginning to the Anthropocene, cites the rise of sedentary societies and the Neolithic Revolution in the period between twelve thousand and eight thousand years ago, and the long-term consequences for animals, vegetation, and soils. According to William Ruddiman, preindustrial farmers actually postponed the next ice age by clearing forests in Eurasia.[17] Bruce Smith and Melinda Zeder point to the role of domestication as a massive change in the earth's ecosystems. In this view, the Holocene and the Anthropocene are coeval.[18]

As discussed in the previous chapter, the Neolithic Revolution was a protracted process extending over thousands of years in which landscapes were transformed and the gene pools of domesticated animals and plants were altered. Food production led to an increase in population sizes and induced migrations, which in turn spread new technologies through globalization effects, as described in chapter 11. The so-called Early Anthropocene Hypothesis argues that the intensification and

15.3. J. Robert Oppenheimer (*left*) and Leslie R. Groves visiting the remains of a test tower after detonation at the Trinity site, New Mexico, 1945. Major General Leslie Groves of the US Army Corps of Engineers directed the Manhattan Project from 1942 to 1946. Nuclear physicist Robert Oppenheimer was the director of the Los Alamos Laboratory that designed the actual bombs. US Army Corps of Engineers.

globalization of agricultural practices deeply influenced terrestrial ecosystems as well as the global climate.

Neolithization, and later the Bronze and Iron Ages, left numerous traces in the archeological and geological records, including traces of charcoal used in metal smelting in glacial ice and indications of a rise in greenhouse gases in the Middle Holocene (seven thousand to five thousand years ago).[19] In later historical times, much of the terrestrial surface was shaped by anthropogenic changes of soils. Such changes to the pedosphere have also been discussed as marking the onset of the Anthropocene.[20]

In chapter 11, I examined the massive global impact of the collision between the ecosystems of the Old World and the New World as a result of the Columbian turn. The globalization of ecosystems in the early modern period and the attendant demographic, economic, and political changes are being debated as another milestone on the pathway to the Anthropocene.[21]

Another proposal, originally favored by the proponents of the Anthropocene concept, Crutzen and Stoermer, identifies the Industrial Revolution and thus the beginning of industrial capitalism in the late eighteenth century as the crucial turning point in which some of the geological signals serving as indicators of the Anthropocene emerged for the first time. Crutzen later revised this idea. In collaboration with the environmental historian John McNeill and Earth system scientist Will Steffen, Crutzen argued for a more protracted process, pointing to the role of atmospheric carbon dioxide as the single most important indicator of the progress into the Anthropocene.[22] Other indicators of this progress are the large-scale transformation of landscapes by agriculture, the massive rise of urbanization and energy consumption, and in particular the development of the fossil fuel industry in Britain between 1800 and 1850 and its atmospheric consequences.

In the case of the fossil fuel industry, the drivers of accelerated technical and socioeconomic development are well known. Earlier industrialization, mainly based on the cotton industry, served as a plateau and a model for what industrialized capitalism could mean. Technological innovations on the order of the steam engine, as developed from the late eighteenth century by James Watt and others, as well as the large-scale exploitation, production, and use of coal and iron, mutually reinforced each other, forming what one might call a "resource couple." Extensive mining and coal use goes back to the Middle Ages. It was particularly prominent in England because of coal resources close to the surface and favorable transport via canals and coastal routes.[23] Owing to specific technological challenges, such as the need to drain water from deeper pits by using mechanical pumps, British coal mines developed an individual local niche in which a feedback mechanism (characteristic of the Industrial Revolution) that combined the use of fossil fuels with further mechanization based on iron and steel could emerge—and from which it would later spread.

As the economic historian Edward Anthony Wrigley writes,

> Coal had been very widely used as a source of heat energy. It overcame the bottleneck in providing heat energy which was inherent in dependence on wood. But without a parallel breakthrough in the provision of mechanical energy to solve the comparable problem associated with dependence on human or animal muscle to supply motive power in industry and transport, energy problems would have continued to frustrate efforts to raise manpower productivity. The link between the drainage problem in mines and the development of a reliable steam engine was intimate. . . .
>
> Moreover, only where coal was available at a very low cost could the early engines have been put to economic use since both the Savery and the Newcomen engines were very inefficient in their use of coal. Only a tiny fraction of the potential energy was harnessed. The cost of providing coal on the scale

needed to operate such engines was prohibitive away from a coalfield, given the transport costs involved. At a pithead, however, even a very inefficient coal-fired engine could be employed since the coal was consumed at the place of production. . . .

It is, of course, impossible to be certain that an effective steam engine would not have been developed outside the peculiar context of coal mining and its drainage problems, but in the English industrial revolution it is evident that there was a close association between the problems experienced in mining coal and the discovery of a satisfactory solution to the problem of providing mechanical energy on a greatly enlarged scale and at much reduced cost.[24]

New iron smelting techniques emerged in the middle of the nineteenth century, leading to use of the metal in a plethora of products, and in particular in the construction of fire-resistant machines—a crucial condition for the further development of the steam engine and later of coal-fired, steam-powered locomotives.

The use of coal contributed to an increase in the production of iron and steel, which in turn allowed for an extension of the railway network, thus facilitating the transport of coal and increasing the demand for and production of iron. As a consequence of this mutual reinforcement, the economic and social structure first of Europe and then of many other regions was profoundly transformed, resulting in a shift in the balance of power in favor of the industrialized nations—a circumstance that in turn promoted further industrialization in ways direct and indirect.[25] Industrialization also brought with it major environmental changes, including a significant increase of carbon dioxide concentrations in the atmosphere. Industrial capitalism was also accompanied, from its inception, by a public awareness of these changes, as well as of its devastating social consequences.[26]

Since the early twentieth century, humans have changed Earth system cycles by introducing industrially produced chemical substances. Examples include the massive human production of reactive nitrogen through the Haber-Bosch process (discussed later in this chapter) and the chlorofluorocarbons and other halocarbons used as propellants and solvents and for refrigeration (also a theme to which I return later).

A further proposal for the onset of the Anthropocene, and one that has received much support, is the so-called Great Acceleration starting after World War II, which, in connection with the advent of artificial nuclear materials mentioned above, provides the most clearly identifiable stratigraphic signals of a new geological epoch. This moment is characterized by a dramatic rise in many parameters of the Earth system, representable by curves resembling a hockey stick and representing a transition from linear to exponential growth of both Earth system parameters and parameters of socioeconomic growth, for example, the increase of gross national product, the use of fertilizers, the accelerated consumption of resources such as oil or water, or international tourism.[27] Clearly, whatever the details of the underlying dynamics may be, the Great Acceleration is inconceivable without the self-reinforcing processes engendered by the first and second industrial revolutions preceding it.

The dynamics driving us into the Anthropocene are closely related to the growth of the human population, which has more than doubled since the first half of the past

15.4. Young miner working at the Brown Mine in West Virginia, 1908. US National Archives.

century. This growth was coupled with similarly dramatic increases in resource and (in particular) energy consumption, agriculture and food production, industrial development, transport, and commerce.[28] Changes in the parameters of the Earth system, such as the rise of greenhouse gases, closely parallel the changes in these socioeconomic parameters. The Earth system and the global dynamics of human societies are evidently linked, for instance, by the necessity to feed a growing number of people with the intensification of agriculture, which in turn requires more energy and fertilizer. The entry into the Anthropocene is a dynamic phenomenon involving feedback loops coupling the growth of human societies with environmental changes favoring that growth in turn—at least until planetary boundaries are reached.[29] "Planetary boundaries," as introduced by Johan Rockström and Will Steffen, refer to those thresholds or tipping points in certain planetary subsystems, such as the nitrogen cycle or biodiversity integrity, beyond which there is a risk of abrupt and irreversible environmental change.[30] Planetary boundaries is a concept characteristic of the borderline problems arising between studies of the Earth system and of global human society in the Anthropocene. As mentioned in chapter 10, the discussion of such limits to growth goes back at least to the 1950s and culminated in the 1970s (in books ranging from Fairfield Osborn's *Our Plundered Planet* (1948) to the famous 1972 report of the Club of Rome), only to resurface in recent years with the growing awareness of the character of the earth as a complex, dynamic system. Network analysis is also central.[31] Given the fact that the Anthropocene represents a situation with no analogue in earth history, we must apply a precautionary principle that takes into account the increased complexity of a coupled human Earth system at risk of displaying unpredictable fluctuations with devastating consequences for humankind.[32] If we want to retain at least some of the comfort of the Holocene, we may have to apply the dialectics of Lampedusa's "Leopard," without, however, restricting this comfort to the privileged: "If we want things to stay as they are, things will have to change."[33]

Looking back at the knowledge economies discussed earlier, we can recognize that the dynamic coupling of human societies and the Earth system not only depended on the generation of knowledge, but enhanced, in turn, the significance of that generation.[34] In a process extending over millennia, scientific knowledge eventually became a crucial component of economic growth; it is now becoming no less important in coping with its consequences. Substantial anthropogenic change on a global scale was evidently already characteristic of the Holocene and the late Pleistocene, from the human-induced extinction of great mammals via Neolithization and urbanization to the Scientific and Industrial Revolutions. What characterizes the Anthropocene is the need to actively prevent the Earth system from transgressing planetary boundaries. The emergence of this need is closely related to the cascade of evolutionary processes described above. Just as cultural evolution started as a sideshow of biological evolution before it became the *conditio humana* (the essential condition of human life), epistemic evolution, that is, the growing dependence of human societies on knowledge economies producing scientific knowledge, was, for a long time, no more than a tangential aspect of cultural evolution. But with the onset of the Anthropocene, the survival of human culture as we know it hinges on this.[35]

Neither the onset of epistemic evolution nor that of the Anthropocene can easily be assigned a singular date, cause, or origin. From this perspective, the primary question is less what or who caused the Anthropocene but how humanity can live with it. Each potential genesis proposed above catches important aspects of the long-term historical development that drove us and the planet we inhabit into this new stage. There is no question that power, violence, exploitation, and oppression were crucial factors. In examining the dark side of history, we see indeed that not all cats are gray; we may well distinguish the oppressors and the oppressed.[36] But identifying causes and culprits does not necessarily put us into a better position for predicting which solutions are equal to the challenges of the Anthropocene. We cannot simply bail out or excuse ourselves from the evolution of knowledge, but we may attempt to understand it better, including the detailed pathways that have brought us to where we now stand—and whence we will be launched into the future.

The Nitrogen Story

If we wish to understand the dynamics catapulting us into the Anthropocene, we must explore the historical pathways that led to an increased dependency of certain parameters of the Earth system on human societies. One of the hallmarks of the Anthropocene is the coupling of Earth system cycles with human activities. Such a coupling makes these cycles ever more dependent on the management of human intervention and thus on human knowledge, in particular on scientific knowledge.

Consider, for instance, the biogeochemical cycle of nitrogen, essential to life on this planet and profoundly modified by modern industrial agriculture and other interventions. The global production of reactive nitrogen has been doubled by the cultivation of legumes, the burning of fossil fuels, and, in particular, by the chemical production of fertilizers with the help of the so-called Haber-Bosch process—named in recognition of the fundamental scientific and technical insights into ammonia synthesis associated with the work of Fritz Haber and Carl Bosch.[37]

At the beginning of the twentieth century, the ability of the agricultural industry to feed the world's population (which had grown significantly during the previous century) was dependent on natural, nitrogen-based fertilizers made mainly from Chile saltpeter. This sodium nitrate from South America was limited in quantity and, in addition, created a precarious dependency on a single supplier. The situation changed dramatically with the advent of synthetic ammonia production. The ability to provide enough foodstuffs to sustain the world's current population, whose sheer size is a determining condition for the Anthropocene, is now dependent on the industrial production of fertilizers.

I want to consider this example in more detail, not only because of its intrinsic importance but also because it illustrates how our framework helps to understand historical pathways into the Anthropocene. As we will see, this example makes use of such notions as the interaction between practical and theoretical knowledge, challenging objects, and borderline problems, as discussed in earlier chapters.

At first glance, the synthesis of ammonia (NH_3) from its elements—nitrogen (N) and hydrogen (H)—appears to be the solution to a seemingly simple chemical question analogous to the synthesis of water from the elements oxygen and hydrogen. The problem was already known at the beginning of the nineteenth century, but the solution in large-scale industrial form was not found until over a century later, between 1904 and 1910.

It may seem as though ammonia synthesis was simply a triumph of chemistry with unexpected consequences for agriculture. But as it turns out, without practical knowledge, agricultural observations, and an understanding of the role of fertilizers in general—and of nitrogen and the nitrogen cycle in particular—essential chemical insights would not have been possible. During the nineteenth century, agriculture became an important area of investigation for the thriving science of chemistry. Given the key economic role of agriculture and the prior history of modern science (namely, its close ties to practical concerns) this comes as no surprise. Yet a sense of desperation was also palpable, as formulated by British economist Thomas Robert Malthus in his warnings about feeding an ever-growing population at the beginning of the nineteenth century. Industrialization and the continuing optimization of agrarian production were able to divert this menace before the century was out.

Agriculture, and in particular the role of fertilizers, eventually became a challenging object for chemical research. One of the pioneers of this research was Justus von Liebig, born in 1803. He deeply influenced education in the field of chemistry by establishing a school that had a lasting impact on the development of chemists for generations. In 1840, he published a book known by its abbreviated title *Agricultural Chemistry*.[38] Liebig made essential contributions to the understanding of agriculture, which would later have an impact on farming practices. Addressing issues of nutri-

15.5. German postcard from 1915 demonstrating the impact of fertilizers on crops: "Who does not amply sprinkle fertilizer into Earth's pleats, will produce potatoes as small as peas. Thus if you want to fill with giants the potato sack, use as fertilizer sulfuric acid ammoniac."

tion and public health with the help of science, he was not just a university professor but also a chemist who combined science with technologically innovative projects.

He claimed that the fertility of farmland can only be maintained if the mineral nutrients removed with the harvest are replaced. He also formulated a "law of the minimum," which states that when a specific element or compound in arable land is limited, the addition of other minerals will have no effect on the growth of crops on that land. But it did not occur to Liebig that ammonia could play one of these crucial limiting roles—he thought it only facilitated plants' uptake of other nutrients.

Liebig also introduced the concept of a cycle of elements. He analyzed, in particular, the nitrogen cycle, thus identifying important chemical and biochemical questions. It turned out that nitrogen in air was chemically inert and that another chemical form was present when the nitrogen was absorbed by plants. In other words, atmospheric nitrogen was transformed into an as yet unknown form and entered the soil via precipitation. There it was absorbed by plants. The nitrogen later reentered the atmosphere as N_2 through an unknown mechanism. Although Liebig correctly claimed that ammonia was somehow involved in the nitrogen cycle, he underestimated its importance. Further investigations were needed before the key role of NH_3 was eventually understood.

The nitrogen cycle posed fundamental questions about nitrogen fixation, specifically about ammonia synthesis: Which mechanisms of nitrogen transformation was nature capable of that chemistry could not yet mimic? It eventually became clear that crops required fixed nitrogen, in which the triple bond between two atmospheric nitrogen atoms was broken and the nitrogen was bound to other elements, as is the case both for ammonia (NH_3) and for Chile saltpeter ($NaNO_3$). Bacteria turned out to be responsible for converting the fixed nitrogen in the soil back into diatomic N_2. With these insights, the complete nitrogen cycle could be understood.

Understanding the key role played by ammonia in agriculture had far-reaching economic and technological consequences. It also impacted chemistry as a scientific field. The main source of fixed nitrogen in the middle of the nineteenth century was saltpeter from Chile and guano—seabird excrement—from Peru. Commercial imports began around the mid-nineteenth century after Alexander von Humboldt had made the subject well known in Europe and after Liebig expounded the importance of chemical fertilizers.

According to the environmental historian Gregory Cushman, the period between 1820 and 1914 was a crucial phase for the onset of the Anthropocene, a phase that he describes as a new "fertility regime." In the context of an emerging global capitalist economy, Western societies and eventually humanity began to depend on the industrial exploitation of substances such as nitrogen and phosphorus, both crucial ingredients in fertilizers.[39] At the end of the nineteenth century, the finite supply of fixed nitrogen in South America was identified as a bottleneck for further socioeconomic development. Replacing it became a political, economic, scientific, and technological challenge. In a speech to the British Association for the Advancement of Science in 1898, the chemist William Crookes expressed the concern that the civilized nations were in danger of producing inadequate quantities of foodstuffs and pointed to the fixation of atmospheric nitrogen as a promising solution.[40] The mysterious process of direct ammonia synthesis thus began to gain attention as the need for a nitrogen fertilizer intensified.

Ammonia synthesis immediately became a jointly economic and scientific challenge—a coupling that, as we have seen, inevitably spurs further developments. The resource transformation that began with the identification of ammonia synthesis as a challenging problem was clearly not the result of a single ingenious breakthrough but rather part of a comprehensive process that eventually enhanced the dependency of humanity on science, while strengthening the marriage of science, technology, industry, and agriculture.

Raising a problem and solving it are, however, different matters. The realization of the economic urgency of ammonia synthesis did not mean that chemistry was actually in the position to confront this challenge. The means to tackle the problem of ammonia synthesis only emerged as a link in a long chain of empirical investigations and also presupposed a new theoretical perspective, that of thermodynamics and physical chemistry. This perspective was in turn shaped by the increasing industrial relevance of chemistry during the nineteenth century.[41]

As a consequence, the understanding of the physical parameters governing a chemical reaction attained increasing practical and economic relevance. Chemical reactions came to be seen as borderline problems between chemistry and physics in the sense that their quantitative aspects were recognized as depending in a precise way on thermodynamic parameters such as temperature and pressure. Like the lines of latitude and longitude on a map, thermodynamics set up a closely spaced grid allowing scientists to find the point at which a reaction runs in a particular direction

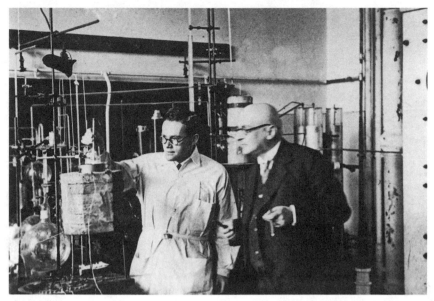

15.6. Fritz Haber smoking a cigar in the laboratory at the Kaiser Wilhelm Institute for Physical Chemistry and Electrochemistry with Ladislaus Farkas around 1930. Courtesy of the Archive of the Max Planck Society, Berlin-Dahlem.

and takes place with a particular efficiency. This new understanding of chemical reactions ushered in a new era of chemistry.

It is against this background that the possibility of synthesizing ammonia from the elements gained a theoretical plausibility. The problem was attacked, around the turn of the nineteenth century, by leading physical chemists such as Wilhelm Ostwald, Walter Nernst, and Fritz Haber. Haber's eventual breakthrough in 1908, which earned him the Nobel Prize, and its later industrial implementation realized by Carl Bosch and Alwin Mittasch were the results of a cooperative effort whose success was, however, in no way a necessary consequence of its premises.[42]

In particular, the technical implementation of ammonia synthesis, for which Haber had delivered the proof of principle, crucially depended on the identification and economic viability of a catalyst to raise the reaction rate. Given the limited knowledge about the exact workings of a catalyst in a chemical reaction, there was no a priori guarantee that an appropriate and affordable material could actually be identified. It took an industrial effort (conducted by the German Badische Anilin- und Sodafabrik (BASF) firm and involving a gigantic experimental program testing thousands of compounds) before a suitable catalyst could be found.

Even after the successful technical and industrial implementation of the "Haber-Bosch process," as the synthesis of ammonia under high pressure came to be called, its economic impact and its role in providing artificial fertilizers for a growing world population may have been uncertain. The rapid establishment of the Haber-Bosch process as the anthropogenic driver of the global nitrogen cycle was, however, soon

15.7. The first tanks of liquid ammonia leaving the plant in Leuna, Germany, on April 28, 1917. Re-flecting the times, *Glück auf!* (good luck) and *Franzosen-Tod!* (death to the French) are chalked on the sides of the wagons. Landesarchiv Sachsen-Anhalt, I 525, FS Nr. G 786.

catalyzed by an external event: World War I, which thus left its own mark on the path into the Anthropocene.

When the war began, the nitrogen market was thrown into a tumult. Suddenly the nitrogen sources in South America were no longer accessible to the Central Pow-ers, and their only alternative was to boost industrial capacities. Gigantic ammonia production facilities were built in Germany. Ammonia synthesis was propelled to the forefront of the industry, serving to produce both "bread from air," as Max von Laue put it,[43] and components for military explosives. Germany soon changed from a nitrogen-importing nation into a net exporter. "Losing the war but gaining ground" is how Haber's biographer Margit Szöllösi-Janze described the paradoxical situation of the German nitrogen industry.[44]

World War I created a feedback mechanism in the sense that the new chemical technology helped to prolong the conflict, while the war fostered in turn the expan-sion of the industrial basis for the new technology. As a consequence, the coupling between science, technology, and industry mentioned above was consolidated with important consequences for further industrial development after the war. Through this feedback loop, the synthesis of ammonia became a scientific discovery with con-tradictory global consequences: without this achievement, a large portion of the world's population could not survive, but it also affected the biosphere and increased its dependence on human interventions.

Given the energy required for the industrial implementation of the Haber-Bosch process, the agricultural system as a whole changed from one that essentially accu-mulated energy from the sun stored in plants into a subsystem of fossil energy use.

Overfertilized farmland all over the world can no longer absorb further nitrate sup-
plements. Owing to eutrophication (the enrichment of water by dissolved nutrients,
leading to a high growth of algae and low oxygen levels), coastal regions, oceans, and
lakes are afflicted with growing pockets devoid of animal life.[45] Power plants and
automobiles produce further large quantities of ammonia. Dealing with the environmental consequences of the human-induced changes in the nitrogen cycle requires
political and economic measures that must rely on further scientific knowledge to
address these unintended consequences.

The example thus shows that the creation of scientific knowledge may have irre-
vocable consequences on the Earth system as well as on the larger material and epistemic economies. The dramatic consequences of this change may become even more
strikingly clear if we include the possibilities of genetic engineering into our discussion of industrialized agriculture. Knowledge involved in the age-old cycle of seeding
and harvesting used to be public. But as seeds become products of genetic engineering, agricultural production increasingly depends on privatized knowledge subject to
a market economy. Farmers lose the seeds from their own plants and are increasingly
forced to purchase them from seed providers—who are part of an ever more concentrated industry. The hope that genetically modified crops will solve the problems of
poverty, hunger, and climate change have so far been unfulfilled, to say nothing of
the unintended consequences of the new technologies for a sustainable economy of
knowledge.[46]

The Threat of Unintended Consequences

Some of these consequences are quite problematic but probably not life-threatening
for the entire species on a short timescale. There are, however, other, more dramatic
examples of the need to cope rapidly with unintended consequences through political
interventions requiring new scientific knowledge. One concerns an episode from the
period of the Great Acceleration that has been analyzed by historians of science: the
opening of the so-called ozone hole in the atmosphere.[47] The ozone hole was an unintended consequence of industrial development, and in particular of the widespread
use of chlorofluorocarbons. It might have become a life-threatening consequence for
the entire biosphere had it not been for the intervention of concerned scientists, who
used their newly gained knowledge about atmospheric chemistry to warn the public
and trigger rapid political and economic measures to prevent a disaster.

The existence of ozone in the upper atmosphere has been known since the end of
the nineteenth century. It later became clear that the ozone layer plays an important
role in protecting the earth from ultraviolet radiation potentially dangerous to life on
its surface. Since the 1930s, chlorofluorocarbons (or CFCs, as they came to be called)
were used for the growing needs of commercial refrigeration and as aerosols in spray
cans, solvents, and foaming agents. This seemed reasonable: CFCs are nontoxic and
nonreactive. But it later turned out that they accumulate in the stratosphere, which
leads, through a sequence of chemical reactions, to a depletion of the protective ozone
layer, with potentially catastrophic consequences for life on Earth. The recognition of
this danger led to a ban of the use of CFCs, codified in the Montreal Protocol of 1987.[48]

Such a matter-of-fact account tends to downplay the dramatic character of this episode, which was actually a close race between the dynamics of industrial development, the production of knowledge capable of coping with its unintended consequences, and political measures implementing this knowledge. The potential impact of human interventions on the ozone layer was first discussed in the 1960s, then connected with the effects of high-altitude flights in proposals of supersonic transport. The realization of this potential danger enhanced public awareness of the global implications of scientific and technological progress. In any case, the further unraveling of the connection between the use of CFCs and the widening of the ozone hole over the Antarctic was driven by a combination of detailed chemical research and investigations of the Earth system.

One of the first to think about the role of nitrogen oxides in causing ozone depletion was Paul Crutzen. In 1970, he showed that nitrogen oxides accelerate the rate of reduction of the ozone content. Harold Johnston's work on nitrogen compounds in 1971 raised questions about the threats to the ozone layer of by-products of jet engines, motivating substantial investments into stratospheric research. But while supersonic transport soon turned out to be economically unviable, similar issues were also raised by NASA's space shuttle.[49]

A first turning came in 1974, when Mario Molina and Sherwood Rowland published a *Nature* paper in which they pointed to the fact that the widespread CFCs were releasing large quantities of chlorine monoxide into the stratosphere.[50] The political reaction to these findings was quick. In January 1975, a task force was set up by the Ford administration, which turned to the National Academy of Sciences for an assessment. Their activities were constantly challenged by industry campaigns. In September 1976, the Academy concluded, on the basis of additional research, that ozone depletion resulting from CFCs was real. Only two years later, a ban on propellant use became effective. Meanwhile, consumers had already significantly decreased their use of CFC-based propellants.

The real shock came in 1985. According to the state of knowledge in the 1970s, ozone depletion would be relatively small, gradual, and occur near the equator. By the mid-1980s, however, results of the British Antarctic Survey indicated that a much larger depletion had actually occurred over Antarctica, amounting to the infamous ozone hole. The findings were soon confirmed by American satellite data, as well as by further expeditions to Antarctica. New data were generated, in particular, by the Airborne Antarctic Ozone Experiment, which was conducted in the late 1980s and offered further clues to the role of the CFCs.

This discovery constituted a challenging object not only for atmospheric chemistry, in the sense discussed in earlier chapters, but also for the political process that eventually led to the signing of the Montreal Protocol even before the scientific debate was definitely settled. An epistemic community emerged around this challenging problem, refocusing existing research agendas and reaching out to the public and political spheres. Although the discovery of the ozone hole seemed to flatly contradict the original prediction, it did not lead to a rejection but to a further elaboration of the underlying chemical theory on the basis of newly acquired empirical knowledge about the Earth system. Crutzen and colleagues eventually identified chemical reactions on the surface of particles in characteristic clouds forming under extremely low temperatures, the "polar stratospheric clouds" (PSCs), as the crucial mechanism giving rise to the Antarctic ozone hole. In 1995, Paul Crutzen, Mario Molina, and Sherwood Rowland received the Nobel Prize for their research.

Both debates, the scientific as well as the political, took place under conditions of uncertainty. The character of this uncertainty was shaped, however, not only by the incomplete knowledge about the processes involved but also by the prior history of industrial chemistry. As we have seen, the deeper chemical understanding arrived only slightly after important political measures had been initiated on the basis of preliminary knowledge, helping to avoid the worst—a lesson to be taken into account for current debates on climate change, as historians of science Jed Buchwald and George Smith argue. But what may seem to be a close race between industrial interventions, scientific understanding, and political action with a positive outcome and an optimistic message also harbors a deeper warning when we consider more closely the constraints inherent in the evolution of knowledge: there was no a priori guarantee that the knowledge needed to prevent a catastrophe could be produced and implemented in time.

Indeed, had bromoflurocarbons been used instead of chlorofluorocarbons, the results would have been catastrophic: on an atom-by-atom basis, bromine is about one hundred times more effective than chlorine. Nobel laureate Paul Crutzen elaborates on this alternative scenario in his Nobel lecture:

> This brings up the nightmarish thought that if the chemical industry had developed organobromine compounds instead of the CFCs—or alternatively, if chlorine chemistry would have run more like that of bromine—then without any preparedness, we would have been faced with a catastrophic ozone hole everywhere and at all seasons during the 1970s, probably before the atmospheric chemists had developed the necessary knowledge to identify the problem and the appropriate techniques for the necessary critical measurements. Noting that nobody had given any thought to the atmospheric consequences of the release of Cl or Br before 1974, I can only conclude that mankind has been extremely lucky.[51]

Another example of threatening unintended consequences in our interaction with nature—again shaping our predicament in the Anthropocene and making us dependent on the production of ever new knowledge, scientific and otherwise—is the discovery of nuclear fission in 1938, amounting to the discovery of a new way of harvesting energy from matter.[52] Nuclear energy is unique in being the only significant source of energy not of solar origin. This discovery would not have been possible without basic science and its unpredictable consequences. Its economic and military significance today, however, is due to a targeted industrial revolution that gained momentum with the emergency of World War II and the Manhattan Project, which produced the atomic bomb. This revolution created not only a new technology, but also a new kind of socioepistemic complex of technological, political, economic, and knowledge structures that seems impossible to abolish. Even if we were to eliminate the technology, we would still need to generate ever new scientific, technological, and policy-relevant knowledge to deal with the enormous quantities of radioactive materials that remain in the Earth system, as well as with knowledge about nuclear explosives, which can hardly be eradicated or even confined.

A final example is the rise of antibiotic resistance, as studied by historians of biology Robert Bud and Hannah Landecker.[53] In chapter 10, I briefly discussed the emergence of the biomedical complex and the upturn of biomedical research during World War II. But I have not yet discussed the dramatic ecological impact of this socioepistemic complex, which may be illustrated by the case of penicillin.

Penicillin is a mold whose ability to inhibit the growth of bacteria was rediscovered in 1928 by Alexander Fleming, after having already been observed in the nineteenth century. During the war, Howard Florey and a team of researchers at Oxford University—among them Ernst Chain and Norman Heatley—developed penicillin into a drug. Its prospective use for treating wounded soldiers motivated further research and eventually led to its mass production in the United States. Government agencies and the US Army facilitated the manufacturing of penicillin by examining soil samples from all over the world and transferring new knowledge and technologies to the pharmaceutical industry. The primary aim was to create a sufficient supply for the American army—a goal that was reached by 1944. After the war, penicillin was made commercially available in the United States. Penicillin radically transformed medicine and was considered an almost magical cure against common bacterial infections.

Antibiotics originated in soil with an ecology completely different from that of the environment in which they are now being employed against disease. In the past, human bacteria had no opportunity to develop resistance mechanisms against antibiotics. They now have, given the ubiquitous presence of antibiotics in an environment in which they are produced and distributed on an industrial scale. Within a few decades, the production of antibiotics (which are rare in natural soils) has reached millions of metric tons per year, with a massive impact on the biosphere and the global conditions for human health. As Landecker pointedly writes, "Our commensals, our pathogens, our parasites, our domestic animals and fish and their commensals, the pathogens of our parasites, the avian scavengers of our cities and the wildlife—are all now participating in an antibiotic ecology."[54] According to a news release by the World Health Organization from 2014, "This serious threat is no longer a prediction for the future, it is happening right now in every region of the world and has the potential to affect anyone, of any age, in any country. Antibiotic resistance—when bacteria change so antibiotics no longer work in people who need them to treat infections—is now a major threat to public health."[55]

In summary, some of the biogeochemical cycles in the Earth system have been wedded to a decidedly socioepistemic component that plays a critical role for their regulation. What ecological systems achieved in the past (such as the limitation of the amount of reactive nitrogen in soils) or cannot achieve at all (such as the degradation of radioactive materials) must now be regulated by human interventions. Human societies need to commit to the installment, maintenance, and adaptation of political, economic, and epistemic infrastructures that will generate the necessary knowledge and implement it in practical regulations over the coming millennia. There may be limits to growth, but with the realization (in the double sense of the word) of additional coupling between the global dynamics of human societies and the Earth system, there may also be a growth of limits that reflect the growing fragility of this coupled system and that we must consider in our tinkering therewith.

Geoanthropology

Novel forms of analysis, new conceptual frameworks, and new research tools will be required to address the challenges of the Anthropocene. What is needed is basic research in a domain that is strongly shaped by technology, applied science, and political and economic interests. Such research will have to overcome traditional borderlines between the natural sciences, the social sciences, and the humanities. One might call this domain of research—human-Earth interactions within an Earth system perspective—"geoanthropology."[56]

Geoanthropology should study the various mechanisms, dynamics, and pathways that have moved us into the Anthropocene. It should deal with the coevolution of natural, sociotechnical, and symbolic environments in an integrated manner and investigate critical junctions and tipping points that endanger human civilization and much of non-human life. It should look at how key systems such as the energy system, the global flow of materials and information, the system of agriculture and land use, industrial chemistry, and the global transport system interact with one another and with the natural spheres—such as the terrestrial and marine ecosystems, geobiochemical cycles, hydrological cycles, and energy storage and transfer cycles across time and space—and it should investigate the role of knowledge in linking all of these processes.

Since its emergence more than thirty years ago, Earth system science has identified a human-led perturbation of the Earth system and has created mathematical frameworks to incorporate the human component into its biophysical modeling strategies.[57] Cultural and historical dimensions of various scales have also been taken into account by the global network initiative known as "Integrated History and Future of People on Earth (IHOPE)." Geoanthropology should make use of these advances in Earth system modeling and methods of integrated assessment, but it should explore on a more fundamental level the environmental, social, economic, political, and epistemic dynamics of the interactions between human actors and the Earth system from historical, evolutionary, and systemic perspectives. At the same time, it should go beyond traditional historical narratives and case studies by providing causal accounts of systemic changes based on data-rich empirical assessments.

How can the dynamics and transformations of human-Earth systems be conceptualized and modeled? What are the major transitions of human-Earth systems and how can they be characterized? What are the processes that allow such major transformations to emerge? How can future "intelligent" human-Earth systems be devised in compliance with planetary boundaries and humanitarian values? The investigation of these questions will have to begin with a novel, integrated approach that brings together three dimensions in the study of human-Earth systems: the resource dimension (labor, energy, materials), the regulatory dimension (economy, politics, law, knowledge, belief systems, automatization, artificial intelligence), and the ecological and evolutionary dimension (biodiversity, ecological challenges, sociotechnical and symbolic environments).

The characterization of human-Earth interactions requires that all of these dimensions and their respective interactions be taken into account. The novelty of the approach lies, however, not only in the combination of the three dimensions, but in the way in which each of them widens the scope of more traditional perspectives. For instance, instead of limiting analysis to the well-known tensions between economy and ecology, their relation will have to be more broadly conceived as an interaction between different regulatory systems, their resources, and their environments. Regulation will have to be broadly understood as comprising not just economy, politics, and law, but also knowledge and belief systems, as well as new regulatory functions in the wake of increasing automatization and digitization and the rise of artificial intelligence. Similarly, ecology must comprise the understanding of natural as well as anthropogenic environments, including the role of endangered biodiversity and the long-term coevolution of cultural and natural systems.

Geoanthropology should combine simulation, evolutionary theory, history, and design. It should advance current integrated assessment models to deal with rich data on past and present human societies and their interactions with the planetary environment. Such data are quantitatively more substantial than the data typically used in historical case studies and qualitatively more refined and contextualized than the Big Data of many social science studies. It should develop and explore theoretical frameworks and analytical tools for investigating the evolution of systems with strong path-dependencies; it should further address the methodological problems of integrating macro-, meso-, and micro-scale historical investigations. Geoanthropology should also be actively involved in shaping current transformations of human-Earth systems and in dealing with the attendant scientific and technological challenges along with the wider systemic properties. Bringing together these different perspectives may establish geoanthropology as a new transdisciplinary venture with a global impact.

Chapter 16

Knowledge for the Anthropocene

Let us not, however, flatter ourselves overmuch on account of our human victories over nature. For each such victory nature takes its revenge on us. . . . Thus at every step we are reminded that we by no means rule over nature like a conqueror over a foreign people, like someone standing outside nature—but that we, with flesh, blood and brain, belong to nature, and exist in its midst, and that all our mastery of it consists in the fact that we have the advantage over all other creatures of being able to learn its laws and apply them correctly.
—FRIEDRICH ENGELS, "Dialectics of Nature"

Earth provides enough to satisfy every man's need but not for every man's greed.
—MAHATMA GANDHI, quoted in Pyarelal,
Mahatma Gandhi, vol. 10, *The Last Phase*

The Place of Humanity in the Anthropocene

In view of the Anthropocene, proposals for dealing with the threat of unintended consequences and of steering human interventions in the Earth system with the aim to control these consequences are controversial.[1] Some propose global interventions in the sense of geoengineering, of global governance, or of a change in the global economic order. Many point to the futility of such interventions since there is no governing center, and some see the further journey into the Anthropocene as being largely prescribed by systemic constraints.[2]

On the one hand, thinkers such as Bruno Latour claim that a new philosophical attitude is required—implying that it is only possible to cope with the global impact of human actions by admitting that humans share the "destiny" of agents who cannot be divided along the lines of subjectivity and objectivity:

> The point of living in the epoch of the Anthropocene is that all agents share the same shape-changing destiny, a destiny that cannot be followed, documented, told, and represented by using any of the older traits associated with subjectivity or objectivity. Far from trying to "reconcile" or "combine" nature and society, the task, the crucial political task, is on the contrary to *distribute*

16.1. Scientists at a San Francisco Rally for Science in December 2016: Peter Frumhoff, director of science and policy for the Union of Concerned Scientists, and (*left to right*) James Coleman, high school student fellow with the Boulder, Colorado–based Alliance for Climate Education; Kim Cobb, professor at the School of Earth and Atmospheric Sciences at the Georgia Institute of Technology, Atlanta; and Naomi Oreskes, professor of the history of science at Harvard University, Cambridge, Massachusetts. Photo by Randy Showstack / American Geophysical Union / *Eos*.

agency as far and in as *differentiated* a way as possible—until, that is, we have thoroughly lost any relation between those two concepts of object and subject that are no longer of any interest any more except in a patrimonial sense.[3]

In his recent critique of fashionable ecological philosophies, Andreas Malm pointedly remarks: "When Latour writes that, in a warming world, 'humans are no longer submitted to the diktats of objective nature, since what comes to them is also an intensively subjective form of action,' he gets it all wrong: there is nothing intensively subjective but a lot of objectivity in ice melting. Or, as one placard read at a demonstration held by scientists at the American Geophysical Union in December 2016: 'Ice has no agenda—it just melts.'"[4]

The reverse claim is that human interventions have only had such menacing and even fatal consequences for our living conditions within the Earth system because human agency has not yet sufficiently freed itself from its dependence on natural history. This seems to be the conviction behind the "Ecomodernist Manifesto," for instance, which claims that "knowledge and technology, applied with wisdom, might allow for a good, or even great, Anthropocene," and that a good Anthropocene "demands that humans use their growing social, economic, and technological powers to make life better for people, stabilize the climate, and protect the natural world."[5]

In this confrontation, an age-old dualism has assumed a new guise: the attempt to establish a complicity with the forces of destiny—if necessary at the price of surrendering human subjectivity or perhaps involving other forms of self-sacrifice—is juxtaposed with the attempt to achieve human autonomy by subordinating the planet under the superior power of human ingenuity. These two positions, a modernist stance and a position critical of it, are usually considered to represent mutually exclusive alternatives. Actually, however, the two positions have more in common than first meets the eye.

At the beginning of chapter 3, I referred to Greek philosophers who suggested that the best way to protect oneself against the vicissitudes of fate was to learn how to submit oneself to it willingly, sacrificing one's drives and ambitions while expecting, at the same time, that this complicity with destiny would empower one to master worldly challenges. What unites the seemingly opposite positions, more generally speaking, is a shared move away from engagement with the concrete and individual human agency (i.e., with empirical human subjects and with the unequal power distribution in human societies) toward some powerful form of abstraction, be it "to distribute agency" or to use the "growing social, economic, and technological powers" of humanity for a better Anthropocene. I suggest that we take a more systematic look at the role of humanity in the Earth system, taking into account both its material interventions and the knowledge that enabled them.

How are we to conceptualize the place of humanity in the Anthropocene without losing sight of empirical human subjects? Can and should one define an "anthroposphere" in analogy to the biosphere, ascribing to it a similar distinctiveness and resilience? This concept has a long history and is broadly used in the context of Earth system science and industrial ecology.[6] The biosphere has existed for more than 3.5 billion years, more than 80 percent of the earth's existence. It is persistent and pervasive. It has survived several mass extinction events and has penetrated into every space on Earth, from the threshold of its atmosphere to far beneath its surface. The biosphere has profoundly transformed the planet (for instance, by the oxygenation of the earth's surface) and continues to keep it under its influence, regulating, for instance, the chemical composition of the atmosphere and oceans.[7] Humanity's growing role in the dynamics of the Earth system, though occurring on a considerably briefer timescale, may therefore suggest a similar concept capturing its global impact.

In the preceding chapter, I quoted from Richard Buckminster Fuller's influential 1969 book, *Operating Manual for Spaceship Earth*, in which he conceived of Earth as a system with finite resources and the Sun as its only energy supply.[8] In the 1970s, chemist James Lovelock and microbiologist Lynn Margulis developed the Gaia hypothesis.[9] They claimed that life has turned the earth's biosphere into a self-regulative, complex, planetary system. Living organisms, with their biochemical activity, have supposedly created better conditions for their own expansion over the globe. While the Gaia hypothesis is simultaneously a theory, a research program, a philosophy of nature, and a religious perspective, it has played (alongside Fuller's vision) a major role in the rise of Earth system science as a wide-ranging, transdisciplinary research program. The importance and influence of these ideas suggest that we have entered a post-Weberian age of science, in which we can no longer claim—as sociologist Max Weber did in his famous 1917 speech on *Science as a Vocation*—that "only through rigorous specialization can the individual experience the certain satisfaction that he has achieved something perfect in the realm of learning" and that this will remain forever so.[10]

16.2. Chris Burden, *Medusa's Head*, 1990. Plywood, steel, cement, rock, model railroad trains, and tracks. The Museum of Modern Art, New York / Scala, Florence.

Since the early twentieth century, similar proposals for human spheres have been made, with different emphases, by thinkers such as Dmitry Anuchin, Vladimir Vernadsky, Pierre Teilhard de Chardin, and Edouard le Roy.[11] Their interpretations of the "anthroposphere" or "noosphere" differed, in particular, with regard to the continuity they saw between this sphere and the others. While Vernadsky saw collective human thinking as emerging from biogeochemical processes, the appearance of the "Mind" as a geological force represented a new stage in cosmogenesis for Teilhard de Chardin. Later authors found this focus on the purely mental dimension of the global human presence deficient, pointing to the role of technical artifacts and infrastructures in shaping the conditions of the biosphere.

The discussion in the previous chapters suggests that what is so special about human history is indeed the emancipation of cultural from biological evolution owing to the resourceful employment of material culture and technologies, which introduce a fundamentally new type of metabolism with the planet. Humanity is changing the earth, covering it with artifacts that mediate between human societies and their environment, a process that came into being with the very emergence of humanity itself. But what does this imply for the place of humanity in the Anthropocene?[12]

Given the growing role of planetwide technologies in modifying the face of the earth, the concept of a "technosphere," first suggested by the philosopher of technology Friedrich Rapp and the landscape ecologist Zev Naveh, has recently been revived by geoscientist Peter Haff.[13] He argues that the entirety of the material and societal components that humanity has constructed—in more or less unplanned ways—is now turning into a global, technologically based system. He claims that this technological macrosystem "exhibits large-scale appropriation of mass and energy resources, shows a tendency to co-opt for its own use information produced by the environment, and is autonomous."[14]

From the perspective of the technosphere, humans are mere components of a planetwide technological system that they did not intentionally design, that they do not control, and from which they cannot escape. Humans need the technosphere for their survival, but the technosphere is indifferent to the fate of individual humans— as long as humanity as a whole helps to sustain it. Human societies are here primarily conceptualized as aggregations of individuals, each restricted to his or her limited scale and incapable of acting on a structural level. Haff justifies this depiction of the technosphere through an operation borrowed from the physical sciences called "coarse graining," which refers to "the adoption of a particular level of resolution or scale in describing the components of a system" in order to capture that system's behavior at a scale larger than that of its individual components.[15]

The technosphere is described here as if neither history nor social or political dynamics mattered. It does not take into account collective agency or political, economic, and social structures, let alone the evolution of knowledge with its powerful impact on shaping technological systems. Has human technology now reached a stage (or will it any time soon) at which it attains the autonomy of an organism with its own agency—an autopoietic structure reproducing its own organization? Such generalizations tend to overlook some essential features of human interaction with the global environment.

For instance, while the biosphere has proven its resilience over the course of at least 3.5 billion years of evolution, the technosphere may turn out to be a rather fragile scaffolding for human existence. While it is quite conceivable that the sum total of the unintended consequences of our actions has developed its own dynamics, even in the age of the Anthropocene escape routes may still be left to us—an observation, however, that does not imply, vice versa, that there will be a guarantee for the existence of an escape route. It rather appears that the dynamics underlying the Anthropocene might well enhance both the challenges with which we are confronted and our opportunities to react to them, leaving the question open as to whether the latter will always be sufficient to match the former. Is it possible, for instance, that geoengineering can intervene in the planetary system to the point that a new state of the planetary system would be reached in which high carbon dioxide concentration, radioactive pollution, and other unintended consequences of industrialization are no longer challenging problems but can be safely kept under control by novel technologies? Given the fact that macro-scale interventions in the Earth system are beyond anything that human engineering has achieved so far, and given the fact that there are still important gaps in our knowledge about our planetary system, we are certainly on the safer side to prioritize, at least for the time being and to the extent that it is possible at all, the preservation of our existing Earth system and damage control.

The Ergosphere

What assessment of the role of humanity within the Earth system would do justice to this insight into the fragility of our existence and that of the technological shell we have constructed for ourselves? Returning to the crucial role of the resourceful employment of material culture and technologies, I suggest the concept of an "ergosphere"—a sphere of human "work"—characterized by the transformative power of human labor both with regard to the global environment and humanity itself. The Greek word *ergon* means "work" in the transformative material sense, referring not primarily to effort and suffering like the word *ponos*, but also not primarily to the procedural, goal-oriented capability captured by the word *techne*. The ergosphere is, by definition, still open in its evolutionary logic to different ways of shaping the relationship between humanity and its planetary home in terms of the cumulative effects of human interventions embodied in their "works."

Earlier I quoted Marx's statement that "men make their own history, but they do not make it just as they please; they do not make it under circumstances chosen by themselves, but under circumstances directly encountered, given and transmitted from the past."[16] Under the conditions of the Anthropocene, this assessment acquires a new urgency as the room to maneuver shrinks in view of the transgression of planetary boundaries. The concept of the ergosphere is meant to capture both: the transformative power of human interventions beyond their intentions and the planetary limits under which these interventions unfold.

Although I call it a sphere because of its global extent, the ergosphere concept should not invite us to overlook the immense heterogeneity of humanity, its striking inability to act collectively, the basic tensions and conflicts of interest tearing it apart, and its asymmetries of power (e.g., between those who are driving the interventions in Earth system cycles and those suffering their consequences). These asymmetries of power also concern the generation of knowledge and science on a global scale.

The Anthropocene is now a "capitalocene," in the sense that the economic, social, and political structures of financialized capitalism drive and fundamentally shape current anthropogenic environmental changes, forcing nature into the role of a world external to the logic of capital, where it serves as a provider of resources and a dumping ground for waste and emissions.[17] I take issue, however, with the proposal to reject the label "Anthropocene" in order to give the current predicament a name supposedly more faithful to the historical causalities involved. In my view, this amounts to tearing down an important bridge between the natural sciences and the social sciences and humanities.[18] As long as the name Anthropocene does not deceive us into believing that a collective "we" or "the people" are in the driver's seat regarding the massive changes to the Earth system (or, vice versa, that a stratigraphic chronology implies historical causality) the concept of the Anthropocene—and perhaps also that of the ergosphere—will help us to integrate knowledge from all the disciplines concerned.

While the technosphere, as conceived by Haff, is essentially inaccessible to individuals, the ergosphere concept takes the role of a knowledge economy into account. The knowledge economy constitutes, so to speak, the software component that is missing or downplayed in the hardware-centered metaphor of the technosphere. In contrast to the circuits of a computer, we humans are, at least in principle, in the position to understand the superordinate logic through which we are involved in the dynamics of the Earth system. For instance, without the progress of recent years in integrating insights from a variety of disciplines, we would be unable to grasp the global changes that we have caused.

While the technosphere concept stresses that most humans lack the potential to influence the behavior of large technological systems, the ergosphere concept makes this possibility dependent on the existence of appropriate social and political structures and knowledge systems, and also on the individual perspectives of human actors. One cause for hope is that a knowledge economy produces and distributes not only the knowledge a society needs for its functioning (and often less) but, to varying degrees, an excess of knowledge (an "epistemic spillover") that may trigger unexpected developments.

Humans must certainly maintain and preserve their tools, technologies, and infrastructures, but they also change them with each implementation. The material world of the ergosphere consists of borderline objects between nature and culture that may trigger innovations as well as unpredictable consequences.[19] The ergosphere has a plasticity and porousness in which materials and functions are not so tightly interwoven as to exclude the repurposing of existing tools for new applications. In principle, each aspect of the ergosphere can be transformed from an end into a means, which is then available to emerging intentions and functions. Repurposing a given tool is, however, a double-edged sword—it may have disastrous consequences. Thus, the responsibility for using and developing technical systems must always be assumed anew.

The problem is also not, as suggested by the technosphere concept, that human beings are incapable of controlling systems that have a larger range of behaviors than they do themselves. The problem rather lies in the question of what "control" means in the first place. The stewardship of technological systems and infrastructures always depends on their specific nature (in particular, the way they are embedded in natural and cultural environments) as well as on their representation in knowledge and belief systems. Human cognition is always embodied cognition. There are historical examples showing that humans have been able to manage and sustain extremely complex ecologies and infrastructures of their own making over the long term. The systems' potential behaviors always far exceeded those of their human components, but these were typically ecologies and infrastructures in which the relevant regulative structures of human behavior had themselves been coevolving over long periods, including in their representation by knowledge and belief systems.

As recent work on Japanese ecologies during the Tokugawa period (between 1603 and 1868) shows, age-old traditions had accumulated knowledge on how to sustainably manage a complex landscape providing humans with food, shelter, clothing, and energy. The knowledge was implemented through a complex system of governance and material practices ranging from sanitation to publishing.[20] I return to this example below. Remarkably, the effective complexity of knowledge systems does not necessarily grow in proportion to that of the technological and environmental systems steered by them.

Since humans, technological systems, and their environments simultaneously operate on different scales, the ergosphere is a highly coupled, nonlinear system that cannot be stratified into a hierarchy of layers, each limited to its own scale in the sense of the coarse-graining operation suggested by Haff. Instead, these layers are connected by metabolic and other cycles. As a result of these couplings, human actions can potentially result in catastrophic instabilities of the Earth system. Then again, by creating suitable interfaces (for instance, by redesigning energy provision systems or urban landscapes) human societies may also be capable of stabilizing such cycles.

Haff claimed that the technosphere provides humans with an artificial environment enabling their survival and functioning as part of the entire system. This is unrealistic. The ergosphere does not provide for its own sustainability or for that of its human components. It rather exposes all of its components, human and nonhuman, to risks that may endanger its very existence. It does not act as a system whose function can be described as succinctly as that of the technosphere (a system allegedly aimed at sustaining itself by converting energy and other resources). The ergosphere includes no mechanisms per se ensuring its long-term stability within planetary boundaries. We can therefore not assume that the technical systems we have constructed will ensure our survival in the future—at least if we are not prepared to continually adapt them to changing conditions and to the existence of these boundaries.

The metaphor of the technosphere may be helpful in describing local or temporary situations in which human societies create a technological shell for themselves that seems to protect them in a momentarily stable system.[21] There is no indication, however, that this has worked or will work, either for extended periods of time or on a global level. Social conflicts, the unintended consequences of our actions, and the inexorable processes of innovation will inevitably slam into planetary boundaries and shatter such a shell. This also differentiates the concept of the ergosphere from that of the technosphere: it does not present us with the deceptive alternatives of surrendering ourselves to the autonomy of global technological systems, redesigning them from scratch, or rejecting them altogether. Rather, it invites us to see ourselves as part of a coevolutionary process in which our opportunities for action depend on the knowledge that we are able to obtain.

Missing Knowledge

From the vantage point of the ergosphere, with its quintessential plasticity, we recognize the predicament of humanity in the Anthropocene as a challenge of knowledge. This is not a technocratic perspective; just producing new scientific and engineering knowledge within the current knowledge economies will not suffice to cope with the Anthropocene. Much of the necessary knowledge does not fall within these categories. It may rather be described as a combination of system knowledge, transformation knowledge, and orientation knowledge.[22] At first blush, one might identify these three types of knowledge with the natural sciences, the social sciences, and the humanities, but this falls short of realizing the major transformation of the disciplinary structure of science envisaged here, as well as the critical role of civic, nonacademic knowledge.

System knowledge is required to understand the Earth system with its human components. It presupposes an integration of knowledge currently fragmented along disciplinary boundaries. *Transformation knowledge* primarily concerns the role of human societies as part of the Earth system and raises the question of how human collective action can affect its dynamics in such a way as to ensure sustainable development and ultimately the survival of the species. *Orientation knowledge* refers to the reflective dimension of the other forms of knowledge, connecting them to ethics, politics, and belief systems for individuals and collectives, and thus to questions of individual and collective identities and values.[23]

System knowledge alone tends to favor technocratic visions; transformation knowledge by itself may encourage blind activism; orientation knowledge without system and transformation knowledge is idle. But even the combination of these types of knowledge will be useless as long as they are not implemented within a suitable knowledge economy, comprising research, education, public discourse, and political action.

Much of the needed knowledge is unavailable—either because it is inaccessible, suppressed, or unimplemented, or because it does not yet exist or has been lost. Even in the face of global challenges, there will not be just one way into the future but various modes of bringing together the richness and diversity of human experiences. New forms of knowledge, as well as new forms of individual and social life ready to meet the challenges of the Anthropocene (including new strategies for knowledge production and energy provision, for dealing with social justice and the flow of materials, for health care and traffic, etc.) will not simply follow from a radical "paradigm shift." They will rather result from exploration processes that may eventually form a matrix, thus giving birth to new insights and forms of life that we could not have anticipated.

The Vulnerable Power of Local Knowledge

New insights may well come from the margins of powerful centers or may be hidden in long-buried experiences. I therefore return, for a moment, to the globalization of knowledge in history discussed in chapter 11, in order to illustrate the role of local knowledge and the consequences of its loss and suppression—for therein may be missed opportunities for finding answers to global problems. The globalization of knowledge was accompanied by the emergence and disappearance of many kinds of local knowledge. Local knowledge must, of course, be understood as a relational concept, to be defined with regard to a particular historical situation as well as to its genesis and development. It describes knowledge that was or did not (or has not yet) become part of a globally dominant constellation.

Nevertheless, local knowledge can be quite relevant to the global challenges of the future, from knowledge concerning pharmacological and health issues, to experiences with environmental management, to insights into human sociality. We should also not forget the relevance of epistemic diversity (an aspect of biocultural diversity) in understanding the evolution of knowledge itself.[24] Given the global character and path dependency of the evolution of knowledge, the destruction of local knowledge may have unpredictable consequences for its future course.

Many kinds of local knowledge have been lost in human history. One of the greatest losses worldwide was due to European colonization and its consequences.[25] From the beginning of the twentieth century, the presence of Western military power, capital, products, science, technology, and ideologies could be felt in virtually every part of the world. Local economies and societies were destroyed or reshaped according to Western priorities in global political and economic competition, which was focused almost exclusively on the exploitation of primary resources. Local knowledge seemed to be confined to niches soon to be eradicated by the expansion of Western economic, political, and epistemic power. The repression of local knowledge corresponded to the disempowerment and pauperization of the majority of people outside Western countries.

Local populations still relied in part on traditional knowledge for their subsistence. At the same time, however, they were affected by modernization processes such as urbanization, monoculture farming, large-scale cattle breeding, the industrial exploitation of primary resources, and militarization or colonial warfare, all of which induced the loss of traditional knowledge. In many instances (e.g., the case of the Americas discussed in chapter 11) local resources were drained further by the spread of epidemic diseases, which reduced the size of human and animal populations and weakened their resistance to newly expanding economies and political regimes. As a result, people became increasingly dependent on a globalized economy, which (in contrast to their traditional methods of subsistence) did not offer them sustainable living conditions unless they belonged to the class of elites who profited from such partial modernization processes.

Yet even in the age of colonialism, there was room for the unfolding of multiple modernities, to use an expression of the sociologist Shmuel Eisenstadt, who conceived "modernity" (often understood as a one-way street in European and Western traditions) as one of several coevolving cultures in the world.[26] During the Tokugawa period in Japan mentioned above, major advances were made in social cohesion and the sustainable management of resources.[27] This was achieved on an island country largely (but never completely) isolated from the rest of the world, with no developed capitalist system or use of fossil fuels. Japan followed pathways different from those that prevailed in Western countries, some of which continued even after the reforms of the Meiji period between 1868 and 1912. One example is the "night soil system," that is, the collection of human feces as a resource, which continued to work more or less until the 1920s.

In the domains of agriculture, forestry, food production, architecture, urbanization, and waste disposal, striking forms of alternative modernity had emerged in Tokugawa Japan, some of which may be models for addressing current problems.

Although only 15 percent of its land was arable, Japan had a population of around twenty-six million people, larger than that of all European countries except Russia. Edo (today's Tokyo) was one of the largest cities in the world, with a population of over a million in 1800. Even more remarkable, this population size remained nearly stable from the beginning of the eighteenth century to the middle of the nineteenth century, owing to measures of deliberate population control. A number of factors contributed to this development, among them a combination of centralized control and decentralized political structures, which created resilience on different ecological scales, an intense local relationship with natural resources, and widespread awareness of resource scarcity and of the need for sustainability. This setting was itself the result of both a dramatic political history and special ecological conditions, favoring the resilience of forests, for example.

The relatively stable constellation of the Tokugawa period followed almost 150 years of civil wars. When the new Tokugawa regime was set up in 1603, the *shogun*, invested by the emperor, directly administered a quarter of Japan's territory while firmly ruling over the local *daimyo* and retaining the monopoly of military force. The ensuing time of peace and prosperity, along with the self-imposed isolation from the rest of the world, cutting trade connections and banning foreign intrusions, led to an increase of population and a depletion of local resources—for example, by deforestation resulting from ambitious building and urbanization projects. The large fire in 1657, which destroyed half of the capital and killed one hundred thousand people, served as a warning shot and eventually led to top-down measures enforced to improve the sustainability of local resources. A culture of sustainability thus emerged in Japan, independently from developments in Germany, where, at about the same time, the very concept of sustainability had been introduced by Hans Carl von Carlowitz, also in the context of silviculture.[28] As Jared Diamond writes in his book *Collapse: How Societies Choose to Fail or Succeed*, "Beginning in that same decade, Japan launched a nationwide effort at all levels of society to regulate use of its forests, and by 1700 an elaborate system of woodland management was in place."[29] Clearly, learning processes are possible, even on the scale of an entire country. But they evidently require special conditions. In the case of Japan, one of them was the consciousness of living within a closed and stable economy, or, in the words of Jared Diamond, "Living in a stable society without input of foreign ideas, Japan's elite and peasants alike expected the future to be like the present, and future problems to be solved with present resources."[30] In fact, however, widespread experience of the fragility of such a closed world eventually changed behaviors at all levels, strengthening, in particular, local subsistence strategies.

In Japan, the local relationship with nature was shaped by a village economy in which people had learned—through wars and famines—to cope with limited resources.[31] As Julia Adeney Thomas and other historians of Japan have emphasized, this included sustainable forest use, focusing, for instance, on renewable materials such as weeds for building purposes. Instead of heating their rooms, Japanese relied on simple and easily reusable clothes for keeping their bodies warm. Houses were designed in a deliberately plain way. Simple customs such as taking one's shoes off before entering a house helped to keep germs out. Public health and life expectancy acquired high standards, also because peasants' dietary practices did not rely to the same extent as in Europe on domestic animals or imported crops liable to suffer from or to spread diseases but on local foodstuffs such as mixed grains or tofu. Population control was also exerted locally; relevant measures ranged from late marriage to infanticide.[32] In periods of crisis, the leading families of a village were expected to take responsibility for families of a lesser social status.

But I am not speaking of an exclusively rural society. As mentioned above, Tokugawa Japan had one of the highest degrees of urbanization in the early modern world. Since the end of the seventeenth century, it had developed a thriving printing culture as well as a high degree of literacy, sustaining a thoroughly regulated society.[33] Its resilience was not just a local affair but depended, quite literally, on metabolic exchanges between urban and rural areas. They involved sophisticated water management as well as a waste management system in which human excrements were collected in the city and traded to be used as fertilizers on the land.[34] In turn, the cities were served with fresh vegetables from the surrounding land. Both systems remained in operation well into the twentieth century, accompanying Japan's special way into modernity.

But Japanese resilience also had its limits, and it had a price. Its limits were indicated by regular severe famines until the 1830s.[35] For example, the "Great Famine of Tempo" (1833–39) claimed up to a million lives. Japan was, moreover, never completely isolated. Famines were tempered by the introduction of sweet potatoes (probably from China) as a relief crop in the seventeenth century and seem to have disappeared only in the second half of the nineteenth century with the Meiji reforms and the adoption of Western economic models, such as an open nationwide market.[36] Furthermore, resilience was achieved under a repressive political regime, heavily restricting individual freedoms such as mobility and birthrates. The Japanese example nevertheless illustrates the kind of potentials harbored by historical experiences that are not part of the standard narrative of modernization dominated by a Western perspective. Even after the Meiji reforms, Japan was not simply a "successful example" for adopting the Western model but retained an independent agency in the process of modernization.

While all in all there was little room for the emergence and spread of different perspectives under the conditions of colonialism, the situation changed when Western political, military, and economic competition escalated into the world wars of the twentieth century. The consequences of these wars substantially weakened Western political and military hegemony and opened up space for the global spread of and experimentation with a variety of models for societal development. Examples include the regionally differentiated pathways or "creolizations" that socialism and capitalism have taken, such as African socialism or the versions of capitalism developed by the "Four Asian Tigers" (Hong Kong, Singapore, South Korea, and Taiwan) between the 1960s and the 1990s.[37] They were still mostly rooted in traditions of modern Western thought but illustrate that models of European origin could be reinterpreted and reinvented.

The spread of such models of modernization was driven at least in part by a globalization of knowledge—in particular, knowledge about the political and economic options of developing countries in response to the external pressure of industrialized nations and of the world market. These models were often adapted to local circumstances, giving rise to autonomous solutions to developmental problems. These localized models shared high hopes that young nations might become independent actors in a globalizing world. Socialism in Tanzania, for example (as interpreted by Julius Nyerere, who led the country to independence in 1961), centered on localized ideas about self-reliance, communal living, and the African family.[38]

Cuba after the revolution of 1959 provides an example of the development of an advanced scientific system in an underdeveloped country along a partially self-determined pathway.[39] From the 1970s onward, Cuban development was strongly shaped by collaborations with the Soviet Union and other nations of the Warsaw Pact, but the establishment of new educational and scientific systems had actually started much earlier on the basis of internal discussions within the local scientific community immediately after the revolution. It also involved collaboration with Western scientists and institutions, through strategies that were not simply imported from more advanced countries but which aimed at new ways of integrating science with local economic and social development.

In most cases, however, external pressures rendered local solutions unsustainable. In fact, the spread of alternative models of modernization was more often than not a vehicle of neocolonialism and global imperialism, particularly during the period of the Cold War. The immunization of models of modernization against the modifications of local knowledge was evidently bolstered by the role of these models to defend claims to power in a postcolonial world—from the ideological confrontations of the Cold War to the religious fundamentalisms and the "war against terrorism" in our times. Under these external pressures, the capability of such models to integrate local knowledge into a global learning experience remained limited.

The dynamics of the world market, and of the world financial market in particular, radically affected the field of policy options in ways that tended to counteract locally sustainable developments.[40] In the 1970s, developing countries were struggling with the consequences of the international oil crisis and the global economic recession. Practically all expectations concerning generic patterns of development were ultimately disappointed. Socialist models did not work as conduits toward a more equal distribution of wealth (either internally or with industrialized nations), nor did protectionism. A reorientation from the world market toward South-South collaborations also did not necessarily foster an autonomous modernization of developing countries. A special case was the rapid economic development of the People's Republic of China, first made possible by the massive transfer of industrial capacities from the Soviet Union in the 1950s and then again unleashed, after the depredations of the Cultural Revolution, by the economic reforms led by Deng Xiaoping from 1978.[41]

In contrast, neoliberalism, with its goal to deregulate local economies and open them toward the world market, did not bring developing countries into alignment with the industrialized world. Its beginnings go back to the so-called Nixon shock of 1971, when US President Richard Nixon unilaterally decided to abandon the guarantee to redeem, within certain limits, dollars for gold. Eventually, in 1973, it led to the collapse of the Bretton Woods system of international monetary management, which had maintained exchange rates within a narrow corridor. The collapse went along with a liberalization of markets and increasingly also of the financial markets, undermining national regulation and control. At the same time, the price of oil increased, with dramatic consequences in particular for the poorer nations, aggravating their balance-of-payment problems.

In the 1980s, interest rates were going up, and capital gravitated to the United States as a consequence of its neoliberal politics, which included tax reductions and increased military spending financed by credit. Many developing countries were thus unable to pay their debts. As a consequence, poor countries now became even more dependent on support from the West—for instance, in the form of financial aid from the International Monetary Fund and the World Bank, institutions dominated by rich countries. Such aid was aimed at macroeconomic stabilization and at paying off national debts.

Western donor countries connected developmental aid with requests for the implementation of Western standards, from human rights to the fight against corruption and the economic viability of aid projects. Such globalized standards for development and structural reform typically also included requests for currency devaluation, deregulation, and the reduction of trade barriers; the privatization of state-owned enterprises; the reduction of public spending for health care, education, and housing programs; and accountability requirements. In Tanzania, for instance, such policies led to a deterioration of the public health-care system, imposed unbearable medical care costs on the poorest strata of the population, and made the country's health-care system increasingly dependent on external funding.[42]

Nevertheless, in the long term, local knowledge as well as other locally diverse conditions did play a role in the differential development of non-Western countries in the decades after the 1970s. Substantial economic development was spurred by sundry conditions, from exploiting local control over oil as a key global resource by Arab feudal regimes to the reorientation of local economies toward global exports by Asian states. These developments reflect the dependence of particular initiatives on specific local conditions, as well as the capability of entire countries to learn not only from their own historical experiences but also from those of others. In fact, the rapid economic development of certain Asian countries in recent decades could have hardly happened without the existence of ancient local traditions of cultivating knowledge as a method of self-improvement (such as Confucianism) nor without some countries (Hong Kong, Taiwan, Singapore, and South Korea) serving as models for others (Indonesia, Malaysia, and Thailand).[43]

Local knowledge connected with traditional techniques for mastering primary living requirements (e.g., food production, medicine, architecture, and mobility) may seem, owing to economic, political, and cultural globalization processes, to be generally waning, confronted as it is with the economic and technological powers of global capitalism and the universalizing tendencies of science. But as I have argued in chapter 11, globalized capitalism and science have themselves, in the course of their emergence, critically benefited from the local knowledge that they now appear to repress. As Marshall Sahlins succinctly states, "Rather than a planetary physics this is a *history* of world capitalism—which, moreover, in a double fashion will testify to the authenticity of other modes of existence. First by the fact that modern global order has been decisively shaped by the so-called peripheral peoples, by these diverse ways they have culturally articulated what was happening to them. Second, and despite the terrible losses that have been suffered, the diversity is not dead; it persists in the wake of Western domination."[44] The place of local knowledge in the global community was and is therefore not a niche but a matrix, a substratum of all other forms of knowledge—one that generates diversification and change. Without residual traditions, without the creative appropriation of globalized knowledge, and without new local responses to global challenges—including the adaption of new foods to traditional eating habits or the recycling of waste—survival for many would be impossible.

Local images of knowledge have often proven to be more resistant to the challenges of globalization than first-order local knowledge, being more removed from immediate experiences. Local social structures such as families, ethnic groups, and religious affiliations, all of which play a role in the regulation of the production, authorization, and transmission of knowledge, may survive and even spread when other, larger-scale social structures (such as political institutions) have long become victims of globalization.

There is, in any case, no spread of knowledge without local efforts to make sense of it. Local knowledge remains the underground from which all other forms of knowledge emerge, not in a primordial sense, but in terms of the inevitably local appropriation of shared knowledge, globalized or not. The immense variability of local conditions hence continues to act as a driving force for the further diversification of knowledge, even in the presence of globalization.

Usually, scientific knowledge is conceived as being produced locally and valid universally, an image of knowledge that Yehuda Elkana has characterized as "local universalism."[45] As a matter of fact, however, beneath the common standards, methodologies, and widely accepted results of science, there is still a great variety of local tradition—of ways to choose problems, to interpret their solutions, and to integrate scientific knowledge into belief systems and societal processes. Speaking of a "global contextualism"[46] (Elkana) may therefore give a more adequate image of the actual diversification of science.

But even if we embrace this new image of science, the actual impact of the diversity of knowledge on our capacity to address the challenges of the Anthropocene remains limited. This dilemma is clearly due to asymmetric power relations, but it is sharpened by a lack of efficient external representations that would allow this multiform knowledge to be reflected as a collective human experience—a critical perspective if we are to shape the forces driving us into the Anthropocene with an awareness of local conditions and consequences.

Dark Knowledge

The loss or lack of access to the hidden treasures of local knowledge is not the only problem facing the current knowledge economy. As stated in an assessment by Jonathan Jeschke, Sophie Lokatis, Isabelle Bartram, and Klement Tockner, it also suffers from a major gap between potential and actual public knowledge. The authors refer to the content hiding in this gap as "dark knowledge."[47] They also point to a lack of research on key topics because of political and economic interests; the public inaccessibility of scientific knowledge because of its specialized character, or because of political, military, or economic interests; the loss of previously acquired knowledge; and the spread of biased information. Examples of the spread of biased information are the attempts of the pharmaceutical industry to influence public opinion on issues such as the health risks connected with tobacco, sugar, or pharmaceutical products, or political and economic strategies that deny the scientific evidence for climate change.

In a seminal study, Naomi Oreskes and Erik Conway have reconstructed the attempts of a handful of scientists, aligning themselves with corporate and political interests, to obscure scientific findings and limit their impact on democratic decision-making processes by spreading doubt in the public sphere.[48] An essential ploy was to distract people's attention from the core facts to marginal issues, for example, by discussing the role of volcanos rather than the effects of anthropogenic air pollution. This strategy (involving massive lobbying activities as well as corporate-funded research) was first successfully practiced in the 1950s with regard to the health risks of tobacco and was then reiterated and refined for other issues such as acid rain, the ozone hole, and climate change. It cynically exploits the open-ended nature of scientific discourse to create the impression that it is always too early to make such drastic decisions as prohibiting smoking in certain environments, stopping the use of CFCs, or limiting the emission of greenhouse gases.

But why can this strategy work at all? It profits, first of all, from the highly differentiated division of labor in modern societies. Science is a complex societal subsystem, seemingly of little direct concern to most of the rest of society. One of its visible interfaces with society is an oversimplified image of knowledge shaped by past experiences, as well as by ideologies. In the United States, in particular, it has been influenced by the experience of the Cold War and neoliberal convictions: society-at-large relies on science when it comes to military challenges, to conquering new frontiers, and to securing commercial superiority. Science, essentially, is an asset in competitive situations, where it is expected to produce technical solutions.

From the perspective of this image of knowledge, the warnings of critical scientists of the unintended side-effects of industrial, technological, or scientific developments appear to be a transgression of their natural sphere of activity; they mingle with politics and create problems, rather than delivering tangible solutions, optimally, in the form of technology. Against this background, it becomes easy for their opponents to cast doubt on their results, and even on their personal integrity, and to call for further, "more serious" research before any actions are taken that do not belong to the sphere of technology but to the sphere of societal regulations. Counteracting the strategic spread of misinformation therefore requires critical engagement with problematic images of science as well as with the economy of knowledge through which scientific knowledge (and misinformation) are shared within society, as has recently also been argued in a contribution to the journal *Nature Climate Change*:

> As science continues to be purposefully undermined at large scales, researchers and practitioners cannot afford to underestimate the economic

influence, institutional complexity, strategic sophistication, financial motivation and societal impact of the networks behind these campaigns. The spread of misinformation must be understood as one important strategy within a larger movement towards post-truth politics and the rise of "fake news." Any coordinated response to this epistemic shift away from facts must both counter the content of misinformation as it is produced and disseminated, and (perhaps more importantly) must also confront the institutional and political architectures that make the spread of misinformation possible in the first place.[49]

As it turns out, only a small fraction of the funds invested into research and development worldwide are dedicated to the augmentation of public knowledge. And even publicly funded research may suffer from constraints and path dependencies imposed by political or economic interests, or by the academic system itself—encouraging, for instance, a concentration of research on mainstream topics, with the danger that precisely the knowledge required to deal with the challenges of the Anthropocene may fail to be generated or publicly shared. What currently prevails is an "oligopolization of knowledge: when few know much, and many know little."[50] This oligopolization of knowledge is, of course, conditioned by the oligopolization of power—and vice versa. The risk is that the innovations necessary to meet the challenges of the Anthropocene are stymied, in particular when they present themselves in forms that may require locally adapted solutions rather than universal prescriptions.

One example of the lack of research on key topics is the way in which the global challenges of disease are being addressed.[51] Diseases are not only part of biological evolution; they are also part of cultural evolution, and they are becoming a challenge of epistemic evolution as well. They have emerged, for instance, from contact between humans and animals in domestication processes. One such example is smallpox, which was transferred from rodents to humans some millennia ago in the age of the Neolithic Revolution. Today global traffic, global nutrition chains, and global inequality of living conditions have set a new stage for the emergence and spread of bacterial and viral diseases. Diseases may constitute challenges that affect societies and economies on a global scale, even if they do so in extremely different ways in different parts of the world. While human health is probably better now than at any other time in history, this progress may have come at the price of degrading the environment and thus at the cost of future generations; and it may now be threatened by the consequences of our interventions in the Earth system. Knowledge produced in the traditional mode (as a by-product of cultural evolution through basic research and market-driven innovations) may turn out to be inadequate to cope with these challenges.[52]

The global pharmaceutical market, for example, is still dominated by the production of drugs for the "First World." The challenges of the major diseases of developing countries (e.g., tuberculosis, malaria, and HIV/AIDS) are not only economic: they are also challenges for the production and transmission of knowledge. Pharmacological research and industry has failed for decades to produce urgently needed drugs that might help to eradicate the major diseases of the developing countries. Among the 1,393 drugs licensed in the last decades of the twentieth century, just 3 were for treating tuberculosis, 4 for malaria, and 13 for all the neglected tropical diseases together, whereas 179 were approved to treat cardiovascular diseases.[53]

The great challenges of humanity confront us with a structural deficit of knowledge. The existing modes of knowledge production and dissemination may well be insufficient to cope with these problems. Take the example of energy transformations mentioned at the beginning of this book.[54] In order to mitigate climate change, we urgently need to defossilize the energy system and use renewable energy instead. From nature, we could take the idea of the carbon cycle as a way of storing solar energy, emulating natural photosynthesis by using physical energy conversion through photovoltaics and wind turbines, and natural respiration by using chemical energy conversion, for instance, to create synthetic fuels.[55] From a historical perspective, a global transformation of the energy system of this magnitude would constitute a major transition on a geological timescale, comparable to other great energy transitions of the Earth system such as the emergence of photosynthesis in nature.[56] But we have no geological timescales to realize this transition.

We are facing, in any case, not only a technological transition but also major socioeconomic and political changes that will affect political and economic power structures. These political and economic structures, and such basic value systems as those related to human rights and democracy, have in fact been deeply shaped by the availability of fossil resources, as Timothy Mitchell and others argue.[57] Mastering the energy transition will therefore require a new understanding of how resource regimes, societal configurations, and technological structures interact. A transformation of energy systems will inevitably affect international trade and global politics. It may favor the decentralized over the centralized production of electric energy, and it will certainly change supply and distribution structures. It will require a reorganization of power grids and induce new patterns of consumption, because consumers may become producers as well.

In short, the challenges of climate change and the necessity of a transformation of the energy system result from our current knowledge and market economies, powered by a fossil energy regime; it is not clear whether they can be successfully addressed without the significant improvement of both. As Timothy Mitchell pointedly writes, "In introducing technical innovations, or using energy in novel ways, or developing alternative sources of power, we are not subjecting 'society' to some new external influence, or conversely using social forces to alter an external reality called 'nature.' We are reorganising socio-technical worlds, in which what we call social, natural and technical processes are present at every point."[58]

Realizing transformations in infrastructure such as the energy system (in its current, fossil form, the main driver of anthropogenic climate change) requires, in any case, new forms of public deliberation and political decision making. It will not succeed unless this transformation acquires the social, economic, and technological momentum that turned earlier innovations of infrastructure into self-accelerating developments, eventually on a global scale, like the spread of microelectronics or computing.

The problem is that we are dealing here with a system that is coupled in complex ways with many other dimensions of society, from food production to mobility, and that cannot be compartmentalized, either into different energy sectors, or into national structures within which such decisions are typically taken. On the other hand, as the discussion in chapter 10 has shown, major transformations of technological systems often require massive state interventions and investments before they gather the necessary technological momentum, in particular when the infrastructure concerned involves large hardware components, such as that of energy provision. The necessary conversion of the global energy system thus amounts to an encompassing societal transformation process, riddled with conflicting interests and complex decision-making processes as well as immense technological and logistic challenges. These challenges should also be conceived of as challenges of knowledge production, requiring not only system knowledge but also transformation and orientation knowledge.

A future energy system will have to regulate, for instance, not only the flow of energy (taking into account knowledge about resource scarcity and locally varying conditions) but also the effects of energy transformations on the global environment, as well as the flow of knowledge itself so that energy provision and the limitation of greenhouse gas emissions are optimized not only locally but globally. In other words, energy provision, like other global infrastructures with anthropocenic consequences, will probably not only have to introduce new market mechanisms such as carbon pricing but also have to make use of the rapidly increasing powers of artificial intelligent systems.[59]

Knowledge as a Common Good

The future of humanity will depend on its ability to preserve and develop its shared resources: natural resources such as clean air and water, sources of food and energy, and cultural resources like infrastructure for mobility, communication, and the dissemination of knowledge. Many of these are common-pool resources and are thus endangered by the "tragedy of the commons," which is to say their overuse and potential destruction owing to the tendency of individuals to maximize their own benefits at the cost of the common good.[60] Common-pool resources are, according to Nobel Prize–winning economist Elinor Ostrom, goods for which it is difficult to exclude users and for which one person's consumption subtracts from the quantity available to others—unlike, for instance, the sunset, which is a public good that can be enjoyed by many without such interference.[61]

While, according to Garrett Hardin, the tragedy of the commons can only be addressed by "either socialism or the privatism of free enterprise,"[62] Ostrom and her colleagues found many different models for the successful governance of common-pool resources in their empirical studies. While the solutions display a great variety dependent on the specific nature of the resources, Ostrom identified a number of generic "design principles" guiding their successful implementation.

She observed, for instance, that it is important to identify the boundaries of a resource to facilitate the exclusion of "free riders"; to impose internal rules within a community that should be respected by higher tiers of governance; to flexibly adapt rules in order to cope with sociocultural variety; to include monitoring and enforcement mechanisms; to strengthen a moral commitment to rules; and to establish procedures for dispute resolution. She also found that rules are more effective when the communities interested in overcoming common-pool resource problems are directly involved in making the regulations. The latter insight in particular, along with the role of monitoring in successfully governing a commons, illustrates the crucial interrelation between institutions and knowledge discussed in chapter 8.

The Anthropocene increasingly confronts us with the need to govern global commons. This need challenges some of Ostrom's design principles: The much larger number of people involved and their cultural diversity make it more difficult to create, agree on, and enforce rules. Different common-pool resources such as arable land, forests, oceans, biodiversity, and the atmosphere are all interacting parts of the complex Earth system and cannot be managed as if they were separate entities. Changes occur much more rapidly than in traditional commons governance, and, in contrast to these traditional cases, there is no room for experimentation. Still, preserving institutional diversity may be crucial for finding appropriate local solutions. Ostrom and her collaborators comment: "These new challenges clearly erode the confidence with which we can build from past and current examples of successful management to tackle the CPR [common-pool resources] problems of the future." They discuss possible institutional and technological responses to these challenges and conclude: "In the end, building from the lessons of past successes will require forms of communication, information, and trust that are broad and deep beyond precedent, but not beyond possibility."[63]

In his comparative study *Collapse: How Societies Choose to Fail or Succeed*, Jared Diamond develops a framework for assessing the factors driving societies into collapse. One critical factor is a society's ability to recognize challenges and respond to its perceived problems, in particular to growing environmental problems. He considers two strategies crucial for the success or failure of societal developments: long-term planning and the readiness to adapt core values. He also emphasizes that today we have unique opportunities to monitor our global situation and to learn from the past.[64]

All these insights point to the need for a new global economy of knowledge. In coping with the challenges of the Anthropocene, knowledge may well be the most important common good. As long as its external representations were subtractable—rival goods, as when the use of a book in a library by one person reduces the accessibility to its contents for others—they constituted a common-pool resource. But in the digital world, external representations of knowledge are no longer subtractable and hence come closer to reflecting the true properties of knowledge as a common good. Yet the infrastructures enabling the knowledge economy are still common-pool resources, and they need to be governed in ways responsive to the needs of this future global economy of knowledge. Here the design principles developed by Ostrom and her school may turn out to be useful.

One such lesson is the need to take into account the specificity of the resources to be governed. This suggestion gets easily lost when talking about data highways or the information society instead of dealing with knowledge as the underlying resource that matters. Data and information are derivative concepts: data are units of information—information is encoded knowledge (without regard to meaning). More than mere "information" is necessary to cope with the challenges of the Anthropocene; it requires specific forms of knowledge, not only about the Earth system but also about human behaviors and values. Whenever new infrastructures for communication have been created, they have raised hopes for improving conditions for the spread of knowledge by creating a greater symmetry between senders and receivers. But such hopes will remain futile as long as we focus simply on improving the technical functions of these infrastructures without taking responsibility for governing knowledge as a common good.

Digitization, Intelligent Systems, and Data Capitalism

Digitization is an accelerating process of transformation with pervasive consequences. It changes our daily lives, the world of work, industrial production, the market, and our social relations.[65] Humans become increasingly involved in an intelligent technological environment that can exert control over them, for instance, by wristbands guiding the movements of workers on the job. And worse is yet to come if one thinks about cerebral integration technologies such as brain implants. Algorithmic decision-making aids are being used to judge the likelihood of recidivism of criminal offenders. The use of automated decision-making of online companies increasingly shapes the knowledge economy of current societies, for instance when news-feed algorithms influence which articles people focus on and read.[66]

Digitization of information sources changes the social perception of reality. Even in affluent Western societies, many people suffer from a feeling of being left behind, not to speak of the growing digital divide between the digital haves and have-nots of this world. Individualism turns into egoism and egoism into a rejection of communal values. Populists construct origin myths to produce a sense of social cohesion, while at the same time sowing seeds of discord. Self-righteousness replaces the quest for justice, fake news takes the place of publicly binding knowledge. These developments are also fostered by an unprecedented abundance of information that is hard to assess. Social scientists speak of social fragmentation and analyze the self-reinforcing echo chambers and filter bubbles forming in the digital universe.[67]

Digitization is closely linked to other global transformation processes, but these links are as yet hardly understood.[68] Digitization goes back to the protracted information revolution beginning with the development of information theory, logical computer design, semiconductor physics, and cybernetics in the second half of the 1940s. What is clear is that without the new information and communication technologies, the rapid economic growth after the end of World War II, enabled by the transformation of wartime economies into consumer societies and the transition from coal to oil, and eventually leading to the Great Acceleration in all domains of human productivity and resource extraction, would have been unthinkable.[69] Early computers such as ENIAC, for instance, were critical for the development of the hydrogen bomb. In short, the nuclear age, the onset of digitalization, and the period that is now being considered as marking the start of the stratigraphic Anthropocene are all closely intertwined.

As for the future, there is little doubt that the expansion of information technologies and the equally expanding rates of production and consumption will continue to mutually reinforce each other and that the digitalized economy will further accelerate the fossil economy. The power of digital technologies to enhance economic effectivity (e.g., by automatization and synchronization of industrial production and distribution, or by creating a planetary labor market for micro tasks) will almost inevitably lead to the production of even more goods and services. It will be hard to escape from William Jevons's paradox, according to which increasing efficiency leads to an increase in consumption in response to lower prices.[70]

The digital turn also affects science in multiple ways: science is subjected to it, it tries to understand it, and it contributes to shaping it. For many sciences, digitization creates great opportunities because it fills a gap between experimentalization, modeling, and theory. Global climate change, for example, can only be investigated because of the availability of large quantities of data, adequate computing facilities, and sophisticated modeling.[71] But digitization not only makes science more effective, it also affects its criteria, for instance, when it comes to issues of reproducibility, trustworthiness, and causal explanations. And it poses novel challenging questions:[72] Which are actually the tasks that intelligent machines can handle better than humans? Where does human judgment play a role? How does machine learning affect decision making? How can machines best assist humans in their decisions? Where do biases creep in? What do optimal interfaces between human and artificial intelligence look like?

Here some of the insights into the dynamics of cultural evolution from chapter 14 may come in handy. After all, machine learning algorithms, for example, are simply a new form of the externalization of human thinking, even if they are a particularly intelligent form. As did other external representations before them, such as calculating machines, for example, they partly take over—in a different modality—functions of the human brain. Will they eventually supersede and even displace human thinking? The crucial point in answering this question is not that their overall intelligence still lags far behind human and even animal intelligence, but that they can play out their full potential only within the cycle of internalization and externalization that, as was discussed in chapter 14, is the hallmark and driving force of cultural evolution.

This cycle is crucial for the human capacity to extract structures from huge amounts of data by reflecting on experiences with external representations and then making them explicit. When the results of this reflection are articulated in a form that is understandable to humans, they can then serve as the next level of external

16.3. Pakistani boy at work in illegally discarded e-waste. According to the United Nations Environment Programme, 90 percent of the five million tons of electronic waste produced each year is illegal trade or dumped in countries where children work in the poisonous waste. KeenforGreen (2018).

representation, engendering the process of iterative abstraction leading to higher-order cognitive structures that was analyzed in chapter 3. Natural and symbolic languages or other rule-based symbolic systems accessible to human communication will hence probably remain among the most efficient external representations of such higher-order structures, in particular in view of the need for cultural transmission. Interfaces between human and artificial intelligence should therefore be optimized with regard to the dynamics of these processes of iterative abstraction and cultural transmission, rather than just relying on algorithmic black boxes, which may be highly efficient but whose functioning is entirely opaque to human understanding.

Digitization, Big Data, artificial intelligence, and machine learning may play a crucial role in managing our environmental challenges; they offer unprecedented insights into the dynamics of the Earth system. But where saving powers are, there danger will grow as well, to ironically paraphrase the poet Friedrich Hölderlin.[73] Electronic devices consume ever-growing amounts of energy, more chemical elements than any prior technology, and produce huge amounts of waste.[74] The personal data available on the Web open up immense possibilities for misuse and manipulation, as is illustrated by the case of a private firm (Cambridge Analytica) inappropriately gathering the personal information of around eighty million Facebook users with the intention to influence the formation of political opinion. One aim of the economic and political forces currently driving digitization is an increased effectiveness in the intelligent control of societal processes. The problem is that this control often focuses on only a few parameters of an attention economy and is geared, for

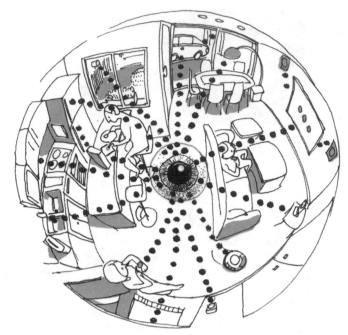

16.4. In a connected world: In whose mind's eye are we? Drawing by Laurent Taudin.

instance, to the time individuals spend on a Facebook page, with the aim to maximize the effects of advertising.

Such a one-sided orientation toward commercially usable effectiveness leaves little room for thinking about possible side effects and may risk generating instability by creating single points of failure. If companies like Google or Uber change their search ranking algorithms, the livelihoods of many people may be affected. Social interactions are increasingly shaped by rules and instructions inscribed in technology and controlled by platform-operating companies. Competition in the market is stiff, and the resulting time pressures are evidently too high for thinking in the long term. Many are convinced that they have to keep up with a global development as if there were only one possible direction in which to run. But this conviction only reinforces in turn the already dominant herd behavior.

Currently, such initiatives as the Internet of Things and the Industrial Internet are pushed by strong economic interests and enabled by new technologies and, in particular, the decreasing cost and size of electronic components. The Internet of Things envisages a global infrastructure in which physical objects are coupled with embedded ubiquitous computing facilities and virtual representations within an electronic network, allowing for new forms of "intelligent" interaction among these objects and with humans.[75] The Industrial Internet of Things, or "Industry 4.0" as it is called in Germany, refers to the use of artificial intelligence for the automatization of manufacturing by self-organizing processes.[76]

The idea of an Internet of Things seems to have started in 1999, with a presentation by the British entrepreneur Kevin Ashton. He claimed, "We need to empower computers with their own means of gathering information, so they can see, hear and smell the world for themselves, in all its random glory. RFID [Radio Frequency Identification] and sensor technology enable computers to observe, identify and understand the world—without the limitations of human-entered data."[77] The idea of ubiquitous computing is older and goes back to pioneering developments in the 1990s at the Xerox Palo Alto Research Center (PARC) in California.[78] According to one of its pioneers, Mark Weiser, the idea was to shift the place of the computer from the desktop or the server room to the human surroundings: "The most profound technologies are those that disappear. They weave themselves into the fabric of everyday life until they are indistinguishable from it."[79]

These developments may become immensely useful when adapting global infrastructure to the challenges of the Anthropocene. But they may just as well lead, in combination with the data capitalism discussed below, to a surveillance society of unprecedented reach; they may even constitute a step in the direction of a self-organizing, intelligent technosphere with complete control over human societies. In this way, the sum total of the unintended consequences of our actions would indeed finally assume its own, autonomous dynamics.

But how can we overcome the dilemma that, on the one hand, we need to cultivate control capabilities and, on the other hand, we are increasingly subdued by the control instruments we have created? What could a future coevolution of human societies and technical systems look like—a coevolution in which human values do not have to be abandoned? Answering these questions requires further investigation. As far as our current understanding of the digital transformation is concerned, we are only at the level that climate research was thirty years ago, that is, at the beginning of Earth system science.

To what historical experiences may the digital transformation be compared, if any? Fortunately, taking complete control over human societies is not that easy; our physical reality is still far from being embedded in an all-encompassing Internet of Things. A more imminent but closely related development shaping humanity's latitude for addressing its challenges is data capitalism. In this respect, are our times comparable to the beginnings of industrial capitalism? We should remember: only after more than a century of class struggles and world wars were those newly unleashed economic powers locally and temporarily somewhat restrained by such instruments of social balance as the social market economy (whose success, however, was also contingent on the affluence of fossil energy sources, which enabled particular forms of "carbon democracy" in some parts of the world).[80] Must we repeat these experiences, or can we identify ways of avoiding such violent struggles? Can we identify ways to profit from the lessons of the past, perhaps in order to establish something like a global "social economy of knowledge"?

Ubiquitous computing and the Internet of Things have given rise to egalitarian visions of a sharing economy—an economy in which humans are empowered with computational capabilities previously reserved only for large data centers, and in which things serve not only as smart service providers but even as intelligent partners.[81] Computation, however, has become both more distributed *and* more centralized.[82] The reason is that distributed computing generates Big Data, which has become a key economic driving force for a new form of capitalism: "data capitalism." Google, Facebook, Amazon, Microsoft, and Apple control more than 80 percent of the traffic generated by end-user devices, and they have turned the collection of data into a key business model that is now being taken over by other drivers of the Internet of Things. As Bruce Sterling succinctly puts it, "The Internet of Things is basically a recognition by other power-players that the methods of the Big Five have won, and that they should be emulated."[83] The impact of data capitalism on the current knowledge economy may be compared to the original rise of capitalism. This rise was accompanied by a "primitive accumulation" of riches, eventually resulting in a class separation between the owners of the means of production and those having nothing to sell but their labor force.

Similarly, recent decades have seen an increasing separation between the owners of Big Data and those forced to generate data in any information transaction over the Internet. As discussed in chapter 8, "data" is an abstract category in an Internet-based circulation sphere similar to the concept of "exchange value" in the traditional material economy. The relationship between the two concepts is established by the cost of data generation, acquisition, storage, and transfer, as well as the production processes of the knowledge represented by the data.

In the historical rise of capitalism, money (serving as an external representation of exchange value in a market economy) assumed a new role as capital, acting as a regulative structure that subjected the development of material production to the conditions of the market. While the accumulation of capital became a purpose in itself, production processes were now subjected to the new, more abstract logic described in chapter 8: the "commodity-money-commodity" cycle was turned into a "money-commodities-more money" cycle, with an inbuilt, self-reinforcing feedback loop that takes no account of global boundary conditions.[84]

Similarly, while electronic data initially served merely to externally represent information as a medium of storage and transfer, the private or state-run accumulation of Big Data, in particular from the social Web, has increasingly become a means of acquiring and transforming control over societal and economic processes. When concepts such as the Industrial Internet of Things are put into practice, the production sphere will also be increasingly controlled by Big Data. As discussed in chapter 8, the information-data-information cycle of the pre-Web market economy is thus effectively changing into a "data-information-more data" cycle enabled by the Internet and controlled by those owning the means and infrastructure for accumulating and using that data—be they private companies or the state. The self-reinforcing processes of capital and data accumulation are not only similar but actually coupled to each other, since the accumulated, privately or state-owned Big Data may serve as "data capital": a novel means for steering and controlling not only the market economy but society as a whole.[85]

In her recent book *The Age of Surveillance Capitalism: The Fight for a Human Future at the New Frontier of Power*, Shoshana Zuboff speaks of "digital expropriation" and describes its far-reaching impact on human self-determination:

[Larry] Page [CEO and cofounder of Google] grasped that human experience could be Google's virgin wood, that it could be extracted at no extra cost online and at very low cost out in the real world, where "sensors are really cheap." Once extracted, it is rendered as behavioral data, producing a surplus that forms the basis of a wholly new class of market exchange. Surveillance capitalism originates in this act of *digital dispossession.* . . . In this new logic, *human experience is subjugated to surveillance capitalism's market mechanisms and reborn as "behavior."* . . . In this future, we are exiles from our own behavior, denied access or control over knowledge derived from its dispossession by others for others. Knowledge, authority, and power rest with surveillance capital, for which we are merely "human natural resources." We are the native peoples now whose tacit claims to self-determination have vanished from the maps of our experience.[86]

These developments affect a wide range of domains: the consumer market (from fitness trackers via mobility and retail services, to smart clothing and ambient assisted living systems), health services, the insurance business, the Industrial Internet mentioned above, basic infrastructure (from the Smart Grid for energy flow to the intelligent control of the water supply or traffic), and agroindustrial production. Since data are generated and may be accumulated in all of these domains, we are faced not only with immense challenges of cybersecurity but also with a rapidly growing means of control over basically all aspects of human existence. As Evgeny Morozov puts it, "Thanks to sensors and internet connectivity, the most banal everyday objects have acquired tremendous power to regulate behaviour."[87] Following Tim O'Reilly, Morozov therefore also speaks of the potential of "algorithmic regulation" competing with traditional forms of regulation through laws and government interventions.[88]

But we should also not forget that Big Data, artificial intelligence, the Internet of Things, and Industry 4.0 may offer opportunities for realizing intelligent systems and knowledge infrastructures (Paul N. Edwards) that could become effective (or even indispensable) in handling global challenges, for instance by enabling new accounting practices aimed at sustainability. How can one reconcile these potential advantages with the evident dangers of such intelligent systems, taking into account conflicting demands on massive infrastructure, the impact on environmental conditions and other societal frameworks, supra-national considerations, diverging economic and political interests, and the quest for democratic legitimacy?

This will hardly work without a global awareness of the necessity of radical change in addressing the manifold challenges of the Anthropocene. This awareness must draw on globally shared knowledge—knowledge that concerns, for instance, historical resource and infrastructure transformations, and that makes the alternatives and their implications as transparent as possible. I readily admit that knowledge alone will hardly suffice to engender the necessary changes, even if it includes knowledge about what might trigger behavioral changes on the required scale. However, knowledge, while failing to be a sufficient condition for these changes on its own, may well be a necessary one.

Toward a Web of Knowledge

The intelligent systems described above fail to provide an answer to the fundamental question of how to enhance the role of knowledge in these changes, since they do not create or lead to the creation of a shared representation of global human knowledge in a digital world. To deal with this question, it is worthwhile to return to the origins of the World Wide Web and to review some of its basic characteristics briefly discussed in chapter 10. The Web is a recent phenomenon, but belongs in a long chain of knowledge representation technologies. The Web has rapidly developed from a small tool used by a specialized research community to a technology with almost four billion users. It was envisaged in the late 1960s with Ted Nelson's idea of a global hypertext to potentially represent the collective knowledge of humanity in a new way: as mutually linked texts.[89] It was realized only in the late 1980s, when the World Wide Web was developed—initially as a communication platform for physicists. Only then did the general idea meet the technical competency and an epistemic community ready to realize it.

The Web offers a high-impact potential to an unprecedented number of people. Its collaborative scalability is enormous: thousands of people (or more) can collaborate in the creation of such products as an open-source operating system or an encyclopedia. The Web promises nearly universal interconnectivity: discrete documents participate in a vast network of connections to other documents. The Web exhibits exceptional plasticity: it can readily accommodate new ways of organizing content, as well as new types of content, and that content can be changed rapidly and frequently. The Web allows rapid ambient findability: amid the vast stockpiles of information, desired knowledge can be located almost instantaneously from anywhere in the world. The Web potentially provides extremely low latency: news—real or fake—spreads worldwide within seconds of an event. Data with radically disparate lifetimes converge: today's news story finds its place in the encyclopedia immediately.

However, in spite of its rapid development, none of the Web's distinctive potential for representing and sharing global human knowledge has yet been systematically realized. The present Web remains a prototype of what the Web might become, and of what its founders envisioned. The present economy of scientific knowledge on the Web remains strikingly atavistic, incorporating anachronistic features of print culture that stretch back to Gutenberg and indeed to the medieval scriptorium. The integration of old and new knowledge is still hindered by the fact that knowledge is fragmented across various media and protected by access control measures that restrict its availability and connectivity. The complex and dynamic structures of links between documents on the Web represent relations between different domains of knowledge, and in themselves constitute an important kind of knowledge. Yet the present Web lacks the means for accessing and annotating these structures and thus for creating new knowledge about them; the structures themselves remain largely invisible to human agents.

A future knowledge economy more adequate to respond to the challenges of the Anthropocene is, of course, not just a technological matter. It has to involve, as I stressed at the beginning of this book, new kinds of rewards that help to overcome the current focus of the academic system on the acquisition of symbolic and real capital, new interfaces between the economy of scientific knowledge and the grand challenges of humanity, and new incentives to implement the knowledge available. The transformation of the current knowledge economy is a political, moral, and intellectual task. It is also, however, a technological and economic challenge.

The lack of open access to scientific knowledge constitutes an important obstacle to the full exploitation of the potential of the Web to support the recursive character of research and scholarship. But while the actual content in the form of digital objects is moving more and more into the public domain (spurred along by the open access movement), the networks they span and hence the semantics inherent in their relations are becoming increasingly privatized. Today's social networking sites and knowledge platforms function as data silos into which contributions can be pumped, but only extracted with difficulty—if at all. In the age of data capitalism, large companies control the relations between documents and the users interacting with them and thus limit the potential for generating the new knowledge engendered by these relations.

New thinking is needed to transform the Web into a technology that facilitates the coproduction and sharing of knowledge for a global society in the Anthropocene. What we need is, I believe, a Web of Knowledge, an "Epistemic Web" optimized for a future knowledge economy. In such a future Web, public access to both content and connectivity would be crucial, as well as a means of controlling the quality and reliability of the represented knowledge. What is published on the Web is browsed—a term that signifies a casual association of documents. As Malcolm Hyman has argued, browsing could be supplemented by the purposeful federation of documents.[90] According to this idea, a group of documents is brought together by means of a federating document. For example, a collection of geographical data sets may be federated into a *mappa mundi*. Or several editions, translations, and commentaries on a literary work may be federated into a synoptic edition.

The development of knowledge in new areas will necessitate, however, new epistemic models for federating documents.[91] Current models (such as the encyclopedia model and the geospatial model) are powerful communal structures for organizing a large amount of information, but they are ultimately only incremental improvements on epistemic models that have been in use for more than a millennium. Overcoming the current compartmentalization of scientific knowledge and its separation from local contexts necessitates a flexible knowledge representation technology that accommodates both data and models and allows for their recursive improvement by a global community. New epistemic models could make use of the deep knowledge structures analyzed in the preceding chapters.

Whereas the structural links between documents in the traditional Web are mostly hidden and do not allow for annotation, in the Epistemic Web these structures would be exposed as federating documents containing enriched links (e.g., incoming as well as outbound links; multidirectional links; transitive and intransitive links; links with attached semantic labels; links with specified behaviors). In turn, such federating documents could be annotated or recursively federated. Librarians ordinarily conceive of metadata as a canonical, structured vocabulary that describes the content and form of a certain knowledge representation. In an Epistemic Web, all data becomes meta-data, and all documents are perspectives into the universe of knowledge. By encourag-ing dynamically enrichable links between documents, the Epistemic Web will allow documents to describe one another.

Since any document can refer to any other set of documents, a document may be understood as a projection of the universe of knowledge instantiated on the Web. Each document serves as a perspective into the entire universe of available knowl-edge; the grandeur of the vista from each perspective is a function of the document's degree of connectivity. Thus documents resemble Leibniz's monads, which "are noth-ing but aspects [*perspectives*] of a single universe. . . ."[92] Any document that is con-nected to other documents is in some sense *about* those other documents and can be construed as metadata.

Users will (in accord with their interests and needs) choose which documents to view together; which documents they wish to select as entryways into the universe of knowledge; and which documents should serve as master documents, controlling the views of secondary documents. These decisions do not have to remain private (like annotations in books kept at home); rather, they may result in the creation of public, shareable knowledge. One person's views can be made available to, and serve as potential starting points for the explorations of others.

On the current Web, user behavior is subject to surreptitious methods of infor-mation capture; the Epistemic Web, by making federation an explicit activity, will give users control over the information they produce. To increase interactivity and reflexivity, the browser concept should be replaced by a technology enabling users to play a more active role. The knowledge consumer and knowledge producer might merge in the knowledge "prosumer," a term that describes individuals who "co-innovate and coproduce the products they consume." According to Malcolm Hyman, an "interagent" should replace the traditional browser and allow the future Web prosumer to annotate existing documents and create new documents as easily as the current Web user can browse.[93] Of course, such novel technologies will have to rely on trustworthy infrastructure that constitutes a global commons in the sense discussed above. Here, the experiences of managing public infrastructure (as gath-ered by pioneer ventures such as Wikipedia) may provide important guidance.

Left to develop in a haphazard fashion, the Web will not spontaneously evolve in a utopian direction. Indeed, the alternative to an Epistemic Web will probably be a Web in which there is a commercial monopoly on content or on the platforms for its dissemination, a lack of open standards and infrastructure, restrictions on innova-tion, and ultimately a forking into two Webs: the Web of slick, mainstream content for the many and an alternative Web for the few. Neither will allow us to muster the knowledge we need in order to confront the challenges of the Anthropocene.

It is therefore important to realize that, just as the original Web was created for and by an epistemic community interested in its development and use, an Epistemic Web will come into being only when driven by a strong public interest in and politi-cal struggle for changing the current knowledge economy, enabling it to benefit from the global diversity of local knowledge and to disclose the dark knowledge rel-evant to deal with anthropocenic challenges. Changing the current knowledge economy is, as I have emphasized, not a technocratic issue but can only be accomplished by a community effort and in the context of a transformation with scientific, technological, economic, and political dimensions.

Yet even in its current state, the Web has become an important medium for driving digitally enabled social change and activism. It has considerably lowered the cost and changed the form of participation in collective social actions, which no longer demand the co-presence of its protagonists in time and space.[94] The Web has also enabled fundamentally new forms of social interaction, in particular new ways of spontaneously organizing and coordinating collective actions and ways of reacting to their results. The Web's impact on collective action addressing the global challenges of the Anthropocene will, however, critically depend not only on the social opportunities that the new technologies afford but also on the knowledge it makes available to these actions. In other words, the Web's impact will depend not so much on its agility as a social Web but on its qualities as an epistemic one.

Current economic and social transformation processes have taken their lead from the social Web and have only just begun to reshape the economy of knowledge as far as the representation of scientific knowledge on the Web as well as its interlacing with other forms of knowledge is concerned. As mentioned above, the representation of scientific knowledge on the Web is, however, still atavistic in part. But even under the conditions of an increasing privatization and commercialization of the Web, it has been possible to establish (at least in certain domains of science, such as high energy physics, astronomy, and climate science) large knowledge infrastructures that allow for the global sharing of vast amounts of data in order to cope with burning issues such as climate change.[95] In view of the dynamics of digital capitalism, such knowledge infrastructures and representations should not only be protected as an epistemic "Almende"—as a common good necessary for our survival in the Anthropocene— but should be further developed into constituents of a Web optimized for the coproduction, open sharing, and public appropriation of knowledge.

In chapter 13, we saw how epistemic communities may come into being by processes of self-organization involving epistemic networks. Clearly, even in its current form, the Web has immensely improved the conditions for such self-organizing networks. The challenges of the Anthropocene might act as a catalyst for the emergence of a global epistemic community beyond disciplinary trenches, for refocusing the Web on problems of knowledge, and for creating new bridges between the academic world and civil society. In approaching this challenge, one should always keep in mind that "being human is not about individual survival or escape. It's a team sport. Whatever future humans have, it will be together."[96]

Chapter 17

Science and the
Challenges of Humanity

Natural science will in time incorporate into itself the science of man, just as the science of man will incorporate into itself natural science: there will be *one* science.
—KARL MARX, *Economic and Philosophic Manuscripts of 1844*

The voice of the intellect is a soft one, but it does not rest till it has gained a hearing. Finally, after a countless succession of rebuffs, it succeeds. This is one of the few points on which one may be optimistic about the future of mankind, but it is in itself a point of no small importance.
—SIGMUND FREUD, "The Future of an Illusion"

We need, I believe, to realign science with the challenges of humanity. This request is, however, neither new nor entirely safe because of the risks it entails. It might be seen as a revival of the early modern alliance between science and the quest to re-center the position of humanity in the world. Then and now, challenging problems required new forms of knowledge integration. In early modern Europe, the challenging objects of science were generated by the great engineering ventures, which made it necessary to assemble all available knowledge resources, from Greek mathematics to the practical experience of contemporary engineers.[1]

In our period, the grand challenges are encountered in their aftermath. They no longer concern the local fate of city-states but all of global society under the conditions of the Anthropocene. Accordingly, our perspective is no longer one of infinite horizons and new worlds. Rather, it is focused on the limits, the intrinsic complexities, and the historical dynamics of systems, be they ecological, societal, or cognitive. The main concern of the present is no longer one of universalizing the local (as it was in the early modern period) but of localizing and contextualizing the supposedly universal. We can no longer categorically segregate culture from nature but must face the fact that these spheres are inescapably mingled.

In the early modern period, artisans, artists, engineers, scientists, and philosophers struggled to overcome traditional boundaries. Today we require no less courage in removing boundaries of a new kind—on both the individual and institutional levels—when we explore spheres of knowledge hitherto divided in order to exploit

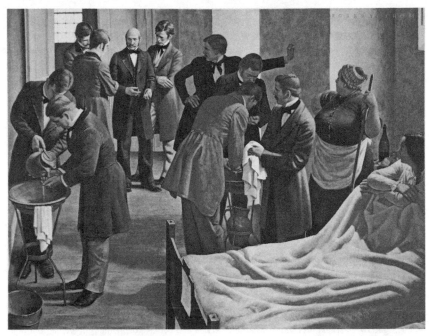

17.1. Robert Thom, *Semmelweis—Defender of Motherhood*. Working in the first obstetrical clinic of Vienna's General Hospital in 1847–48, Ignaz Philipp Semmelweis recommended that physicians and medical students wash their hands with chlorinated lime solutions before examining patients to reduce the deaths of women in childbed caused by puerperal fever (also known as "childbed fever"). His recommendation was rejected by the contemporary medical community. In 1865, Semmelweis suffered a nervous breakdown and died at age forty-seven in an asylum. From Bender and Thom (1961).

the potential of science as a medium of reflection in a globalized world. A reflective approach to scientific practice will be crucial for regaining control of the seemingly insuperable dynamic that is bringing about the Anthropocene—a dynamic that may threaten the very conditions of our existence.

The early modern engineer-scientists' tradition of engagement with practical challenges was, in a way, continued in the eighteenth and nineteenth centuries by hybrid experts producing useful knowledge with the intention of improving the human condition. Figures such as Alexander von Humboldt were not only engaged in solving societal problems by contributing to scientific knowledge and technical innovation; they also appreciated the political and, in his case, global dimension of that engagement.[2]

There are, however, other examples that may raise skepticism as to the wisdom of bending science toward a broader engagement with civil and political issues. The engagement of scientists in problematic and ultimately murderous political movements, such as the eugenics movement in the late nineteenth and early twentieth centuries, highlights the dangers of crossing the boundary between science and politics—to say nothing of the crimes against humanity committed by scientists in

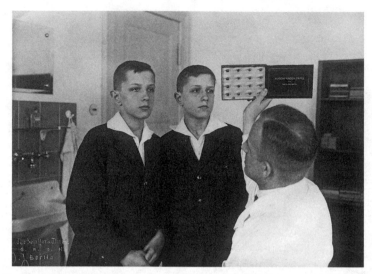

17.2. The human biologist and geneticist, Otmar von Verschuer determining eye color in twelve-year-old twins, February 1928. Von Verschuer was a leading eugenicist who during the first half of the twentieth century advocated compulsory sterilization programs. One of his students was Josef Mengele, who in 1937 also became his assistant and collaborator. In 1938, Mengele joined the SS and in 1943, encouraged by Verschuer, applied to transfer to service in a concentration camp. There he performed deadly human experiments on prisoners, in particular also on twins, and selected victims to be killed in the gas chambers. See Posner and Ware (1986); Sachse (2003); Schmuhl (2005). Courtesy of the Archive of the Max Planck Society, Berlin-Dahlem.

the name of Nazi or other ideologies. Can we, in light of this, truly claim that science may still serve as a beacon of life and the survival of the human species?

In addressing the ambiguities of such requests (which fall within the tradition of the Enlightenment) we may learn from the history of religions—not a simple task, and one that may be easily misunderstood. But I do not wish to erase the dividing lines between science and religion. Knowledge, critique, and self-criticism—often believed to be hallmarks of science—are not alien to religious traditions. Religions, however, are not primarily concerned with the institutionalized production and critical evaluation of knowledge; in the sense used in this book, science is. Traditionally, religions have offered life orientation to large communities and have even claimed to do so for all of humanity. Religions transmit basic human experiences and offer individuals participation not only in a community providing collective identity but, in a sense, also in the fate of humanity as a whole.[3] Is it possible that science could offer such a participatory perspective on the fate of humanity as well?

In the face of the complexity of human experience in a modernized and globalized world, traditional religions are increasingly incapable of offering orientation in so broad a sense, in particular one based on comprehensive knowledge about this world. This incapability may lead to a withdrawal from the challenges of this world or to violent attempts at organizing life itself according to the relatively narrow

17.3. Science piling up mountains of facts. Drawing by Laurent Taudin.

orientation that traditional religions frequently offer, based on the more limited stock of experiences that constitute their core. From a secular point of view, this shortcoming suggests reducing religions to a historically obsolete source of ethical reflection, thus ignoring their power to create and maintain strong communities held together by common practices, beliefs, and knowledge.

From an Enlightenment perspective, we have become accustomed to measuring religion by the standard of science. Religion is thereby seen, at best, as a cultural catchall, responsible for the final questions of life and death and for preserving traditional identities, values, customs, and ways of coping with the challenges of life. What is easily overlooked from this perspective, however, is the millennia of guidance that religions have offered individuals and societies across the globe—might this not help us formulate questions about the potential of science to orient humanity? For example: Does science employ accumulated human experience to address questions concerning humanity's survival in the Anthropocene? Does it contribute to making human societies self-aware? Does it enable individuals to contribute to shaping the fate of humanity (which is determined in no small measure by science itself)?

In Christian theology, eschatology is concerned with the ultimate destiny of humanity. Today, as the destiny of humanity cannot be separated from the knowledge of science and technology, we may seek out the eschatological dimensions of science itself and cultivate its role as a guide in a fragile world whose future depends on it.

There can be no constructive answers to these questions as long as we view science as a mere mountain of facts whose peaks offer no vistas, and as long as we do

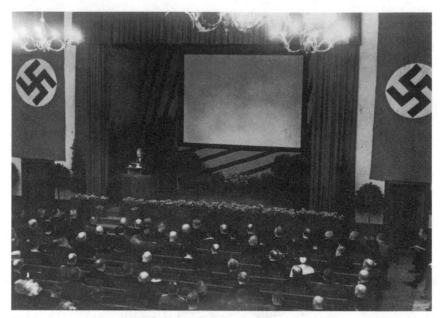

17.4. Continuing his tenure as president of the Kaiser Wilhelm Society, also at the beginning of the Nazi era, Max Planck intended to protect pure science from ideology, but in doing so helped to place it in the hands of the regime. The image shows him speaking in 1936 at the Harnack House on the society's twenty-fifth anniversary. Planck believed that he could save German science by clinging to his office as president despite the purges by the Nazi regime of Jewish or otherwise undesired colleagues. See Renn, Castagnetti, and Rieger (2001). Courtesy of the Archive of the Max Planck Society, Berlin-Dahlem.

not lift the heavy veil of inevitability that sometimes seems to weigh it down with the illusion of blind instrumental rationality. On the other hand, the preceding chapters have shown us that science is just the tip of an iceberg whose substance is the knowledge of the world and whose shape is subject to constant change. We have come to identify reflection (thinking about thinking) as the essential mechanism that makes this structural change of knowledge possible. The categories that shape our understanding of the world are therefore not stuck in the straitjacket of a predetermined logic—they are, at least in principle, constantly subject to change under our evaluation. Meanwhile, scientific knowledge has become the bottom of the iceberg as well, first because it has become relevant to virtually all domains of society and, more important, because its capacity to deliver the knowledge society needs has become a matter of survival.

In some periods in world history, science or scholarship and erudition have served as a "quasi-religion"—as an elitist religion for experts, for instance, or as an all-encompassing salvific enterprise, or as an institutionalized church. Let us briefly consider these different options one by one. Religions for experts are typically the constructions of elites who see themselves as an intellectual aristocracy, aloof from the masses and their trivial beliefs. This type of quasi-religious image of knowledge

Ales vt à primis producit in aëra nidis
Iam iam plumantes certo modulamine foetus,
Hortaturque sequi, breuibusque insurgere pennis;

Nature au monde met l'homme pour trauailler,
Ainsi qu'elle y produit tout oyseau à voller.

Sic genus humanum rerum Natura nouatrix
Mollibus è cunis, grauidaq, parentis ab aluo,
Ducit ad aerumnias, & duros cauta labores.

Natura brengt den mensch ter werelt vantsoch der wieghen,
Tot moeyte en arbeyt, als den voghel tot vlieghen.

17.5. In this 1572 engraving by Marten van Heemskerck from the series *The Reward of Labour and Diligence*, the world appears under three different perspectives: as an "ergosphere" covered by tools and instruments, as Diana of Ephesos nurturing a human child at one of her many breasts, and as a cultivated landscape. Can changing the world through human work and entrusting oneself to the care of nature be reconciled? In the early modern period, this was the hope. See Bredekamp (1984; 2003). © The Trustees of the British Museum.

is as old as the separation of intellectual and manual labor; it was already found among the Mesopotamian and Egyptian scribes, and later (as we saw in chapter 3) among Greek philosophers. The view of science as a mystery to be revealed only to a few chosen people often goes hand in hand with the prospect of abandoning those who do not understand it—an attitude that has wed scientists to criminal regimes in the past.

In contrast, in the early modern period we see science being pursued within a "workshop of hopes," as part of an all-encompassing, inner-worldly salvific enterprise directed at all of humanity, in parallel and often in competition with the great religious upheavals of the time. Early modern science was not only characterized by epistemic and economic transformations as I have discussed them; it also stirred numerous hopes for improving life, and not just for the scientists. The philosopher, pedagogue, and theologian Jan Amos Comenius, for instance, engaged himself in universal education and proposed a "General Consultation on the Improvement of Human Affairs," which is also the title of his principal work.[4] Among science's eschatological characteristics—in the sense of the ultimate hopes it raised—were its claim to productivity (i.e., that its value should be judged by its fruits); its inductive character (i.e., the individual is conceived as a representative of the whole, by which the latter than stands or falls); its immanence (i.e., its claim to explain the world of its own accord); and its inclusivity (i.e., the claim that its knowledge is able, as Descartes put it, "to secure the general welfare of mankind"[5]).

The period of awakening and high hopes was eventually followed by historical processes familiar from the history of religions: dogmatic rigidity and progressive institutionalization led to a loss of immediacy in the original promise of salvation. In the face of the disappointments in the wake of the French Revolution, enlightened hopes for an emancipation of humanity as a whole from the bonds of immaturity and irrationality began to wane. Hopes for an overarching integration of knowledge from an emancipatory perspective still survived in such philosophical programs as the natural philosophy of the so-called German idealism associated with Georg Wilhelm Friedrich Hegel, Friedrich Hölderlin, and Friedrich Schelling. One of their original questions was about the implications of the existence of moral beings for the constitution of the physical world: what must a world look like in which moral beings with the freedom to think and act are possible at all?[6] Today, the Anthropocene brings us back in dramatic ways to such a crossroad of natural and moral philosophy.

As discussed in chapter 10, however, science and technology of the nineteenth century began to evolve much more rapidly than philosophers could anticipate or incorporate into their frameworks. This expansion led to a different type of science, namely, what one might call "science as a church" owing to the institutionalization of scientific practices and the strict epistemic norms imposed on it.[7] From the perspective of the individual, the belief in science as a church was to a large degree determined by the increasing dominance of organizational structures, procedures, and hierarchies at universities, academies, and research organizations—a dominance that could hardly be questioned. The original promise of salvation—which, at least in principle, was still made to all of humanity—was thus ultimately diluted to an institutional mission celebrated on ceremonial occasions in praise of the unity of science.

The awareness that scientific unity not only entails a unity of method or epistemology but also an actual integration of content, fragmented as it may be over a variety of disciplines, was nevertheless preserved in certain scientific traditions, such as in those of geography and evolutionary biology, as the works of Alexander von Humboldt and Charles Darwin make clear. In the five volumes of his famous *Cosmos*, Humboldt tried to give an overview of contemporary scientific knowledge, showing "nature as one great whole, moved and animated by internal forces."[8] As discussed in chapter 7, Darwin integrated knowledge from areas as diverse as morphology and geography. The tradition of knowledge integration was preserved in later efforts as well. Prominent examples from the nineteenth century include Hermann von Helmholtz, Henri Poincaré, and Ernst Mach.

An integrative outlook on science and its role in society was also encouraged by its popularization. An outstanding example of the influence these popular notions might have on an individual is personified by the young Albert Einstein, who devoured Aaron Bernstein's *Naturwissenschaftliche Volksbücher*.[9] These small volumes popularized the recent advancements of science with an emphasis on its international,

17.6. Albert Einstein during a television speech in February 1950 denouncing the US government's decision to develop the hydrogen bomb. Agence France-Presse.

cooperative character; its affinity with social and political progress; and a critical attitude with regard to authorities. They gave the young Einstein an understanding of science as a human undertaking that he could not only admire but in which he could also actively participate himself—with well-known consequences not only for his scientific achievements but also for his views on the moral and political implications of science. Toward the end of his life, Einstein still remembered the impact that his reading of popular science books had in overcoming his prior religious beliefs and more generally on his outlook on the world: "Mistrust of every kind of authority grew out of this experience, a skeptical attitude toward the convictions that were alive in any specific social environment—an attitude that has never again left me, even though, later on, it has been tempered by a better insight into the causal connections."[10] Einstein's example underlines the significance of both a scientific worldview and its alliance with reflective and critical thinking about the world in all of its dimensions. As Bonneuil and Fressoz write: "To strive for decent lives in the Anthropocene therefore means freeing ourselves from repressive institutions, from alienating dominations and imaginaries. It can be an extraordinary emacipatory experience."[11]

To conclude, the very possibility of addressing our current global challenges continues to depend on processes of knowledge integration based on globally shared experiences and mediated by critical reflection that includes the societal conditions for the evolution of knowledge as well as its eschatological dimension. But we must maintain a careful balance, aware of the temptations, pitfalls, and transgressions of which the history of science reminds us.

On the one hand, in confronting the Anthropocene, we should reorient the current knowledge economy toward global responsibility. We should integrate local perspectives and find new ways of combining problem-oriented research with teaching and learning inside and outside of academia. Civic engagement and courage will also be imperative in the unavoidable global transformation processes that will follow. This is a complex challenge that requires the careful identification of cognitive and institutional structures capable of sustaining such a reorientation. The transformation will have to include the development of new curricula and research agendas, the interlocking of multiple knowledge dimensions, and critical engagement with the entanglement of knowledge with political, economic, and moral issues—these are the hallmarks of science in the twenty-first century. Societies require mechanisms and institutions to help them ensure that their policies use and comply with the available knowledge on global challenges.

On the other hand, given the long-term innovation cycles of science and its quintessential need for autonomy, it remains necessary to muster societal support for high-trust exploratory research free of utilitarian constraints, while maintaining the methodological standards and the stability of knowledge traditionally offered by the disciplines. The new scientific knowledge needed to address the global challenges of the Anthropocene can only be generated if the freedom of science to self-organize (which is at the roots of its innovative potential) is strengthened rather than weakened through further layers of control or short-term expectations; if problem choice in science can be fostered by reflection and trust, instead of being enforced by external pressure or formalized career patterns; if the necessary reality checks of intellectual ventures are not used as a pretext for the confinement of science to a blind race for "progress" or economically profitable applications; if the social and institutional structures of science encourage intellectual mobility and recruitment from all strata of a global society; and if new ways of accessing scientific information are not blocked by its transformation into a commodity.

Some of the knowledge that remains in the dark can, however, only be generated, shared, and implemented by direct participation in the struggle to change the human condition—that is, in practical attempts and local initiatives to find concrete responses to the problems on the ground with which the Anthropocene confronts us. Whatever way we encounter these problems as individuals, realizing that science is part of a global, comprehensive evolution of knowledge (and not just an elitist pursuit for and by experts) may help to restore the life-orienting dimension that was a hallmark of early modern science and its emancipatory legacy. Thereby, in searching for that elusive, shadowy knowledge, scientists may again become collaborators in a workshop of hopes, including humanity's hope for survival.

Glossary

Abstraction and Reflection

abstract concepts: Concepts such as those of number, space, causality, matter, gene, atom, chemical element, state, mind, color, action, proof, fact, or beauty, typically covering a wide range of experiences and having a large domain of reference with semantic relations to many other concepts. It is difficult to distinguish, just by looking at words, however, whether abstract concepts are part of an elementary classification of the world (e.g., the distinction between living versus nonliving things as learned by children) or part of a rich and powerful conceptual system (e.g., the concept of space in geometry). In the latter case, abstract concepts are typically the result of a long chain of iterative abstractions.

abstraction: The cognitive and historical processes by which abstract concepts are generated, such as generalization from specific cases, or reflective and iterative abstraction.

cultural abstraction: Applies the general term abstraction specifically to shared abstract concepts serving regulatory functions in a society. Examples are concepts such as honor, value, capital, or time.

iterative abstraction: The historical process through which a hierarchy of reflective abstractions is formed. This hierarchy is based on the possibility of generating a new level of abstraction by a reflection on material practices typically involving external representations such as, for instance, the constructed diagrams of Euclidean geometry, which themselves result from reflection on more elementary and historically antecedent experiences (such as the use of ruler and compass in surveying practices).

recursive blindness: A side effect of iterative abstractions—the seeming independence of abstract concepts from the specific experiences and concrete actions from which they originated.

reflection: Thinking about thinking, wherein thinking typically involves material actions and, in particular, external representations such as speaking, writing, or calculating. "Thinking about thinking" thus often becomes "thinking about the coordination of actions involving external representations of thinking," leading to reflective abstraction.

reflective abstraction: A concept introduced by Jean Piaget, here used to describe the cognitive process by which abstract concepts emerge from reflection on the coordination of material actions (e.g., the actions involved in ordering sticks by their length). Higher-order forms of knowledge may seem to be decoupled from the primary objects of experience, but they actually remain related to them by historically specific knowledge transformations. The resulting abstract concepts serve to "integrate" numerous experiences within a conceptual system (e.g., the conceptual system of thermodynamics or of modern biology). All of these abstract concepts remain, nevertheless, liable to further change.

reflective expansion: The process in which reflection on knowledge representations produces higher-order knowledge systems. Compare extensive expansion under "Cognitive Psychology and Cognitive Science" heading.

Cognitive Psychology and Cognitive Science

abstraction: *See under* "Abstraction and Reflection" heading.

accommodation: The revision of extant cognitive structures in response to new content.

assimilation: The incorporation of new content into extant cognitive structures.

chunking: The cognitive ability to memorize several items, such as the steps of a procedure, as a single unit.

cognitive psychology: The branch of psychology dealing with processes of cognition in humans and nonhuman primates.

cognitive science: The interdisciplinary study of cognition. The field became popular in the 1960s, stimulated by cybernetics and later by studies of artificial intelligence.

cognitive structures: Mental structures used to process and comprehend experiences. Cognitive structures may be composed of other cognitive structures.

collective intentionality: According to Michael Tomasello, the human capacity for creating cultural conventions, norms, and institutions based not on personal but cultural common ground. It presupposes joint attention.

concepts: Cognitive structures represented by language (sometimes by single words) that can be described in terms of their intrinsic meaning (as defined by their relations to other concepts, thus constituting semantic networks) as well as their extrinsic meaning (referring to the set of referents to which they apply, i.e., their domain of reference).

constructivist structuralism: An approach (such as that of Piaget) which assumes that cognitive structures are constructed in the course of development.

cultural memory (also external or institutional memory): The long-term transmission of habits, knowledge, or norms involving the transmission of external representations.

default settings (or assumptions): Plausible expectations resulting from prior experience that can be used to fill the variables or slots in frames or mental models and that can be replaced when new input becomes available.

descriptor: A label, such as a word or phrase, that refers to the result of chunking (e.g., the name of a procedure composed of various steps).

embodied cognition: A perspective of cognitive science that emphasizes the constitutive role of the body's interaction with its environment in cognitive processes (e.g., perception, thought, emotions), contrary to the approach of former research that modeled cognition as a kind of computation in the brain (as in robotics or artificial evolution).

enactivism: An approach within cognitive science emphasizing the interaction of mind, organism, and environment in cognitive processes. Their interrelations are understood as closely intertwined so that dualistic models of cognition are circumvented and the activity of the organism is emphasized.

equilibration: The process in which cognitive structures react to challenges that reveal insufficiencies, and through which (according to Piaget) cognitive balance is reestablished through an interplay of assimilation and accommodation.

extensive expansion: The growth of a domain of knowledge through the widening of its borders (e.g., by new experiences and the consequent incorporation of previously foreign elements). Compare reflective expansion.

frame: A concept used by the cognitive scientist Marvin Minsky to describe internal knowledge representations of typical situations. A frame comprises variables (or slots) that can be filled with various inputs such as sensual experience or prior knowledge. A frame offers information about what to expect in a particular situation. Such information may be supplied by the default settings of the relevant slots.

genetic epistemology: An approach to cognitive psychology (developed by Jean Piaget and his Geneva School of the 1920s) whose key idea is that cognitive structures are successively built up during child development through an internalization of the coordination of actions. Children thus form an active intelligence by adaptation to new experiences through assimilation and accommodation.

gestalt psychology: A direction in psychology (dating back to the Berlin School founded by Max Wertheimer, Wolfgang Köhler, and Kurt Koffka at the beginning of the twentieth century) that emphasizes the holistic and self-organizing character of cognitive structures and their changes.

internal knowledge representation: The form in which given knowledge is stored in the mind, which may be described in terms of concepts, frames, mental models, procedures, or other cognitive structures.

joint attention: The capacity to share attention with others, regarded by the developmental and comparative psychologist Michael Tomasello as one of the distinctive features of human cognition, alongside the capacity for intentional action, the capacity to imitate others, and collective intentionality.

mental model: *See under* "Mental Models" heading.

non-monotonic logic: formal logic that allows reasoning by default and the retraction of conclusions if further evidence becomes known—in contrast to monotonic logic, in which the set of logical consequences cannot be altered by the addition of new information.

noun-status: The state achieved by a chunked procedure when it can be intelligibly referred to by its descriptor (e.g., a named recipe [*see* recipe knowledge *under* "Knowledge Types" heading]).

object permanence: The cognitive capacity to realize that an object still exists even if it is outside the sensory range.

operational thinking: The latter stage in Piaget's sequence of developmental stages, separated into concrete-operational and formal-operational thinking, depending on whether mental operations remain tied to concrete contexts or can be separated from them.

operations: Reversible thinking processes (according to Piaget) in which the effects of a change can be cancelled out by imagining the opposite change (e.g., pouring water from one vessel into another and then pouring it back).

preoperational thinking: The second stage in Piaget's sequence of developmental stages, extending from about the age of two to the age of seven. In this stage, children learn to form internal representations, thus acquiring the ability to use symbols and language.

procedure: A relatively stable series of repeatable actions that can be encoded as a set of instructions allowing a person to execute a number of actions in a definitive sequence even in new situations.

real-time synthesis: The process in which an external representation is constructed through reasoning in a given situation.

reflection: *See under* "Abstraction and Reflection" heading.

retrieval: The process in which a knowledge representation structure is called up from memory.

sensorimotor intelligence: The first stage in Piaget's sequence of developmental stages, starting at birth and completed roughly by the age of two. It refers to the increasing coordination of reflexes and action schemes (e.g., sucking or grasping) that then become the bases for subsequent developmental steps.

situative action coordination: The context-dependent interaction of individuals challenged to master a situation requiring their cooperation.

zone of proximal development: A concept introduced by the Soviet psychologist Lev
Vygotsky that characterizes the difference between what a learner can achieve with and
without favorable circumstances and support (e.g., when acquiring language).

Complex Systems Theory and Evolution

action plateaus: Opportunities for action, planning, and innovation provided by given
material means constituting preconditions that may themselves be the sediment of prior
actions. They become platforms when they act as an infrastructure with regulatory functions that determine a specific space of possibilities.

adaptation: As a process, this refers to the gradual correspondence of functional traits to
their selective challenges; as a trait, it is the product of an adaptive process.

artificial evolution: An optimization procedure based on the evolutionary mechanisms of
mutation and selection. It is used to optimize the design of technical systems and is also
relevant to machine learning.

autopoietic structure: According to social theorist Niklas Luhmann, this is the quality of a
system's architecture that allows it to be self-organizing, to reproduce its own elements,
and thus to reproduce itself.

biological evolution: A special case of evolution that is based on the effects of random variation at the level of the internal genetic and developmental system of individual living
organisms and heritability across generations. The key regulative structure is selective
reproduction.

bootstrapping: A process in which a system generates the conditions for developing itself
into a more complex system.

coevolution: The phenomenon that a system does not evolve independently of other systems. As such, all evolutionary processes are in principle coevolutionary processes
that also affect the many mutual interdependencies of a nested hierarchy of regulatory
networks.

complex regulatory networks, structures, or systems: Nested hierarchies of structures that
govern the operation of a complex system. Examples include gene-regulatory networks
that control the development of organisms and institutions that govern the operation of
societies.

complex systems: Networks of components interacting on multiple levels to form a functioning and dynamic whole.

cultural evolution: An extension of biological evolution that emerges with the transmission
of material practices and knowledge in the context of social and cultural learning. Generally, cultural systems have multiple inheritance paths; the dynamics of cultural evolution are therefore more complex and operate on a multitude of timescales.

emergence: The appearance in complex systems of phenomena that could not have been
predicted or deduced from analysis of the simple systems within the complex system.
Attributes of complex systems that integrate several components (genetically or spatially) are said to be "resultant" if they are the sum of or could be determined from the
attributes of the components. They are "emergent" if they are more than their sum and
could not be determined from the parts.

endaptation: The result of an internalization whereby certain external conditions become
part of the complex regulatory structures that govern the behavior of systems. An example from cultural evolution is the Neolithization process, in which accidental ecological conditions became an intrinsic feature of human societies' production of food from
domesticated plants and animals.

epistemic evolution: A process emerging from cultural evolution in which the knowledge economy of science has transformed from an accidental into a necessary condition for preserving, sharing, and developing the achievements of cultural evolution, and possibly even the survival of the human species on a global scale.

evolution: An iterative, dynamic process of change in systems where agents and environmental factors in each (generational) step generate similar agents and environmental factors, allowing for variation. The reproduction of a system is governed by regulative structures; variants are subject to selective (and other) forces affecting their representation in the next generation.

exaptation: A term introduced by Stephen Jay Gould and Elisabeth Vrba to describe the phenomenon in which a trait that originally served a certain evolutionary function is then "repurposed" to serve another function (e.g., when feathers—according to one hypothesis—may have evolved for temperature regulation and were then repurposed for flying). An example from cultural evolution might be the invention of the wheel for the purpose of locomotion and its subsequent repurposing as a pottery wheel or vice versa.

extended evolution: A theoretical framework for understanding phenotypical, social, cultural, and epistemic evolution that is based on the integration of regulatory network and niche construction perspectives. It comprises internalization and externalization processes as fundamental components.

externalization: A process in extended evolution whereby the internal functional states or behaviors of systems transform not only the systems but also their external environment, resulting in niche construction. Examples include beaver dams or the material culture of human societies. Such constructed niches can subsequently be incorporated by internalization into extended regulatory structures that govern the behavior of systems.

feedback mechanism: A system feature through which a system's output is applied to its input, thus creating a loop in which cause and effect are difficult to discern, as shifts in either may create changes in the other, thus engendering more changes.

fitness landscape: The distribution of the fitness values of each variant for any given population of agents. Fitness landscapes allow for the prediction of evolutionary trajectories insofar as the mean fitness of a population will increase under selection. The structure of a fitness landscape also determines the possibilities for local or global optimization.

internalization: A process in extended evolution whereby elements of the environment (e.g., constructed niches or external representations in human thinking) are incorporated into the regulatory structures governing the behavior and function of systems. The result of this process is an endaptation.

major transitions: A term introduced by John Maynard Smith and Eörs Szathmáry to characterize events in evolution that involve the major reorganization of regulatory functions corresponding to the emergence of large-scale phenotypic novelty. Examples include the origins of life, multicellularity, and eusociality.

niche construction: The phenomenon wherein organisms (or systems more generally) actively construct their own functionally relevant environments. As part of this process, they externalize internal functional states, thus shaping their relevant niche; they may also internalize elements of the environment into their own regulatory structures. Niche constructions are the sediments of prior actions and may act as action plateaus for future actions.

path-dependency: The property of the evolution of complex systems that any prior system state may facilitate or constrain future states. History (as opposed to time) is a consequence of path-dependency.

platform: An environment (or regulatory niche) for the execution of functions opening up a specific state space of possibilities. The term was introduced by Manfred Laubichler to characterize major transitions in evolutionary processes.

regulative (or regulatory) structures: Rules whose functions emerge in the evolution of complex systems where the coordination of elements is a major source of novelty and innovation.

resonance effects: The stimulation of a system's internal dynamics by external forces and contexts.

scaffolding: A concept widely used across disciplines in the developmentalist tradition of conceptualizing evolutionary change (also designated as "evo-devo"). The term was introduced by the American cognitive psychologist Jerome Bruner in the late 1950s to describe support structures of learning processes within the zone of proximal development (e.g., the guidance of adult speakers in the language acquisition of a child). It has been used by the historian of science Michel Janssen to describe an intermediate stage of theory development acting as the matrix for a new system of knowledge.

sediment and *sedimentation of prior actions*: The process by which societies leave material traces in their environments. The sediment, whether physical (e.g., toxic waste) or otherwise (e.g., language spread in the wake of immigration), alters the environment for subsequent generations.

self-reinforcing mechanisms: Feedback loops within a system that perpetuate that system. For example, the integration of science into the knowledge economy of early modern Europe created a self-reinforcing mechanism in which increased knowledge production led to economic or political advantages, which in turn led to an increased demand for knowledge production.

stability and instability: The ability (or inability) of a system to accommodate internal or external stimuli and thus the system's relative susceptibility to change (evolution, endaptation, tipping points, major transitions, emergence, etc.).

symbolic evolution: The emergence and ongoing development of human capacities for external representation (speech, the creation of images, reading, writing, counting, etc.). In a more general sense, all forms of code-based representations (such as genetic code and the evolution of regulatory control and communication systems) can be seen as instances of symbolic evolution.

tipping point: The threshold beyond which a system risks abrupt and irreversible change, leading to a new state of equilibrium. In the Anthropocene (a post–tipping point state), the new equilibrium of the Earth system remains unpredictable.

XXL transitions: Major transitions that fundamentally affect the position and role of humans in their environment, as in the cascade from biological to cultural to epistemic evolution. This term is differentiated from major transitions in that XXL transitions refer specifically to humans.

Earth System

Anthropocene: A proposed new geological epoch following the Holocene in which the impact of human activities on the Earth system is the defining characteristic. Its existence and onset is currently being evaluated by the stratigraphic community in order to become formally added to the official geological timescale.

biogeochemical cycles: The pathways and transformations of substances through the various components and spheres of the Earth system, such as the carbon, oxygen, nitrogen, phosphorus, sulfur, and water cycles. These cycles may be affected—and some even induced—by humans.

biosphere: The zone in which life exists, comprising all ecosystems.

Earth system: The complex system constituted by the ensemble of Earth's spheres (atmosphere, hydrosphere, lithosphere, biosphere, etc.).

ergosphere: The Earth sphere shaped by human material interventions and infrastructures; alternative concepts are the anthroposphere and the technosphere. In ancient Greek, *ergon* designates something that has been created by human work.

geoanthropology: The study of the processes, mechanisms, and pathways leading human societies into the Anthropocene.

geoengineering: The conscious attempt to manipulate the Earth system on a systemwide scale.

Great Acceleration: A term used to describe the unprecedented rise in the influence of human activity on the Earth system beginning in the second half of the twentieth century.

Holocene: The interglacial epoch within the so-called Quaternary period in which humanity has been living in remarkable comfort for the past eleven thousand years. The word *Holocene* means "entirely recent."

planetary boundaries: A concept of Earth system science introduced by a group of scientists led by Johan Rockström and Will Steffen to describe those thresholds or tipping points in planetary biogeophysical subsystems beyond which there is a risk of abrupt, irreversible, and unsafe change.

resource couple: A technological innovation and the accompanying resource exploitation that mutually reinforce each other (e.g., the steam engine and the large-scale exploitation, production, and use of coal and iron).

technosphere: The proposed novel Earth sphere constituted by the global ensemble of technical systems and infrastructures created by humans, with an emphasis on its self-organizing dynamics; an alternative concept is the ergosphere.

Epistemic Networks

average shortest path: A network characteristic that measures the average number of edges of the shortest connections of two nodes in a fully connected graph.

epistemic networks: Networks of processes of knowledge storage, accumulation, transfer, transformation, and appropriation; epistemic networks comprise semantic networks, semiotic networks, and social networks.

semantic networks: Networks whose nodes are concepts, mental models, or the elements thereof that obtain meaning from their role in the interpretation of experiences as well as from their relation to other nodes. The edges of a semantic network consist of the thought processes by which these cognitive building blocks relate to one another.

semiotic networks: Networks whose nodes are real objects such as artifacts, material instruments, or external representations and whose edges are actions, physical processes, or other relations by which connections between them are established.

small world: A network where high connectivity between the nodes can be reached by adding only a few new connections. High connectivity is understood here as a small average shortest-path.

social networks: Networks whose nodes are individual or collective actors and whose edges are their interactions or other relations between them.

strong ties: Relationships between the nodes in a network that are designed to hold for the long term and display a high degree of interaction (e.g., long-term business relationships). As a rule, these relationships have to be ensured by high investments of economic and social resources.

weak ties: Connections based on occasional interactions (e.g., relationships within an extended circle of friends or informal meetings at trade fairs or exhibitions).

Epistemic Web

ambient findability: The ability to locate specific information among the collected masses of available information with relative ease.

collaborative scalability: The ability of a network to host collaborations between many people at once.

Epistemic Web: A World Wide Web optimized for the purposes of the global knowledge economy, including open access to both content and enriched link structures, and offering prosumers multiple ways to interact with the knowledge through interagents. The structures of links between documents are exposed as federating documents; all data are metadata; and all documents are perspectives into the universe of knowledge.

federate: To bring together a series of related documents by means of a single, federating document (e.g., geographical data sets federated into a *mappa mundi*).

interagent: A more interactive replacement for the Web browser that allows the Web prosumer to create new documents, to federate documents, and to annotate documents as easily as the current Web user can now browse.

Internet of Things: A global infrastructure in which physical objects are coupled with embedded ubiquitous computing facilities and virtual representations within an electronic network, allowing for "intelligent" interaction among these objects and with humans.

latency: The relative lag between a network's acceptance of information and its dissemination.

open access: The public availability of knowledge and the means to access it on the Web.

open source: The public availability of the source code of computer programs.

prosumer: The merged person of the knowledge consumer and the knowledge producer.

External Representations

enactive representation: The representation of thinking by a material embodiment of an action that is immediately associated with the action itself.

external representation: Any aspect of the material culture or environment of a society that may serve as an encoding of knowledge (landmarks, tools, artifacts, models, rituals, sound, gestures, language, music, images, signs, writing, symbol systems, etc.). External representations can be used to share, store, transmit, appropriate, or control knowledge, but also to transform it. The handling of external representations is guided by the cognitive structures represented by them and may give rise to novel cognitive structures. The use of external representations may also be constrained by semiotic rules characteristic of their material properties and their employment in a given social or cultural context, such as orthographic rules or stylistic conventions in the case of writing.

generative ambiguity: The paradoxical quality of external representations of knowledge as agents of transmission and fidelity as well as of innovation and change.

horizon of possibilities: The shifting, often expanding potential of a knowledge system, as opened up by the action space given by a specific physical environment and material culture, as well as by the external representations of that knowledge system.

iconic representation: A representation that refers to an action, object, or idea but that also mimics the properties of that which it represents.

individual or *subjective meaning*: The individual's internal representations as related to possible usages of given external representations and informed by that individual's unique experiences.

knowledge representation technology: A technology available for creating external representations of knowledge.

orders of external representations: Levels of representations of knowledge as defined by the iterative abstractions required for their generation, though not necessarily for their perpetuation or use.

paper tool: A concept introduced by the historian of science Ursula Klein to denote an external representation of knowledge manifested in symbols, diagrams, or formulas passed down and manipulated in writing.

social meaning: The action potentials of external representations (such as the meanings of words in a language or the possible usages of an instrument) that may not be independently perceived or anticipated by every individual but that are nevertheless realized in societal practice.

symptomatic meaning: The ambivalent capture and preservation of conflictual experiences by cultural practices such as art or religion and their related external representations.

transcendence: The phenomenon in which something acquires meaning pointing beyond itself.

Historical Epistemology

crisis: A term used by the historian of science Thomas Kuhn to describe an untenable dissonance within a scientific community when a particular paradigm seems to be breaking down or belied.

epistemic things: A term used by historian of science Hans-Jörg Rheinberger to describe objects and processes that come into the focus of a research endeavor, displaying unsuspected behaviors that may become the targets of an experimental inquiry.

epistemic virtues: A concept meant to facilitate the analysis of a scientific ethos. In essence, it was introduced by sociologist of science Robert K. Merton when he proposed basic normative principles (or ethical values) of science. It was later adopted by the historians of science Lorraine Daston and Peter Galison when they explored the shared normative standards of scientists (e.g., truth-to-nature, objectivity, or trained judgment) as collective or individual virtues, connecting the concept with a "history of scientific selves." Epistemic virtues are an important aspect of knowledge economies.

historical epistemology: A historical understanding of knowledge aimed at questioning ahistorical epistemological claims.

images of knowledge: A term introduced by the historian of science Yehuda Elkana to describe shared views about knowledge in a society or within an epistemic community. Images of knowledge are a form of second-order knowledge and are an important component of knowledge economies.

incommensurability: A term used by the historian of science Thomas Kuhn to describe the property of mutual incompatibility or untranslatability among different scientific theories and worldviews.

local universalism: A term introduced by the historian of science Yehuda Elkana to describe a specific image of knowledge based on the illusion that inferences drawn from local conditions are sufficient to create concepts of universal validity.

normal science: A term introduced by the historian of science Thomas Kuhn to describe the regular work of scientists in resolving standard problems within an accepted, established framework.

paradigm: A term introduced by the historian of science Thomas Kuhn to describe a set of typical problems, theories, methods, and practices constituting an exemplary framework within which normal contributions to a given field occur.

paradigm shift: A term introduced by the historian of science Thomas Kuhn to describe the sudden, disruptive transformations in knowledge systems and scientific communities occurring during scientific revolutions when established frameworks are given up.

scientific community: A term used by the historian of science Thomas Kuhn to describe a group of scientists and researchers focused on the same paradigm and operating within a common domain or discipline.

scientific revolutions: A term used by the historian of science Thomas Kuhn to describe transformations of scientific knowledge that give rise to far-reaching reorganizations of both the architecture of knowledge and the relevant scientific communities. According to Kuhn, scientific revolutions can be described as paradigm shifts.

thought collective: A term used by the historian of science Ludwik Fleck to describe a group of researchers constituted by their interchange of thoughts and ideas as well as by a common thought style.

thought style: A term used by the historian of science Ludwik Fleck to describe the conditioning of perception, thinking, and practices of researchers through shared tacit presuppositions that may change over time. For Fleck, agreement is only possible within a shared style of thought.

trading zone: A term used by the historian of science Peter Galison to describe a sphere of limited exchange and communication among different thought collectives (e.g., among theoretical and experimental physicists) that can function in spite of disagreements about the objects and objectives involved.

universal contextualism: A term introduced by the historian of science Yehuda Elkana to describe the necessity of taking into account a variety of local contexts when choosing problems, interpreting their solutions, and integrating scientific knowledge into belief systems and societal processes.

Knowledge

data: Information encoded in a specific but universally applicable external representation (typically as a sequence of symbols in digital form in an electronic medium, e.g., the ones and zeros of the binary number system) that may also serve as a universal standard and measure of information.

epistemic vacuum: The hypothetical (and impossible) state in which all default assumptions about reality are abandoned.

information: A measure of surprise in the transmission of messages between a sender and a receiver via a communication channel, conceptualized and quantified in information theory going back to Claude Shannon. As information marks the distance from randomness, it is closely related to the physical concept of entropy. Information is used here, just as are other abstract concepts (such as the concept of time), as part of a conceptual framework applicable to a wide range of natural and societal phenomena. In the context of a knowledge economy, information is knowledge encoded in external representations for exchange purposes.

knowledge: The capacity of an individual agent or of a group to solve problems and mentally anticipate or perform corresponding actions. Knowledge is internally represented by cognitive structures, allowing for the connection of past and current experiences.

Knowledge is not just experience encoded in individual mental structures; it also has material and social dimensions, allowing human societies to acquire and encode experiences and thereby to share and transmit them over generations. Without materially embodied external representations of knowledge, it cannot be communicated between individuals and transmitted across generations. Without being produced and shared within the knowledge economy of a society, individual learning processes would be limited.

knowledge development: *See under* "Knowledge Development" heading.

knowledge representation: *See* internal knowledge representation *under* "Cognitive Psychology and Cognitive Science" heading; *also entries under* "External Representations" heading.

knowledge system: Knowledge amalgamated by the connectivity of its elements within their mental, material, and social dimensions. Knowledge systems are typically part of the shared knowledge of a community. Knowledge systems may vary according to their distributivity (the extent to which they are shared within the community); their systematicity (their degree of organization, determining how closely the components are interwoven); and their reflexivity (their distance from primary material actions expressed by their order within a chain of iterative abstraction).

knowledge types: *See under* "Knowledge Types" heading.

package of knowledge: A knowledge system with low systematicity.

Knowledge Development

borderline problems: Challenging objects or problems that belong to multiple distinct systems of knowledge. Borderline problems put these systems in contact (and sometimes into direct conflict) with each other, potentially triggering their integration and reorganization.

challenging objects: Any material objects, problems, processes, or practices that fall into the domain of an explanatory framework and trigger experiences that cannot easily be assimilated to it but rather require its expansion, transformation, or capitulation. The knowledge system of early modern mechanics, for example, was characterized by the investigation of challenging objects such as the pendulum, the flywheel, and projectile motion, provoking traditional explanatory frameworks.

disintegration: The process by which sublevel knowledge systems secede from or are extracted from the parent systems in which they arose.

epistemic islands: Clusters of insights, results, or activities that develop at a certain distance from the core of a knowledge system (i.e., the part of a knowledge system that is characterized by high systematicity at the outset of its development, such as the axioms and some core theorems of a deductive system).

exploration: The process in which knowledge systems and their representations are investigated and probed, leading inevitably to encounters with and breaches of boundaries and to the expansion of horizons.

historiogenesis: The historical processes by which knowledge systems or other aspects of human culture emerge.

integration: Any process by which initially foreign elements of knowledge are subsumed into a given knowledge system.

internal challenges: Ambiguities, inconsistencies, or contradictions unearthed within a knowledge system through its exploration, thus triggering reorganization.

matrix: A set of results generated by the exploration of a given system which typically constitutes an *epistemic island* and serves as an intermediate stage in the transition between knowledge systems. It provides a scaffold for the emergence of a new system by supporting bootstrapping processes within a zone of proximal development (e.g., protolinguistic communication systems, protolanguages, predomestication cultivation, protoarithmetic, preclassical mechanics).

ontogenesis: The development of an individual organism.

perspective (on a knowledge system): The specific set of experiences and cognitive structures connected with the appropriation of a knowledge system in an individual learning process.

phylogenesis: The biological development of a taxon, such as that of the human species.

re-centering: The shift of an intellectual center of gravity (e.g., the thematic or conceptual focus of a theory) attendant to the transition from an old knowledge system to a new one.

recontextualization: The transfer of a system or representation to a new domain, often leading to changes in default assumptions and an expansion of horizons.

triggers of change: Catalytic effects in knowledge development. Triggers of change are inescapable disturbances and destabilizations of existing knowledge systems. They are induced by internal challenges or external challenges, such as borderline problems or challenging objects.

Knowledge Economy

appropriation: The act (on the part of an individual or a society) of actively integrating knowledge into cognitive structures or systems of knowledge, with the inevitable result of transforming both the received knowledge and the preexisting structures.

circulation: Knowledge transfer within an organized knowledge economy (e.g., the circulation of manuals in early modern Europe that spread technical knowledge).

codification: The act of creating durable external representations of knowledge (such as printed books) that may then circulate within a knowledge economy.

connectivity: The degree to which a knowledge representation is connected to other knowledge, and how explicitly so.

control procedures: Requisite societal practices that guarantee that a transmitted knowledge system satisfies expectations incorporated in certain images of knowledge (e.g., its verifiability, stability, or corrigibility).

cultural refraction: The transformation of knowledge through recontextualization when it is imported into another cultural context.

dissemination: The intentional spread of knowledge (e.g., through commercial processes).

durability: The capability of a given knowledge representation technology to endure without significant deterioration.

emancipatory reversal: The reversal of the relation between problems and solution strategies, characteristic of the emergence of separate teaching institutions. Initially, real-world problems determine the methods to be taught. In teaching, however, the methods become central and the choice of problems depends on the methods to be transmitted.

epistemic community: A group (or thought collective) concerned with the preservation, accumulation, transmission, and improvement of a particular body of shared knowledge.

epistemic spillover: An excess of knowledge beyond the knowledge specifically produced and distributed in a knowledge economy for the functioning of a society. Epistemic spillover may trigger unexpected developments.

extrinsic dynamics of knowledge globalization: The spread of knowledge through political, economic, military, or religious expansions, fellow traveling with other transfer processes, such as commerce, conquest, or missionary activity.

fellow traveling: The process in which knowledge participates in other transmission processes without governing them or being their purpose.

globalization of knowledge: The spread of knowledge that is no longer bound to local contexts and hence, in principle, is capable of participating in globalization processes.

hybridization: The process of generating new forms of knowledge by integrating knowledge systems of different cultural backgrounds.

immunization: A knowledge system's ability to resist external influences.

interactivity: The relative flexibility with which a knowledge representation can be accessed (e.g., a book can be skipped through or reread, but an electronic text can be comprehensively searched).

intrinsic dynamics of knowledge globalization: The development of knowledge that is no longer bound to specific local circumstances. Knowledge is unbound from locale through an ever-broadening experiential pool as well as by processes of iterative abstraction and hence in principle capable of global spread.

knowledge economy: All societal processes pertaining to the production, preservation, accumulation, circulation, and appropriation of knowledge mediated by its external representations.

knowledge economy of science: A knowledge economy specifically dedicated to the generation of scientific knowledge, allowing for the corrigibility of this knowledge, and involving appropriate control procedures.

modes of integration: The means by which new knowledge is absorbed into an existing knowledge system.

ownership: Access to the means of production of a knowledge representation technology.

portability: The ease with which a knowledge representation can travel.

recursiveness: The degree to which higher-order knowledge about a knowledge representation (resulting from reflection on its usage) can in turn be represented and integrated with the existing representation.

rivalry: The degree to which an individual's use of a knowledge representation decreases the value of that representation for others.

self-reinforcing dynamics: The tendency of transferred knowledge to stimulate further knowledge transmission.

social accessibility of knowledge: The relative availability or unavailability of knowledge within a society (e.g., the oligopolization of knowledge, in which "few know much, and many know little").

spread (or *diffusion*): The ensemble of individual knowledge transfer processes referring to the same body of knowledge and involving a multitude of recipients.

stimulus transfer: Mediated transmission of knowledge by an external representation not intentionally designed to represent knowledge.

transaction costs: A term introduced by Douglass North, here used to designate the economic costs of the acquisition, reproduction, appropriation, and transfer or transmission of knowledge and the relevant control procedures in a given knowledge economy.

transfer or *transmission*: The movement of bodies of knowledge from one place to another or from one individual or group to another (e.g., the movement of linguistic practices or dietary habits through migration). Knowledge transfer is characterized according to three dimensions: mediation (the employment of external representations), directness, and intentionality. Processes of knowledge transfer always involve agency on the part of the receiver as well (e.g., rejection or appropriation) and usually also have repercussions on the sender. Transferred knowledge is typically transformed in the transmission process.

vehicles: Persons or external representations by which knowledge travels.

Knowledge Types

dark knowledge: A term introduced by Klemens Tockner (et al.) to designate the content hiding in the gap between potential and actual public knowledge. It may also refer to a lack of research on key topics owing to political, military, or economic interests, the public inaccessibility of scientific knowledge because of its specialized character, the loss of previously acquired knowledge, and the spread of biased information.

globalized knowledge: Knowledge no longer bound to local contexts and hence, in principle, capable of participating in globalization processes.

implicit knowledge: Often used as a characterization of intuitive or practical knowledge because verbal expressions represent only a very limited aspect of its transmission.

intuitive knowledge: The broadly shared knowledge that results from the interaction of humans with their natural and cultural environments in the process of ontogenesis. Intuitive does not mean "without mediation" but points to the necessarily embodied and situated aspects of cognition (*see also* embodied cognition). Certain aspects of intuitive knowledge may have a universal character.

local knowledge: Knowledge emergent from and dependent on the specific experiences of locally definable groups (cultures, societies, communities).

missing knowledge: The unproduced knowledge demanded of humanity for maintaining the achievements of cultural evolution—and perhaps for its survival in the Anthropocene.

normative knowledge: The knowledge allowing individuals or groups to act in accordance with the regulations and forms of cooperation germane to institutions.

orientation knowledge: The reflective element that connects forms of knowledge to ethics, politics, and belief systems for individuals and collectives, and thus to questions of individual and collective identities and values.

practical knowledge: The knowledge resulting from the experiences of specially trained practitioners. It is generated from the pursuit of a specific task or the use of specific tools and is characteristic of all kinds of craftsmanship (e.g., architecture, medicine, etc.). It has been transmitted, for long historical periods, as part of the transmission of professional skills. When practical knowledge is habitualized, it becomes "automated" and thus may resemble implicit knowledge.

recipe knowledge: Knowledge comprising procedures with a relatively stable series of repeatable actions that can be encoded as a set of instructions allowing a person to execute a number of actions in a definitive sequence, each time in a new situation.

scientific knowledge: Knowledge resulting from the exploration of the potentials inherent in the material or symbolic culture of a society within a knowledge economy specifically dedicated to the generation of such knowledge, allowing for its corrigibility and involving appropriate control procedures (i.e., a knowledge economy of science).

second-order knowledge: Knowledge resulting from the reflection of other knowledge, such as images of knowledge or concepts like proof or objectivity.

shared knowledge: The knowledge distributed within a society or group, constituting a common ground for the intellectual activities of its members.

system knowledge: Knowledge required to understand complex systems like the spheres of the Earth system and its human components.

theoretical knowledge: Knowledge systems with high degrees of systematicity and reflexivity, typically represented by texts in which abstract concepts are represented by controlled vocabularies or symbol systems understandable only with prior knowledge.

transformation knowledge: Knowledge concerned with the collective and individual actions required to ensure a sustainable life in the Anthropocene.

Language

glottography: According to the linguist Malcolm Hyman, the subsystem of a writing system that can be read aloud (i.e., that represents linguistic content that can be spoken).

lexicalization: The representation of concepts in language.

lingua franca: A language common across cultures and linguistic groups (e.g., Koine Greek in late antiquity and English in the present).

lingua sacra: A language used exclusively or mostly in sacred or liturgical practice (e.g., Coptic or Slavonic in some Orthodox Christian churches).

multilingualism: The phenomenon in which multiple languages are employed by a cultural, geographic, or professional society, indicating (and facilitating) knowledge transmission.

translation: The process in which a text is transferred from one language to another, a process in which the text is inescapably altered in both form and meaning.

Mental Models

mental model: An internal knowledge representation structure similar to a frame, but typically referring to (real or ideal) objects and processes that display a rich internal structure or dynamic. Mental models can often be externally represented by material models. They allow conclusions to be drawn from incomplete (or excessive) information, in particular by running a mental model. A mental model comprises variables that can be filled with inputs from different sources, including default settings. Some of these variables are critical in the sense that they need to be filled for the model to be operable. Mental models are context-specific and therefore are in general not universally valid. Mental models link present with past experiences by embedding new experiences in a cognitive network of previous experiences.

acceleration-implies-force model: A cornerstone model of classical physics in which it is not motion that requires a force but the change of motion (i.e., acceleration).

arch model: A model based on the practical knowledge that an opening in a building can be spanned by an arch, which only begins to bear its own weight when the keystone is inserted.

balance-lever model: A combination of the lever model and the equilibrium model emerging with the invention of balances with unequal arms.

center-of-gravity model: A generalization of the model of a balance, according to which every heavy body can, in principle, be regarded as a balance in equilibrium with the center of gravity acting as its fulcrum.

equilibrium model: The model of weight determination in which the force that keeps a balance in equilibrium is equal to the weight of the body.

field theory model: A model of classical physics according to which a source generates a field in a way that is described by the field equation, while the field prescribes the motion of a body according to the equation of motion.

impetus model: The model stating that a force causing a motion is an entity that can be transferred from the mover to the object moved.

lever model: A specific version of the mechanae model, which recognizes the lever as a force-saving device.

mechanae model: A model in which a mechanical instrument achieves, with a given force, an "unnatural" effect that could not have been achieved without the instrument.

model of isostasy: A model of geology according to which mountains or continental blocks are in hydrostatic equilibrium with their substrate.

motion-implies-force model: The model in which motions are perceived as being caused by a force exerted by a mover and in which a greater force must also cause a stronger motion.

running a mental model: The operation in which a mental model is deployed or brought to bear on a particular situation, mentally exploring the consequences of varying its input.

stability model: A model according to which a heavy object, not supported or stabilized by another body, falls downward.

Society and Institutions

academic capitalism: An economy of scientific knowledge shaped according to models of a capitalist economy by introducing market mechanisms and evaluating the outcomes of research and teaching according to economic standards.

administrations: Second-order institutions whose task consists in controlling or steering the operation of existing institutions.

anthropological gamut: A set of the key properties of humans and human societies that should constitute a common basis for investigations in the humanities and social sciences. Among these key properties are biological features such metabolic needs (breathing, nutrition, etc.) and sexual reproduction, as well as cultural aspects like the transmission of material culture and knowledge.

belief systems: Mental states shared within a group or society, typically comprising packages of knowledge and often playing a part in normative systems.

collective action potential: The capability of a society to perform a function (e.g., to produce infrastructure, police citizenry, or fight a war).

data capitalism: The self-reinforcing process of data accumulation leading to an ever more comprehensive subjugation of knowledge under private or state interests. Under the conditions of data capitalism, the information-data-information cycle of the pre-Web knowledge economy is changing into a data-information-more data cycle controlled by those owning the means and infrastructure for accumulating and using Big Data—be they private companies or the state.

engineer-scientists: A group of intellectuals emerging in fifteenth-century Europe who combined practical and theoretical knowledge in addressing contemporary technological challenges.

entrapment by mutual limitation: A feedback loop in which individual capabilities and perspectives are blunted by the limited experiential horizon a society offers individuals, while institutional regulations are themselves confined by the bounded mental capabilities and perspectives of the individuals for whom they exist. An alternative feedback loop would be a stimulation of cooperative actions by the mutual enablement of individuals and societal institutions.

external representations (or material embodiments) of institutions: The tangible symbols, tools, and infrastructures of an institution that, along with internal representations in the minds of individuals, structure behaviors in accordance with institutional regulations. When the material embodiments of an institution have an operative function (such as money in a market economy) we also speak of social technologies.

fractalization: The generation of similar complex patterns of social structures on varying scales of societies through globalization.

globalization: The rise of global interdependencies by the global spread of means of survival and social cohesion—ranging from goods and technologies to language, knowledge, and normative and belief systems—as well as of political and economic institutions. The term applies even if the spread is part of a long-term development.

homogenization: The tendency toward uniformity and standardization in a globalized world.

hybrid experts: Technical specialists who are also recognized as scientists.

institutional persona: A type of person within an institution who is the characteristic result of socialization processes shaped by his or her specific position within that institution. The person then acts as a role model for that position.

institutions: Regulatory structures of human interactions, enabling human societies to cope with recurring problems. They encode collective experience and mediate cooperative actions, relying on distributed knowledge, on the material means and external representations of institutional regulations, and on their internal representation in the minds of individuals in terms of values and beliefs belonging to normative systems. *See also* triggers of institutional change.

material means: The tools, machinery, infrastructures, production forces, and external representations available to a given society and its institutions.

normative systems: Systems that comprise values, beliefs, and knowledge concerned with the functioning, rationale, and meaning of institutions for the actors involved, allowing them to act in compliance with the behavioral norms implicit in institutional regulations.

objective interests: The pattern goals (according to an institutional persona) of an individual in a specific societal position. Objective interests are realized as individual, subjective interests and may therefore take diverse forms.

overcritical theories: Theories that assume more than the anthropological gamut.

power: The potential of an individual or a group to direct and control the actions of other individuals or groups, typically by controlling the conditions of their actions.

retardation: The slowing of globalization processes after the realization of their initial purpose.

second-order institutions: Institutions that enable the functioning of existing institutions (e.g., supervisory authorities) and impose a superordinate logic on individuals and alter the conditions of cooperative action.

social cohesion: The ability of a community to bond and to avoid fragmentation and dissolution.

social fragmentation: The loss of social cohesion.

socialization: The sum of the processes that interface concrete individuals and their needs, desires, and cognitive abilities with the demands of a society and its institutions. This often takes the form of education, learning, and training in the contexts of family, apprenticeship, and schooling.

social technologies: The material means of institutions employed in order to operatively implement institutional regulations, such as written laws applied to a violation of norms, the use of money in the market, or the use of time clocks in a factory.

societal reflexivity: A mechanism of hierarchy formation whereby new institutions (which rely on external representations) emerge to steer existing institutions or practices.

socio-epistemic complexes: Large-scale societal structures (e.g., of the chemical industry, nuclear technology, genetic engineering, the global networks of mobility, or of information and communication technologies) that enhance the dependence of human society on the production of scientific knowledge.

subjective interests: The actual, subjective, individualized goals of a particular individual in a specific institutional position and that individual's perception of the corresponding objective interests.

triggers of institutional change: External or internal stimuli or challenges (conflicts, the extension of experience, or the introduction of new material means) that illicit shifts in social frameworks. Institutions constitute dynamic equilibria in societal interactions. They vary constantly as a consequence of the fact that individuals are different and hence embody and enact social frameworks in ways that tend to differ in time and space. A critical condition for the stability of institutions when faced with challenges is the capacity to break down solutions to collective problems into individual actions.

undercritical theories: Theories that assume less than the anthropological gamut.

Notes

The Story of This Book

1. For early programmatic publications, see Renn (1994), published as Renn (1995). See also Renn (1996). See also the detailed discussions of ongoing work in the biannual research reports of the Max Planck Institute for the History of Science, appearing regularly since 1994: https://www.mpiwg-berlin.mpg.de/research-reports.
2. Damerow (1996a). For a broader account of Peter Damerow's contributions, see Renn and Schemmel (2019).
3. For the history of *Drosophila* as one of the most productive of all laboratory animals, see Kohler (1994). Kohler argues that fly laboratories constituted a special ecological niche in which the wild fruit fly was transformed into a laboratory tool for genetic research.
4. Damerow, Freudenthal, McLaughlin, et al. (2004).
5. Büttner, Damerow, and Renn (2001); Renn (2001b); Renn and Valleriani (2001); Büttner (2008); Valleriani (2017b).
6. The Historical Epistemology of Mechanics (four-volume series): Schemmel (2008); Valleriani (2010); Feldhay, Renn, Schemmel, et al. (2018); Büttner (2019). See also Renn and Damerow (2012); Valleriani (2013).
7. See in particular Renn (2007b).
8. See Duncan and Janssen (2018).
9. See, among others, Klein (1994; 2003; 2015b); Klein and Lefèvre (2007).
10. See Schemmel (2016a; 2016b).
11. See Schlimme (2006); Renn, Osthues, and Schlimme (2014a).
12. Renn (2012c).
13. See Renn and Schemmel (2006); Zhang and Renn (2006); Zhang, Tian, Schemmel, et al. (2008).
14. See Abattouy, Renn, and Weinig (2001a); Brentjes and Renn (2016c).
15. See Renn, Wintergrün, Lalli, et al. (2016).
16. See Bödeker (2006).
17. See Renn and Schemmel (2000); Damerow, Renn, Rieger, et al. (2002).
18. See Laubichler and Renn (2015; 2019); Renn and Laubichler (2017).
19. See the Anthropocene Curriculum Project (2015–18), https://www.anthropocene-curriculum.org. See also Klingan, Sepahvand, Rosol, et al. (2014); Renn and Scherer (2015a).
20. See Renn, Schlögl, and Zenner (2011); Nelson, Rosol, and Renn (2017); Renn, Schlögl, Rosol, et al. (2017).
21. For a critical analysis of recent discussions, see Omodeo (2018, in press a). See also Engler and Renn (2018).

Chapter 1: History of Science in the Anthropocene

1. This chapter is based on the assessments of the Anthropocene given in Steffen, Sanderson, Tyson, et al. (2004); Steffen, Grinevald, Crutzen, et al. (2011); Schwägerl (2014);

Davies (2016); Trischler (2016). It makes use of Renn, Laubichler, and Wendt (2014); Renn and Scherer (2015b); and Rosol, Nelson, and Renn (2017).

2. For further references, see Costanza, Graumlich, and Steffen (2007); Zalasiewicz, Williams, Steffen, et al. (2010); Zalasiewicz, Williams, Waters, et al. (2014); Steffen, Broadgate, Deutsch, et al. (2015); Steffen, Richardson, Rockström, et al. (2015); Zalasiewicz, Waters, Williams, et al. (2015); Moore (2016).

3. See Brooke (2018).

4. Davies (2016, 42).

5. Davies (2016, 43). See also Crutzen and Stoermer (2000).

6. For historical overviews, see Trischler (2016), and Bonneuil and Fressoz (2016) for excellent surveys of the current discussion. See also Vernadsky ([1938] 1997); Moiseev (1993). On Vernadsky's noosphere and its difference from Teilhard de Chardin's and Le Roy's interpretations, see Levit (2001).

7. Quoted after Trischler (2016, 311). Buffon (1778, 237): "enfine la face entière de la Terre porte aujourd'hui l'empreinte de la puissance de l'homme."

8. The discussion started with Crutzen and Stoermer's publication of their proposal in the newsletter of the International Biosphere-Geosphere Program in 2000: Crutzen and Stoermer (2000). It was immediately picked up by the Earth system science community. See, for example, Falkowski, Scholes, Boyle, et al. (2000).

9. Waters, Zalasiewicz, Williams, et al. (2014); Steffen, Leinfelder, Zalasiewicz, et al. (2016); Voosen (2016); Zalasiewicz, Waters, Summerhayes, et al. (2017); Zalasiewicz, Waters, Williams, et al. (2018).

10. Marx ([1844] 1970, 143).

11. For a discussion of the challenges of the Anthropocene for historical scholarship, see also Nelson (2001); Oreskes (2004); Chakrabarty (2009; 2012; 2015; 2017); Bonneuil and Fressoz (2016); Haber, Held, and Vogt (2016); Omodeo (2017); Szerszynski (2017); Lewis and Maslin (2018); Steininger (2018).

12. On the concept of the "Great Acceleration," see Hibbard, Crutzen, Lambin, et al. (2007); Steffen, Crutzen, and McNeill (2007); McNeill and Engelke (2014); Steffen, Broadgate, Deutsch, et al. (2015).

13. For a recent overview of the history of capitalism, see Kocka (2016). For an analysis of "fossil capitalism," see Malm (2016).

14. For an exemplary study of the coevolution of solution strategies and unintended consequences, see Leeuw (2012). Bonneuil and Fressoz (2016) rightly point to the "environmental reflexivity" of modern societies.

15. The following is based on Schlögl (2012); Renn, Laubichler, and Wendt (2014); Renn, Schlögl, Rosol, et al. (2017). See also Fischer-Kowalski, Krausmann, and Pallua (2014).

16. Bourdieu ([1984] 2000, 291).

17. See Mayr (1971, 185), as well as the discussion in Tomlinson (2018, 185).

18. For a critical discussion of science and chemical warfare, see Friedrich, Hoffmann, Renn, et al. (2017).

19. For further discussion, see Chakrabarty (2015); Delanty and Mota (2017).

20. "Die Welt ist uns nicht gegeben, sondern aufgegeben." This is often attributed to Kant, among others, by Einstein in his *Autobiographical Notes*; see Gutfreund and Renn (forthcoming 2020). Actually, the formulation is found in Cohn (1907, 96).

21. See the various approaches discussed in Losee (2004).

22. The relation between the history of science and the history of knowledge has been the subject of much recent discussion; see, for example, Burke (2000; 2016); Lefèvre (2000); Vogel (2004); Ammon, Heineke, and Selbmann (2007); Renn (2015b); Adolf and Stehr (2017); Östling, Sandmo, Larsson Heidenblad, et al. (2018).

23. Bacon ([1620] 2004, 193) [NO. 1, aphorism 129]. "Moreover, improvement of political conditions seldom proceeds without violence and disorder, whereas discoveries enrich and spread their blessings without causing hurt or grief to anybody." For discussion, see Vickers (1992).

24. Condorcet (1796). For historical discussion, see Meier and Koselleck (1975); Rohbeck (1987).

25. See Klein (2012b; 2015b).

26. Lefèvre (2003).

27. For an exemplary study of the history of objectivity as an image of knowledge and control procedure of scientific practice, see Daston and Galison (2007). For the notion of the "classical image of science," see Renn, Schoepflin, and Wazeck (2002).

28. For a discussion of philosophical attempts in the early twentieth century to use science as a model of rationality for the rest of society, see Engler and Renn (2018).

29. See Renn (2001b); Valleriani (2010). See also Smith (2004); Valleriani (2017b). The role of prescientific knowledge has also been stressed much earlier; see, for example, Schumpeter (1949); Zilsel ([1976] 2000).

30. Galilei (1638; 1974).

31. See Renn and Valleriani (2001).

32. Galilei (1974, 11).

33. See Elman (2005).

34. See Popper ([1959] 2002).

35. Elkana (1981, 15–19). See also Elkana (1975).

36. This definition is based on Damerow (1996b, 398), where Damerow and Lefèvre state that we "may speak of science if the goal of a certain social activity consists in elaborating the potentials of the material tools of mental labor, which are otherwise used in the planning of work, apart from such goals solely for the purpose of gaining knowledge about the possible outcome."

37. Accounts of long-term developments have become rare. For examples, see, e.g., Dijksterhuis ([1950] 1986); Cohen (2015); Høyrup (2017b). For reflections on the case study approach, see Forrester (1996).

38. See Chakrabarty (2007).

39. Collins and Pinch (1993, 1).

40. See, for example, Mach ([1905] 1976; 1910; [1910] 1992). For historical reviews, see Bayertz (1987); Bayertz, Gerhard, and Jaeschke (2007). See Marx ([1867] 1990), 324, n. 4).

41. See Sternberger (1977), in particular chap. 4.

42. There have been some exceptions, see, e.g., Toulmin (1972); Hull (1988; 2001); Thagard (1993); Fangerau (2013).

43. Darwin (1859).

44. See Cavalli-Sforza and Bodmer (1971); Bodmer and Cavalli-Sforza (1976); Dawkins (1976; 1982); Boyd and Richerson (1985); Cavalli-Sforza, Menozzi, and Piazza (1994); Boyd and Silk (1997); Richerson and Boyd (2005).

45. For a succinct summary and an original view of the role of niche construction for the emergence of human culture, see Tomlinson (2018).

46. For more recent discussions of cultural evolution, see Mesoudi, Whiten, and Laland (2006); Richerson and Christiansen (2013); Gray and Watts (2017); Laland (2017); Tomlinson (2018).

47. See Damerow and Lefèvre (1981).

48. See Tomlinson (2018).

49. Kuhn (1970).

50. Frege ([1892] 1997).

51. For "Papert's Principle," see Minsky (1986, 102).
52. For a call for cooperation in the context of the Anthropocene, see Brondizio, O'Brien, Bai, et al. (2016).

Chapter 2: Elements of a Historical Theory of Human Knowledge

1. This chapter presents an overview of the argument developed in this book. It explains how the theoretical analysis of knowledge, the investigation of episodes of the history of science, and the theme of the Anthropocene are related to each other. It argues that the concept of evolution of knowledge is necessary in tackling the question of the beginning of the Anthropocene. This concept can only be filled with meaning, on the other hand, on the basis of a history of knowledge that takes long-term developments into account. The specific examples serving to illustrate such long-term developments have been chosen mostly against the background of the author's own research foci.
2. For discussion, see Gutfreund and Renn (forthcoming 2020).
3. Klein (2001).
4. Piaget (1970, 17).
5. Elkana (2012, 611–612).
6. For the notion of ecological footprint, see Wackernagel (1996).

Chapter 3: The Historical Nature of Abstraction and Representation

1. This chapter is based on an investigation of various approaches to the concepts of "abstraction" and "reflection," including philosophical, psychological, and historical traditions. The views developed here bring together insights from pragmatism, neo-Kantianism, semiotics, and Marxism, along with perceptions from cognitive psychology in the traditions of Piaget, Vygotsky, Lurija, and Leontiev. It owes much to the pivotal work of Peter Damerow; see his book *Abstraction and Representation* (1996a), on which part of the following is based, but also to the critical examination of the role of abstraction in religious and philosophical traditions by Klaus Heinrich (1986; 1987; 1993; 2000; 2001).
2. Piaget (1970).
3. Crowe (1999, 205–8).
4. Sagan (1985).
5. See Wöhrle (2014). The following is based on Heinrich (2001).
6. Aristotle (1933, 19) [*Metaphysics*, I. 3, 983 b 5].
7. Plato (1997c, 193) [*Theaetetus* 174 a–b].
8. Aristotle (1932, 55–57) [*Politics* I. 4, 1259 a 5–6].
9. Plato (1997b, 1135) [*Republic* 517 b–c].
10. The following argument is based on Damerow (1996a).
11. Aristotle (1935, 67–69) [*Metaphysics* XI 3, 1061 a 7].
12. Hume ([1748] 2000, 37).
13. Hume ([1748] 2000, 24).
14. Kant ([1781] 1998, 247) [*Critique of Pure Reason* 8, B 132].
15. See Stadler (2001). See also Engler and Renn (2018).
16. For the following, see Damerow (1994). See also Gutfreund and Renn (2017, 118).

17. Wertheimer ([1959] 1978, 213).
18. Wertheimer ([1959] 1978, 6).
19. Wertheimer ([1959] 1978, 9).
20. Piaget (1970, 15–16).
21. The example is from Piaget (1982, 217).
22. For a definition of the concept of schema, see, for instance, Piaget (1983, 180–85).
23. See, e.g., Piaget ([1947] 1981, 109–10); Newcombe and Huttenlocher (2003, 53–71). The discussion here is based on Newcombe and Huttenlocher (2003); Schemmel (2016a, 10–12).
24. Piaget (1970, 21ff.).
25. Piaget (1970, 29, 53).
26. For the following, see Piaget (1970, 28–31).
27. Damerow (1996a, 99–100).
28. Piaget and Inhelder (1956, 455).
29. In the history and philosophy of science, such insights took hold only with delay. Early examples are the work of Yehuda Elkana on "concepts in flux" and the work of Kostas Gavroglu and Yorgos Goudarouli on the history of low-temperature physics. See Elkana (1970; 1974); Gavroglu and Goudaroulis (1989); see also the discussion in Arabatzis, Renn, and Simoes (2015). Gavroglu and Goudaroulis specifically investigated how scientific concepts could become relatively autonomous with regard to their original theoretical context when describing and explaining novel phenomena.
30. Piaget (1970, 16–18).
31. Piaget (1970, here 41ff.).
32. See in particular Elias ([1984] 2007).
33. For the following, see Schemmel (2016a, 24–25). On the Eipo, see Schiefenhövel (1991). For their spatial language and practices, see also Thiering and Schiefenhövel (2016).
34. Marx ([1867] 1990, 62) [Capital, vol. 1, pt. 1, chap. 1, sec. 4, 41].
35. Marx ([1867] 1990, 62) [Capital, vol. 1, pt. 1, chap. 1, sec. 4, 42–43].
36. Marx ([1867] 1990, 62) [Capital, vol. 1, pt. 1, chap. 1, sec. 4, 43].
37. For the following, see Englund (2012, 427–58); Cripps (2014); Ritt-Benmimoun (2014); Schaper (2019).
38. Marx ([1867] 1990, 62) [Capital, vol. 1, pt. 1, chap. 1, sec. 4, 42].
39. See Piaget and Garcia (1989). For a critical assessment of Piaget's approach from a historical perspective, see Damerow (1998).
40. For historical reviews, see Yasnitsky (2011; 2018a; 2018b); Hyman (2012). See also Leontiev (1978; 1981); Lurija (1979); Vygotskij (1987–99).
41. Vygotskij and Lurija (1994).
42. See Freudenthal (2005); Freudenthal and McLaughlin (2009b).
43. Marx ([1867] 1990, 62) [Capital, vol. 1, pt. 1, chap. 1, sec. 4, 42].
44. See Freudenthal (2005).
45. Cassirer (1944, 24).
46. For the following, see Schemmel (2016c).
47. This was Peter Damerow's essential idea in approaching the historiogenesis of cognition. It emerged in Berlin during the 1970s in the context of vibrant discussions with Wolfgang Lefèvre, Gideon Freudenthal, Peter Ruben, Peter Furth, Klaus Holzkamp, and others about Marx, Cassirer, Hessen, Grossmann, and psychological traditions ranging from Piaget to Leontiev. See, for example, Holzkamp (1968); Holzkamp and

Schurig ([1973] 2015); Furth (1980); Freudenthal (1986); Freudenthal and McLaughlin (2009b); Hedtke and Warnke (2017). These and other texts were central to the discussions of a regular colloquium headed by Peter Damerow and Wolfgang Lefèvre that was held at the Max Planck Institute for Human Development from the 1970s to the 1990s and that brought together what later became a core group of the Max Planck Institute for the History of Science, among them also Ursula Klein and Hans-Jörg Rheinberger.

48. See, for example, Klein (2003); Kaiser (2005).

49. See Piaget (1951); Cassirer (1944; 1955–96); Simmel (2004). For historical discussion, see Freudenthal (2002).

50. For an overview, see Nöth (1995).

51. Tomlinson (2018, 73–74).

52. Peirce (1967, 415).

53. Galilei (1623, 25; 1960, 183–84).

54. My discussion of semiotics and its role for the phylogenesis of culture relies on Tomlinson (2018, 153).

55. Tomlinson (2018, 69).

56. Tomlinson (2018, 107).

57. For the first two episodes, see Freudenthal and McLaughlin (2009b, 1–40). For the third episode, see Engler and Renn (2016). See also Engler and Renn (2018). For the broader context, see Omodeo and Badino (forthcoming in 2019).

58. See Hessen ([1931] 2009).

59. Cited after Freudenthal and McLaughlin (2009b, 28–29). For the bell story, see Chilvers (2015, 80).

60. Bernal (1949, 336).

61. Bernal (1949, 338).

62. Grossmann (1935; 2009).

63. Borkenau ([1934] 1976).

64. Freudenthal and McLaughlin (2009a, 2–3).

65. Freudenthal (2005).

66. See Engler and Renn (2016); Engler, Renn, and Schemmel (2018).

67. Fleck ([1935] 1980). Published in English as Fleck ([1935] 1979). See Engler and Renn (2016, 140).

68. Fleck ([1935] 1979, 98ff.).

69. Engler and Renn (2016, 140–41).

70. Cf. Freudenthal (2012, 34ff.).

71. Hegel ([1837] 1942, 33).

72. Dewey ([1925] 1981, 146), cited after Klein (2001, 13).

73. For Husserl, see Husserl (1939, 27). See also Husserl (2001). Damerow and Lefèvre introduced the term in the sense used here in Damerow and Lefèvre (1981). Freudenthal (2005, 167).

74. Kirk and Raven (1957, 197–98, §218).

75. For a broader discussion of such iterative processes, see Sperber (2000); Tomlinson (2018).

76. Damerow (1996a, 46 ff.).

77. This is somewhat similar to the ambiguities connected with "boundary objects." See Star (2010).

78. "Whose thinking is abstract? That of the uneducated rather than the educated person." Hegel ([1807] 1996, 577ff.), quoted after Damerow (1996a, 77).

79. This includes all forms of knowledge embodiment. See Overmann (2017).

Chapter 4: Structural Changes in Systems of Knowledge

1. This chapter brings together insights from various studies of transformations of knowledge systems, integrating them into a broader framework for understanding scientific change. This framework makes, in turn, use of a variety of different theoretical traditions. An important starting point was the investigation of the transition from preclassical to classical mechanics; see Damerow, Freudenthal, McLaughlin, et al. (2004). An outline of the larger research program, which is integrated into the current text, is found in Renn (1995). See also Renn (1993), where the concept of borderline problems is introduced. The concept of a "challenging object" is discussed in Renn (2001a); Renn, Damerow, and Rieger (2001). In contrast to the concept of "epistemic things" (introduced by Hans-Jörg Rheinberger) in a tradition going back to Gaston Bachelard, the notion of challenging objects specifically emphasizes that which is problematic or irritating for a given system of knowledge, rather than generally "that which is still unknown" (Rheinberger (1997, 70).

   The use of concepts from cognitive science for a historical theory of knowledge is discussed in Renn and Damerow (2007); for a revised English translation, see Renn, Damerow, Schemmel, et al. (2018). This chapter is partly based on that text.

   Basic ideas of mental models as explanatory tools go back to Craik (1943). The idea of a frame has been discussed in Minsky (1975). A pioneering publication on non-monotonic logic is McDermott and Doyle (1980). Further milestone publications on the subject are McCloskey, Caramazza, and Green (1980); Gentner and Stevens (1983); Johnson-Laird (1983); McCloskey (1983). For an overview of this tradition, on which the following account is based, see Davis (1984); Lattery (2016). There is a largely independent tradition of discussing models in the philosophy and history of science; see Morgan and Morrison (1999).

2. See Valleriani (2017b).

3. See Schiefenhövel (2013).

4. For tacit knowledge, see Polanyi (1983).

5. See Valleriani (2017b).

6. See Høyrup (2007; 2009).

7. ". . . ea quae sunt necessaria talibus ad sciendum non traduntur secundum ordinem disciplinae, sed secundum quod requirebat librorum expositio. . . ." Aquinas (2006, 2).

8. See Archimedes's *On the Equilibrium of Planes* in Heath (2009).

9. For a collection of classical papers on cognitive science, see Johnson-Laird and Wason (1977). The relation between cognitive psychology and cognitive science is also discussed in Aebli (1980–81). Aebli gives a broad overview of theories of thinking that relate thinking to action. For mental models, see Gentner and Stevens (1983); Davis (1984) on which the following is largely based.

   The implications of cognitive science for the philosophy of science are explored in Giere (1992). For the role of language in cognitive development, see Bowerman and Levinson (2001). A more recent approach to cognitive science, which takes up many of the classical insights expounded in the following and which may hold some promise also for a historical theory of knowledge, is Gärdenfors (2004). For the role of models in science education, including a helpful survey of the literature on the subject, see Lattery (2016). See also Geus and Thiering (2014).

10. Specifically, see Davis (1984, 37ff.).

11. For the following, see Lefèvre (2009a).

12. For historical discussion, see Camardi (1999, 540). See also Hooykaas (1963, 32ff.).

13. Darwin ([1876] 1958, 119–20).
14. See Clement (1983, 326ff.). See also Bödeker (2006, 22ff.).
15. Charniak (1972). Marvin Minsky used the paradigm in 1974 in a memo for the Artificial Intelligence Laboratory at MIT to explain the concept of a frame (published electronically: http://web.media.mit.edu/~minsky/papers/Frames/frames.html). The paradigm became known through the later publication of the memo: Minsky (1975, 241–47). See also Minsky (1986). Here I use the version and the interpretation of the paradigm by Davis (1984). See also Renn, Damerow, Schemmel, et al. (2018).
16. Minsky (1986, 230).
17. See Büttner, Damerow, and Renn (2001); Büttner (2019).
18. See Renn (2007a).
19. For the following, see Freudenthal (2000)—and based on it, Renn and Damerow (2012).
20. See Leibniz ([1686] 1989; [1695] 1989). For historical discussion, see Garber (1994); Smith (2006, 34ff.).
21. See Janssen (2019).
22. See Renn (1993). For Lorentz, see Janssen (2019, 22ff.). For Bohr and Sommerfeld, see Janssen (2019, 3ff.). For Einstein and Grossmann, see Janssen and Renn (2015, 31).
23. See Blum, Renn, and Schemmel (2016).

Chapter 5: External Representations at Work

1. The role of visual and graphical representations for scientific thinking has long been noted and discussed. See, for example, Hankins (1999); Netz (1999); Lefèvre, Renn, and Schoepflin (2003); Kusukawa and Maclean (2006); Bredekamp, Dünkel, and Schneider (2015); Leeuwen (2016).
2. This chapter owes much to the research of Peter Damerow and collaborators on the origin of writing and arithmetic. A key publication is Nissen, Damerow, and Englund (1993). The notion of external representations relevant to this work has been elaborated in Damerow (1996a), based on work that was first published in German: Damerow (1981). A succinct version of Damerow's account the origin of writing and arithmetic is Damerow (2012). The chapter is based on my own summary; see Renn (2015c; 2015d). The chapter also refers to Ursula Klein's research on the use of paper tools as external representations in chemistry: Klein (2003). The use of medieval diagrams of change as external representations in analyzing accelerated motion has been discussed in Damerow, Freudenthal, McLaughlin, et al. (2004). The presentation here relies on the systematic account given in Schemmel (2014).
3. See Damerow (1981; 2012); Damerow and Englund (1987); Nissen, Damerow, and Englund (1993); Bauer, Englund, and Krebernik (1998); Englund (1998; 2006); Høyrup and Damerow (2001); Høyrup (2015; 2017a).
4. For a magisterial study of the history of Babylonian mathematics and its context, see Robson (2008). See also Damerow (2010).
5. All dates are approximations. From Woods (2015, 13).
6. All dates are approximations. Adapted after Damerow (1999, 52) by Robert Middeke-Conlin.
7. See Schmandt-Besserat (1992a; 1992b).
8. I here follow the conventional "Middle Chronology" that gives 2000 BCE as the approximate date of the end of the Ur III period, although there is solid evidence that place value sexagesimal notations were used earlier. See Ouyang and Proust (forthcoming).

9. For the history and decipherment of Mayan script, see Coe (1992).

10. For seminal works on the history of Chinese mathematics, see Chemla and Shuchun (2004). See also Chemla (2006; 2012; 2017).

11. See Lisheng (2017). For the Mesopotamian abacus, see Woods (2017).

12. See Ritter (2000). See also Damerow (1996a, 157–58).

13. Plato (1997b, 1142–43) [*Republic* 7, 525b–526c].

14. For the following, see Englund (1991).

15. For the Babylonian case, see Damerow (2006; 2012). For the Egyptian case, see Baines (1983; 2001; 2007). For an overview of the origin of writing, also of other systems, see Houston (2004); Woods (2015). For the Chinese case, see Boltz (1986). For the development of the Greek alphabet, see Woodard (1997).

16. For an anthropological view on issues of orality and literacy, see, e.g., Goody and Watt (1963); Goody (1986; 2010a).

17. Euclid (1956, book 8). See the online edition: https://mathcs.clarku.edu/~djoyce/java/elements. The following is based on Lefèvre (1981); Damerow (1996a, 123–24).

18. For a history of knowledge encoded in cuneiform, see Rochberg (2017).

19. This follows Klein (2003).

20. Berzelius (1813; 1814).

21. Proust (1794, 341; 1799, 31); Dalton (1810, 329).

22. Klein (2003, 18–20).

23. The use of this graphical representation in preclassical mechanics is discussed in Damerow, Freudenthal, McLaughlin, et al. (2004). The following draws on Schemmel (2014).

24. See Oresme ([1350s] 1968). For historical discussion, see Maier (1949–58; 1964; 1982).

25. See Clagett (1968, 15–19).

26. See Damerow, Freudenthal, McLaughlin, et al. (2004). See also Schemmel (2008).

27. See Peirce (1967, 415).

Chapter 6: Mental Models at Work

1. This chapter is based on research on the history of mechanics pursued jointly with Peter Damerow, Peter McLaughlin, Gideon Freudenthal, Matthias Schemmel, Jochen Büttner, and Matteo Valleriani. Some of the materials discussed go back to Damerow, Freudenthal, McLaughlin, et al. (2004) and subsequent publications such as Büttner, Damerow, and Renn (2001); Renn (2001b); Renn and Damerow (2012); Valleriani (2017b). A comprehensive account of the history of preclassical mechanics is found in The Historical Epistemology of Mechanics (a four-volume series in Boston Studies in the Philosophy and History of Science): Schemmel (2008); Valleriani (2010); Feldhay, Renn, Schemmel, et al. (2018); Büttner (2019). The text is essentially based on Renn and Damerow (2007); an extended English version has been published as Renn, Damerow, Schemmel, et al. (2018).

2. Aristotle (1934; 1957). See, e.g., Prantl (1881).

3. See McCloskey (1983). See also Piaget (1999).

4. The reasons for the emergence of impetus theory were complex, and its application was not restricted to physical phenomena. For an overview of the various contexts of the theory, see Wolff (1978). See also Feldhay, Renn, Schemmel, et al. (2018).

5. For a discussion of Galileo, Descartes, and Beeckman, see Damerow, Freudenthal, McLaughlin, et al. (2004). For a discussion of Harriot, see Schemmel (2008).

6. See Valleriani (2013); Büttner (2017).

7. For the following, see Büttner, Damerow, and Renn (2001); Damerow, Freudenthal, McLaughlin, et al. (2004); Büttner (2019).

8. See Renn, Damerow, and Rieger (2001).

9. See, e.g., Lefèvre, Renn, and Schoepflin (2003); Valleriani (2013).

10. For a detailed discussion, see Renn, Damerow, and Rieger (2001).

11. See the discussion in Wertheimer ([1912] 2012). Neutral motion, that is, motion that is neither natural nor forced, is discussed in Galileo's early treatise *De motu* (1968a, 299–300). See also Damerow, Freudenthal, McLaughlin, et al. (2004, 163).

12. See Renn, Damerow, and Rieger (2001).

13. See Galilei (1968b, 272–73) [*Discorsi* 4, theorema 1, prop. 1]; English translation: Galilei (1974, 221).

14. See Ms. Gal. 72, folio 175v (http://www.imss.fi.it/ms72); and the discussion in Schemmel (2001a; 2006, 366–68); Damerow, Freudenthal, McLaughlin, et al. (2004, 216ff.).

15. See, for example, Torricelli (1919, 156) as discussed in Damerow, Freudenthal, McLaughlin, et al. (2004, 284–85).

16. Newton ([1687] 2016).

17. See definitions 3 and 4 in Newton ([1687] 2016, 404–5).

18. Toynbee (1954, 4); see also Goody (2010b).

19. The mutual construction of recipient culture and reference culture has been called "allelopoiesis" and developed into a central conceptual tool in the DFG Collaborative Research Center "Transformations of Antiquity." See Böhme (2011, esp. 9–15).

20. The discussion of the *Querelle des anciens et des modernes* is based on Lehner and Wendt (2017), while the following is indebted also to helpful comments by Mordechai Feingold.

21. See Force (1999); Guicciardini (2002); Buchwald and Feingold (2013).

22. For discussion, see Meli (1993).

23. Lagrange ([1811] 1997, 7).

24. See also Pulte (2005).

Chapter 7: The Nature of Scientific Revolutions

1. This chapter is based on the results of collaborative research projects dedicated to the understanding of major transformative processes of systems of knowledge. In addition to the emergence of classical mechanics, which has been discussed in previous chapters, and recent work on the Copernican revolution by Pietro D. Omodeo, the major transformations under investigation comprised the so-called chemical revolution studied by Ursula Klein, the Darwinian revolution reconstructed by Wolfgang Lefèvre, and the relativity and quantum revolutions of modern physics, which have been analyzed extensively by larger research teams. All projects were centered at Department 1 of the Max Planck Institute for the History of Science. The relevant publications are cited in the notes to this chapter.

2. The protracted character of such transformation processes was emphasized, for the case of the so-called probability revolution, also in Hacking (1987). For the case of the so-called quantum revolution, see Schweber (2015).

3. See Klein (2016a); Omodeo (2016), both in Blum, Gavroglu, Joas, et al. (2016). See also the discussion of Kuhn's position in Engler and Renn (2018).

4. For a synthesis of Kuhn's philosophy of science, see Hoyningen-Huene (1993). For historical studies of the work of Thomas S. Kuhn, see, e.g., Gattei (2008). For recent assessments, see Devlin and Bokulich (2015); Blum, Gavroglu, Joas, et al. (2016). For a

discussion of the notion of incommensurability in Kuhn, see Buchwald (1992); Buchwald and Smith (2001b).

5. See Engler and Renn (2016); Engler, Renn, and Schemmel (2018).
6. Kuhn (2000, 283).
7. Kuhn (1979, x).
8. See Gattei (2016, 127ff.).
9. Fleck ([1960] 1986, 154).
10. Kuhn's conception of scientific revolutions can be seen as a generalization from the Copernican revolution as depicted by Alexandre Koyré in works such as Koyré (1939), English translation: Koyré ([1939] 1978). Cf. Omodeo (2016). On Kuhn on Copernicus, see Swerdlow (2004).
11. See Kuhn (1959).
12. For a thorough analysis, see Swerdlow and Neugebauer (1984).
13. See Feldhay and Ragep (2017).
14. See Hyman (1986); Genequand (2001). On their relevance for Renaissance astronomy, see Omodeo (in press b).
15. See Swerdlow (1973).
16. Peuerbach (1473).
17. See Malpangotto (2016).
18. On Copernicus from the angle of early modern celestial physics, cf. Regier and Omodeo (in press). On the intricacies of the reception of Copernican work up to the early seventeenth century, cf. Omodeo (2014). On Giordano Bruno's natural philosophy and its contexts, see Hufnagel and Eusterschulte (2013).
19. See Klein (2015a). See also Klein and Lefèvre (2007).
20. Lavoisier, Guyton de Morveau, Berthollet, et al. (1787, 32).
21. Lavoisier (1778, 536).
22. Lavoisier, Guyton de Morveau, Berthollet, et al. (1787, 31).
23. Klein and Lefèvre (2007, 125–26, 191).
24. The following is based on Lefèvre (2009a).
25. Darwin ([1876] 1958, 124).
26. Huxley (1948, 22).
27. For a history of biological inheritance, see Müller-Wille and Rheinberger (2012).
28. Lefèvre (2009b, 314). See also Lefèvre (2003; 2007).
29. Several elements of this process are described in a series of papers: Laubichler (2009); Laubichler and Maienschein (2013); Laubichler, Prohaska, and Stadler (2018).
30. See Laubichler (2009); Laubichler and Maienschein (2013); Laubichler, Prohaska, and Stadler (2018); Laublichler and Renn (2019).
31. For overviews, see Harman (1982); Jungnickel and McCormmach (1986a); Nye (2003); Renn (2007a); Staley (2009). For contemporary discussion of scientific worldviews, see Gutfreund and Renn (forthcoming 2020).
32. See Planck (1900), published in English as Planck ([1900] 1967). For overviews, see Planck ([1920] 1967); Badino (2015). The following is based on Renn (1993). See also Büttner, Renn, and Schemmel (2003).
33. See Einstein (1905; 1989a). For historical discussion, see Norton (2014); Renn and Rynasiewicz (2014).
34. See Lorentz (1892; 1895; 1899). For historical discussion, see Janssen (1995; 2002).
35. Einstein (1905; 1989a).
36. Einstein (1905; 1989a).
37. Lorentz (1895).

38. Lorentz (1899).
39. Lorentz (1904).
40. Poincaré (1905, 1505).
41. Lorentz (1892).
42. Lorentz ([1892] 1937, 221).
43. Michelson (1881); Michelson and Morley (1887). For historical discussion, see Janssen and Stachel (2004); Staley (2009).
44. See, e.g., Wien (1894); Planck ([1900] 1967).
45. For the following, see Renn and Rynasiewicz (2014).
46. The role of reading Hume and Mach for "solving the problem" is mentioned in Einstein (1991, 53).
47. See Renn and Sauer (2007); Janssen and Renn (2015).
48. For the following, see Renn (2007b).
49. Einstein (1907; 1989b).
50. Einstein (1915b; 1996a).
51. Einstein (1915a; 1996b).
52. Dyson, Eddington, and Davidson (1920).
53. Einstein (1995).
54. Einstein and Grossmann (1913; 1995).
55. Janssen and Renn (2015).
56. Gutfreund and Renn (2017).
57. Eisenstaedt (1986; 1989; 2006).
58. Will (1986; 1989). See also Blum, Lalli, and Renn (2015; 2016); Blum, Giulini, Lalli, et al. (2017).
59. The following is based on Blum, Jähnert, Lehner, et al. (2017). See also Renn (2013b).
60. Heisenberg (1925).
61. Planck ([1900] 1967).
62. Bohr (1913).
63. Born and Jordan (1925, 859).
64. Heisenberg (1925, 880, 886, 893).
65. Von Neumann (1927a, 1927b, 1927c).
66. See, also for the following, Bub (2016, 2).
67. Feynman (1967, 129).
68. For these three criteria, see Blum, Renn, and Schemmel (2016).
69. For a discussion of more examples, see Janssen (2019).

Chapter 8: The Economy of Knowledge

1. This chapter presents some reflections on the role of knowledge in society and on the social and institutional conditions of science that have emerged in the context of a research project on the history of the Max Planck Society, pursued jointly with Jürgen Kocka, Carsten Reinhardt, Florian Schmaltz, and a team of historians and historians of science. Parts and preliminary versions have been published in Renn (2014c; 2015a; 2016). The concept of knowledge economy in the sense used here has been introduced in Renn and Hyman (2012b).
2. I use the term differently from the sociological and economic literature; see, e.g., Powell and Snellman (2004). For a history of the recent "knowledge society," see Reinhardt (2010).

3. Shapin and Schaffer (1985, 15).
4. See Jasanoff (2004, 2–3).
5. For an overview, see Misak (2013). See in particular Dewey (1897).
6. For a helpful overview, see Reckwitz (2003). See also Renn (2016, 98). For the following, see also Joas and Beckert (2001); Beckert (2003).
7. Durkheim ([1895] 1966).
8. Parsons (1949, 710–11).
9. What I propose in the following is close to the "constructivist structuralism or structuralist constructivism" of Bourdieu. See Bourdieu (1989). See also Bourdieu (2001).
10. For an innovative look at action as a borderline problem between social theory and psychology, see Prinz (2012).
11. For an overview of classical approaches, see Schülein (1987). For a more recent overview with an emphasis on different philosophical conceptions and research traditions, see Miller (2014).
12. See, in contrast, Douglas (1986). The role of shared knowledge in institutions is also related to the issue of collective intentionality raised in Searle (1995). The historical perspective on shared knowledge offered here may provide an escape from the problematic aspects of methodological individualism notoriously plaguing these fundamental sociological questions. For stimulating remarks pointing in this direction, see Arrow (1994).
13. Otherwise, I closely follow Streeck and Thelen (2005). See also Thelen and Steinmo (1992); Thelen (2004). Compare also the role of the concept of "charter" in Malinowski (1947, 157). For a broader discussion, see Schelsky (1970). For the relation between formal and informal rules, see also Meyer and Rowan (1977).
14. See Elkana (1981, 15–19). See also Elkana (2012, 608).
15. An essay on which much subsequent work is based is Elkana (1981). See also Elkana (1986a). For a specific study of the emergence and historical transformations of such a second-order concept and pertinent practices in particular historical contexts, see Daston and Galison (2007).
16. Transaction costs are, according to Douglass North, costs that are caused by information asymmetries in economic transactions. They may be search and information costs required to find out whether goods are available on the market. They may be bargaining costs required to establish agreement; and they may be policing and enforcement costs to make sure that contracts are being honored. For North, institutions play a central role in determining transaction costs. See North (1981; 1992).
17. See the discussion in Raj (2013), as well as Markovits, Pouchepadass, and Subrahmanyam (2006), where "circulation" is used to emphasize the bidirectional character of knowledge exchange, as well as its social and cultural framing conditions, which are assumed here to be general features underlying *all* knowledge exchange processes. For a recent overview, see Östling, Sandmo, Larsson Heidenblad, et al. (2018). On the historiography of the notions of center and periphery, see Gavroglu, Patiniotis, Papanelopoulou, et al. (2008).
18. For a systematic analysis of modes of transmission and transformation, see Böhme, Bergemann, Dönike, et al. (2011).
19. See Kroeber (1940); Smith (1977). See also the discussion in Potts (2012, 107).
20. This section follows Hyman and Renn (2012a).
21. Marx ([1867] 1990, 29–135) [*Capital*, vol. 1, pts. 1–2].
22. Marx ([1867] 1990, 132) [*Capital*, vol. 1, pt. 2, chap. 4].
23. The *homo oeconomicus* is supposed to act rationally and in his or her own self-interest, maximizing utility as a consumer and profit as a producer.

24. Shannon and Weaver (1949, 31).
25. Actually, the concept of "data" has a much older history, reaching back to Greek mathematics; see Taisbak (2003). For a historical survey of the long-term development of "cultures of information in the sciences," see Aronova, Oertzen, Sepkoski (2017).
26. The concept generalizes the notion of "scientific persona." See Daston and Sibum (2003).
27. For a historical review of psychological research on moral socialization, see Keller (2007). See also Keller and Edelstein (1991). Piaget's groundbreaking work is Piaget (1965).
28. See Hughes (1994).
29. Hayek (1945, 526–27). See also Hayek (1937).
30. The role of representations for the functioning of society has been discussed in theoretical terms at least since the Enlightenment. See Freudenthal (2012).
31. Karl August Wittfogel even characterized such cultures as "hydraulic civilizations" in Wittfogel (1957).
32. See Elias ([1984] 2007, esp. 99–100).
33. See Freud ([1930] 1962). The interpretation of symptoms as simultaneously concealing and revealing societal conflicts follows Heinrich (2001).
34. Agee and Evans (1941, 310).
35. For a discussion of techno fossils as part of this sediment, see Zalasiewicz, Williams, Waters, et al. (2014).
36. For the following, see Marin (1988).
37. I here follow Heinrich (2000, 146; 2001, 32–33).
38. See, e.g., the discussion in Rahwan and Cebrian (2018).

Chapter 9: An Economy of Practical Knowledge

1. This chapter presents some of the results of a major research project dedicated to the epistemic history of architecture, which have been published in three volumes: Renn, Osthues, and Schlimme (2014b; 2014c; 2014d). The project was jointly initiated with Elisabeth Kieven and Peter Damerow, and pursued together with Wilhelm Osthues, Hermann Schlimme, and a team of historians of art and architecture, archaeologists, and experts in Egyptology and Near Eastern cultures. The following text is based on the introduction to the three volumes, written jointly with Matteo Valleriani and the authors of the single contributions: Renn and Valleriani (2014). For construction history, see Becchi, Corradi, Foce, et al. (2002); Huerta (2003).
2. See Bührig, Kieven, Renn, et al. (2006).
3. Schlimme, Holste, and Niebaum (2014, 102).
4. Schlimme, Holste, and Niebaum (2014, 5).
5. For the history of architecture in antiquity, see, e.g., Frontinus Gesellschaft (1988); Heisel (1993); Lamprecht (1996); Cech (2012).
6. Renn and Valleriani (2014, 51).
7. Renn and Valleriani (2014, 8); Schlimme, Holste, and Niebaum (2014, 336).
8. See Bührig (2014, 348–49, 361).
9. See Rapoport (1969); Bernbeck (1994); Kurapkat (2014, 107, 113ff.).
10. See Fauerbach (2014, 112); Renn and Valleriani (2014, 52–53).
11. See Belli and Belluzzi (2003); Gargiani (2003); Schlimme, Holste, and Niebaum (2014, 332).
12. See Osthues (2014b, 396).
13. See Kurapkat (2014, 118–19); Sievertsen (2014, 250, 267).

14. See Kroeber (1940).
15. See Schlimme, Holste, and Niebaum (2014, 279).
16. See Hilgert (2014, 283–84); Kurapkat (2014, 114).
17. See Fauerbach (2014, 98); Osthues (2014a, 234–36).
18. Barbaro (1567, 162–66 [*I dieci libri dell'architettura di M. Vitruvio*, bk. 4, chap. 1], 325 [*Vitruvio*, bk. 7, chap. 12]); Osthues (2014b, 390–91); Schlimme, Holste, and Niebaum (2014, 100–101, 131).
19. Binding (2014, 30); Schlimme, Holste, and Niebaum (2014, 327). Cf. Becchi (2014).
20. Kurapkat (2014, 104).
21. Fauerbach (2014, 13); Renn and Valleriani (2014, 13–24, esp. 23, 52–53); Sievertsen (2014, 135, 153, 155).
22. Bührig (2014, 336, 338–39, 343, 361); Fauerbach (2014, 10 esp. n.17, 103, 109–10); Osthues (2014a, 133); Sievertsen (2014, 152).
23. See Binding (2014, 33); Fauerbach (2014, 28); Osthues (2014a, 127); Schlimme, Holste, and Niebaum (2014, 99, 103); Sievertsen (2014, 158).
24. See Binding (2014, 13); Kurapkat (2014, 100); Osthues (2014a, 141); Schlimme, Holste, and Niebaum (2014, 206).
25. See Osthues (2014a, 174, 177, 225ff.); Schlimme, Holste, and Niebaum (2014, 102–3).
26. For the emergence of architecture as a profession, see Binding (2004); Nerdinger (2012). For the emergence of the engineer, see Kaiser and König (2006).
27. See Kurapkat (2014, 115); Osthues (2014a, 128); Sievertsen (2014, 148–49).
28. See Osthues (2014a, 144ff.).
29. Osthues (2014a, 229).
30. See Osthues (2014a, 127, 145, 227, 232).
31. See Kurapkat (2014, 60–61).
32. See Kurapkat (2014, 75, 100, 106).
33. See Kurapkat (2014, 73–75).
34. See Kurapkat (2014, 63–64); Osthues (2014a, 196); Renn and Valleriani (2014, 14).
35. See Osthues (2014b, 270–71, 294–96, 300, 405).
36. Fauerbach (2014, 54); Renn and Valleriani (2014, 27).
37. Binding (2014, 83); Osthues (2014a, 158–59; 2014b, 354); Schlimme, Holste, and Niebaum (2014, 296, 325–26).
38. Schlimme, Holste, and Niebaum (2014, 102).
39. Binding (2014, 25–26); Dubois, Guillouët, Van den Bossche, et al. (2014).
40. Binding (1985–86; 2014, 32, 34, 54).
41. Schlimme, Holste, and Niebaum (2014, 98–100, 111–12, 161).
42. Schlimme, Holste, and Niebaum (2014, 102–3).
43. Schlimme, Holste, and Niebaum (2014, 129, 139, 142–43, 151–52, 328–29).
44. Schlimme, Holste, and Niebaum (2014, 105–6).
45. See Haines (2015b, II 2 1, c. 177). This was probably the "invention" of the *capomaestro* and vaulting expert, Giovanni di Lapo Ghini, "circa armaturam fiendam de volta cupole" who was awarded a prize of 115 florins in 1371 (Guasti 1887, doc. 231). This circumstance confirms Manetti's and Vasari's accounts that Brunelleschi had to convince his interlocutors in the Opera that building such a huge structure on wood armature would be impossible before proceeding to demonstrate how it could instead be erected as a self-supporting dome. I am indebted to Margaret Haines for these remarks.
46. Schlimme, Holste, and Niebaum (2014, 188–92, 196, 325–26, 333).
47. Schlimme, Holste, and Niebaum (2014, 102–3, 111–12).
48. Schlimme, Holste, and Niebaum (2014, 104–5).

49. Approximate measurements are taken from Galluzzi (1996a, 94). The following is mainly based on Schlimme, Holste, and Niebaum (2014) as well as on the work by Margaret Haines (1989; 2011–12), who also gave very helpful commentaries on this text. Further basic references are Saalman (1980); Galluzzi (1996a); Ippolito and Peroni (1997); Di Pasquale (2002); Fanelli and Fanelli (2004); Corazzi and Conti (2011).

50. For early biographical sources on Brunelleschi, see Vasari (1550; 1878–85); for an English translation, see Manetti (1970; 1976); Vasari (1987). See also Ghiberti (1948–67); Bartoli (1998). For Brunelleschi's work, see Battisti (1981); Gärtner (1998).

51. Alberti ([1436] 1970, 39–40). Quoted from I. Hyman (1974, 27).

52. Galluzzi (1996a, 98). Cf. Fanelli and Fanelli (2004, 28, 184).

53. See Prager and Scaglia (1970); Galluzzi (1996a; 1996b; 2005). For the general context of Renaissance machine technology, see also Lefèvre (2004); Lefèvre and Popplow (2006–9); and Popplow (2015).

54. The state of studies on the undocumented Arnolfo cathedral plan is treated in Peroni (2006). For the alterations in plan documented in the course of the following century, see Saalman (1980, 32–57), and for the administrative strategies used to build the knowledge and consensus necessary to implement the expanding program, see Haines (1989, 89–125).

55. For the institutional background, see Grote (1959).

56. Haines (2011–12). In-depth studies of two major aspects of the worksite administration are published online at "The Years of the Cupola": Becattini (2015); Terenzi (2015). The biannual rolls of workers authorized for employment show fluctuating numbers averaging sixty-five master workers in the period of construction of the cupola and demonstrate the important contribution of the so-called core group of expert, long-standing masons who represented the majority (62.6 percent) presences over the whole period.

57. For the following, see Haines (1996).

58. Fabbri (2003).

59. A first selection of the administrative documents of the Opera was published by Guasti (1857; 1887). "The Years of the Cupola, 1417–1436," a complete digital edition of the extant documentation of the Opera for the period of the final planning and construction of the dome, prepared by Margaret Haines and her collaborators (2015a), is available online at the website of the Opera http://www.operaduomo.firenze.it/cupola and that of the Max Planck Institute for the History of Science http://duomo.mpiwg-berlin.mpg.de. The electronic edition was realized by Jochen Büttner and Klaus Thoden in the context of a joint project of the Opera del Duomo and the Max Planck Institute for the History of Science.

60. Haines (2008).

61. Haines (2015b, II 1 77, c. 34).

62. Haines (2015a).

63. Haines and Battista (2014). Other examples in the Opera workforce are examined in Terenzi (2015). For the broader context, see also Pinto (1984; 1991).

64. For the general context, see Goldthwaite (1980).

65. See Schlimme, Holste, and Niebaum (2014, vol. 3). See also Lepik (1994; 1995).

66. Guasti (1887, 100–103, 199–205).

67. The document, dated August 19, 1418, was published in Guasti (1857, n.11), and is now available in Haines (2015a), under date, archival location [Haines (2015b, II 1 74, c. 9)], and various subjects.

68. See Saalman (1980).

69. This precious written program, lost from the Opera archives, is known in copies passed along by Brunelleschi's biographers as synthesized in Guasti (1857, doc. 51, 28–30). Subsequently, another official copy was discovered in the Wool Guild archives and published by Doren (1898).

70. Haines (2015b, II 2 1, c. 107v).

71. The following is based on Renn and Valleriani (2001); Valleriani (2010).

72. On Galilei's science of strength of materials, see Portz (1994).

73. Galilei (1974, 127 [Day 2, Proposition IX]).

74. On the conservation of the cupola, see Rossi (1982); Dalla Negra (1995); Di Pasquale (1996); Corazzi and Conti (2006); Rocchi Coopmans De Yoldi (2006); Como (2010); Ottoni and Blasi (2014).

Chapter 10: Knowledge Economies in History

1. See Schumpeter ([1934] 2008). For historical discussion, see McCraw (2007); Kocka (2017b). See also Komlos (2016).

2. See Hansen and Renn (2018).

3. This chapter is based on and closely follows the typology of knowledge systems and their description by Damerow and Lefèvre (1998). It extends their account by drawing on further literature, as well as on related research pursued jointly with them and other colleagues such as Markus Popplow, in particular making use of the studies collected in Renn (2012c), and other sources indicated in the following notes. For the more recent history, extensive use has also been made of the essays collected in Krige and Pestre (1997), as well as on an insightful essay by Joel Mokyr (1999).

4. Barth (2002, 3).

5. See Barth (1975; 1987).

6. Barth (2002, 4).

7. See, for example, the work of Bronislaw Malinowski (2002).

8. For the Guarani, see Silva da Silva and Arantes Sad (2012). For the Eipo, see Schiefenhövel, Heeschen, and Eibl-Eibesfeldt (1980); Eibl-Eibesfeldt, Schiefenhövel, and Heeschen (1989); Thiering and Schiefenhövel (2016).

9. See Silva da Silva and Arantes Sad (2012, 536).

10. See Gladwin (1974); Hutchins (1996); Holbrook (2012).

11. See Eibl-Eibesfeldt, Schiefenhövel, and Heeschen (1989, 30–33); Thiering and Schiefenhövel (2016).

12. See Thiering and Schiefenhövel (2016, 41–42).

13. For the following, see Eibl-Eibesfeldt, Schiefenhövel, and Heeschen (1989).

14. There is still a lack of understanding of the so-called proto-Elamite texts from ca. 3100–2900 BCE, which may reflect speech. For discussion, see Englund (2004).

15. See Rochberg (2004); Cancik-Kirschbaum (2005).

16. More precisely, glottography is the subsystem of a writing system that can be read aloud (i.e., that represents linguistic content that can be spoken), although written texts differ in a number of structural ways from speech. See Hyman (2006). For historical overviews of Babylonian literature, see Sasson (1995); Alster (2005).

17. See Scharf and Hyman (2012).

18. See Krebernik (2007a; 2007b).

19. For the early dynastic period, see Bauer (1998); Krebernik (1998). For the Ur III period, see Sallaberger (1999).

20. For the Old Assyrian and Old Babylonian periods, see Charpin, Edzard, and Stol (2004); Veenhof and Eidem (2008).
21. For historical discussion, see Hooker, Walker, Davies, et al. (1990). See also Krebernik (2007a).
22. For the Old Akkadian period, see Westenholz (1999).
23. See Lambert (1957).
24. See Graßhoff (2012).
25. See Heisel (1993); Bagg (2011); Bührig (2014).
26. See Kirkhusmo Pharo (2013).
27. Jaspers ([1949] 1953).
28. For a discussion of this context, see Dittmer (1999).
29. See Eisenstadt (1982; 1986). For the beginnings of the thesis's reception by the historical disciplines, see Graubard (1975).
30. Elkana (1986b).
31. See Assmann (1992; 2001; 2008; 2018).
32. See Elkana (1986b).
33. See Assmann (1989).
34. Jaspers ([1949] 1953, 3). "Im *spekulativen Gedanken* schwingt er sich auf zu dem Sein selbst, das ohne Zweiheit, im Verschwinden von Subjekt und Objekt, im Zusammenfallen der Gegensätze ergriffen wird." Jaspers ([1949] 1983, 22).
35. This is also emphasized in Eisenstadt (1988).
36. See Dittmer (1999).
37. For an overview, see Maier (1998).
38. See Heinrich (1986).
39. Xenophanes (2016, 33–35) [LM D16–D19/DK 21 B23–B26]. See also Asper (2013).
40. For a classical overview, see Kirk, Raven, and Schofield (1983).
41. For historical discussion, see Dijksterhuis (1956).
42. For an introduction, see Lloyd (1964).
43. See, e.g., Most (1999); Freudenthal (2015).
44. For an introduction, see Barnes (1995); Russo (2004).
45. See Pedersen (1993); Lloyd (1970; 1973); Cuomo (2000); Schiefsky (2007).
46. For Judaism, see Levine (2014). For Buddhism, see Braarvig (2012).
47. The following history of medieval and early modern technology is based on and includes passages from Popplow and Renn (2002). See also Popplow (2015).
48. See Lucas (2005).
49. For early historical studies, see Olschki (1919, 1922, 1927); Zilsel ([1976] 2000); Smith (2004); Freudenthal and McLaughlin (2009b). See also Klein (2016c; 2017).
50. See Tsuen-Hsuin (1987). See also Bloom (2017).
51. The following is based on conversations with Matteo Valleriani. For more detailed treatments of the early modern book culture, see Febvre and Martin (1900); Giesecke (1990, 1991); Nuovo (2013).
52. See Nuovo (2013).
53. See Buringh and van Zanden (2009).
54. For an example of alliances between educational institutions and great international academic publishers, see Pantin and Renouard (1986); Pantin (1998, 2006). For the establishment of canonic teaching texts during the early modern period, see Valleriani (2017c) and, for a more general overview, MacLean (2009, 2012).
55. See Chartier (1994); Walsby and Constantinidou (2013).
56. See Vecce (2017).

57. For an investigation of the process of codification of practical knowledge in form of texts and drawing, see Hall (1979). For a more general overview, see Valleriani (2017b).

58. See Valleriani (2017a).

59. For Hero of Alexandria, see Schiefsky (2007).

60. See Rüegg (1996).

61. See Renn (2013a).

62. See Jack (1976); Schlimme (2009). See also Schlimme, Holste, and Niebaum (2014, sec. 2.4.3).

63. See Omodeo (2019).

64. Popplow and Renn (2002, 269).

65. See the discussion in Chalmers (2017, 176ff.).

66. Popplow and Renn (2002, 268–69).

67. Popplow and Renn (2002, 265, 271–72).

68. Galilei (1967, 103 [Day 1]).

69. Bruno ([1584] 1995; [1584] 1999); Galilei (1967); Descartes (1985b).

70. Kepler ([1619] 1997); Newton ([1687] 2016, 390 [*Principia*, vol. 2, sec. 9, prop. 52, theorem 40, case 3, cor. 4]; [1730] 1952, 399 [*Opticks*, vol. 3, pt. 1, qu. 31]); Kant ([1746–49] 2012; [1786] 2002); Lamarck ([1809] 1963).

71. For a history of the mechanical worldview, see Dijksterhuis ([1950] 1986). For an overview of its demise, see Harman (1982). For a more recent account, see Buchwald and Fox (2013).

72. Giovanni Botero, the political thinker, stressed the importance of territorial policy and industry over bare military expansion as the most apt "reason of state" in memorable works such as *Della ragion di stato* (1589). See Keller (2015, 35–45).

73. The relationship between the emerging sovereign state and a new economy of knowledge is masterfully discussed in Feldhay (2018).

74. For a striking example, see Omodeo and Renn (2019).

75. For the following, see Grimm (2015).

76. Bodin (1576; 1606).

77. Elias ([1969] 2006).

78. Hobbes (1651, 66–67) [*Leviathan*, pt. 1, chap. 10, sec. 41–42]. For historical discussion of the Leviathan image, see Bredekamp (2006).

79. Locke (1689, bk. 2, chaps. 10 and 11).

80. Rousseau ([1762] 1964, 360–61). Cited from Scott (2006, 121).

81. This is shown in two classic works on the history of literature and art, respectively: Benjamin (1928); Marin (1988).

82. For a discussion of the limitations of this external representation, see Marx ([1859] 1987).

83. Appleby (2013).

84. For a history of the emergence of scientific disciplines, see Stichweh (1984).

85. See Renn, Schoepflin, and Wazeck (2002).

86. On the foundation of Humboldt University, see, e.g., Humboldt ([1809/10] 1993). For historical discussion, see Bruch (2010).

87. For historical discussion, see Thomas (1985); Stearns (1993); Jacob (1997; 2014); Cohen (2015).

88. Mokyr (2002, 29).

89. Jacob (2014, 221).

90. Mokyr (2002, 103).

91. Mokyr (2002, 74–75).
92. Mokyr (2002, 36).
93. Mokyr (2002, 65). Note, however, that knowledge and ingenuity was not all that was needed to propel the Industrial Revolution, which also crucially depended on environmental and material circumstances, see Albritton Jonsson (2012).
94. See Cushman (2013).
95. The following is based on Mokyr (1999).
96. See Jungnickel and McCormmach (1986b); Agar (2012).
97. Klein (2012a; 2015b; 2016c; 2017).
98. For the following, see Steinberg ([1955] 2017); Austrian (1982); Nunberg (1996); Huurdeman (2005); Poe (2011); Hyman and Renn (2012b).
99. See, for example, Cahan (2011); Hoffmann, Kolboske, and Renn (2015).
100. See Simon, Knie, Hornbostel, et al. (2016). See also Kaldeway and Schauz (2018).
101. The following closely follows Mokyr (1999). John Desmond Bernal was one of the first historians of science to discuss a Second Industrial Revolution and described it as an event in which "science is playing a much larger and more conscious part than in the first," see Bernal (1939, 392).
102. Mowery and Rosenberg (1999, chap. 2).
103. See Carlson (1997, 214–15).
104. See Hughes (1983).
105. See Carlson (1997); Collet (1997); Mokyr (1999).
106. Maxwell (1865).
107. See, for example, Hughes (1983, 91). For the notion of technological momentum, see Hughes (1994).
108. Hertz (1888; 1889).
109. Collet (1997); Huurdeman (2005).
110. For the following, see Carlson (1997); Collet (1997).
111. See Mokyr (1999).
112. See Cahan (2011); Hoffmann, Kolboske, and Renn (2015).
113. See Cohen (1997).
114. See Collet (1997, 256).
115. See Collet (1997, 258).
116. For a comprehensive survey of scientific and technological experts in the twentieth century (with a focus on Europe), see Kohlrausch and Trischler (2014). For an analysis of the role of the wars of the twentieth century in entering into the Anthropocene, see Bonneuil and Fressoz (2016).
117. Mokyr (1999).
118. See Pestre (1997, 67–68).
119. See Krementsov (1997, 791–92).
120. See Agar (2012).
121. See Heim, Sachse, and Walker (2009).
122. See Preston (1996, 36); Hiltzik (2016).
123. Hahn and Strassmann (1939).
124. Eisenhower (1961, 1038–39).
125. Rheinberger (2012, 739).
126. National Human Genome Research Institute (n.d.).
127. See Pestre (1997); Galambos and Sturchio (1998); Agar (2012, 433–65).
128. For the following, see Collet (1997).
129. Collet (1997).
130. Moore (1965, 115). For discussion, see Mody (2016).

131. See Hafner and Lyon (1996, 247–54); Spicer, Bell, Zimmerman, et al. (1997).

132. See Gillies and Cailliau (2000).

133. See Minsky (1986); McCorduck (2004); Nilsson (2010); Lenzen (2018); Rosol, Steininger, Renn, et al. (2018).

134. Garfield (2006). For the history of peer reviews, see also Lalli (2014; 2016).

135. Abir-Am (1997). See also Pestre (1997).

136. See Oldroyd (1996); Doel (1997); Weart (2003); Zalasiewicz and Williams (2009); Rudwick (2014); Lax (2018a; 2018b). For the emergence of modern meteorology, see Friedman (1993).

137. Wegener (1912; 1915).

138. The following is based on the seminal study by Naomi Oreskes: *The Rejection of Continental Drift* (1999).

139. See Hallam (1989); Lewis (2000); Dalrymple (2004).

140. Wegener (1912; 1915). For the English translation, see Wegener (1966).

141. Suess (1904–24).

142. Dana (1873).

143. For the following, see Oreskes (1999, 23–48). See also Airy (1855); Pratt and Challis (1855).

144. Holmes (1926).

145. See Oreskes (1999, chap. 10).

146. Doel (1997, 412ff.).

147. Carson (1962). For a historical account of the emergence of Earth system science from climate science since the 1980s, see Dahan (2010). See also Paillard (2008).

148. See Porter (1997).

149. See Berkhout (1997).

150. Based on Renn and Hyman (2012a, 578), which follows Oreskes and Conway (2010). See also Oreskes, Conway, Karoly, et al. (2018).

151. Oreskes and Conway (2010); Proctor (2011).

152. Lane (2000).

153. Slaughter and Leslie (1997) discusses the increasing dependence of higher education in the English-speaking countries on an extra-academic market in the period between ca. 1970 and 1990, with the consequence that research becomes less "curiosity-driven" and more market-driven. See also Nelson (1959; 1986); Heller and Eisenberg (1998); Mirowski and Sent (2007); Pagano and Rossi (2009); Münch (2016).

154. See Slaughter and Leslie (1997, 50ff.).

155. Flexner ([1939] 1997, 34–35).

156. Flexner ([1939] 1997, 36).

157. Ordine (2017, 107). The quotation is from Henri Poincaré (1907), who in turn quotes a famous hexameter in Juvenal's *Satires*: "summum crede nefas animam praeferre pudori / et propter vitam vivendi perdere causas" ("I hold it to be the greatest infamy to prefer life to honor / and for the love of life to lose the very reason for life itself").

Chapter 11: The Globalization of Knowledge in History

1. This chapter is based on joint work with several colleagues on the globalization of knowledge in history. These investigations began with studies of the transmission and transformation of science in late antiquity and the Middle Ages, pursued with Mohammed Abattouy, Peter Damerow, and Paul Weinig, and later with Mark Geller and Sonja Brentjes. Some of the results reported here have already been published: Abattouy, Renn, and Weinig (2001b); Renn and Damerow (2012); Geller (2014); Brentjes and Renn (2016a; 2016c). Studies of the globalization of knowledge in the Iberian world

have been initiated in the context of a collaboration with the Fundación Canaria Orotava de Historia de la Ciencia, at the time under the directorship of José Montesinos,
and then pursued jointly with Helge Wendt and Angelo Baracca, with particular attention given to the Latin American world. See, e.g., Montesinos Sirera and Renn (2003);
Baracca, Renn, and Wendt (2014b); Wendt (2016b). Investigations of transfer and
transformation processes between European and Chinese science have been pursued
jointly with Matthias Schemmel and are discussed in the next chapter. A comprehensive
survey of the globalization of knowledge in history was undertaken jointly with Peter
Damerow, Malcolm and Ludmilla Hyman, Kostas Gavroglu, Eva Cancik-Kirschbaum,
Gerd Graßhoff, and Yehuda Elkana, among many other colleagues, and was published as
The Globalization of Knowledge in History; see Renn (2012c). Its introductory essays, in
part written jointly with Malcolm Hyman, as well as the contributions by Jens Braarvig,
Dan Potts, Arie Krampf, Hansjörg Dilger, Circe Mary Silva da Silva, Ligia Arantes Sad,
and Mark Schiefsky have been essential for the present chapter; see Braarvig (2012);
Dilger (2012); Hyman and Renn (2012a); Krampf (2012); Potts (2012); Renn (2012a;
2012b); Renn and Hyman (2012a); Schiefsky (2012); Silva da Silva and Arantes Sad
(2012). The important role of multilingualism in globalization processes has meanwhile
been treated more systematically in Braarvig and Geller (2018).

2. See Basalla (1967); Wendt and Renn (2012). For critical assessments, see Cunningham
and Williams (1993); Raina (1999); Lyth and Trischler (2003); Secord (2004); Arnold
(2005); Werner and Zimmermann (2006); Edgerton (2007); Selin (2008); Elshakry
(2010); Goody (2010c); Sivasundaram (2010a; 2010b); Günergun and Raina (2011);
Howlett and Morgan (2011); Lipphardt and Ludwig (2011); Wallerstein (2011); Zemon
Davis (2011); Patiniotis (2013); Raj (2013); Cancik-Kirschbaum and Traninger (2016);
Duve (2016); Medick (2016); Kocka (2017a); Östling, Sandmo, Larsson Heidenblad,
et al. (2018).

3. See Balter (2012); Fu, Rudan, Pääbo, et al. (2012). For discussion and further references, see Hansen and Renn (2018, 14).

4. See Diamond (2005b).

5. On the role of multilingualism, see Braarvig and Geller (2018).

6. For the following, see Mignolo (2000); Bayly (2004, 312–13); Osterhammel (2009,
1142–46); Tilley (2010); Wallerstein (2011).

7. See Hoffmann (2012).

8. For the IPCC, see IPCC (2018). See also Krige (2006); Edwards (2010); Klingenfeld
and Schellnhuber (2012).

9. IPCC and Watson (2001, 37).

10. For a critical assessment, see Uhrqvist and Linnér (2015). For the role of transnational
private regulations in globalization processes and the political construction of market
institutions, see Bartley (2007). See also Bartley and Child (2014); Bertelsen (2014);
Bartley (2018).

11. See Ezrahi (1990, 13).

12. European Parliament (2000).

13. Friedman (2007).

14. For the following see Krampf (2012).

15. See Edgerton (2006).

16. For the following, see Rottenburg (2012). See also Porter (2006).

17. For the following, see Dilger (2012).

18. Streeck (2014).

19. See, e.g., Frank (1998).

20. Pacey (1990, vii–viii).

21. See Hublin, Ben-Ncer, Bailey, et al. (2017).
22. For a recent survey of the relation between migrations and climate, see Mauelshagen (2018b).
23. See Rahmsdorf (2011); Parzinger (2014).
24. See Braarvig and Geller (2018).
25. See Hansen and Helwing (2018).
26. See Gronenborn (2010); Fuller (2011); Özdoğan (2014; 2016).
27. Özdoğan (2014, 33).
28. See, e.g., Beckwith (2009, 50–52); Potts (2011).
29. See Rosińska-Balik, Ochał-Czarnowicz, Czarnowicz, et al. (2015).
30. Diamond (2005b, chap. 10).
31. For a discussion of the conditions for the spread of writing, see the survey by Cancik-Kirschbaum (2012).
32. See Potts (2007).
33. The following is based on Renn (2014b).
34. Glick (2005, 6–7).
35. See Burnett (2001, 260).
36. See Po-chia Hsia (2015).
37. Edgerton (2006, 28ff.).
38. Casas ([1552] 1992). For historical discussion, see Abril Castelló (1987).
39. For the emergence of "Western science" as a historiographical construct, see Elshakry (2010).
40. The following is based on Plofker (2009).
41. See, e.g., Elkana (1986b).
42. See Schiefsky (2012). See also Ostwald (1992).
43. For the following, see Høyrup (2012).
44. See Szabó (1978); Lefèvre (1981).
45. The following is based on Gutas (1998); Abattouy, Renn, and Weinig (2001b); Speer and Wegener (2008); Brentjes and Renn (2016c).
46. See Braarvig (2012).
47. See Brentjes and Renn (2016b).
48. See Moody and Clagett (1960).
49. See Jordanus de Nemore (1960).
50. See Renn and Damerow (2012).
51. See Lemay (1977).
52. See Abattouy, Renn, and Weinig (2001b).
53. See Lemay (1963).
54. See Fraenkel, Fumo, Wallis, et al. (2011).
55. See Brentjes and Renn (2016a).
56. See Lemay (1962); Cronin (2003); Speer and Wegener (2008).
57. See Russell-Wood (1998); Kamen (2003).
58. See Bethell (1984–2008); Stern (1988).
59. See Crosby (2003).
60. See Yule (1903); Jackson and Morgan (1990); Larner (1999); Rossabi (2010); Francopan (2016).
61. See McNeill (1993).
62. See Needham (1986, 572–74); Pacey (1990, 45–46).
63. For the following, see Crosby (1993).
64. See, e.g., Bushnell (1993).
65. See Diamond (2005b).

66. See Bray, Coclanis, Fields-Black, et al. (2015).
67. See Wendt (2016b).
68. See Gruzinski (2004).
69. See, e.g., Müller-Wille (1999). See also Montesinos Sirera and Renn (2003).
70. Altshuler and Baracca (2014, 58).
71. McNeill (1994); Grove (1995); Haynes (2001); Klein (2016b).
72. Darwin ([1865] 2002, 238).

Chapter 12: The Multiple Origins of the Natural Sciences

1. "Da Galileo, come già ho scritto altre volte, desidererei molto vivamente [ricevere] il calcolo delle eclissi, in particolare di quelle solari che si possono conoscere dalle sue molte osservazioni; tutto questo, infatti, ci è estremamente utile per la correzione del calendario; e se c'è qualche scritto su cui possiamo fare affidamento, perché non ci caccino da ogni parte del regno, è solo questo." Iannaccone (1998, 67–68).
2. Pomeranz (2000).
3. The study of Greek science in this chapter is based on collaboration with Peter Damerow, Peter McLaughlin, Markus Asper, Mark Schiefsky, and Istvan Bodnar. The study of Chinese science, including its comparison with Greek science, is based on collaboration with Matthias Schemmel, Zhang Baichun, Tian Miao, and Bill Boltz. The text incorporates and extends passages from earlier joint publications: Renn and Schemmel (2006; 2012; 2017); Büttner and Renn (2016); McLaughlin and Renn (2018). Important results on the Jesuit transmission used in this chapter have been published in Schemmel (2012). The treatment of early Chinese science is furthermore indebted to Graham (1978). The history of the calendar reform is based on Elman (2005). The history of the Protestant mission in the nineteenth century relies on Hu (2005). For the history of the Jesuit order, I have made use of Friedrich (2016).
4. Adapted from Diogenes Laertius (1925, 43) [DL 6, 40]. (Original translation: "Plato had defined Man as an animal, biped and featherless, and was applauded. Diogenes plucked a fowl and brought it into the lecture-room with the words, 'Here is Plato's man.' In consequence of which there was added to the definition, 'having broad nails.'")
5. For Darwin's argument for the descent of humans from an "early progenitor," see Darwin (1877, 160).
6. See Pääbo (2014); Reich (2018). See also Rogers Ackermann, Mackay, and Arnold (2016).
7. For the Moroccan findings, see Hublin, Ben-Ncer, Bailey, et al. (2017); for the sapient paradox, see Renfrew (2008).
8. See Chiaroni, Underhill, and Cavalli-Sforza (2009); Coop, Pickrell, Novembre, et al. (2009), and the discussion in Tomlinson (2018).
9. Aristotle (1939).
10. For the following, see Damerow, Renn, Rieger, et al. (2002); Renn, Damerow, and McLaughlin (2003); McLaughlin and Renn (2018).
11. Aristotle (1939, 353) [*Mechanica*, 850 a 30–33].
12. See Schemmel (2001b); Boltz, Renn, and Schemmel (2003); Renn and Schemmel (2006); Schemmel and Boltz (2016).
13. For the following, see also Graham (1978), for the textual history, Graham (1978, 68–69).
14. Graham (1978, 30ff.)
15. Translation taken from Schemmel (2001b, 7).

16. Schemmel and Boltz (2016).
17. See Ostwald and Lynch (1994).
18. Graham (1978, 19–22).
19. Graham (1978, 6–8).
20. Graham (1978, 23).
21. This has been argued in Schemmel and Boltz (2016, 143).
22. For the role of abstraction in ancient Chinese mathematics, see Chemla (2006). For a general reflection on abstraction in ancient science, see Schemmel (2019). The relation between ancient Greek and Chinese science has been a central subject in the work of Geoffrey E. R. Lloyd: see, e.g., Lloyd (1996; 2002); Lloyd and Sivin (2003).
23. Needham (1954–2015). For critical reassessments of the Needham problem, see O'Brien (2009); Goody (2012, chap. 5). For an attempt to answer the Needham question from a comparative point of view, emphasizing the uniqueness of the combination of Greek philosophy and science, Roman law, and Christian theology in Europe, and more broadly the role of social and institutional structures, see Huff (2017, in particular chap. 8).
24. See Feldhay (2006; 2011). For a general history of the Jesuit order, see Friedrich (2016).
25. See Feldhay (2011).
26. For a history of the calendar reform in the context of European and Chinese relations, see Elman (2005).
27. See Elman (2005, chap. 2).
28. For an overview of the translation and compilation projects, see Elman (2005, 90–106).
29. See Schemmel (2013).
30. See Schemmel (2012, 269–70).
31. Schemmel (2012, 276–77).
32. Schemmel (2012, 280).
33. For the history of the Protestant mission in the nineteenth century, see Elman (2005, chaps. 8–10); Hu (2005).
34. Hu (2005, 16–18).
35. See, for example, Fung (1922, 249–50, 261); Hu (1922, 8–9, 59 ff.); Su (1996).

Chapter 13: Epistemic Networks

1. This chapter is based on a collaboration that began with Malcolm Hyman and was then pursued with Manfred Laubichler, Roberto Lalli, Matteo Valleriani, and Dirk Wintergrün. It includes and extends text passages from Renn and Hyman (2012b); Renn, Wintergrün, Lalli, et al. (2016); and various publications by Alexander Blum, Roberto Lalli, and Jürgen Renn on the renaissance of general relativity (Blum, Lalli, and Renn (2015; 2016; 2018)). Of key importance to this presentation is Hyman (2007); Lalli (2017); Valleriani (2017c); Wintergrün (forthcoming 2019).
2. See Lazega and Snijders (2016).
3. See Renn, Wintergrün, Lalli, et al. (2016).
4. See Granovetter (1973; 1983); Watts (2003).
5. See Valente (1995); Wasserman and Faust (1997); Weyer (2012).
6. See, e.g., Padgett and Ansell (1993). See also Reinhard (1979); Vleuten and Kaijser (2005); Malkin (2011).
7. Preiser-Kapeller (2015). See also Laubichler, Maienschein, and Renn (2013); Lemercier (2015); Düring (2017).

8. See Bettencourt, Kaiser, Kaur, et al. (2008); Bettencourt, Kaiser, and Kaur (2009); Bettencourt and Kaiser (2015); Herfeld and Doehne (in press 2018).

9. See Barabási and Albert (1999). See also Barabási (2016).

10. See Price (1965).

11. See Merton (1968); Price (1976).

12. See Granovetter (1973; 1983).

13. See Milgram (1967); Watts (2003, 70ff.).

14. For the following, see Hyman (2007). See also Kintsch (1998).

15. For the following, see Russo (2004); Malkin (2011); Hyman and Renn (2012a); Schiefsky (2012). See also Ossendrijver (2011).

16. For historical discussion, see Di Pasquale (2005; 2007; 2010; 2012; 2013).

17. Kühn (1828, 606–7). See Blum (1977).

18. See Russo (2004).

19. For historical discussion, see, e.g., Graßhoff (1990); Schiefsky (2007); Hankinson (2008).

20. Valleriani (2017c).

21. Harris (2006).

22. The following is based on personal communication with Matteo Valleriani. For the history of the calendar and the Gregorian reform, see Poole and Poole (1934); Pedersen (1983); Richards (1998); Wallis (1999); Glick, Livesey, and Wallis (2005).

23. See Rochot (1967); Nellen (1990); Mauelshagen (2003); Grosslight (2013). For an online database on early modern correspondence, see *Early Modern Letters Online* by Cultures of Knowledge, http://emlo.bodleian.ox.ac.uk.

24. Fleck ([1935] 1979, 39); Kuhn (1996, 4).

25. See Kräutli and Valleriani (2018).

26. For historical analysis of a text similar to the *Sphere,* ascribed to Prochlos, circulating in particular at Protestant universities, see Biank (2019).

27. Sacrobosco and Nunes (1537).

28. Sacrobosco, Vinet, Valerianus, et al. (1556).

29. Kuhn (1996, 122–23).

30. See Lalli (2017).

31. See Eisenstaedt (1986; 1989).

32. See Will (1986; 1989). Some of the key publications of the renaissance are Kerr (1963); Matthews and Sandage (1963); Penrose (1965); Penzias and Wilson (1965); Hawking and Penrose (1970). First possible explanations of Penzia's and Wilson's findings: Dicke, Peebles, Roll, et al. (1965); Hewish, Bell, Pilkington, et al. (1968); Ruffini and Wheeler (1971).

33. See Thorne (1994, 258–99).

34. The historical discussion here is based on Blum, Lalli, and Renn (2015; 2016; 2018).

35. See Kaiser (2005).

36. Lalli (2017, 150–51).

37. Lalli (2017, 55).

38. See Mercier and Kervaire (1956).

Chapter 14: Epistemic Evolution

1. This chapter is based on various collaborative projects. The speculation about the onset of a new form of evolution goes back to the collaboration with Malcolm Hyman on *The Globalization of Knowledge in History*. Texts from the introduction to that volume and

its Survey 4 have been reused for this chapter; see Renn and Hyman (2012a; 2012b). The concept of an extended evolution, combining niche construction with complex regulatory networks, goes back to joint work with Manfred Laubichler. Texts from our publications have been used and extended: Laubichler and Renn (2015); Renn and Laubichler (2017). The development of spatial thinking relies entirely on the work of Matthias Schemmel and his group: Schemmel (2016a; 2016b). The example of weighing technology relies on joint research with Peter Damerow, Matthias Schemmel, and Katja Bödeker, as well as with Jochen Büttner and his group. Texts from publications resulting from this collaboration have been used and extended; see Damerow, Renn, Rieger, et al. (2002); Büttner and Renn (2016). The treatment of language evolution owes much to discussions with Peter Damerow, Stephen Levinson, and Caroline Rowland. The treatment of the Neolithic and urban revolutions closely follows the work of Hans Nissen, Dorian Fuller, and James Scott. The examples of extended evolution have been previously presented in the following publications, from which various texts have been taken and reworked for the current purpose: Renn (2015b; 2015c, 2015d).

2. For the following, see Lefèvre (2003).
3. For the concept of niche construction, see Odling-Smee, Erwin, Palkovacs, et al. (2013). The following is based on Laubichler and Renn (2015); Renn and Laubichler (2017). For like-minded discussions of cultural evolution, see Gray and Watts (2017); Tomlinson (2018).
4. Toepfer (2011, 523–27).
5. Toepfer (2011, 40–41)
6. Marx ([1852] 1979, 103).
7. See Laubichler and Renn (2015); Peter and Davidson (2015); Tomlinson (2018, 34, 113).
8. The following is adapted from Büttner and Renn (2016); Büttner (2018); Büttner, Renn, and Schemmel (2018); McLaughlin and Renn (2018). For a classical account, see Basalla (1988); for an overview of the more recent discussion, see Ziman (2000).
9. Bödeker (2006); Senft (2016). See also Malinowski (2002).
10. Bödeker (2006, 322ff.).
11. See Bödeker (2006, 199–205).
12. For the following, see Büttner and Renn (2016); Damerow, Renn, Rieger, et al. (2002).
13. Aristophanes (1998, 587).
14. Schemmel (2016a).
15. For the relation between language and cognition in spatial thinking, see Levinson (2003).
16. Euclid (1956).
17. See Vogel (1995).
18. See Piaget (1969). The approach to the history of time sketched in this box was jointly developed with Matthias Schemmel, following the example of his studies of space: Schemmel (2016a; 2016b).
19. See Szabó and Maula (1982, 23–24, 46–47).
20. For the integration of Babylonian astronomical observations into Greek astronomical models, see Graßhoff (1990, 214).
21. For the interpretation of Euclidean geometry as a science of space, see De Risi (2015).
22. See Elias ([1984] 2007, 33–34).
23. The origin of language has been long discussed; among the early milestones are the works of Herder, Grimm, and Humboldt: Herder (1772); Humboldt ([1820] 1994); Grimm (1852). Later discussions have related the origin of language to the general issue of symbol use and its role in the context of evolution; see, e.g., Bickerton (1990; 1995); Deacon (1997). My discussion here starts from the considerations in Damerow (2000);

Lock (2000); Sperber (2000); Levinson and Holler (2014); Tomlinson (2018). Other, more recent discussions relevant to the approach taken here are Corballis (2003; 2011); Tallerman (2005). For the current discussion, see also Coolidge and Wynn (2009); Richerson and Boyd (2010); Steels (2011); Dor, Knight, and Lewis (2014); Friederici (2017). For the views of Michael Tomasello, see the following discussion.

24. See Tomasello, Kruger, and Ratner (1993). The authors identify cultural learning "with those instances of social learning in which intersubjectivity or perspective-taking plays a vital role, both in the original learning process and in the resulting cognitive product"; Tomasello, Kruger, and Ratner (1993, 495). They characterize cultural learning as "a uniquely human form of social learning that allows for a fidelity of transmission of behaviors and information among conspecifics not possible in other forms of social learning, thereby providing the psychological basis for cultural evolution"; Tomasello, Kruger, and Ratner (1993, 495). See also Tomasello (1999; 2003). Tomasello (2014) develops the hypothesis that, in order to survive, humans had to develop what he calls "shared intentionality" (chap. 1), including the ability to see the world from multiple social perspectives. For a different view on the evolutionary process by which such achievements were reached, see Levinson (2014). Levinson differs from Tomasello in thinking that we need recursive, reciprocal feedback to evolve speech, as well as its supportive neuroanatomy.

25. Tomasello (2014). For Tomasello's work on primate cognition, see Tomasello and Call (1997).

26. See Maynard Smith and Szathmáry (1995).

27. Tomasello (2014, 141).

28. Tomasello (2014, 127).

29. Dediu and Levinson (2013, 1). See also Levinson and Dediu (2018).

30. This relation between the organization of human actions, which comprises networks and hierarchies, and the structure of language envisaged here does not seem to be incompatible with recent insights into the neuronal foundations of language: see Friederici (2017). There she argues that "it is conceivable that the ability to process structural hierarchies is what should be considered as a crucial step toward the language faculty" (206). In another paper, she argues that "language is best described as a biologically determined computational cognitive mechanism that yields an unbounded array of hierarchically structured expressions"; see Friederici, Chomsky, Berwick, et al. (2017, 713). For hints at deeper evolutionary roots of the underlying cognitive abilities, see Krause, Lalueza-Fox, Orlando, et al. (2007); Milne, Mueller, Männel, et al. (2016).

31. Levinson and Holler (2014, 2); Levinson (in press).

32. For a review of the discussion on protolanguages, see Tomlinson (2015).

33. See Dediu and Levinson (2013); Levinson and Holler (2014).

34. Vygotskij (1978, 86). See also Damerow (2000); Lock (2000); Pradhan, Tennie, and van Schaik (2012); Bickerton (2014, 334).

35. See Savage-Rumbaugh, Sevic, Rumbaugh, et al. (1985); Savage-Rumbaugh, Murphy, Sevic, et al. (1993); Sevic and Savage-Rumbaugh (1994); Savage-Rumbaugh, Fields, Segerdahl, et al. (2005); Lyn, Greenfield, Savage-Rumbaugh, et al. (2011).

36. Tomlinson (2018, 156).

37. Tomlinson (2018, 157).

38. Tomlinson (2018, 160).

39. Tomlinson (2018, 108–10).

40. I am grateful to Caroline Rowland for drawing my attention to the case of Nicaraguan Sign Language discussed below and for reading the first draft of this box.

41. For pioneering work on Nicaraguan Sign Language from a linguistic point of view, see Senghas (1995; 2000); Senghas (1997); Senghas and Coppola (2001); Senghas, Senghas, and Pyers (2005). The present review is based on and follows the arguments in Slobin (2004) and Blunden (2014).
42. Polich (2005).
43. Senghas and Coppola (2001, 328).
44. For the first aspect, see Senghas and Coppola (2001), and for the second, see Morford (2002, 333).
45. This point was stressed in Blunden (2014).
46. Pyers, Shusterman, Senghas, et al. (2010).
47. Vygotskij (1978, 86).
48. See Polich (2000, 298–99).
49. The following is based on Renn (2015b; 2015c). The interpretation and the description of many of the details given in the following closely follow Scott (2017) as well as the work of Dorian Fuller: Fuller, Allaby, and Stevens (2010); Fuller (2011; 2012); Fuller, Willcox, and Allaby (2011). I have furthermore made use of Kennett and Winterhalder (2006); Pinhasi and Stock (2011); Watkins (2013); Krause and Haak (2017); Kavanagh, Vilela, Haynie, et al. (2018). See also Gronenborn (2010); Özdoğan (2014; 2016).
50. See the discussion in Scott (2017, 10, 107ff.).
51. Marx and Engels ([1845–46] 1976, 47).
52. See Gowlett (2016).
53. See Ruddiman (2003).
54. See Scott (2017, 3).
55. See Zeder (2009, 32–33); Asouti (2010); Fuller, Willcox, and Allaby (2011).
56. For a discussion of predomestication, see, e.g., Willcox, Fornite, and Herveux (2008).
57. See Scott (2017, 43–44).
58. See the discussion in Kavanagh, Vilela, Haynie, et al. (2018).
59. See Fuller, Allaby, and Stevens (2010, 15ff.). For a broader discussion of the traps connected with the evolution of sedentism and agriculture, see Scott (2017).
60. For an approach to Neolithization through niche construction theory, see also Sterelny and Watkins (2015).
61. See Bogaard (2005).
62. For a discussion of the role of symbolic constructions in the context of the Neolithic Revolution, see Watkins (2010).
63. See Scott (2017, 7). The following is based on this book.
64. Zeder (2011, 230–31).
65. See Nissen (1988, 33, 59–60, 66ff.); Thompson (2006, 171–72).
66. Scott (2017, 21–23).
67. Scott (2017, 96).

Chapter 15: Exodus from the Holocene

1. See Ward and Kirschvink (2015).
2. See Zalasiewicz (2008); Zalasiewicz and Williams (2012); Schellnhuber, Serdeczny, Adams, et al. (2016).
3. See Boulding (1966); Fuller (1969); Poole (2010); Höhler (2015).
4. This chapter is based on joint work with Christoph Rosol, Benjamin Johnson, Benjamin Steininger, Thomas Turnbull, Helge Wendt, Manfred Laubichler, Sara Nelson, Bernd Scherer and Giulia Rispoli. It uses and extends texts elaborated jointly with

these colleagues, in particular Renn, Laubichler, and Wendt (2014); Renn and Scherer (2015a); Renn, Johnson, and Steininger (2017); Rosol, Nelson, and Renn (2017).

5. See Brooke (2018).

6. Davies (2016, 5).

7. For discussions of Earth system analysis, see Steffen, Sanderson, Tyson, et al. (2004); Huber, Schellnhuber, Arnell, et al. (2014); Rockström, Brasseur, Hoskins, et al. (2014); Ghil (2015); Donges, Lucht, Müller-Hansen, et al. (2017); Donges, Winkelmann, Lucht, et al. (2017); Steffen, Rockström, Richardson (2018).

8. For the historical development of climate science, see Weart (2003; 2004); Edwards (2010); Rosol (2017); Heymann and Achermann (2018); Lax (2018a; 2018b); Mauelshagen (2018a).

9. For recent discussions, see Zalasiewicz, Williams, Steffen, et al. (2010); Davies (2016); Yusoff (2016); Zalasiewicz, Waters, Williams, et al. (2018).

10. The following is based on Nelson, Rosol, and Renn (2017). See also Bonneuil and Fressoz (2016).

11. Zalasiewicz, Waters, Williams, et al. (2015); Zalasiewicz, Waters, Summerhayes, et al. (2017).

12. See, e.g., Certini and Scalenghe (2011); Edgeworth, Richter, Waters, et al. (2015).

13. See Barnosky, Matzke, Tomiya, et al. (2011); Ceballos, Ehrlich, and Dirzo (2017); Ceballos and Ehrlich (2018).

14. See Davies (2016, 86); Trischler (2016, 316). For recent discussion, see Waters, Zalasiewicz, Summerhayes, et al. (2018).

15. See Zalasiewicz, Waters, Williams, et al. (2015); Waters, Zalasiewicz, Summerhayes, et al. (2018).

16. See, e.g., Sandom, Faurby, Sandel, et al. (2014); Johnson, Alroy, Beeton, et al. (2016).

17. See Ruddiman (2005; 2013); Kaplan, Krumhardt, Ellis, et al. (2011).

18. See Zeder, Bradley, Emshwiller, et al. (2006); Zeder and Smith (2009); B. Smith and Zeder (2013). See also Bellwood (2004); Fuller (2010); Fuller, van Etten, Manning, et al. (2011).

19. See Petit, Jouzel, Raynaud, et al. (1999).

20. See Edgeworth, Richter, Waters, et al. (2015).

21. See Lewis and Maslin (2015; 2018).

22. See Steffen, Grinevald, Crutzen, et al. (2011). See also Crutzen and Stoermer (2000); Crutzen (2002).

23. The following is based on Renn, Laubichler, and Wendt (2014). For a discussion of the ecological conditions of the Industrial Revolution, see Sieferle (2001).

24. Wrigley (2010, 44–45).

25. See Landes (1969); Hobsbawm (1975); Wrigley (2010); Malm (2016). An analysis of this connection in other historical contexts is given in Evans and Rydén (2005). See also Bayly (2004); Wendt (2016a).

26. See Buffon (1778, 237), cited after Trischler (2016, 311).

27. See Steffen, Broadgate, Deutsch, et al. (2015). See also McNeill and Engelke (2014).

28. For a comprehensive historical survey, see McNeill (2000).

29. For a discussion of how to conceptualize the dynamic interaction between the Earth system and human behaviors, see, e.g., Müller-Hansen, Schlüter, Mäs, et al. (2017).

30. See Rockström, Steffen, Noone, et al. (2009); Steffen, Richardson, Rockström, et al. (2015).

31. See Osborn (1948); Meadows, Meadows, Randers, et al. (1972); Warde and Sörlin (2015).

32. For discussion, see Steffen, Sanderson, Tyson, et al. (2004); Donges, Winkelmann, Lucht, et al. (2017); Müller-Hansen, Schlüter, Mäs, et al. (2017).

33. Lampedusa ([1958] 1963, 29).

34. See Falkowski, Scholes, Boyle, et al. (2000); Klingenfeld and Schellnhuber (2012); Fischer-Kowalski, Krausmann, and Pallua (2014).

35. See Steffen, Rockström, Richardson, et al. (2018).

36. See, e.g., Bonneuil and Fressoz (2016); Malm (2016); Rich (2019).

37. Based on Renn, Johnson, and Steininger (2017). See also the contributions to Ertl and Soentgen (2015). See also Steininger (2014).

38. Liebig (1840).

39. Cushman (2013, 346).

40. See Crookes (1898).

41. See, e.g., Baracca, Ruffo, and Russo (1979); Travis (1993).

42. See Nernst (1907); Haber and Le Rossignol (1913); Ostwald (1926–27, 278–98); Mittasch (1951); Holdermann (1953); Farbwerke Hoechst AG (1964; 1966); Szöllösi-Janze (1998, chap. 4, esp. 175ff.). See also Steininger (2014); Friedrich, Hoffmann, Renn, et al. (2017).

43. Laue (1934).

44. Szöllösi-Janze (2000).

45. See Singh and Verma (2007); Gorman (2015); Wissemeier (2015).

46. See Mulvany (2005); McNeill, Rangarajan, and Padua (2009); Zeller (2016).

47. The following is based on Buchwald and Smith (2001a) as well as Oreskes and Conway (2010, chap. 4). See also Grundmann (2002); Buchwald (2017); Lax (2018a; 2018b).

48. Ozone Secretariat (1987).

49. See Crutzen (1970); Johnston (1971).

50. Molina and Rowland (1974).

51. Crutzen (1997, 214).

52. The following is based on Baracca (2012).

53. The following is based on Bud (2007); Landecker (2016).

54. Landecker (2016, 41). See also D'Abramo and Landecker (2019).

55. See World Health Organization (2014).

56. For an illustration, see, e.g., Turner, Matson, McCarthy, et al. (2003).

57. For the history of Earth system science, see Uhrqvist (2014); Lax (2018a; 2018b).

Chapter 16: Knowledge for the Anthropocene

1. This chapter is based on the collaboration with Christoph Rosol and Sara Nelson on the concept of the technosphere; it uses and extends passages from Rosol, Nelson, and Renn (2017), as well as from Renn (2017). The discussion of different interpretations of the Anthropocene heavily relies on Davies (2016); Trischler (2016). The history of the sphere concept is indebted to suggestions made by Giulia Rispoli and Christoph Rosol. The discussion of local knowledge uses and extends passages from Renn (2012b); Baracca, Renn, and Wendt (2014a). The concept of the Epistemic Web has been jointly developed with Malcom Hyman; its presentation makes use of Hyman and Renn (2012b).

2. The futility and the danger of geoengineering interventions and environmental geo-technologies is argued, for instance, by Hamilton (2013).

3. Latour (2014, 15). See also Trischler (2016, 318).

4. Malm (2018, 97).

5. Asafu-Adjaye, Blomqvist, Brand, et al. (2015), cited after Trischler (2016, 322).

6. Russian geographer and anthropologist Dmitry Anuchin spoke about the "Anthropo-sphere" in his lecture at the opening of the Geographical Department of the Moscow Pedagogical Society on March 9, 1902. He described the anthroposphere as the sphere of human culture, which includes its material products. See Anuchin ([1902] 1949, 99–100). For recent uses, see Lucht (2010). See also Baccini and Brunner (2002).
7. See Lovelock ([1979] 2000).
8. Fuller (1969).
9. Lovelock and Margulis (1974); Lovelock ([1979] 2000).
10. Weber ([1917/19] 2004, 7).
11. On Vernadsky's noosphere and its difference from Teilhard de Chardin and Le Roy's interpretations, see Levit (2001).
12. The following is based on Nelson, Rosol, and Renn (2017); Rosol, Nelson, and Renn (2017).
13. Rapp (1981, 123, 154); Naveh (1982, 207); Haff (2014b, 301), quoted after Rosol, Nelson, and Renn (2017, 3); Haff and Renn (2019).
14. Haff (2014b, 301). See also Rosol, Nelson, and Renn (2017, 3).
15. Haff (2014a, 129). See also Haff (2014b).
16. Marx ([1852] 1979, 103).
17. See Altvater (2016).
18. See also Brondizio, O'Brien, Bai, et al. (2016); Delanty and Mota (2017).
19. The ergosphere may also be conceived as the sphere of energy conversion resulting from human interventions. See, e.g., Cipolla (1978); Fischer-Kowalski, Krausmann, and Pallua (2014); Kleidon (2016); Judson (2017).
20. The reference to Japanese history as an example was inspired by a talk given by Julia Adeney Thomas at the Max Planck Institute for the History of Science in Berlin in October 2017; see Thomas (2017), see also her forthcoming book with Princeton University Press: *The Historian's Task in the Anthropocene: Theory, Practice, and the Case of Japan*. The more extensive discussion of the Japanese example below is also based on Ishikawa (2000b), English translation: Ishikawa (2000a); Diamond (2005a); Ochiai (2007); Niles (2018). See also Parker (2013, 484–506); Niles and Leeuw (2018).
21. The technosphere has meanwhile even become the subject of sociotechnological ex-periments within self-contained ecospheres, see Höhler (2018).
22. For this typology of knowledge, see Hirsch Hadorn and Pohl (2007); Vilsmaier and Lang (2014, 87–113). On orientation knowledge, see Schmieg, Meyer, Schrickel, et al. (2017); Lucht and Pachauri (2004). On transformation knowledge, see Kollmorgen, Wagener, and Merkel (2015). An example of missing knowledge are the true global costs and benefits of food production: Sukhdev, May, and Müller (2016).
23. See, e.g., Raffnsøe (2016).
24. See Mauelshagen (2016). For the loss of biodiversity, see Kolbert (2014).
25. See, e.g., Fanon ([1961] 2005); Chambers and Gillespie (2000); Dirks (2001); Diawara (2004); Selin (2008).
26. Eisenstadt (2000; 2002).
27. See Beasley ([1969] 2001); Howell (1992); Inkster (2001, esp. chaps. 2–5); Diamond (2005a, chap. 9); Thomas (2017); Niles (2018).
28. Carlowitz (1713).
29. Diamond (2005a, 300–301).
30. Diamond (2005a, 305).
31. For a discussion of the broader context, see Smitka (1998); Isett and Miller (2016). On knowledge of nature in Japan, see Marcon (2015).
32. For the Japanese practice of infanticide (*mabiki*), see Drixler (2013).

33. See Brokaw and Kornicki (2013).
34. Hanley (1987). For the night soil system, see Tajima (2007). From a global perspective, this practice is discussed in Ferguson (2014).
35. See Saito (2002).
36. For the introduction of sweet potatoes into Japan, see O'Brien (1972).
37. See Vogel (1991). For the concept of creolization and its relation to technological knowledge, see Edgerton (2007).
38. See Nyerere (1968).
39. See Baracca, Renn, and Wendt (2014b).
40. See Mosley (2003).
41. See Zhang, Zhang, and Fan (2006).
42. See Anderson (1999); Easterly (2006); Collier (2007); Dilger (2012); Rottenburg (2012).
43. See Dirlik (1995); Kim (2000).
44. Sahlins (1994, 414).
45. Elkana (2012, 610–12).
46. Elkana (2012, 610–12).
47. Jeschke, Lokatis, Bartram, et al. (2018, 2).
48. Oreskes and Conway (2010). See also Oreskes, Conway, Karoly, et al. (2018).
49. Farrell, McConnell, and Brulle (2019, 194).
50. Jeschke, Lokatis, Bartram, et al. (2018, 12).
51. This paragraph draws on Kaufmann (2009). See also Benatar, Daar, and Singer (2005).
52. For the challenges of the Anthropocene for global health issues, see Whitmee, Haines, Beyrer, et al. (2015).
53. See Kaufmann and Parida (2007, 301).
54. See Schlögl (2012); Renn, Schlögl, Rosol, et al. (2017).
55. See Schlögl (2012).
56. For overviews, see Kleidon (2016); Judson (2017).
57. See Mitchell (2011).
58. Mitchell (2011, 239).
59. See Edwards (2017).
60. See Hardin (1968).
61. See Ostrom, Burger, Field, et al. (1999); Hess and Ostrom (2003); Ostrom (2012; 2015). The use of the notion "common good" follows Romer (1990); Engel (2002); Hess and Ostrom (2007).
62. Hardin (1998, 683).
63. Ostrom, Burger, Field, et al. (1999, 282).
64. Diamond (2005a, 10–15, 522–25). For a history of the new possibilities of global climate monitoring, see Harper (2008); Edwards (2010).
65. See, e.g., Graham (2018); Morozov (2018).
66. See Kenney and Zysman (2016); Srnicek (2017).
67. See Flaxman, Goel, and Rao (2016).
68. For the following, see Rosol, Steininger, Renn, et al. (2018).
69. See Beniger (1986).
70. See Jevons (1865).
71. See Edwards (2010); Gabrys (2016); Rosol (2017); Schneider (2017).
72. I am grateful to Krishna Gummadi, Max Planck Institute for Software Systems, for a very helpful conversation on these issues. Some of the following draws on his input.

73. "Where there is danger some Salvation grows there too." (*Wo aber Gefahr ist, wächst das Rettende auch*). From the poem "Patmos" by Friedrich Hölderlin (1990, 54).

74. See Rosol, Steininger, Renn, et al. (2018).

75. See Evans (2011).

76. The following is based on Sprenger and Engemann (2015b), esp. the introduction by Florian Sprenger and Christoph Engemann (2015a). The following quotations and references are also taken from this work.

77. Ashton (2009). For a history of RFID technology, see Rosol (2007); Hayles (2009).

78. See Want (2010). See also Hiltzik (1999).

79. Weiser (1991, 94). Quoted after Sprenger and Engemann (2015a, 10).

80. Mitchell (2011).

81. For the following, see again Sprenger and Engemann (2015b), as well as for the quotations and references.

82. Spindler (2014).

83. Sterling (2015, 7). Quoted after Sprenger and Engemann (2015a, 19).

84. Marx ([1867] 1990, 91). [Capital, vol. 1, pt. 1, chap. 3, sec. 2a, 78].

85. See also Moulier-Boutang (2012).

86. Zuboff (2019, 99–100).

87. Morozov (2014b). Quoted from Sprenger and Engemann (2015a, 21).

88. Morozov (2013; 2014a); O'Reilly (2013).

89. See Gillies and Cailliau (2000, 104–5). See also Gromov (1995–2011); Abbate (1999). The following is based on Hyman and Renn (2012b).

90. Hyman and Renn (2012b, 833).

91. See Renn (2014a).

92. Leibniz ([1714] 1889, 248) [§57].

93. Hyman and Renn (2012b, 834).

94. See Earl and Kimport (2011).

95. See Edwards (2010); Hoffmann (2012).

96. See Rushkoff (2018).

Chapter 17: Science and the Challenges of Humanity

1. This chapter is based on Renn (2005); Renn and Hyman (2012a).

2. See Klein (2015b).

3. See Heinrich (1986; 1987; 1993; 2000).

4. See Comenius ([1644] 1966). For historical discussion, see, e.g., Sadler (2014). Even today, his original pedagogical ideas have not lost their topicality and are being practiced, for instance, in the form of an experimental cooperation between historians of science, pedagogues, and children in the Berlin Comenius Garten: see Vierck (2001). For a historical discussion of the hopes for emancipation raised by early modern science, see Lefèvre (1978); Heinrich (1987).

5. Descartes (1985a, esp. 142).

6. See Jamme and Schneider (1984, 21–78).

7. I am thinking here, e.g., of Ringer (1990).

8. Humboldt (1849–58, author's preface, ix).

9. Bernstein (1897).

10. Einstein (1991, 3, 5).

11. Bonneuil and Fressoz (2016, 291).

References

Abattouy, Mohammed, Jürgen Renn, and Paul Weinig, eds. (2001a). "Intercultural Transmission of Scientific Knowledge in the Middle Ages: Graeco-Arabic-Latin." Special issue of *Science in Context* 14 (1–2).

———(2001b). "Transmission as Transformation: The Translation Movements in the Medieval East and West in a Comparative Perspective." In "Intercultural Transmission of Scientific Knowledge in the Middle Ages: Graeco-Arabic-Latin," ed. M. Abattouy, J. Renn, and P. Weinig. Special issue of *Science in Context* 14 (1–2): 1–12.

Abbate, Janet (1999). *Inventing the Internet*. Cambridge, MA: MIT Press.

Abir-Am, Pnin G. (1997). "The Molecular Transformation of Twentieth-Century Biology." In *Science in the Twentieth Century*, ed. J. Krige and D. Pestre, 495–524. Amsterdam: Harwood Academic.

Abril Castelló, Vidal (1987). "Las Casas contra Vitoria, 1550–1552: La revolución de la duodécima réplica—Causas y consecuencias." *Revista de Indias* 47 (179): 83–101.

Adolf, Marian, and Nico Stehr (2017). *Knowledge: Is Knowledge Power?* 2nd ed. Abingdon: Routledge.

Aebli, Hans (1980–81). *Denken: Das Ordnen des Tuns*. 2 vols. Stuttgart: Klett-Cotta.

Agar, John (2012). *Science in the Twentieth Century and Beyond*. London: Polity Press.

Agee, James, and Walker Evans (1941). *Let Us Now Praise Famous Men: Three Tenant Families*. Boston: Houghton Mifflin.

Airy, George Biddell (1855). "III. On the Computation of the Effect of Attraction of Mountain-Masses, as Disturbing the Apparent Astronomical Latitude of Stations in Geodetic Surveys." *Philosophical Transactions of the Royal Society of London* 145:101–4.

Albert, Réka, and Albert-László Barabási (2002). "Statistical Mechanics of Complex Networks." *Reviews of Modern Physics* 74 (1): 47–97.

Alberti, Leon Battista ([1436] 1970). *On Painting*. Translated by John R. Spencer. New Haven, CT: Yale University Press.

Albritton Jonsson, Frederik (2012). "The Industrial Revolution in the Anthropocene." *The Journal of Modern History* 84 (3): 679–696.

Alster, Bendt (2005). *Wisdom of Ancient Sumer*. Bethesda, MD: CDL Press.

Altshuler, José, and Angelo Baracca (2014). "The Teaching of Physics in Cuba from Colonial Times to 1959." In *The History of Physics in Cuba*, ed. A. Baracca, J. Renn, and H. Wendt, 57–106. Dordrecht: Springer.

Altvater, Elmar (2016). "The Capitalocene, or, Geoengineering against Capitalism's Planetary Boundaries." In *Anthropocene or Capitalocene? Nature, History, and the Crisis of Capitalism*, ed. J. W. Moore, 138–52. Oakland, CA: PM Press.

Ammon, Sabine, Corinna Heineke, and Kirsten Selbmann, eds. (2007). *Wissen in Bewegung: Vielfalt und Hegemonie in der Wissensgesellschaft*. Weilerswist: Velbrück Wissenschaft.

Anderson, Mary B. (1999). *Do No Harm: How Aid Can Support Peace—or War*. Boulder, CO: Lynne Rienner.

Anuchin, Dmitry Nikolaevich ([1902] 1949). "O prepodavanii geografii i o voprosakh s nim svyazannykh" [On the teaching of geography and on issues related to it]. Reprinted in *Izbrannye geograficheskie raboty* [Selected geographical works], 99–110. Moscow: Gosudarstvennoe izdatel'stvo geograficheskoj literatury.

Appleby, Joyce (2013). *Shores of Knowledge: New World Discoveries and the Scientific Imagination*. New York: W. W. Norton.

Aquinas, Thomas (2006). *Summa Theologiae: Latin Text and English Translation, Introductions, Notes, Appendices and Glossaries*. Vol. 1: *Christian Theology (1a. 1)*. Cambridge: Cambridge University Press. https://aquinas.cc.

Arabatzis, Theodore, Jürgen Renn, and Ana Simoes, eds. (2015). *Relocating the History of Science: Essays in Honor of Kostas Gavroglu*. Boston Studies in the Philosophy and History of Science 312. Dordrecht: Springer.

Aristophanes (1998). "Peace." In *Clouds—Wasps—Peace*, 417–602. Loeb Classical Library 488. Cambridge, MA: Harvard University Press.

Aristotle (1932). "Book 1." In *Politics*, 3–67. Loeb Classical Library 264. Cambridge, MA: Harvard University Press.

———(1933). "Book 3." In *Metaphysics. Vol. 1.: Books 1–9*, 17–25. Loeb Classical Library 271. Cambridge, MA: Harvard University Press.

———(1934). *Physics*. Vol. 2: *Books 5–8*. Translated by P. H. Wicksteed and F. M. Cornford. Loeb Classical Library 255. Cambridge, MA: Harvard University Press.

———(1935). "Book 11." In *Metaphysics*. Vol. 2: *Books 10–14—Oeconomica—Magna Moralia*, 53–121. Loeb Classical Library 287. Cambridge, MA: Harvard University Press.

———(1939). "Mechanical Problems." In *Minor Works*, 327–412. Loeb Classical Library 307. Cambridge, MA: Harvard University Press.

———(1957). *Physics*. Vol. 1: *Books 1–4*. Translated by P. H. Wicksteed and F. M. Cornford. Loeb Classical Library 228. Cambridge, MA: Harvard University Press.

Arnold, David (2005). "Europe, Technology, and Colonialism in the 20th Century." *History and Technology* 21 (1): 85–106.

Aronova, Elena, Christine von Oertzen, and David Sepkoski, eds. (2017). *Data Histories*. Osiris 32. Chicago: University of Chicago Press.

Arrow, Kenneth J. (1994). "Methodological Individualism and Social Knowledge." *American Economic Review* 84 (2): 1–9.

Asafu-Adjaye, John, Linus Blomqvist, Stewart Brand, Barry Brook, Ruth DeFries, Erle Ellis, Christopher Foreman, David Keith, Martin Lewis, Mark Lynas, Ted Nordhaus, Roger Pielke Jr., Rachel Pritzker, Joyashree Roy, Mark Sagoff, Michael Shellenberger, Robert Stone, and Peter Teague (2015). "An Ecomodernist Manifesto." Ecomodernist. Accessed March 28, 2018. http://www.ecomodernism.org/manifesto-english.

Ashton, Kevin (2009). "That 'Internet of Things' Thing: In the Real World, Things Matter More Than Ideas." *RFID Journal*. Accessed February 5, 2019. https://www.rfidjournal.com /articles/view?4986.

Asouti, Eleni (2010). "Beyond the 'Origins of Agriculture': Alternative Narratives of Plant Exploitation in the Neolithic of the Middle East." In *Proceedings of the 6th International Congress of the Archaeology of the Ancient Near East, Rome, May 5–10 2009*. Vol. 1, ed. P. Matthiae, F. Pinnock, L. Nigro, and N. Marchetti, 189–204. Wiesbaden: Harrassowitz Verlag.

Asper, Markus (2013). "Explanation between Nature and Text: Ancient Commentaries on Science." *Studies in History and Philosophy of Science Part A* 44 (1): 43–50.

Assmann, Aleida (1989). "Jaspers' Achsenzeit, oder Schwierigkeiten mit der Zentralperspektive in der Geschichte." In *Karl Jaspers: Denken zwischen Wissenschaft, Poliktik und Philosophie*, ed. D. Harth, 187–205. Stuttgart: J. B. Metzler.

Assmann, Jan (1992). *Das kulturelle Gedächtnis: Schrift, Erinnerung und politische Identität in frühen Hochkulturen*. Munich: C. H. Beck.

———(2001). *Ma'at: Gerechtigkeit und Unsterblichkeit im Alten Ägypten*. Munich: C. H. Beck.

———(2008). *Of God and Gods: Egypt, Israel, and the Rise of Monotheism*. Madison: University of Wisconsin Press.

———(2018). *The Invention of Religion: Faith and Covenant in the Book of Exodus*. Princeton, NJ: Princeton University Press.

Austrian, Geoffrey D. (1982). *Herman Hollerith: Forgotten Giant of Information Processing*. New York: Columbia University Press.

Baccini, Peter, and Paul H. Brunner (2002). *Metabolism of the Anthroposphere: Analysis, Evaluation, Design*. Cambridge, MA: MIT Press.

Bacon, Francis ([1620] 2004). *Novum organum*. In *The Oxford Francis Bacon*. Vol. 11: *The "Instauratio magna," Part II: "Novum organum" and Associated Texts*, ed. G. Rees and M. Wakely, 48–447. Oxford: Clarendon Press.

Badino, Massimiliano (2015). *The Bumpy Road: Max Planck from Radiation Theory to the Quantum (1896–1906)*. SpringerBriefs in History of Science and Technology. Cham: Springer.

Bagg, Ariel M. (2011). "Mesopotamische Bauzeichnungen." In *The Empirical Dimension of Ancient Near Eastern Studies: Die empirische Dimension altorientalischer Forschungen*, 543–86. Wiener offene Orientalistik 6. Vienna: G. J. Selz.

Baines, John (1983). "Literacy and Ancient Egyptian Society." *Man New Series*, 18 (3): 572–99.

———(2001). *The Earliest Egyptian Writing: Development, Context, Purpose*. Preprint 180. Berlin: Max Planck Institute for the History of Science.

———(2007). *Visual and Written Culture in Ancient Egypt*. Oxford: Oxford University Press.

Balter, Michael (2012). "Ancient Migrants Brought Farming Way of Life to Europe." *Science* 336 (6080): 400–401.

Barabási, Albert-László (2016). *Network Science*. Cambridge: Cambridge University Press. http://networksciencebook.com.

Barabási, Albert-László, and Réka Albert (1999). "Emergence of Scaling in Random Networks." *Science* 286 (5439): 509–12.

Baracca, Angelo (2012). "The Global Diffusion of Nuclear Technology." In *The Globalization of Knowledge in History*, ed. J. Renn, 669–711. Studies 1. Berlin: Edition Open Access. http://edition-open-access.de/studies/1/31/index.html.

Baracca, Angelo, Jürgen Renn, and Helge Wendt (2014a). "A Short Introduction to This Volume." In *The History of Physics in Cuba*, ed. A. Baracca, J. Renn, and H. Wendt, 3–7. Boston Studies in the Philosophy and History of Science 304. Dordrecht: Springer.

———, eds. (2014b). *The History of Physics in Cuba*. Boston Studies in the Philosophy and History of Science 304. Dordrecht: Springer.

Baracca, Angelo, Stefano Ruffo, and Arturo Russo (1979). *Scienza e industria, 1848–1915: Gli sviluppi scientifici connessi alla seconda rivoluzione industriale*. Rome: Editori Laterza.

Baran, Paul (1964). "On Distributed Communications: I. Introduction to Distributed Communications Networks." Rand Corporation, Research Memoranda. Accessed January 31, 2019. https://www.rand.org/pubs/research_memoranda/RM3420.html.

Barbaro, Daniele (1567). *I dieci libri dell'architettura di M. Vitruvio, Tradotti & commentati da Mons: Daniel Barbaro eletto Patriarca d'Aquileia, da lui riveduti & ampliati; & hora in piu commoda forma ridotti*. Venice: Francesco de'Franceschi Senese & Giovanni Chrieger Alemano Compagni. http://echo.mpiwg-berlin.mpg.de/MPIWG:2D11R617.

Barnes, Jonathan, ed. (1995). *The Cambridge Companion to Aristotle*. Cambridge: Cambridge University Press.

Barnosky, Anthony D., Nicholas Matzke, Susumu Tomiya, Guinevere O. U. Wogan, Brian Swartz, Tiago B. Quental, Charles Marshall, Jenny L. McGuire, Emily L. Lindsey, Kaitlin C. Maguire, Ben Mersey, and Elizabeth A. Ferrer (2011). "Has the Earth's Sixth Mass Extinction Already Arrived?" *Nature* 471 (7336): 51–57.

Barth, Fredrik (1975). *Ritual and Knowledge among the Baktaman of New Guinea*. New Haven, CT: Yale University Press.

————(1987). *Cosmologies in the Making: A Generative Approach to Cultural Variation in Inner New Guinea.* Cambridge: Cambridge University Press.

————(2002). "An Anthropology of Knowledge." *Current Anthropology* 43 (1): 1–18.

Bartley, Tim (2007). "Institutional Emergence in an Era of Globalization: The Rise of Transnational Private Regulation of Labor and Environmental Conditions." *American Journal of Sociology* 113 (2) :297–351.

————(2018). *Rules without Rights: Land, Labor, and Private Authority in the Global Economy.* Transformations in Governance Series. Oxford: Oxford University Press.

Bartley, Tim, and Curtis Child (2014). "Shaming the Corporation: The Social Production of Targets and the Anti-sweatshop Movement." *American Sociological Review* 79 (4): 653–79.

Bartoli, Lorenzo, ed. (1998). *Lorenzo Ghiberti: I commentarii.* Florence: Giunti.

Basalla, George (1967). "The Spread of Western Science." *Science* 156 (3775): 611–22.

————(1988). *The Evolution of Technology.* Cambridge History of Science. Cambridge: Cambridge University Press.

Battisti, Eugenio (1981). *Brunelleschi: The Complete Work.* London: Thames & Hudson.

Bauer, Josef (1998). "Der Vorsargonische Abschnitt der Mesopotamischen Geschichte." In *Mesopotamien: Späturuk-Zeit und Frühdynastische Zeit,* ed. P. Attinger and M. Wäfler, 431–585. Annäherungen 1. Orbis Biblicus et Orientalis 160. Göttingen: Vandenhoeck & Ruprecht.

Bauer, Josef, Robert K. Englund, and Manfred Krebernik, eds. (1998). *Mesopotamien: Späturuk-Zeit und Frühdynastische Zeit.* Annäherungen 1. Orbis Biblicus et Orientalis 160. Göttingen: Vandenhoeck & Ruprecht.

Bayertz, Kurt (1987). "Wissenschaftsentwicklung als Evolution? Evolutionäre Konzeptionen wissenschaftlichen Wandels bei Ernst Mach, Karl Popper und Stephen Toulmin." *Zeitschrift für allgemeine Wissenschaftstheorie* 18 (1/2): 61–91.

Bayertz, Kurt, Myriam Gerhard, and Walter Jaeschke, eds. (2007). *Der Darwinismus-Streit.* Weltanschauung, Philosophie und Naturwissenschaft im 19. Jahrhundert 2. Hamburg: Felix Meiner Verlag.

Bayly, Christopher A. (2004). *The Birth of the Modern World, 1780–1914: Global Connections and Comparisons.* Malden, MA: Blackwell Publishing.

Beasley, William G. ([1969] 2001). "Japan and the West in the Mid-Nineteenth Century: Nationalism and the Origins of the Modern State." In *The Collected Writings of Modern Western Scholars on Japan.* Vol. 5, 127–44. Collected Writings of W. G. Beasley. Tokyo: Edition Synapse.

Becattini, Ilaria (2015). "Dalla Selva alla Cupola: Il trasporto del legname dell'Opera di Santa Maria del Fiore e il suo impiego nel cantiere brunelleschiano." The Years of the Cupola— Studies. http://duomo.mpiwg-berlin.mpg.de/STUDIES/study003/study003.html.

Becchi, Antonio (2014). "Fokus: Architektur und Mechanik." In *Wissensgeschichte der Architektur.* Vol. 3: *Vom Mittelalter bis zur Frühen Neuzeit,* ed. J. Renn, W. Osthues, and H. Schlimme, 397–428. Studies 5. Berlin: Edition Open Access. http://edition-open-access .de/studies/5/6/index.html.

Becchi, Antonio, Massimo Corradi, Federico Foce, and Orietta Pedemonte, eds. (2002). *Towards a History of Construction: Dedicated to Eduardo Benvenuto.* Basel: Birkhäuser.

Beckert, Jens (2003). "Economic Sociology and Embeddedness: How Shall We Conceptualize Economic Action?" *Journal of Economic Issues* 37 (3): 769–87.

Beckwith, Christopher I. (2009). *Empires of the Silk Road: A History of Central Eurasia from the Bronze Age to the Present.* Princeton, NJ: Princeton University Press. http://hdl.handle .net/2027/fulcrum.fn106z494.

Belli, Gianluca, and Amedeo Belluzzi (2003). *Il Ponte a Santa Trinita.* Florence: Edizioni Polistampa.

Bellwood, Peter (2004). *First Farmers: The Origins of Agricultural Societies*. Malden, MA: Blackwell Publishing.

Benatar, Solomon R., Abdallah S. Daar, and Peter A. Singer (2005). "Global Health Challenges: The Need for an Expanded Discourse on Bioethics." *PLOS Medicine* 2 (7): 587–89.

Bender, George, and Robert A. Thom, eds. (1961). *A History of Medicine in Pictures*. 3 vols. Detroit, MI: Parke, Davis & Co.

Beniger, James R. (1986). *The Control Revolution: Technological and Economic Origins of the Information Society*. Cambridge, MA: Harvard University Press.

Benjamin, Walter (1928). *Ursprung des deutschen Trauerspiels*. Berlin: Rowohlt.

———(2003). *Selected Writings*. Vol. 4: *1938–1940*. Translated by Edmund Jephcott et al. Cambridge, MA: Belknap Press of Harvard University Press.

Berkhout, Frans (1997). "Science in Public Policy: A History of High-Level Radioactive Waste Management." In *Science in the Twentieth Century*, ed. J. Krige and D. Pestre, 275–99. Amsterdam: Harwood Academic.

Bernal, John D. (1939). *The Social Function of Science*. London: Routledge.

———(1949). *The Freedom of Necessity*. London: Routledge & Kegan Paul.

Bernbeck, Reinhard (1994). *Die Auflösung der Häuslichen Produktionsweise: Das Beispiel Mesopotamiens*. Berliner Beiträge zum Vorderen Orient 14. Berlin: Reimer.

Bernstein, Aaron (1897). *Naturwissenschaftliche Volksbücher*. 21 vols. 5th ed. Berlin: Ferd. Dümmlers Verlagsbuchhandlung.

Bertelsen, Rasmus Gjedssø (2014). "American Missionary Universities in China and the Middle East and American Philanthropy: Interacting Soft Power of Transnational Actors." *Global Society* 28 (1): 113–27.

Bertolini, Lucia, ed. (2011). *Leon Battista Alberti: De pictura (redazione volgare)*. Florence: Edizioni Polistampa.

Berzelius, Jöns Jacob (1813). "Experiments on the Nature of Azote, of Hydrogen, and of Ammonia, and upon the Degrees of Oxidation of Which Azote Is Susceptible." *Annals of Philosophy* 2:276–84, 357–68.

———(1814). "Essay on the Cause of Chemical Proportions, and on Some Circumstances Relating to Them: Together with a Short and Easy Method of Expressing Them." *Annals of Philosophy* 3:51–62, 93–106, 244–57, 353–64.

Bethell, Leslie (1984–2008). *The Cambridge History of Latin America*. 11 vols. Cambridge: Cambridge University Press.

Bettencourt, Luís M. A., and David I. Kaiser (2015). "Formation of Scientific Fields as a Universal Topological Transition." Cornell University, arXiv. org. Accessed March 5, 2018. https://arxiv.org/abs/1504.00319.

Bettencourt, Luís M. A., David I. Kaiser, and Jasleen Kaur (2009). "Scientific Discovery and Topological Transitions in Collaboration Networks." *Journal of Informetrics* 3 (3): 210–21.

Bettencourt, Luís M. A., David I. Kaiser, Jasleen Kaur, Carlos Castillo-Chávez, and David Wojick (2008). "Population Modeling of the Emergence and Development of Scientific Fields." *Scientometrics* 75 (3): 495–518.

Biank, Johanna (2019). *Pseudo-Proklos' Sphaera und die Sphaera-Gattung im 15. bis 17. Jh*. Berlin: Edition Open Sources.

Bickerton, Derek (1990). *Language and Species*. Chicago: University of Chicago Press.

———(1995). *Language and Human Behavior*. Seattle: University of Washington Press.

———(2014). *More Than Nature Needs: Language, Mind, and Evolution*. Cambridge, MA: Harvard University Press.

Binding, Günther (1985–86). "Zum Kölner Stadtmauerbau: Bemerkungen zur Bauorganisation im 12./13. Jahrhundert." *Wallraf-Richartz-Jahrbuch* 46–47:7–17.

————(2004). *Meister der Baukunst: Geschichte des Architekten- und Ingenieurberufes*. Darmstadt: Wissenschaftliche Buchgesellschaft.

————(2014). "Bauwissen im Früh- und Hochmittelalter." In *Wissensgeschichte der Architektur* Vol. 3: *Vom Mittelalter bis zur Frühen Neuzeit*, ed. J. Renn, W. Osthues, and H. Schlimme, 9–94. Studies 5. Berlin: Edition Open Access. http://edition-open-access.de/studies/5/3/index.html.

Blagdon, Francis William (1813). "A Brief History of Ancient and Modern India, from the Earliest Periods of Antiquity to the Termination of the Late Mahratta War." In *The European in India: From a Collection of Drawings, by Charles Doyley, Esq., Engraved by J. H. Clark and C. Dubourg, with a Preface and Copious Descriptions, by Captain Thomas Williamson, Accompanied with a Brief History of Ancient and Modern India, from the Earliest Periods of Antiquity to the Termination of the Late Mahratta War by F. W. Blagdon, Esq.*, ed. C. D'Oyly, 61–149. London: Edward Orme.

Bloom, Jonathan M. (2017). "Papermaking: The Historical Diffusion of an Ancient Technique." In *Mobilities of Knowledge*, ed. H. Jöns, P. Meusburger, and M. Heffernan, 51–66. Cham: Springer.

Blum, Alexander S., Kostas Gavroglu, Christian Joas, and Jürgen Renn, eds. (2016). *Shifting Paradigms: Thomas S. Kuhn and the History of Science*. Proceedings 8. Berlin: Edition Open Access. http://edition-open-access.de/proceedings/8/index.html.

Blum, Alexander S., Domenico Giulini, Roberto Lalli, and Jürgen Renn (2017). "Editorial Introduction to the Special Issue 'The Renaissance of Einstein's Theory of Gravitation.'" *European Physical Journal H* 42 (2): 95–105.

Blum, Alexander S., Martin Jähnert, Christoph Lehner, and Jürgen Renn (2017). "Translation as Heuristics: Heisenberg's Turn to Matrix Mechanics." *Studies in the History and Philosophy of Modern Physics* 60:3–22.

Blum, Alexander S., Roberto Lalli, and Jürgen Renn (2015). "The Reinvention of General Relativity: A Historiographical Framework for Assessing One Hundred Years of Curved Space-Time." *Isis* 106 (3): 598–620.

————(2016). "The Renaissance of General Relativity: How and Why It Happened." *Annalen der Physik* 528 (5): 344–49.

————(2018). "Gravitational Waves and the Long Relativity Revolution." *Nature Astronomy* 2:534–43.

Blum, Alexander S., Jürgen Renn, and Matthias Schemmel (2016). "Experience and Representation in Modern Physics: The Reshaping of Space." In *Spatial Thinking and External Representation: Towards a Historical Epistemology of Space*, ed. M. Schemmel, 191–212. Studies 8. Berlin: Edition Open Access. http://edition-open-access.de/studies/8/8/index.html.

Blum, Rudolf (1977). *Kallimachos und die Literaturverzeichnung bei den Griechen: Untersuchungen zur Geschichte der Biobibliographie*. Archiv für Geschichte des Buchwesens 18. Frankfurt am Main: Buchhändler-Vereinigung.

Blunden, Andy (2014). "The Invention of Nicaraguan Sign Language." Ethical Politics. Accessed November 28, 2018. https://ethicalpolitics.org/ablunden/works/nsl.htm.

Bödeker, Katja (2006). *Die Entwicklung intuitiven physikalischen Denkens im Kulturvergleich*. Münster: Waxmann Verlag.

Bodin, Jean (1576). *Les six livres de la republique*. Paris: Iacques du Puys.

————(1606). *The Six Bookes of a Commonweale*. Translated by Richard Knolles. London: G. Bishop.

Bodmer, Walter F., and Luigi L. Cavalli-Sforza (1976). *Genetics, Evolution, and Man*. San Francisco: W. H. Freeman and Co.

Bogaard, Amy (2005). "'Garden Agriculture' and the Nature of Early Farming in Europe and the Near East." *World Archaeology* 37 (2): 177–96.

Böhme, Hartmut (2011). "Einladung zur Transformation." In *Transformation: Ein Konzept zur Erforschung kulturellen Wandels*, ed. H. Böhme and Sonderforschungsbereich Transformationen der Antike, 7–37. Munich: Wilhelm Fink Verlag.

Böhme, Hartmut, Lutz Bergemann, Martin Dönike, Albert Schirrmeister, Georg Toepfer, Marco Walter, and Julia Weitbrecht, eds. (2011). *Transformation: Ein Konzept zur Erforschung kulturellen Wandels*. Munich: Wilhelm Fink Verlag.

Bohr, Niels (1913). "I. On the Constitution of Atoms and Molecules." *London, Edinburgh, and Dublin Philosophical Magazine and Journal of Science: Series 6* 26 (151): 1–25.

Boltz, William G. (1986). "Early Chinese Writing." *Early Writing Systems* 17 (3): 420–36.

Boltz, William G., Jürgen Renn, and Matthias Schemmel (2003). *Mechanics in the Mohist Canon and Its European Counterpart: Texts and Contexts*. Preprint 241. Berlin: Max Planck Institute for the History of Science.

Bonneuil, Christophe, and Jean-Baptiste Fressoz (2016). *The Shock of the Anthropocene*. London: Verso Books.

Borkenau, Franz ([1934] 1976). *Der Übergang vom feudalen zum bürgerlichen Weltbild: Studien zur Geschichte der Philosophie der Manufakturperiode*. Reprint, Darmstadt: Wissenschaftliche Buchgesellschaft.

Born, Max, and Pascual Jordan (1925). "Zur Quantenmechanik." *Zeitschrift für Physik* 34 (1): 858–88.

Botero, Giovanni (1589). *Della ragion di stato libri dieci, con tre libri delle cause della grandezza, e magnificenza delle città di Giovanni Botero Benese*. Venice: Gioliti.

Boulding, Kenneth E. (1966). "The Economics of the Coming Spaceship Earth." In *Environmental Quality in a Growing Economy*, ed. H. Jarrett, 3–14. Baltimore, MD: Johns Hopkins University Press. http://www.zo.utexas.edu/courses/thoc/Boulding_SpaceshipEarth.pdf.

Bourdieu, Pierre ([1984] 2000). *Distinction: A Social Critique of the Judgement of Taste*. Translated by Richard Nice. Reprint, Cambridge, MA: Harvard University Press.

———(1989). "Social Space and Symbolic Power." *Sociological Theory* 7 (1): 14–25.

———(2001). *Science of Science and Reflexivity*. Chicago: University of Chicago Press.

Bowerman, Melissa, and Stephen C. Levinson, eds. (2001). *Language Acquisition and Conceptual Development*. Cambridge: Cambridge University Press.

Bowie, William (1927). *Isostasy*. London: E. P. Dutton.

Boyd, Robert, and Peter J. Richerson (1985). *Culture and the Evolutionary Process*. Chicago: University of Chicago Press.

Boyd, Robert, and Joan B. Silk (1997). *How Humans Evolved*. New York: W. W. Norton.

Braarvig, Jens (2012). "The Spread of Buddhism as Globalization of Knowledge." In *The Globalization of Knowledge in History*, ed. J. Renn, 245–67. Studies 1. Berlin: Edition Open Access. http://edition-open-access.de/studies/1/14/index.html.

Braarvig, Jens, and Markham D. Geller, eds. (2018). *Multilingualism, Lingua Franca and Lingua Sacra*. Studies 10. Berlin: Edition Open Access. http://edition-open-access.de/studies/10/index.html.

Bray, Francesca, Peter A. Coclanis, Edda Fields-Black, and Dagmar Schäfer, eds. (2015). *Rice: Global Networks and New Histories*. Cambridge: Cambridge University Press.

Bredekamp, Horst (1984). "Der Mensch als Mörder der Natur: Das 'Iudicium Iovis' von Paulus Niavis und die Leibmetaphorik." In *All Geschöpf ist Zung' und Mund: Beiträge aus dem Grenzbereich von Naturkunde und Theologie*, ed. H. Reinitzer, 261–83. Hamburg: Friedrich Wittig Verlag.

———(2003). "Kulturtechnik zwischen Mutter und Stiefmutter Natur." In *Bild–Schrift–Zahl*, ed. S. Krämer and H. Bredekamp, 117–41. Munich: Wilhelm Fink Verlag.

————(2006). *Thomas Hobbes: Der Leviathan—Das Urbild des modernen Staates und seine Gegenbilder, 1651–2001*. 3rd ed. Berlin: Akademie Verlag.

————(2010). *Theorie des Bildakts*. Berlin: Suhrkamp.

Bredekamp, Horst (2019). *Galileo's Thinking Hand: Mannerism, Anti-Mannerism and the Virtue of Drawing in the Foundation of Early Modern Science*. Berlin: De Gruyter.

Bredekamp, Horst, Vera Dünkel, and Birgit Schneider, eds. (2015). *The Technical Image: A History of Styles in Scientific Imagery*. Chicago: University of Chicago Press.

Brentjes, Sonja, and Jürgen Renn (2016a). "A Re-evaluation of the 'Liber de canonio.'" In *Scienze e rappresentazioni: Saggi in onore di Pierre Souffrin*, ed. P. Caye, R. Nanni, and P. D. Napolitani, 119–50. Florence: Leo S. Olschki.

————(2016b). "Contexts and Content of Thābit ibn Qurra's (Died 288/901) Construction of Knowledge on the Balance." In *Globalization of Knowledge in the Post-Antique Mediterranean, 700–1500*, ed. S. Brentjes and J. Renn, 67–99. New York: Routledge.

————, eds. (2016c). *Globalization of Knowledge in the Post-Antique Mediterranean, 700–1500*. New York: Routledge.

Brierley, Chris, Katie Manning, and Mark Maslin (2018). "Pastoralism May Have Delayed the End of the Green Sahara." *Nature Communications* 9 (2018): 1–9.

Brokaw, Cynthia, and Peter Kornicki, eds. (2013). *The History of the Book in East Asia*. Abingdon: Routledge.

Brondizio, Eduardo S., Karen O'Brien, Xuemei Bai, Frank Biermann, Will Steffen, Frans Berkhout, Christophe Cudennec, Maria Carmen Lemos, Alexander Wolfe, Jose Palma-Oliveira, and Chen-Tung Arthur Chen (2016). "Re-conceptualizing the Anthropocene: A Call for Collaboration." *Global Environmental Change* 39:318–27.

Brooke, John L. (2018). "The Holocene." In *The Palgrave Handbook of Climate History*, ed. S. White, C. Pfister, and F. Mauelshagen, 175–82. London: Palgrave Macmillan.

Bruch, Rüdiger vom (2010). "Aufbrüche und Zäsuren: Stationen der Berliner Wissenschaftsgeschichte." In *Weltwissen: 300 Jahre Wissenschaften in Berlin*, ed. J. Hennig and U. Andraschke, 22–33. Munich: Hirmer Verlag.

Bruno, Giordano ([1584] 1995). *La cena de le ceneri: The Ash Wednesday Supper*. Translated by Edward A. Gosselin and Lawrence S. Lerner. Renaissance Society of America Reprint Texts 4. Toronto: University of Toronto Press.

————([1584] 1999). *La cena de le ceneri—De la causa, principio et uno—De l'infinito, universo et mondi*. Opere italiane 2. Florence: Leo S. Olschki.

Bub, Jeffrey. 2016. *Bananaworld: Quantum Mechanics for Primates*. Oxford: Oxford University Press.

Buchwald, Jed Z. (1992). "Kinds and the Wave Theory of Light." *Studies in History and Philosophy of Science Part A* 23 (1): 39–74.

————(2017). "Politics, Morality, Innovation, and Misrepresentation in Physical Science and Technology." In *The Romance of Science: Essays in Honour of Trevor H. Levere*, ed. J. Z. Buchwald and L. Stewart, 201–18. Archimedes 52. Cham: Springer.

Buchwald, Jed Z., and Mordechai Feingold (2013). *Newton and the Origin of Civilization*. Princeton, NJ: Princeton University Press.

Buchwald, Jed Z., and Robert Fox (2013). *The Oxford Handbook of the History of Physics*. Oxford: Oxford University Press.

Buchwald, Jed Z., and George E. Smith (2001a). "An Instance of the Fingerpost." Review of *The Ozone Layer: A Philosophy of Science Perspective*, by Maureen Christie. *American Scientist* 89 (6): 546–49.

————(2001b). "Incommensurability and the Discontinuity of Evidence." *Perspectives on Science* 9 (4): 463–98.

Bud, Robert (2007). *Penicillin: Triumph and Tragedy.* Oxford: Oxford University Press.

Buffon, Georges-Louis Leclerc de (1778). *Histoire naturelle, générale et particulière.* Suppl. 5: *Des époques de la nature.* Paris: Imprimerie Royale.

Bührig, Claudia (2014). "Fokus: Bauzeichnungen auf Tontafeln." In *Wissensgeschichte der Architektur.* Vol. 1: *Vom Neolithikum bis zum Alten Orient,* ed. J. Renn, W. Osthues, and H. Schlimme, 335–407. Studies 3. Berlin: Edition Open Access. http://edition-open-access .de/studies/3/8/index.html.

Bührig, Claudia, Elisabeth Kieven, Jürgen Renn, and Hermann Schlimme (2006). "Towards an Epistemic History of Architecture." In *Practice and Science in Early Modern Italian Building: Towards an Epistemic History of Architecture,* ed. H. Schlimme, 7–12. Milan: Electa.

Bullard, Edward, J. E. Everett, A. Gilbert Smith, Patrick Maynard Stuart Blackett, Edward Bullard, and Stanley Keith Runcorn (1965). "The Fit of the Continents around the Atlantic." *Philosophical Transactions of the Royal Society A: Mathematical, Physical and Engineering Sciences* 258 (1088): 41–51.

Buringh, Eltjo, and Jan Luiten van Zanden (2009). "Charting the 'Rise of the West': Manuscripts and Printed Books in Europe; A Long-Term Perspective from the Sixth through Eighteenth Centuries." *Journal of Economic History* 69 (2): 409–45.

Burke, Peter (2000). *A Social History of Knowledge: From Gutenberg to Diderot.* Cambridge: Polity Press.

———(2016). *What Is the History of Knowledge?* Malden, MA: Polity Press.

Burnett, Charles (2001). "The Coherence of the Arabic-Latin Translation Program in Toledo in the Twelfth Century." In "Intercultural Transmission of Scientific Knowledge in the Middle Ages: Graeco-Arabic-Latin," ed. M. Abattouy, J. Renn, and P. Weinig. Special issue of *Science in Context* 14 (1–2): 249–88.

Bushnell, Oswald A. (1993). *The Gifts of Civilization: Germs and Genocide in Hawai'i.* Honolulu: University of Hawai'i Press.

Büttner, Jochen (2008). "Big Wheel Keep on Turning." *Galilaeana* 5:33–62.

———(2017). "Shooting with Ink." In *The Structures of Practical Knowledge,* ed. M. Valleriani, 115–88. Cham: Springer.

———(2018). "Waage und Wandel: Wie das Wiegen die Bronzezeit prägt." In *Innovationen der Antike,* ed. G. Graßhoff and M. Meyer, 60–78. Darmstadt: Philipp von Zabern.

———(2019). *Swinging and Rolling: Unveiling Galileo's Unorthodox Path from a Challenging Problem to a New Science.* Cham: Springer.

Büttner, Jochen, Peter Damerow, and Jürgen Renn (2001). "Traces of an Invisible Giant: Shared Knowledge in Galileo's Unpublished Treatises." In *Largo campo di filosofare: Eurosymposium Galileo 2001,* ed. J. Montesinos and C. Solís, 183–201. La Orotava: Fundación Canaria Orotava de Historia de la Ciencia.

Büttner, Jochen, and Jürgen Renn (2016). "The Early History of Weighing Technology from the Perspective of a Theory of Innovation." In "Space and Knowledge." Special issue of *eTopoi: Journal for Ancient Studies* 6:757–76.

Büttner, Jochen, Jürgen Renn, and Matthias Schemmel (2003). "Exploring the Limits of Classical Physics: Planck, Einstein and the Structure of a Scientific Revolution." *Studies in History and Philosophy of Modern Physics* 34 (1): 37–59.

———(2018). "The Early History of Weighing Technology from the Perspective of a Theory of Innovation." In *Emergence and Expansion of Pre-classical Mechanics,* ed. R. Feldhay, J. Renn, M. Schemmel, and M. Valleriani, 81–137. Boston Studies in the Philosophy and History of Science 333. Cham: Springer.

Cahan, David (2011). *Meister der Messung: Die Physikalisch-Technische Reichsanstalt im Deutschen Kaiserreich.* Bremerhaven: Wirtschaftsverlag NW.

Camardi, Giovanni (1999). "Charles Lyell and the Uniformity Principle." *Biology and Philosophy* 14 (4): 537–60.

Cancik-Kirschbaum, Eva (2005). "Beschreiben, Erklären, Deuten: Ein Beispiel für die Operationalisierung von Schrift im alten Zweistromland." In *Schrift: Kulturtechnik zwischen Auge, Hand und Maschine*, ed. G. Grube, W. Kogge, and S. Krämer, 399–411. Munich: Wilhelm Fink Verlag.

———(2012). "Writing, Language and Textuality: Conditions for the Transmission of Knowledge in the Ancient Near East." In *The Globalization of Knowledge in History*, ed. J. Renn, 125–51. Studies 1. Berlin: Edition Open Access. http://edition-open-access.de/studies/1/9/index.html.

Cancik-Kirschbaum, Eva, and Anita Traninger, eds. (2016). *Wissen in Bewegung: Institution—Iteration—Transfer*. Episteme in Bewegung 1. Wiesbaden: Harrassowitz Verlag.

Carlowitz, Hans Carl von (1713). *Sylvicultura oeconomica oder Hauswirthliche Nachricht und Naturgemäße Anweisung zur Wilden Baum-Zucht*. Leipzig: Johann Freidrich Braun.

Carlson, W. Bernard (1997). "Innovation and the Modern Corporation: From Heroic Invention to Industrial Science." In *Science in the Twentieth Century*, ed. J. Krige and D. Pestre, 203–26. Amsterdam: Harwood Academic.

Carson, Rachel (1962). *Silent Spring*. Boston: Houghton Mifflin.

Casas, Bartolomé de las ([1552] 1992). *A Short Account of the Destruction of the Indies*. Translated by Nigel Griffin. London: Penguin Books. http://www.columbia.edu/~daviss/work/files/presentations/casshort.

Cassirer, Ernst (1944). *An Essay on Man: An Introduction to a Philosophy of Human Culture*. New Haven, CT: Yale University Press.

———(1955–96). *The Philosophy of Symbolic Forms*. Translated by Ralph Manheim and J. M. Krois. Reprint, New Haven, CT: Yale University Press.

———(1996). "From the Introduction to the First Edition of *The Problem of Knowledge in Modern Philosophy and Science*." *Science in Context* 9 (7): 195–215.

Cavalli-Sforza, Luigi L., and Walter F. Bodmer (1971). *The Genetics of Human Populations*. San Francisco: W. H. Freeman.

Cavalli-Sforza, Luigi L., Paolo Menozzi, and Alberto Piazza (1994). *The History and Geography of Human Genes*. Princeton, NJ: Princeton University Press.

Ceballos, Gerardo, and Paul R. Ehrlich (2018). "The Misunderstood Sixth Mass Extinction." *Science* 360 (6393): 1080–81.

Ceballos, Gerardo, Paul R. Ehrlich, and Rodolfo Dirzo (2017). "Biological Annihilation via the Ongoing Sixth Mass Extinction Signaled by Vertebrate Population Losses and Declines." *Proceedings of the National Academy of Sciences of the United States of America* 114 (30): E6089–96.

Cech, Brigitte (2012). *Technik in der Antike*. Darmstadt: Wissenschaftliche Buchgesellschaft.

Certini, Giacomo, and Riccardo Scalenghe (2011). "Anthropogenic Soils Are the Golden Spikes for the Anthropocene." *Holocene* 21 (8): 1269–74.

Chakrabarty, Dipesh (2007). *Provincializing Europe: Postcolonial Thought and Historical Difference*. Princeton, NJ: Princeton University Press. http://hdl.handle.net/2027/heb.04798.0001.001.

———(2009). "The Climate of History: Four Theses." *Critical Inquiry* 35 (2): 197–222.

———(2012). "Postcolonial Studies and the Challenge of Climate Change." *New Literary History* 43 (1): 1–18.

———(2015). "The Human Condition in the Anthropocene." University of Utah. Accessed August 27, 2018. https://tannerlectures.utah.edu/lecture-library.php.

————(2017). "The Future of the Human Sciences in the Age of Humans: A Note." *European Journal of Social Theory* 20 (1): 39–43.

Chalmers, Alan F. (2017). *One Hundred Years of Pressure: Hydrostatics from Stevin to Newton.* Archimedes: New Studies in the History of Science and Technology 51. Cham: Springer.

Chambers, David Wade, and Richard Gillespie (2000). "Locality in the History of Science: Colonial Science, Technoscience, and Indigenous Knowledge." In "Nature and Empire: Science and the Colonial Enterprise." Special issue of *Osiris* 15:221–40.

Charniak, Eugene (1972). "Toward a Model of Children's Story Comprehension." Ph.D. diss., Massachusetts Institute of Technology.

Charpin, Dominique, Dietz Otto Edzard, and Marten Stol (2004). *Mesopotamia: The Old Assyrian Period.* Annäherungen 4; Orbis Biblicus et Orientalis 160. Göttingen: Vandenhoeck & Ruprecht.

Chartier, Roger (1994). *The Order of Books: Readers, Authors, and Libraries in Europe between the Fourteenth and Eighteenth Centuries.* Stanford, CA: Stanford University Press.

Chemla, Karine (2006). "Documenting a Process of Abstraction in the Mathematics of Ancient China." In *Studies in Chinese Language and Culture: Festschrift in Honor of Christoph Harbsmeier on the Occasion of His 60th Birthday*, ed. C. Anderl and H. Eifring, 169–94. Oslo: Hermes Academic Publishing.

————(2012). *The History of Mathematical Proof in Ancient Traditions.* Cambridge: Cambridge University Press.

————(2017). "Changing Mathematical Cultures, Conceptual History, and the Circulation of Knowledge: A Case Study Based on Mathematical Sources from Ancient China." In *Cultures without Culturalism: The Making of Scientific Knowledge*, ed. K. Chemla and E. Fox Keller, 352–98. Durham, NC: Duke University Press.

Chemla, Karine, and Guo Shuchun (2004). *Les neuf chapitres: Le classique mathématique de la Chine ancienne et ses commentaires.* Malakoff: Éditions Dunod.

Cheney, Dorothy L., and Robert M. Seyfarth (2018). "Flexible Usage and Social Function in Primate Vocalizations." *Proceedings of the National Academy of Sciences of the United States of America* 115 (9): 1974–79.

Chiaroni, Jacques, Peter A. Underhill, and Luca L. Cavalli-Sforza (2009). "Y Chromosome Diversity, Human Expansion, Drift, and Cultural Evolution." *Proceedings of the National Academy of Sciences of the United States of America* 106 (48): 20174–79.

Chilvers, Christopher A. J. (2015). "Five Tourniquets and a Ship's Bell: The Special Session at the 1931 Congress." *Centaurus* 57 (2): 61–95.

Cipolla, Carlo M. (1978). *The Economic History of World Population.* 7th ed. Sussex: Harvester Press.

Clagett, Marshall (1968). *Nicole Oresme and the Medieval Geometry of Qualities and Motions: A Treatise on the Uniformity and Difformity of Intensities Known as "Tractatus de configurationibus qualitatum et motuum."* Publications in Medieval Science 12. Madison: University of Wisconsin Press.

Clement, John (1983). "A Conceptual Model Discussed by Galileo and Used Intuitively by Physics Students." In *Mental Models*, ed. D. Gentner and A. L. Stevens, 325–40. Hillsdale, NJ: Erlbaum.

Coe, Michael D. (1992). *Breaking the Maya Code.* London: Thames & Hudson.

Cohen, Hendrik Floris (2015). *The Rise of Modern Science Explained: A Comparative History.* Cambridge: Cambridge University Press.

Cohen, Yves (1997). "Scientific Management and the Production Process." In *Science in the Twentieth Century*, ed. J. Krige and D. Pestre, 111–25. Amsterdam: Harwood Academic.

Cohn, Jonas (1907). *Führende Denker: Geschichtliche Einleitung in die Philosophie*. Aus Natur und Geisteswelt: Sammlung wissenschaftlich-gemeinverständlicher Darstellungen 176. Leipzig: Teubner.

Collet, John Peter (1997). "The History of Electronics: From Vacuum Tubes to Transistors." In *Science in the Twentieth Century*, ed. J. Krige and D. Pestre, 253–74. Amsterdam: Harwood Academic.

Collier, Paul (2007). *The Bottom Billion: Why the Poorest Countries Are Failing and What Can Be Done about It*. Oxford: Oxford University Press.

Collins, Harry, and Trevor J. Pinch (1993). *The Golem: What Everyone Should Know about Science*. Cambridge: Cambridge University Press.

Comenius, Johann Amos ([1644] 1966). *Iohannis Amos Comenii: De rerum humanarum emendatione consultatio catholica*. 2 vols. Prague: Academiae Scientiarum Bohemoslovacae.

———(2001). *Pampaedia—Allerziehung*. 3rd ed. Schriften zur Comeniusforschung 18. Sankt Augustin: Academia Verlag.

Como, Mario (2010). *Statica delle costruzioni storiche in muratura: Archi, volte, cupole, architetture monumentali, edifici sotto carichi verticali e sotto sisma*. Rome: Aracne Editrice.

Condorcet, Jean Antoine Nicolas de Caritat de (1796). *Outlines of an Historical View of the Progress of the Human Mind: Being a Posthumous Work of the Late M. de Condorcet*. Philadelphia: M. Carey, H. & P. Rice & Co. J. Ormrod, B. F. Bache, and J. Fellows. http://oll.libertyfund.org/titles/1669.

Coolidge, Frederick L., and Thomas Wynn (2009). *The Rise of Homo Sapiens: The Evolution of Modern Thinking*. Chichester: Wiley-Blackwell.

Coop, Graham, Joseph K. Pickrell, John Novembre, Sridhar Kudaravalli, Jun Li, Devin Absher, Richard Myers, Luigi Luca Cavalli-Sforza, Marcus W. Feldman, and Jonathan K. Pritchard (2009). "The Role of Geography in Human Adaptation." *PLOS Genetics* 5 (6): e1000500.

Corazzi, Roberto, and Giuseppe Conti (2006). *La cupola di Santa Maria del Fiore a Firenze: Il rilievo fotogrammetrico*. Livorno: Sillabe Casa Editrice.

———(2011). *Il segreto della cupola del Brunelleschi a Firenze*. Florence: Angelo Pontecorboli Editore.

Corballis, Michael C. (2003). *From Hand to Mouth: The Origins of Language*. Princeton, NJ: Princeton University Press.

———(2011). *The Recursive Mind: The Origins of Human Language, Thought, and Civilization*. Princeton, NJ: Princeton University Press.

Costanza, Robert, Lisa J. Graumlich, and Steffen Will, eds. (2007). *Sustainability or Collapse? An Integrated History and Future of People on Earth*. Cambridge, MA: MIT Press.

Craik, Kenneth (1943). *The Nature of Explanation*. Cambridge: Cambridge University Press.

Cripps, Eric L. (2014). "Money and Prices in the Ur III Economy of Umma." Review of *Monetary Role of Silver and Its Administration in Mesopotamia during the Ur III Period (c. 2112–2004 BCE): A Case Study of the Umma Province*, by Xiaoli Ouyang. *Wiener Zeitschrift für die Kunde des Morgenlandes* 104:205–32.

Cronin, Michael (2003). *Translation and Globalization*. London: Routledge.

Crookes, William (1898). "Address of the President before the British Association for the Advancement of Science, Bristol, 1898." *Science* 8 (200): 561–75.

Crosby, Alfred W. (1993). "The Columbian Voyages, the Columbian Exchange, and Their Historians." In *Islamic and European Expansion: The Forging of a Global Order*, ed. M. Adas, 141–64. Philadelphia: Temple University Press.

———(2003). *The Columbian Exchange: Biological and Cultural Consequences of 1492*. 30th anniversary ed. Westport, CT: Praeger.

Crowe, Michael J. (1999). *The Extraterrestrial Life Debate, 1750–1900*. Mineola, NY: Dover.

Crutzen, Paul J. (1970). "The Influence of Nitrogen Oxides on the Atmospheric Ozone Content." *Quarterly Journal of the Royal Metrological Society* 96 (408): 320–25.

———(1997). "My Life with O_3, NO_x and Other YZO_xs." In *Nobel Lectures: Chemistry, 1991–1995*, ed. B. G. Malmström, 189–242. Singapore: World Scientific Publishing.

———(2002). "Geology of Mankind." *Nature* 415 (6867): 23.

———(2017). "Foreword: Transition to a Safe Anthropocene." In *Well Under 2 Degrees Celsius: Fast Action Policies to Protect People and the Planet from Extreme Climate Change*, ed. V. Ramanathan, M. J. Molina, D. Zaelke, N. Borgford-Parnell, Y. Xu, K. Alex, M. Auffhammer, P. Bledsoe, W. Collins, B. Croes, F. Forman, Ö. Gustafsson, A. Haines, R. Harnish, M. Z. Jacobson, S. Kang, M. Lawrence, D. Leloup, T. Lenton, T. Morehouse, W. Munk, R. Picolotti, K. Prather, G. Raga, E. Rignot, D. Shindell, A. K. Singh, A. Steiner, M. Thiemens, D. W. Titley, M. E. Tucker, S. Tripathi, and D. Victor, 3–4. Washington, DC: Institute of Governance and Sustainable Development.

Crutzen, Paul J., and Eugene F. Stoermer (2000). "The 'Anthropocene.'" *Global Change Newsletter* 41:17–18.

Cultures of Knowledge (n.d.). "Early Modern Letters Online." Accessed January 31, 2019. http://emlo.bodleian.ox.ac.uk.

Cunningham, Andrew, and Perry Williams (1993). "De-centring the 'Big Picture': The Origins of Modern Science and the Modern Origins of Science." *British Journal for the History of Science* 26 (4): 407–32.

Cuomo, Serafina (2000). *Pappus of Alexandria and the Mathematics of Late Antiquity*. Cambridge: Cambridge University Press.

Cushman, Gregory T. (2013). *Guano and the Opening of the Pacific World: A Global Ecological History*. Cambridge: Cambridge University Press.

D'Abramo, Flavio, and Hannah Landecker. 2019. "Anthropocene in the Cell." *Technosphere Magazine*: https://technosphere-magazine.hkw.de/p/Anthropocene-in-the-Cell-fQjoLLgrE7jbXzLYr1TLNn.

Dahan, Amy (2010). "Putting the Earth System in a Numerical Box? The Evolution from Climate Modeling toward Global Change." *Studies in History and Philosophy of Science Part B: Studies in History and Philosophy of Modern Physics* 41 (3): 282–92.

Dalla Negra, Riccardo (1995). "La cupola del Brunelleschi: Il cantiere, le indagini, i rilievi." In *Cupola di Santa Maria del Fiore: Il cantiere di restauro 1980–1995*, ed. C. Acidini Luchinat and R. Della Negra, 1–45. Rome: Istituto Poligrafico e Zecca dello Stato.

Dalrymple, Gary Brent (2004). *Ancient Earth, Ancient Skies: The Age of Earth and Its Cosmic Surroundings*. Stanford, CA: Stanford University Press.

Dalton, John (1810). *A New System of Chemical Philosophy*. Part 2. Manchester: Russell and Allen.

Damerow, Peter (1981). "Die Entstehung des arithmetischen Denkens: Zur Rolle der Rechenmittel in der altägyptischen und der altbabylonischen Arithmetik." In *Rechenstein, Experiment, Sprache: Historische Fallstudien zur Entstehung der exakten Wissenschaften*, ed. P. Damerow and W. Lefèvre, 11–113. Stuttgart: Klett-Cotta.

———(1994). "Albert Einstein e Max Wertheimer." In *L'eredità di Einstein*, ed. G. Pisent and J. Renn, 43–60. Padua: Il Poligrafo.

———(1996a). *Abstraction and Representation: Essays on the Cultural Evolution of Thinking*. Translated by Renate Hanauer. Boston Studies in the Philosophy and History of Science 175. Dordrecht: Kluwer Academic.

———(1996b). "Tools of Science." In *Abstraction and Representation: Essays on the Cultural Evolution of Thinking*, ed. P. Damerow, 395–404. Dordrecht: Kluwer Academic.

———(1998). "Prehistory and Cognitive Development." In *Piaget, Evolution, and Development*, ed. J. Langer and M. Killen, 247–69. Mahwah, NJ: Erlbaum.

————(1999). *The Material Culture of Calculation: A Conceptual Framework for an Historical Epistemology of the Concept of Number*. Preprint 117. Berlin: Max Planck Institute for the History of Science.

————(2000). "How Can Discontinuities in Evolution Be Conceptualized?" *Cultural Psychology* 6 (2): 155–60.

————(2006). "The Origins of Writing as a Problem of Historical Epistemology." *Cuneiform Digital Library Journal* 1.

————(2010). "From Numerate Apprenticeship to Divine Quantification." Review of *Mathematics in Ancient Iraq: A Social History*, by Eleanor Robson. *Notices of the American Mathematical Society* 57 (3): 380–84.

————(2012). "The Origins of Writing and Arithmetic." In *The Globalization of Knowledge in History*, ed. J. Renn, 153–73. Studies 1. Berlin: Edition Open Access. http://edition-open -access.de/studies/1/10/index.html.

Damerow, Peter, and Robert K. Englund (1987). "Die Zahlzeichensysteme der Archaischen Texte aus Uruk." In *Zeichenliste der Archaischen Texte aus Uruk*, ed. M. W. Green and H. J. Nissen, 117–66. Archaische Texte aus Uruk 2. Berlin: Gebr. Mann Verlag.

Damerow, Peter, Gideon Freudenthal, Peter McLaughlin, and Jürgen Renn (2004). *Exploring the Limits of Preclassical Mechanics: A Study of Conceptual Development in Early Modern Science—Free Fall and Compounded Motion in the Work of Descartes, Galileo, and Beeckman*. 2nd ed. Sources and Studies in the History of Mathematics and Physical Sciences. New York: Springer.

Damerow, Peter, and Wolfgang Lefèvre (1981). "Arbeitsmittel der Wissenschaft: Nachbemerkung zur Theorie der Wissenschaftsentwicklung." In *Rechenstein, Experiment, Sprache: Historische Fallstudien zur Entstehung der exakten Wissenschaften*, ed. P. Damerow and W. Lefèvre, 223–33. Stuttgart: Klett-Cotta.

————(1998). "Wissenssysteme im geschichtlichen Wandel." In *Enzyklopädie der Psychologie: Themenbereich C: Theorie und Forschung*. Series 2: *Kognition*. Vol. 6: *Wissen*, ed. F. Klix and H. Spada, 77–113. Göttingen: Hogrefe Verlag.

Damerow, Peter, Jürgen Renn, Simone Rieger, and Paul Weinig (2002). "Mechanical Knowledge and Pompeian Balances." In *Homo Faber: Studies on Nature, Technology, and Science at the Time of Pompeii*, ed. J. Renn and G. Castagnetti, 93–108. Studi della Soprintendenza Archeologica di Pompei 6. Rome: L'Erma di Bretschneider.

Dana, James Dwight (1873). "On Some Results of the Earth's Contraction from Cooling, Including a Discussion of the Origin of Mountains, and the Nature of the Earth's Interior." *American Journal of Science* 3/5 (105): 423–43.

Darwin, Charles (1859). *On the Origin of Species by Means of Natural Selection, or the Preservation of Favoured Races in the Struggle for Life*. London: John Murray. https://en .wikisource.org/wiki/On_the_Origin_of_Species_(1859).

————([1865] 2002). "To A. R. Wallace 22 September." In *The Correspondence of Charles Darwin*. Vol. 13: *1865—Supplement to the Correspondence 1822–1864*, ed. F. Burkhardt and D. M. Porter, 237–39. Cambridge: Cambridge University Press. http://www .darwinproject.ac.uk/DCP-LETT-4896.

————([1876] 1958). "Recollections of the Development of My Mind and Character." In *The Autobiography of Charles Darwin, 1809–1882*, ed. N. Barlow, 17–145. London: Collins. http://darwin-online.org.uk/content/frameset?itemID=F1497&viewtype=text& pageseq=1.

————(1877). *The Descent of Man: Selection in Relation to Sex*. Rev. and aug. 2nd ed. London: John Murray.

Daston, Lorraine, and Peter Galison (2007). *Objectivity*. New York: Zone Books.

Daston, Lorraine, and H. Otto Sibum, eds. (2003). "Scientific Personae and Their Histories." Special issue of *Science in Context* 16 (1–2).

Davies, Jeremy (2016). *The Birth of the Anthropocene*. Oakland: University of California Press.

Davis, Robert B. (1984). *Learning Mathematics: The Cognitive Science Approach to Mathematics Education*. Norwood, NJ: Ablex Publishing.

Dawkins, Richard (1976). *The Selfish Gene*. Oxford: Oxford University Press.

———(1982). *The Extended Phenotype: The Long Reach of the Gene*. Oxford: Oxford University Press.

Deacon, Terrence W. (1997). *The Symbolic Species: The Co-evolution of Language and the Human Brain*. New York: W. W. Norton.

de Bry, Theodor (1613). *Das vierdte Buch von der Neuwen Welt: Oder neuwe und gründtliche Historien, von dem Nidergängischen Indien, so von Christophoro Columbo im Jar 1492. erstlich erfunden*. Frankfurt am Main: Dietrichs von Bry. https://archive.org/details/dasvierdtebuchvooobenz_1.

Dediu, Dan, and Stephen C. Levinson (2013). "On the Antiquity of Language: The Reinterpretation of Neandertal Linguistic Capacities and Its Consequences." *Frontiers in Psychology* 4 (397): 1–17.

Delanty, Gerard, and Aurea Mota (2017). "Governing the Anthropocene: Agency, Governance, Knowledge." *European Journal of Social Theory* 20 (1): 9–38.

De Risi, Vincenzo (2015). "Introduction." In *Mathematizing Space: The Objects of Geometry from Antiquity to the Early Modern Age*, ed. V. De Risi. Basel: Birkhäuser.

Descartes, René (1664). *Le monde de Mr. Descartes, ou Le traité de la lumière et des autres principaux objets des sens: Avec un discours de l'action des corps, & un autre des fièvres, composez selon les principes du même auteur*. Paris: Theodore Girard. https://gallica.bnf.fr/ark:/12148/bpt6k5534491g.texteImage.

———(1985a). "Discourse on the Method of Rightly Conducting One's Reason and Seeking Truth in the Sciences." In *The Philosophical Writings of Descartes*. Vol. 1, 111–51. Cambridge: Cambridge University Press.

———(1985b). "Principles of Philosophy (1644)." In *The Philosophical Writings of Descartes*, 1:177–291. Cambridge: Cambridge University Press.

———(1986). "Le monde ou Le traité de la lumière." In *Œuvres de Descartes*. Vol. 11, ed. C. Adam and P. Tannery, 11:1–118. Paris: Librairie Philosophique J. Vrin.

Devlin, William J., and Alisa Bokulich (2015). *Kuhn's Structure of Scientific Revolutions: 50 Years On*. Cham: Springer.

Dewey, John ([1895–98] 1972). *The Early Works, 1882–1898*. Vol. 5: *1895–1898: Early Essays*, ed. J. A. Boydston. Carbondale: Southern Illinois University Press.

———(1897). *The Significance of the Problem of Knowledge*. University of Chicago Contributions to Philosophy 1, 3. Chicago: University of Chicago Press.

———([1925] 1981). *The Later Works, 1925–1953*. Vol. 1: *1925: Experience and Nature*, ed. J. A. Boydston. Carbondale: Southern Illinois University Press.

Diamond, Jared M. (2005a). *Collapse: How Societies Choose to Fail or Succeed*. New York: Viking Press.

———(2005b). *Guns, Germs, and Steel: The Fates of Human Societies*. Rev. ed. New York: W. W. Norton.

Diawara, Mamadou (2004). "Colonial Appropriation of Local Knowledge." In *Between Resistance and Expansion: Explorations of Local Vitality in Africa*, ed. P. Probst and G. Spittler, 273–93. Münster: LIT Verlag.

Dicke, Robert H., Phillip James E. Peebles, P. G. Roll, and David T. Wilkinson (1965). "Cosmic Black-Body Radiation." *Astrophysical Journal* 142 (1): 414–19.

Diderot, Denis, and Jean Le Rond d'Alembert, eds. (1751–72). *Encyclopédie ou Dictionnaire raisonné des sciences, des arts et des métiers, par une Société de Gens de lettres*. Paris: Briasson, David, Le Breton, Durand.

Dijksterhuis, Eduard Jan ([1950] 1986). *The Mechanization of the World Picture: Pythagoras to Newton*. Translated by C. Dikshoorn. Princeton, NJ: Princeton University Press.

———(1956). *Archimedes*. Copenhagen: Ejnar Munksgaard.

Dilger, Hansjörg (2012). "The (Ir)Relevance of Local Knowledge: Circuits of Medicine and Biopower in the Neoliberal Era." In *The Globalization of Knowledge in History*, ed. J. Renn, 501–24. Studies 1. Berlin: Edition Open Access. http://edition-open-access.de/studies/1/26/index.html.

Diogenes Laertius (1925). "Book 6.2: Diogenes (404–323 B.C.)." In *Lives of the Eminent Philosophers. Vol.2: Books 6–10*, 22–85. Loeb Classical Library 185. Cambridge, MA: Harvard University Press.

Di Pasquale, Giovanni (2005). "The Museum of Alexandria: Myth and Model." In *From Private to Public: Natural Collections and Museums*, ed. M. Beretta, 1–12. New York: Science History Publications.

———(2007). "Una enciclopedia delle tecniche nel Museo di Alessandria." In *Il giardino antico da Babilonia a Roma*, ed. G. Di Pasquale and F. Paolucci, 58–71. Livorno: Sillabe Casa Editrice.

———(2010). "The 'Syrakousia' Ship and the Mechanical Knowledge between Syracuse and Alexandria." In *The Genius of Archimedes: 23 Centuries of Influence on Mathematics, Science and Engineering—Proceedings of an International Conference, Syracuse, June 8–10, 2010*, ed. S. A. Paipetis and M. Ceccarelli, 289–301. History of Mechanism and Machine Science 11. Dordrecht: Springer.

———(2012). *Le strade della tecnica: Tecnologia e pratica della scienza nel mondo antico*. Florence: Centro Di.

———(2013). "From Syracuse to Alexandria: A Technological Network in the Mediterranean." In *Archimedes: The Art and Science of Invention*, ed. G. Di Pasquale and C. Parisi Presicce, 77–82. Florence: Giunti.

Di Pasquale, Salvatore (1996). *L'arte del construire: Tra conoscenza e scienza*. Venice: Marsilio Editori.

———(2002). *Brunelleschi: La costruzione della cupola di Santa Maria del Fiore*. Venice: Marsilio Editori.

Dirks, Nicholas B. (2001). *Castes of Mind: Colonialism and the Making of Modern India*. Princeton, NJ: Princeton University Press.

Dirlik, Arif (1995). "Confucius in the Borderlands: Global Capitalism and the Reinvention of Confucianism." *boundary 2* 22 (3): 229–73.

Dittmer, Jörg (1999). "Jaspers' 'Achsenzeit' und das interkulturelle Gespräch: Überlegungen zur Relevanz eines revidierten Theorems." In *Globaler Kampf der Kulturen? Analysen und Orientierungen*, ed. D. Becker, 191–214. Stuttgart: W. Kohlhammer.

Doel, Ronald E. (1997). "The Earth Sciences and Geophysics." In *Science in the Twentieth Century*, ed. J. Krige and D. Pestre, 391–416. Amsterdam: Harwood Academic.

Donges, Jonathan F., Wolfgang Lucht, Finn Müller-Hansen, and Will Steffen (2017). "The Technosphere in Earth System Analysis: A Coevolutionary Perspective." *Anthropocene Review* 4 (1): 23–33.

Donges, Jonathan F., Ricarda Winkelmann, Wolfgang Lucht, Sarah E. Cornell, James G. Dyke, Johan Rockström, Jobst Heitzig, and Hans Joachim Schellnhuber (2017). "Closing the Loop: Reconnecting Human Dynamics to Earth System Science." *Anthropocene Review* 4 (2): 151–57.

Dor, Daniel, Chris Knight, and Jerome Lewis, eds. (2014). *The Social Origins of Language*. Oxford: Oxford University Press.

Doren, Alfred (1898). "Zum Bau der Florentiner Domkuppel." *Repertorium für Kunstwissenschaft* 21:249–62.

Douglas, Mary (1986). *How Institutions Think*. Syracuse, NY: Syracuse University Press.

Drixler, Fabian (2013). *Mabiki: Infanticide and Population Growth in Eastern Japan, 1660–1950*. Berkeley: University of California Press.

Dubois, Jacques, Jean Marie Guillouët, Benoît Van den Bossche, and Annamaria Ersek (2014). *Les transferts artistiques dans l'Europe gothique: Repenser les circulations des hommes, des œuvres, des savoir-faire et des modèles (XIIe–XVIe siècle)*. Paris: Picard.

Du Halde, Jean Baptiste (1735). *Description géographique, historique, chronologique, politique et physique de l'empire de la Chine et de la Tartarie chinoise*. Vol. 3 of 4 vols. Paris: P. G. Lemercier. https://gallica.bnf.fr/ark:/12148/bpt6k5699174c.

Duncan, Anthony, and Michel Janssen, eds. (forthcoming 2019). *Constructing Quantum Mechanics*. Vol. 1: *The Scaffold: 1900–1923*. Oxford: Oxford University Press.

Düring, Marten (2017). "Historical Network Research: Network Analysis in the Historical Disciplines." Accessed March 1, 2018. http://historicalnetworkresearch.org.

Durkheim, Émile ([1895] 1966). *The Rules of Sociological Method*. Translated by Sarah A. Solovay and John H. Mueller. 8th ed. New York: Free Press.

du Toit, Alexander Logie (1937). *Our Wandering Continents*. Edinburgh: Oliver and Boyd.

Duve, Thomas (2016). "Global Legal History: A Methodological Approach." Oxford: Oxford University Press. Accessed June 27, 2018. http://www.oxfordhandbooks.com/view/10.1093/oxfordhb/9780199935352.001.0001/oxfordhb-9780199935352-e-25.

Dyson, Frank W., Arthur S. Eddington, and Charles Davidson (1920). "IX. A Determination of the Deflection of Light by the Sun's Gravitational Field, from Observations Made at the Total Eclipse of May 29, 1919." *Philosophical Transactions of the Royal Society A: Mathematical, Physical and Engineering Sciences* 220 (571–81): 291–333.

Earl, Jennifer, and Katrina Kimport (2011). *Digitally Enabled Social Change: Activism in the Internet Age*. Cambridge, MA: MIT Press.

Easterly, William Russell (2006). *The White Man's Burden: Why the West's Efforts to Aid the Rest Have Done So Much Ill and So Little Good*. Oxford: Oxford University Press.

Edgerton, David (2006). *The Shock of the Old: Technology and Global History since 1900*. London: Profile Books.

———(2007). "Creole Technologies and Global Histories: Rethinking How Things Travel in Space and Time." *HoST: Journal of History of Science and Technology* 1:75–112.

Edgeworth, Matt, Daniel D. Richter, Colin Waters, Peter Haff, Cath Neal, and Simon James Price (2015). "Diachronous Beginnings of the Anthropocene: The Lower Bounding Surface of Anthropogenic Deposits." *Anthropocene Review* 2 (1): 33–58.

Edwards, Paul N. (2010). *A Vast Machine: Computer Models, Climate Data, and the Politics of Global Warming*. Cambridge, MA: MIT Press.

———(2017). "Knowledge Infrastructures for the Anthropocene." *Anthropocene Review* 4 (1): 34–43.

Eibl-Eibesfeldt, Irenäus, Wulf Schiefenhövel, and Volker Heeschen (1989). *Kommunikation bei den Eipo: Eine humanethologische Bestandsaufnahme*. Berlin: Dietrich Reimer Verlag.

Einstein, Albert (1905). "Zur Elektrodynamik bewegter Körper." *Annalen der Physik* 322 (10): 891–921.

———(1907). "Über das Relativitätsprinzip und die aus demselben gezogenen Folgerungen." *Jahrbuch der Radioaktivität und Elektronik* 4:411–62.

————(1915a). "Die Feldgleichungen der Gravitation." *Sitzungsberichte der Königlich Preussischen Akademie der Wissenschaften* 48:844–47.

————(1915b). "Zur allgemeinen Relativitätstheorie." *Sitzungsberichte der Königlich Preussischen Akademie der Wissenschaften* 44:778–86.

————(1987). *Letters to Solovine: 1906–1955.* Translated by Wade Baskin. New York: Philosophical Library.

————(1989a). "On the Electrodynamics of Moving Bodies." 1905. In *The Collected Papers of Albert Einstein.* Vol. 2: *The Swiss Years: Writings, 1900–1909; English Translation,* trans. Anna Beck, 140–71. Princeton, NJ: Princeton University Press. https://einsteinpapers .press.princeton.edu/vol2-trans/154.

————(1989b). "On the Relativity Principle and the Conclusions Drawn from It." 1907. In *The Collected Papers of Albert Einstein.* Vol. 2: *The Swiss Years: Writings, 1900–1909; English Translation,* trans. Anna Beck, 252–311. Princeton, NJ: Princeton University Press. https://einsteinpapers.press.princeton.edu/vol2-trans/266.

————(1991). *Autobiographical Notes.* Translated by Paul Arthur Schilpp. 2nd ed. LaSalle, IL: Open Court Publishing.

————(1995). "Research Notes on a Generalized Theory of Relativity." Ca. August 1912. In *The Collected Papers of Albert Einstein.* Vol. 4: *The Swiss Years: Writings, 1912–1914,* ed. M. J. Klein, A. J. Kox, J. Renn, and R. Schulman, 201–69. Princeton, NJ: Princeton University Press. https://einsteinpapers.press.princeton.edu/vol4-doc/223.

————(1996a). "On the General Theory of Relativity." 1915. In *The Collected Papers of Albert Einstein.* Vol. 6: *The Berlin Years: Writings, 1914–1917; English Translation,* trans. Alfred Engel, 98–107. Princeton, NJ: Princeton University Press. https://einsteinpapers.press .princeton.edu/vol6-trans/110.

————(1996b). "The Field Equations of Gravitation." 1915. In *The Collected Papers of Albert Einstein.* Vol. 6: *The Berlin Years: Writings, 1914–1917; English Translation,* trans. Alfred Engel, 117–20. Princeton, NJ: Princeton University Press. https://einsteinpapers.press.princeton .edu/vol6-trans/129.

Einstein, Albert, and Marcel Grossmann (1913). *Entwurf einer verallgemeinerten Relativitätstheorie und einer Theorie der Gravitation.* Leipzig: Teubner. https://einsteinpapers.press .princeton.edu/vol4-doc/324.

————(1995). "Outline of a Generalized Theory of Relativity and of a Theory of Gravitation." 1913. In *The Collected Papers of Albert Einstein.* Vol. 4: *The Swiss Years: Writings, 1912–1914; English Translation,* trans. Anna Beck, 151–88. Princeton, NJ: Princeton University Press. https://einsteinpapers.press.princeton.edu/vol4-trans/163.

Eisenhower, Dwight D. (1961). "Farewell Radio and Television Address to the American People: January 17, 1961." In *Dwight D. Eisenhower: 1960–1961: Containing the Public Messages, Speeches, and Statements of the President, January 1, 1960, to January 20, 1961,* 1035–40. Public Papers of the Presidents of the United States. Washington, DC: Office of the Federal Register, National Archives and Records Service, General Services Administration. http://name.umdl.umich.edu/4728424.1960.001.

Eisenstadt, Shmuel N. (1982). "The Axial Age: The Emergence of Transcendental Visions and the Rise of Clerics." *European Journal of Sociology—Archives Européennes de Sociologie* 23 (2): 294–314.

————, ed. (1986). *The Origins and Diversity of Axial Age Civilizations.* SUNY Series in Near Eastern Studies. Albany: State University of New York Press.

————(1988). "Explorations in the Sociology of Knowledge: The Soteriological Axis in the Construction of Domains of Knowledge." In *Cultural Traditions and Worlds of Knowledge: Explorations in the Sociology of Knowledge,* ed. S. N. Eisenstadt and I. Friedrich-Silber, 1–71.

Knowledge and Society: Studies in the Sociology of Culture Past and Present 7. Greenwich, CT: JAI Press.

——(2000). "Multiple Modernities." *Daedalus* 129 (1): 1–29.

——, ed. (2002). *Multiple Modernities*. New Brunswick, NJ: Transaction.

Eisenstaedt, Jean (1986). "La relativité générale à l'étiage: 1925–1955." *Archive for History of Exact Sciences* 35 (2): 115–85.

——(1989). "The Low Water Mark of General Relativity: 1925–1955." In *Einstein and the History of General Relativity*, ed. D. Howard and J. Stachel, 1–277. Einstein Studies 1. Basel: Birkhäuser.

——(2006). *The Curious History of Relativity: How Einstein's Theory of Gravity Was Lost and Found Again*. Princeton, NJ: Princeton University Press.

Elias, Norbert ([1969] 2006). *The Court Society*. Vol. 2 of *The Collected Works of Norbert Elias*. Translated by Edmund Jephcott. Rev. ed. Dublin: University College Dublin Press.

——([1984] 2007). *An Essay on Time*. Vol. 9 of *The Collected Works of Norbert Elias*. Dublin: University College Dublin Press.

Elkana, Yehuda (1970). "Helmholtz' 'Kraft': An Illustration of Concepts in Flux." *Historical Studies in the Physical Sciences* 2:263–98.

——(1974). *The Discovery of the Conservation of Energy*. Cambridge, MA: Harvard University Press.

——(1975). "Boltzmann's Scientific Research Program and Its Alternatives." In *The Interaction between Science and Philosophy*, ed. Y. Elkana, 243–79. Atlantic Highlands, NJ: Humanities Press.

——(1981). "A Programmatic Attempt at an Anthropology of Knowledge." In *Sciences and Cultures: Anthropological and Historical Studies of the Sciences*, ed. E. Mendelsohn and Y. Elkana, 1–76. Sociology of the Sciences 5. Dordrecht: D. Reidel Publishing.

——(1986a). *Anthropologie der Erkenntnis: Die Entwicklung des Wissens als episches Theater einer listigen Vernunft*. Translated by Ruth Achlama. Wissenschaftsforschung 1. Frankfurt am Main: Suhrkamp.

——(1986b). "The Emergence of Second-Order Thinking in Classical Greece." In *The Origins and Diversity of Axial Age Civilizations*, ed. S. N. Eisenstadt, 40–64. Albany, NY: SUNY Press.

——(2012). "The University of the 21st Century: An Aspect of Globalization." In *The Globalization of Knowledge in History*, ed. J. Renn, 605–30. Studies 1. Berlin: Edition Open Access. http://edition-open-access.de/studies/1/29/index.html.

Elman, Benjamin A. (2005). *On Their Own Terms: Science in China, 1550–1900*. Cambridge, MA: Harvard University Press.

Elshakry, Marwa (2010). "When Science Became Western: Historiographical Reflections." *Isis* 101 (1) :98–109.

Engel, Christoph (2002). *Abfallrecht und Abfallpolitik*. Baden-Baden: Nomos Verlagsgesellschaft.

Engels, Frederick ([1883] 1987). "Dialectics of Nature." In *Karl Marx / Frederick Engels: Collected Works*. Vol. 25: *Engels: Anti-Dühring - Dialectics of Nature*, 311–588. London: Lawrence & Wishart. First published in its entirety in Russian and German in 1925.

Engler, Fynn Ole, and Jürgen Renn (2016). "Two Encounters." In *Shifting Paradigms: Thomas S. Kuhn and the History of Science*, ed. A. Blum, K. Gavroglu, C. Joas, and J. Renn, 139–47. Proceedings 8. Berlin: Edition Open Access. http://edition-open-access.de/proceedings/8/11/index.html.

——(2018). *Gespaltene Vernunft: Vom Ende eines Dialogs zwischen Wissenschaft und Philosophie*. Berlin: Matthes & Seitz.

Engler, Fynn Ole, Jürgen Renn, and Matthias Schemmel (2018). "Creating Room for Historical Rationality." *Isis* 109 (1): 87–91.

Englund, Robert K. (1991). "Hard Work: Where Will It Get You? Labor Management in Ur III Mesopotamia." *Journal of Near Eastern Studies* 50 (4): 255–80.

———(1998). "Texts from the Late Uruk Period." in *Mesopotamien: Späturuk-Zeit und Frühdynastische Zeit*, ed. J. Bauer, R. K. Englund, and M. Krebernik, 13–233. Fribourg: Academic Press.

———(2004). "The State of Decipherment of Proto-Elamite." In *The First Writing: Script Invention as History and Process*, ed. S. D. Houston, 100–149. Cambridge: Cambridge University Press.

———(2006). "An Examination of the 'Textual' Witnesses to Late Uruk World Systems." In "A Collection of Papers on Ancient Civilizations of Western Asia, Asia Minor and North Africa," ed. Y. Gong and Y. Chen. Special issue of *Oriental Studies* (Beijing: University of Beijing): 1–38.

———(2012). "Equivalency Values and the Command Economy of the Ur III Period in Mesopotamia." In *The Construction of Value in the Ancient World*, ed. J. Papadopoulos and G. Urton, 427–58. Los Angeles: Cotsen Institute of Archaeology Press.

Ertl, Gerhard, and Jens Soentgen, eds. (2015). *N: Stickstoff—Ein Element schreibt Weltgeschichte*. Stoffgeschichten 9. Munich: Oekom Verlag.

Euclid (1956). *The Thirteen Books of the Elements*. Translated by Thomas L. Heath. 2nd rev. with add. ed. New York: Dover.

European Parliament (2000). "Lisbon European Council 23 and 24 March 2000: Presidency Conclusions." Accessed February 21, 2018. http://www.europarl.europa.eu/summits/lis1_en.htm.

Evans, Chris, and Göran Rydén (2005). *The Industrial Revolution in Iron: The Impact of British Coal Technology in Nineteenth-Century Europe*. Aldershot: Ashgate.

Evans, Dave (2011). "The Internet of Things: How the Next Evolution of the Internet Is Changing Everything." CISCO white paper. Accessed October 15, 2018. http://www.cisco.com/web/about/ac79/docs/innov/IoT_IBSG_0411FINAL.pdf.

Ezrahi, Yaron (1990). *The Descent of Icarus: Science and the Transformation of Contemporary Democracy*. Cambridge, MA: Harvard University Press.

Fabbri, Lorenzo (2003). "La 'Gabella di Santa Maria del Fiore': Il finanziamento pubblico della cattedrale di Firenze." In *Pouvoir et édilité: Les grands chantiers dans l'Italie communale et seigneuriale*, ed. É. Crouzet-Pavan, 195–244. Collection de l'École française de Rome 302. Rome: École française de Rome.

Falkowski, Paul, Robert J. Scholes, Edward A. Boyle, Josep Canadell, Don Canfield, James Elser, Nicolas Gruber, Kathy Hibbard, Peter Högberg, Sune Linder, Fred T. Mackenzie, Berrien Moore III, Thomas Pedersen, Yair Rosenthal, Sybil Seitzinger, Victor Smetacek, and Will Steffen (2000). "The Global Carbon Cycle: A Test of Our Knowledge of Earth as a System." *Science* 290 (5490): 291–96.

Fanelli, Giovanni, and Michele Fanelli (2004). *La cupola del Brunelleschi: Storia e futuro di una grande struttura*. Florence: Editrice La Mandragora.

Fangerau, Heiner (2013). "Evolution of Knowledge from a Network Perspective: Recognition as a Selective Factor in the History of Science." In *Classification and Evolution in Biology, Linguistics and the History of Science: Concepts, Methods, Visualization*, ed. H. Fangerau, H. Geisler, T. Halling, and W. Martin, 11–32. Kulturanamnesen 5. Stuttgart: Franz Steiner Verlag.

Fanon, Frantz ([1961] 2005). *The Wretched of the Earth*. Translated by Richard Philcox. New York: Grove Press.

Farbwerke Hoechst AG, ed. (1964). *Wilhelm Ostwald und die Stickstoffgewinnung aus der Luft.* Dokumente aus Hoechster Archiven 5. Frankfurt am Main: Farbwerke Hoechst AG.

Farey, John ([1827] 1971). *A Treatise on the Steam Engine: Historical, Practical and Descriptive.* Reprint, Newton Abbot: David & Charles.

Farrell, Justin, Kathryn McConnell, and Robert Brulle (2019). "Evidence-Based Strategies to Combat Scientific Misinformation." *Nature Climate Change* 9 (3): 191–95.

Fauerbach, Ulrike (2014). "Bauwissen im Alten Ägypten." In *Wissensgeschichte der Architektur.* Vol. 2: *Vom Alten Ägypten bis zum Antiken Rom*, ed. J. Renn, W. Osthues, and H. Schlimme, 7–124. Studies 4. Berlin: Edition Open Access. http://edition-open-access.de/studies/4/3/index.html.

Favaro, Antonio, ed. (1964–66). *Le opera di Galileo Galilei: Nuova ristampa della edizione nazionale 1890–1909.* 20 vols. Florence: Barbèra.

Febvre, Lucien Paul Victor, and Henri-Jean Martin (1990). *The Coming of the Book: The Impact of Printing, 1450–1800.* London: Verso Books.

Feldhay, Rivka (2006). "Religion." In *Early Modern Science*, ed. K. Park and L. Daston, 727–55. Cambridge History of Science 3. Cambridge: Cambridge University Press.

———(2011). "The Jesuits: Transmitters of the New Science." In *Il caso Galileo: Una rilettura storica, filosofica, teologica*, ed. M. Bucciantini, M. Camerota, and F. Giudice, 47–74. Florence: Leo S. Olschki.

———(2018). "Pre-classical Mechanics in Context: Practical and Theoretical Knowledge between Sovereignty, Religion, and Science." In *Emergence and Expansion of Pre-classical Mechanics*, ed. R. Feldhay, J. Renn, M. Schemmel, and M. Valleriani, 29–53. Boston Studies in the Philosophy and History of Science 333. Cham: Springer.

Feldhay, Rivka, and F. Jamil Ragep (2017). *Before Copernicus: The Cultures and Contexts of Scientific Learning in the Fifteenth Century.* Montreal: McGill-Queen's University Press.

Feldhay, Rivka, Jürgen Renn, Matthias Schemmel, and Matteo Valleriani, eds. (2018). *Emergence and Expansion of Pre-classical Mechanics.* Boston Studies in the Philosophy and History of Science 333. Cham: Springer.

Ferguson, Dean T. (2014). "Nightsoil and the 'Great Divergence': Human Waste, the Urban Economy, and Economic Productivity, 1500–1900." *Journal of Global History* 9 (3): 379–402.

Feynman, Richard (1967). *The Character of Physical Law.* Cambridge, MA: MIT Press.

Fischer-Kowalski, Marina, Fridolin Krausmann, and Irene Pallua (2014). "A Sociometabolic Reading of the Anthropocene: Modes of Subsistence, Population Size and Human Impact on Earth." *Anthropocene Review* 1 (1): 8–33.

Flaxman, Seth, Sharad Goel, and Justin M. Rao (2016). "Filter Bubbles, Echo Chambers, and Online News Consumption." *Public Opinion Quarterly* 80 (S1): 298–320.

Fleck, Ludwik ([1935] 1979). *Genesis and Development of a Scientific Fact.* Translated by Fred Bradley and Thaddeus J. Trenn. Chicago: University of Chicago Press.

———([1935] 1980). *Entstehung und Entwicklung einer wissenschaftlichen Tatsache: Einführung in die Lehre vom Denkstil und Denkkollektiv.* Reprint, Frankfurt am Main: Suhrkamp.

———([1960] 1986). "Crisis in Science." In *Cognition and Fact: Materials on Ludwik Fleck*, ed. R. S. Cohen and T. Schnelle, 153–58. Dordrecht: D. Reidel.

Flexner, Abraham (1939). "The Usefulness of Useless Knowledge." *Harper's Magazine* 179:544–52.

———([1939] 1997). *The Usefulness of Useless Knowledge: With a Companion Essay by Robbert Dijkgraaf.* Princeton, NJ: Princeton University Press.

Force, James E. (1999). "Newton, the 'Ancients,' and the 'Moderns.'" In *Newton and Religion: Context, Nature, and Influence*, ed. J. E. Force and R. H. Popkin, 237–57. Dordrecht: Springer.

Forrester, John (1996). "If P, Then What? Thinking in Cases." *History of the Human Sciences* 9 (3): 1–25.

Foucault, Michel (1983). *This Is Not a Pipe: With Illustrations and Letters by René Magritte*. Berkeley: University of California Press.

Foucher, Alfred (1950). "Le cheval de Troie au Gandhâra." *Comptes rendus des séances de l'Académie des Inscriptions et Belles-Lettres* 94 (4): 407–12.

Fraenkel, Carlos, Jamie Fumo, Faith Wallis, and Robert Wisnovsky, eds. (2011). *Vehicles of Transmission, Translation, and Transformation in Medieval Textual Culture*. Turnhout: Brepols.

Francopan, Peter (2016). *The Silk Roads: A New History of the World*. London: Bloomsbury.

Frank, Andre Gunder (1998). *ReOrient: Global Economy in the Asian Age*. Berkeley: University of California Press. http://hdl.handle.net/2027/heb.31038.0001.001.

Frege, Gottlob (1892). "Über Sinn und Bedeutung." *Zeitschrift für Philosophie und philosophische Kritik* 100 (1): 25–50.

———([1892] 1997). "On Sinn and Bedeutung." In *The Frege Reader*, ed. M. Beaney, 151–71. Malden, MA: Blackwell Publishing.

Freud, Sigmund ([1927] 1964). "The Future of an Illusion." In *The Standard Edition of the Complete Psychological Works of Sigmund Freud*. Vol. 21: *1927–1931: The Future of an Illusion— Civilization and Its Discontents—and Other Works*, ed. J. Strachey, 1–56. London: Hogarth Press and the Institute of Psycho-Analysis.

———([1930] 1962). *Civilization and Its Discontents*. Translated by James Strachey. New York: W. W. Norton.

Freudenthal, Gideon (1986). *Atom and Individual in the Age of Newton: On the Genesis of the Mechanistic World View*. Dordrecht: D. Reidel.

———(2000). "A Rational Controversy over Compounding Forces." In *Scientific Controversies: Philosophical and Historical Perspectives*, ed. P. Machamer, M. Pera, and A. Baltas, 125–42. New York: Oxford University Press.

———(2002). " 'Substanzbegriff und Funktionsbegriff' als Zivilisationstheorie bei Georg Simmel und Ernst Cassirer." In *Gesellschaft denken: Eine erkenntnistheoretische Standortbestimmung der Sozialwissenschaften*, ed. L. Bauer and K. Hamberger, 251–76. Vienna: Springer.

———(2005). "The Hessen-Grossman Thesis: An Attempt at Rehabilitation." *Perspectives on Science* 13 (2): 166–93.

———(2012). *No Religion without Idolatry: Mendelssohn's Jewish Enlightenment*. Notre Dame, IN: University of Notre Dame Press.

———(2015). "Commentary as Intercultural Practice." In *Wissen in Bewegung: Institution— Iteration—Transfer*, ed. E. Cancik-Kirschbaum and A. Traninger, 49–63. Wiesbaden: Harrassowitz Verlag.

Freudenthal, Gideon, and Peter McLaughlin (2009a). "Classical Marxist Historiography of Science: The Hessen-Grossmann-Thesis." In *The Social and Economic Roots of the Scientific Revolution: Texts by Boris Hessen and Henryk Grossmann*, ed. G. Freudenthal and P. McLaughlin, 1–40. Dordrecht: Springer.

———, eds. (2009b). *The Social and Economic Roots of the Scientific Revolution: Texts by Boris Hessen and Henryk Grossmann*. Boston Studies in the Philosophy and History of Science 278. Dordrecht: Springer.

Friederici, Angela D. (2017). *Language in Our Brain: The Origins of a Uniquely Human Capacity*. Cambridge, MA: MIT Press.

Friederici, Angela D., Noam Chomsky, Robert C. Berwick, Andrea Moro, and Johan J. Bolhuis (2017). "Language, Mind and Brain." *Nature Human Behaviour* 1 (10): 713–22.

Friedman, Robert Marc (1993). *Appropriating the Weather: Vilhelm Bjerknes and the Construction of a Modern Meteorology*. Ithaca, NY: Cornell University Press.

Friedman, Thomas L. (2007). *The World Is Flat: A Brief History of the Twenty-First Century*. Updated and exp. 3rd ed. New York: Picador.

Friedrich, Bretislav, Dieter Hoffmann, Jürgen Renn, Florian Schmaltz, and Martin Wolf, eds. (2017). *One Hundred Years of Chemical Warfare: Research, Deployment, Consequences*. Cham: Springer. https://link.springer.com/book/10.1007/978-3-319-51664-6.

Friedrich, Markus (2016). *Die Jesuiten: Aufstieg, Niedergang, Neubeginn*. Munich: Piper Verlag.

Frontinus Gesellschaft, ed. (1988). *Die Wasserversorgung antiker Städte: Mensch und Wasser—Mitteleuropa—Thermen—Bau/Materialien—Hygiene*. Geschichte der Wasserversorgung 3. Mainz: Verlag Philipp von Zabern.

Fu, Qiaomei, Pavao Rudan, Svante Pääbo, and Johannes Krause (2012). "Complete Mitochondrial Genomes Reveal Neolithic Expansion into Europe." *PLOS ONE* 7 (3): e32473.

Fuller, Dorian Q. (2010). "An Emerging Paradigm Shift in the Origins of Agriculture." *General Anthropology* 17 (2): 1, 8–12.

———(2011). "Pathways to Asian Civilizations: Tracing the Origins and Spread of Rice and Rice Cultures." *Rice* 4 (3–4): 78–92.

———(2012). "New Archaeobotanical Information on Plant Domestication from Macro-Remains: Tracking the Evolution of Domestication Syndrome Traits." In *Biodiversity in Agriculture: Domestication, Evolution, and Sustainability*, ed. P. L. Gepts, T. R. Famula, R. L. Bettinger, S. B. Brush, A. B. Damania, P. E. McGuire, and C. O. Qualset, 110–35. Cambridge: Cambridge University Press.

Fuller, Dorian Q., Robin G. Allaby, and Chris Stevens (2010). "Domestication as Innovation: The Entanglement of Techniques, Technology and Change in the Domestication of Cereal Crops." *World Archeology* 42 (1): 13–28.

Fuller, Dorian Q., Jacob van Etten, Katie Manning, Cristina Castillo, Eleanor Kingwell-Banham, Alison Weisskopf, Ling Qin, Yo-Ichiro Sato, and Robert J. Hijmans (2011). "The Contribution of Rice Agriculture and Livestock Pastoralism to Prehistoric Methane Levels: An Archaeological Assessment." *Holocene* 21 (5): 743–59.

Fuller, Dorian Q., George Willcox, and Robin G. Allaby (2011). "Cultivation and Domestication Had Multiple Origins: Arguments against the Core Area Hypothesis for the Origins of Agriculture in the Near East." *World Archeology* 43 (4): 628–52.

Fuller, Richard Buckminster (1969). *Operating Manual for Spaceship Earth*. New York: Simon & Schuster.

Fung, Yu-Lan (1922). "Why China Has No Science: An Interpretation of the History and Consequences of Chinese Philosophy." *International Journal of Ethics* 32 (3): 237–63.

Furth, Peter (1980). "Arbeit und Reflexion." In *Arbeit und Reflexion: Zur materialistischen Theorie der Dialektik—Perspektiven der Hegelschen Logik*, ed. P. Furth, 71–80. Cologne: Pahl-Rugenstein Verlag.

Gabrys, Jennifer (2016). *Program Earth: Environmental Sensing Technology and the Making of a Computational Planet*. Minneapolis: University of Minnesota Press.

Galambos, Louis, and Jeffrey L. Sturchio (1998). "Pharmaceutical Firms and the Transition to Biotechnology: A Study in Strategic Innovation." *Business History Review* 72 (2): 250–78.

Galilei, Galileo (1623). *Il saggiatore vel quale con bilancia esquisita e giusta si ponderano le cose contenute nella libra astronomica e filosofica di Lotario Sarsi Sigensano* [pseud.] *scritto in forma di lettera all'illmo. et reuermo. monsre. d. Virginio Cesarini acco. linceo mo. di camera di N S dal sig. Galileo Galilei*. Rome: Appresso Giacomo Mascardi. http://lhldigital.lindahall.org/cdm/ref/collection/astro_early/id/10173.

———(1638). *Discorsi e dimostrazioni matematiche, intorno à due nuoue scienze Attenenti alla mecanica & i movimenti locali, del signor Galileo Galilei linceo, filosofo e matematico primario del Serenissimo Grand Duca di Toscana: Con vna appendice del centro di grauità d'alcuni solidi*. Leiden: Appresso gli Elsevirii. https://books.google.de/books?id=E9BhikF658wC.

———(1960). "The Assayer." 1623. In *The Controversy on the Comets of 1618: Galileo Galilei, Horatio Grassi, Mario Guiducci, Johann Kepler*, ed. S. Drake and C. D. O'Malley, 151–336. Philadelphia: University of Pennsylvania Press.

———(1967). *Dialogue concerning the Two Chief World Systems: Ptolemaic & Copernican*. 1632. Translated by Stillman Drake. 2nd rev. ed. Berkeley: University of California Press.

———(1968a). "De motu." In vol. 1 of *Le opere di Galileo Galilei: Nuova ristampa della edizione nazionale*, ed. A. Favaro, 243–420. Florence: Giunti Barbèra.

———(1968b). *Le nuove scienze*. Vol. 8 of *Le opere di Galileo Galilei: Nuova ristampa della edizione nazionale*. Florence: Giunti Barbèra.

———(1974). *Two New Sciences: Including "Centers of Gravity" & "Force of Percussion."* 1638/1655. Translated by Stillman Drake. Madison: University of Wisconsin Press.

Galison, Peter (2003). *Einstein's Clocks and Poincare's Maps: Empires of Time*. New York: W. W. Norton.

Galluzzi, Paolo (1996a). *Gli ingegneri del Rinascimento: Da Brunelleschi a Leonardo da Vinci*. Florence: Giunti.

———(1996b). *Mechanical Marvels: Invention in the Age of Leonardo*. Florence: Giunti.

———(2005). "Machinae pictae: Immagine e idea della macchina negli artisti-ingegneri del Rinascimento." In *Machina: XI Colloquio Internazionale*, ed. M. Veneziani, 241–72. Florence: Leo S. Olschki.

Garber, Daniel (1994). "Leibniz: Physics and Philosophy." In *The Cambridge Companion to Leibniz*, ed. N. Jolley, 270–335. Cambridge: Cambridge University Press.

Gärdenfors, Peter (2004). *Conceptual Spaces: The Geometry of Thought*. Cambridge, MA: MIT Press.

Garfield, Eugene (2006). "The History and Meaning of the Journal Impact Factor." *JAMA* 295 (1): 90–93.

Gargiani, Roberto (2003). *Principi e costruzione nell'architettura italiana del Quattrocento*. Bari: Editori Laterza.

Gärtner, Peter (1998). *Filippo Brunelleschi 1377–1446*. Cologne: Könemann.

Gattei, Stefano (2008). *Thomas Kuhn's "Linguistic Turn" and the Legacy of Logical Empiricism: Incommensurability, Rationality and the Search for Truth*. Aldershot: Ashgate.

———(2016). "Science, Criticism and the Search for Truth: Philosophical Footnotes to Kuhn's Historiography." In *Shifting Paradigms: Thomas S. Kuhn and the History of Science*, ed. A. Blum, K. Gavroglu, C. Joas, and J. Renn, 123–38. Proceedings 8. Berlin: Edition Open Access. http://edition-open-access.de/proceedings/8/10/index.html.

Gavroglu, Kostas, and Yorgos Goudaroulis (1989). *Methodological Aspects of the Development of Low Temperature Physics 1881–1956: Concepts out of Context(s)*. Science and Philosophy 4. Dordrecht: Kluwer Academic.

Gavroglu, Kostas, Manolis Patiniotis, Faidra Papanelopoulou, Ana Simões, Ana Carneiro, Maria Paula Diogo, José Ramón Bertomeu Sánchez, Antonio García Belmar, and Agustí Nieto-Galan (2008). "Science and Technology in the European Periphery: Some Historiographical Reflections." *History of Science* 46 (2): 153–75.

Geller, Markham J., ed. (2014). *Melammu: The Ancient World in an Age of Globalization*. Proceedings 7. Berlin: Edition Open Access. http://edition-open-access.de/proceedings/7/index.html.

Genequand, Charles, ed. (2001). *Alexander of Aphrodisias on the Cosmos*. Islamic Philosophy, Theology and Science 44. Leiden: Brill.

Gentner, Dedre, and Albert L. Stevens, eds. (1983). *Mental Models*. Cognitive Science. Hillsdale, NJ: Erlbaum.

Geoffroy, Etienne François (1777). "Table des différents rapports observés en chymie entre différentes substances." In *Histoire de l'Académie royale des sciences: Avec les mémoires de mathématique & de physique, pour la même année—Tirés des registres de cette Académie*, 256–69. Paris: Imprimerie Royal.

Geus, Klaus, and Martin Thiering, eds. (2014). *Features of Common Sense Geography: Implicit Knowledge Structures in Ancient Geographical Texts*. Vienna: LIT Verlag.

Ghiberti, Lorenzo (1948–67). *The Commentaries*. Translated by Julius von Schlosser. London: Courtauld Institute of Art.

Ghil, Michael (2015). "A Mathematical Theory of Climate Sensitivity or, How to Deal with Both Anthropogenic Forcing and Natural Variability?" In *Climate Change: Multidecadal and Beyond*, ed. C.-P. Chang, M. Ghil, M. Latif, and J. M. Wallace, 31–51. World Scientific Series on Asia-Pacific Weather and Climate 6. Singapore: World Scientific Publishing/ Imperial College Press.

Giere, Ronald N., ed. (1992). *Cognitive Models of Science*. Minnesota Studies in the Philosophy of Science 15. Minneapolis: University of Minnesota Press.

Giesecke, Michael (1990.) "Printing in the Early Modern Era: A Media Revolution and Its Historical Significance." *Universitas: A Quarterly German Review of the Arts and Sciences* 32 (3): 219–27.

———(1991). *Der Buchdruck in der frühen Neuzeit: Eine Fallstudie über die Durchsetzung neuer Informations- und Kommunikationstechnologien*. Frankfurt am Main: Suhrkamp.

Gillies, James, and Robert Cailliau (2000). *How the Web Was Born: The Story of the World Wide Web*. Oxford: Oxford University Press.

Ginzburg, Carlo (1999). *History, Rhetoric, and Proof*. The Menahem Stern Jerusalem Lectures. Hanover, NH: University Press of New England.

Gladwin, Thomas (1974). *East Is a Big Bird: Navigation and Logic on Puluwat Atoll*. 3rd ed. Cambridge, MA: Harvard University Press.

Glick, Thomas (2005). *Islamic and Christian Spain in the Early Middle Ages*. The Medieval and Early Mordern Iberian World 27. Leiden: Brill.

Glick, Thomas F., Steven J. Livesey, and Faith Wallis, eds. (2005). *Medieval Science, Technology, and Medicine: An Encyclopedia*. New York: Routledge.

Goldthwaite, Richard A. (1980). *The Building of Renaissance Florence: An Economic and Social History*. Baltimore, MD: Johns Hopkins University Press.

Goody, Jack (1986). *The Logic of Writing and the Organization of Society*. Cambridge: Cambridge University Press.

———(2010a). *Myth, Ritual and the Oral*. Cambridge: Cambridge University Press.

———(2010b). *Renaissances: The One or the Many?* Cambridge: Cambridge University Press.

———(2010c). *The Eurasian Miracle*. Cambridge: Polity Press.

———(2012). *The Theft of History*. Cambridge: Cambridge University Press.

Goody, Jack, and Ian Watt (1963). "The Consequences of Literacy." *Comparative Studies in Society and History* 5 (3): 304–45.

Gorman, Hugh S. (2015). "Wie kann der menschliche Anteil am Stickstoffkreislauf begrenzt werden?" In *N: Stickstoff—Ein Element schreibt Weltgeschichte*, ed. G. Ertl and J. Soentgen, 217–38. Munich: Oekom Verlag.

Gould, Stephen J., and Richard C. Lewontin (1979). "The Spandrels of San Marco and the Panglossian Paradigm: A Critique of the Adaptationist Programme." *Proceedings of the Royal Society B: Biological Sciences* 205 (1161): 581–98.

Gowlett, John A. J. (2016). "The Discovery of Fire by Humans: A Long and Convoluted Process." *Philosophical Transactions of the Royal Society B: Biological Sciences* 371 (1696): 20150–64.

Graham, Angus C. (1978). *Later Mohist Logic, Ethics and Science*. Hong Kong: Chinese University Press.

Graham, Mark (2018). "The Rise of the Planetary Labor Market—and What It Means for the Future of Work." *Technosphere Magazine*. Accessed October 1, 2018. https://www .technosphere-magazine.hkw.de/p/The-Rise-of-the-Planetary-Labor-Marketand-What-It -Means-for-the-Future-of-Work-nyqzMRoxhWycwvwvtvAZVv.

Granovetter, Mark S. (1973). "The Strength of Weak Ties." *American Journal of Sociology* 78 (6): 1360–80.

———(1983). "The Strength of Weak Ties: A Network Theory Revisited." *Sociological Theory* 1:201–33.

Graßhoff, Gerd (1990). *The History of Ptolemy's Star Catalogue*. Studies in the History of Mathematics and Physical Sciences 14. New York: Springer.

———(2012). "Globalization of Ancient Knowledge: From Babylonian Observations to Scientific Regularities." In *The Globalization of Knowledge in History*, ed. J. Renn, 175–90. Berlin: Edition Open Access. http://edition-open-access.de/studies/1/11/index.html.

Graubard, Stephen R. (1975). "Preface to the Issue 'Wisdom, Revelation, and Doubt: Perspectives on the First Millennium B.C.'" *Daedalus* 104 (2): v–vi.

Gray, Russell D., and Joseph Watts (2017). "Cultural Macroevolution Matters." *Proceedings of the National Academy of Sciences of the United States of America* 114 (30): 7846–52.

Grimm, Dieter (2015). *Sovereignty: The Origin and Future of a Political and Legal Concept*. Translated by Belinda Cooper. Columbia Studies in Political Thought / Political History. New York: Columbia University Press.

Grimm, Jacob (1852). "Über den Ursprung der Sprache." 1851. In *Abhandlungen der Königlichen Akademie der Wissenschaften zu Berlin aus dem Jahre 1851*, 103–40. Berlin: Druckerei der Königlichen Akademie der Wissenschaften.

Gromov, Gregory (1995–2011). "History of the Internet and WWW: The Roads and Crossroads of Internet History, 1995–1998." Internet Valley. Accessed January 24, 2019. http://www .internetvalley.com/intval.html.

Gronenborn, Detlef (2010). "Climate, Crises, and the 'Neolithisation' of Central Europe between IRD-Events 6 and 4." In *The Spread of the Neolithic to Central Europe*, ed. D. Gronenborn and J. Petrasch, 61–81. Mainz: Verlag des Römisch-Germanischen Zentralmuseums Mainz.

Grosslight, Justin (2013). "Small Skills, Big Networks: Marin Mersenne as Mathematical Intelligencer." *History of Science* 51 (3): 337–74.

Grossmann, Henryk (1935). "Die gesellschaftlichen Grundlagen der mechanistischen Philosophie und die Manufaktur." *Zeitschrift für Sozialforschung* 4 (2): 161–231.

———(2009). "The Social Foundations of the Mechanistic Philosophy and Manufacture." In *The Social and Economic Roots of the Scientific Revolution: Texts by Boris Hessen and Henryk Grossmann*, ed. G. Freudenthal and P. McLaughlin, 103–56. Dordrecht: Springer.

Grote, Andreas (1959). *Studien zur Geschichte der Opera di Santa Reparata zu Florenz im vierzehnten Jahrhundert*. Munich: Prestel-Verlag.

Grove, Richard H. (1995). *Green Imperialism: Colonial Expansion, Tropical Island Edens and the Origins of Environmentalism, 1600–1860*. Cambridge: Cambridge University Press.

Grundmann, Reiner (2002). *Transnational Environmental Policy: Reconstructing Ozone*. Routledge Studies in Science, Technology and Society. London: Routledge.

Gruzinski, Serge (2004). *Les quatre parties du monde: Histoire d'une mondialisation.* Paris: La Martinière.

Guasti, Cesare (1857). *La cupola di Santa Maria del Fiore, illustrata con i documenti dell'archivio dell'Opera secolare.* Florence: Barbèra, Bianchi & Co.

———(1887). *Santa Maria del Fiore: La costruzione della chiesa e del campanile secondo i documenti tratti dall'Archivio dell'Opera secolare e da quello di stato.* Florence: Tipografia Ricci.

Guicciardini, Niccolò (2002). "Analysis and Synthesis in Newton's Mathematical Work." In *The Cambridge Companion to Newton*, ed. I. B. Cohen and G. E. Smith, 308–28. Cambridge: Cambridge University Press.

Günergun, Feza, and Dhruv Raina, eds. (2011). *Science between Europe and Asia: Historical Studies on the Transmission, Adoption and Adaptation of Knowledge.* Dordrecht: Springer.

Gutas, Dimitri (1998). *Greek Thought, Arabic Culture: The Graeco-Arabic Translation Movement in Baghdad and Early 'Abbasid Society (2nd–4th / 8th–10th Centuries).* London: Routledge.

Gutfreund, Hanoch, and Jürgen Renn (2017). *The Formative Years of Relativity: The History and Meaning of Einstein's Princeton Lectures.* Princeton, NJ: Princeton University Press.

———(forthcoming 2020). *Einstein on Einstein: Autobiographical and Scientific Reflections.* Princeton, NJ: Princeton University Press.

Haber, Fritz, and Robert Le Rossignol (1913). "Über die technische Darstellung von Ammoniak aus den Elementen." *Zeitschrift für Elektrochemie* 19 (2): 53–72.

Haber, Wolfgang, Martin Held, and Markus Vogt, eds. (2016). *Die Welt im Anthropozän: Erkundungen im Spannungsfeld zwischen Ökologie und Humanität.* Munich: Oekom Verlag.

Hacking, Ian (1987). "Was There a Probabilistic Revolution 1800–1930?" In *The Probabilistic Revolution.* Vol. 1: *Ideas in History*, ed. L. Krüger, L. Daston, and L. J. Heidelberger, 45–58. Cambridge: MIT Press.

Haeckel, Ernst (1879). *The Evolution of Man: A Popular Exposition of the Principal Points of Human Ontogeny and Phylogeny.* 2 vols. New York: D. Appleton and Co.

Haff, Peter K. (2014a). "Humans and Technology in the Anthropocene: Six Rules." *Anthropocene Review* 1 (2): 126–36.

———(2014b). "Technology as a Geological Phenomenon: Implications for Human Well-Being." In *A Stratigraphical Basis for the Anthropocene*, ed. C. N. Waters, J. A. Zalasiewicz, M. Williams, M. A. Ellis, and A. M. Snelling, 301–9. Special Publications 395. London: Geological Society.

Haff, Peter K., and Jürgen Renn (2019). " 'Was Menschen wollen,' ist keine Richtschnur dafür, wie die Welt tatsächlich funktioniert." In *Technosphäre*, ed. K. Klingan and C. Rosol, 26–46. Berlin: Matthes & Seitz.

Hafner, Katie, and Matthew Lyon (1996). *Where Wizards Stay up Late: The Origins of the Internet.* New York: Simon & Schuster.

Hahn, Otto, and Fritz Strassmann (1939). "Über den Nachweis und das Verhalten der bei der Bestrahlung des Urans mittels Neutronen entstehenden Erdalkalimetalle." *Naturwissenschaften* 27 (1): 11–15.

Haines, Margaret (1989). "Brunelleschi and Bureaucracy: The Tradition of Public Patronage at the Florentine Cathedral." *I Tatti Studies in the Italian Renaissance* 3:89–125.

———(1996). "L'arte della lana e l'opera del Duomo a Firenze con un accenno a Ghiberti due istituzioni." In *Opera: Carattere e ruolo delle Fabbriche cittadine fino all'inizio dell'età moderna*, ed. M. Haines and L. Riccetti, 267–94. Florence: Leo S. Olschki.

———(2008). "Oligarchy and Opera: Institution and Individuals in the Administration of the Florentine Cathedral." In *Florence and Beyond: Culture, Society and Politics in Renaissance Italy—Essays in Honour of John M. Najemy*, ed. D. S. Peterson and D. E. Bornstein, 153–77. Toronto: Centre for Reformation and Renaissance Studies.

———(2011–12). "Myth and Management in the Construction of Brunelleschi's Cupola." *I Tatti Studies in the Italian Renaissance* 14/15:47–101.

———(2015a). "The Years of the Cupola, 1417–1436: Digital Archive of the Opera di Santa Maria del Fiore." Accessed August 29, 2018. http://duomo.mpiwg-berlin.mpg.de/home_eng.HTML.

———(2015b). "The Years of the Cupola: Sources." Accessed January 25, 2019. http://duomo.mpiwg-berlin.mpg.de/ENG/AR/ARM001.HTM.

Haines, Margaret, and Gabriella Battista (2014). "Fokus: Die Kuppel des Florentiner Doms und ihre Handwerker." In *Wissensgeschichte der Architektur*. Vol. 3: *Vom Mittelalter bis zur Frühen Neuzeit*, ed. J. Renn, W. Osthues, and H. Schlimme, 467–492. Studies 5. Berlin: Edition Open Access. http://edition-open-access.de/studies/5/8/index.html.

Hall, Bert S. (1979). "Der Meister sol auch kennen schreiben und lesen: Writings about Technology ca. 1400–ca. 1600 A.D. and Their Cultural Implications." In *Early Technologies*, ed. D. Schmandt-Besserat, 47–58. Malibu, CA: Undena Publications.

Hallam, Anthony (1989). *Great Geological Controversies*. 2nd ed. New York: Oxford University Press.

Hamilton, Clive (2013). *Earthmasters*. New Haven, CT: Yale University Press.

Hankins, Thomas L. (1999). "Blood, Dirt, and Nomograms: A Particular History of Graphs." *Isis* 90 (1): 50–80.

Hankinson, Robert J., ed. (2008). *The Cambridge Companion to Galen*. Cambridge: Cambridge University Press.

Hanley, Susan B. (1987). "Urban Sanitation in Preindustrial Japan." *Journal of Interdisciplinary History* 18 (1): 1–26.

Hansen, Svend, and Barbara Helwing (2018). "Der Beginn der Landwirtschaft im Kaukasus." In *Gold und Wein: Georgiens älteste Schätze*, ed. L. Glemsch and S. Hansen, 26–41. Mainz: Nünnerich-Asmus Verlag.

Hansen, Svend, and Jürgen Renn (2018). "Technische und soziale Innovationen." In *Innovationen der Antike*, ed. G. Graßhoff and M. Meyer, 8–19. Zaberns Bildbände zur Archäologie. Darmstadt: Verlag Philipp von Zabern.

Hardin, Garrett (1968). "The Tragedy of the Commons." *Science* 162 (3859): 1243–48.

———(1998). "Extensions of 'The Tragedy of the Commons.'" *Science* 280 (5364): 682–83.

Harman, Peter M. (1982). *Energy, Force, and Matter: The Conceptual Development of Nineteenth-Century Physics*. Cambridge: Cambridge University Press.

Harper, Kristine C. (2008). *Weather by the Numbers: The Genesis of Modern Meteorology*. Cambridge, MA: MIT Press.

Harrell, James A. (n.d.). "Turin Papyrus Map from Ancient Egypt." University of Toledo. Accessed January 28, 2019. http://www.eeescience.utoledo.edu/faculty/harrell/egypt/turin%20papyrus/harrell_papyrus_map_text.htm.

Harris, Steven (2006). "Networks of Travel, Correspondence, and Exchange." In *Early Modern Science*, ed. K. Park and L. Daston, 341–62. The Cambridge History of Science 3. Cambridge: Cambridge University Press.

Hawking, Stephen W., and Roger Penrose (1970). "The Singularities of Gravitational Collapse and Cosmology." *Proceedings of the Royal Society A: Mathematical, Physical and Engineering Sciences* 314 (1519): 529–48.

Hayek, Friedrich A. (1937). "Economics and Knowledge." *Economica* 4 (13): 33–54.

———(1945). "The Use of Knowledge in Society." *American Economic Review* 35 (4): 519–30.

Hayles, N. Katherine (2009). "RFID: Human Agency and Meaning in Information-Intensive Environments." *Theory, Culture & Society* 26 (2–3): 47–72.

Haynes, Douglas M. (2001). *Imperial Medicine: Patrick Manson and the Conquest of Tropical Disease*. Philadelphia: University of Pennsylvania Press.

Heath, Thomas L., ed. (2009). *The Works of Archimedes: Edited in Modern Notation with Introductory Chapter*. Cambridge: Cambridge University Press. https://www.cambridge.org /core/books/works-of-archimedes/E5F35917BA320E2B40696056CB6ED610.

Hedtke, Ulrich, and Camilla Warnke (2017). "Peter Ruben: Philosophische Schriften—Online Edition." Accessed August 23, 2018. http://www.peter-ruben.de.

Hegel, Georg Wilhelm Friedrich ([1837] 1942). *The Philosophy of History*. Translated by John Sibree. Rev. ed. Ann Arbor, MI: Edwards Brothers.

———([1807] 1996). "Wer denkt abstrakt?" In *Jenaer Schriften: 1801–1807*. Vol. 2 of *Georg Wilhelm Friedrich Hegel Werke*, ed. E. Moldenhauer and K. M. Michel, 575–81. Frankfurt am Main: Suhrkamp.

Heim, Susanne, Carola Sachse, and Mark Walker, eds. (2009). *The Kaiser Wilhelm Society under National Socialism*. Cambridge: Cambridge University Press.

Heinrich, Klaus (1986). *Anthropomorphe: Zum Problem des Anthropomorphismus in der Religionsphilosophie*. Dahlemer Vorlesungen 2. Basel: Stroemfeld/Roter Stern.

———(1987). *Tertium datur: Eine religionsphilosophische Einführung in die Logik*. 2nd ed. Dahlemer Vorlesungen 1. Basel: Stroemfeld/Roter Stern.

———(1993). *Arbeiten mit Ödipus: Begriff der Verdrängung in der Religionswissenschaft*. Dahlemer Vorlesungen 3. Basel: Stroemfeld/Roter Stern.

———(2000). *Vom Bündnis denken: Religionsphilosophie*. Dahlemer Vorlesungen 4. Basel: Stroemfeld/Roter Stern.

———(2001). *Psychoanalyse Sigmund Freuds und das Problem des konkreten gesellschaftlichen Allgemeinen*. Dahlemer Vorlesungen 7. Basel: Stroemfeld/Roter Stern.

Heisel, Joachim P. (1993). *Antike Bauzeichnungen*. Darmstadt: Wissenschaftliche Buchgesellschaft.

Heisenberg, Werner (1925). "Über quantentheoretische Umdeutung kinematischer und mechanischer Beziehungen." *Zeitschrift für Physik* 33 (1): 879–93.

Heller, Michael A., and Rebecca S. Eisenberg (1998). "Can Patents Deter Innovation? The Anticommons in Biomedical Research." *Science* 280 (1): 698–701.

Herder, Johann Gottfried (1772). *Abhandlung über den Ursprung der Sprache*. Berlin: Christian Friedrich Voß. http://www.deutschestextarchiv.de/book/view/herder_abhandlung_1772 ?p=5.

Herfeld, Catherine, and Malte Doehne (in press 2018). "The Diffusion of Scientific Innovations: A Role Typology." *Studies in History and Philosophy of Science Part A*.

Hertz, Heinrich (1888). "Ueber electrodynamische Wellen im Luftraume und deren Reflexion." *Annalen der Physik und Chemie* 270 (8a): 609–23.

———(1889). "Die Kräfte electrischer Schwingungen, behandelt nach der Maxwell'schen Theorie." *Annalen der Physik und Chemie* 272 (1): 1–22.

Hess, Charlotte, and Elinor Ostrom (2003). "Ideas, Artifacts, and Facilities: Information as a Common-Pool Resource." *Law and Contemporary Problems* 66 (1): 111–46.

———(2007). *Understanding Knowledge as a Commons: From Theory to Practice*. Cambridge, MA: MIT Press.

Hessen, Boris ([1931] 2009). "The Social and Economic Roots of Newton's *Principia*." In *The Social and Economic Roots of the Scientific Revolution: Texts by Boris Hessen and Henryk Grossmann*, ed. G. Freudenthal and P. McLaughlin, 41–101. Dordrecht: Springer.

Hewish, Antony, S. Jocelyn Bell, J.D.H. Pilkington, Paul F. Scott, and R. A. Collins (1968). "Observation of a Rapidly Pulsating Radio Source." *Nature* 217 (5130): 709–13.

Heymann, Matthias, and Dania Achermann (2018). "From Climatology to Climate Science in the Twentieth Century." In *The Palgrave Handbook of Climate History*, ed. S. White, C. Pfister, and F. Mauelshagen, 605–32. London: Palgrave Macmillan.

Hibbard, Kathy A., Paul J. Crutzen, Eric F. Lambin, Diana Liverman, Nathan J. Mantua, John R. McNeill, Bruno Messerli, and Will Steffen (2007). "Decadal Interactions of Humans and the Environment." In *Sustainability or Collapse? An Integrated History and Future of People on Earth*, ed. R. Costanza, L. Graumlich, and W. Steffen, 341–75. Dahlem Workshop Report 96. Cambridge, MA: MIT Press.

Hilgert, Markus (2014). "Fokus: Keilschriftliche Quellen zu Architektur und Bauwesen." In *Wissensgeschichte der Architektur*. Vol. 1: *Vom Neolithikum bis zum Alten Orient*, ed. J. Renn, W. Osthues, and H. Schlimme, 281–96. Studies 3. Berlin: Edition Open Access. http://edition-open-access.de/studies/3/6/index.html.

Hiltzik, Michael A. (1999). *Dealers of Lightning: Xerox PARC and the Dawn of the Computer Age*. New York: HarperCollins.

———(2016). *Big Science: Ernest Lawrence and the Invention That Launched the Military-Industrial Complex*. New York: Simon & Schuster.

Hirsch Hadorn, Gertrude, and Christian Pohl (2007). *Principles for Designing Transdisciplinary Research*. Munich: Oekom Verlag.

The Historical Epistemology of Mechanics. 4-title series within Springer's Boston Studies in the Philosophy and History of Science (*see* entries under each author for full details).

    No. 1. Schemmel (2008). *The English Galileo: Thomas Harriot's Work on Motion as an Example of Preclassical Mechanics*.

    No. 2. Valleriani (2010). *Galileo Engineer*.

    No. 3. Büttner (2019). *Swinging and Rolling: Unveiling Galileo's Unorthodox Path from a Challenging Problem to a New Science*.

    No. 4. Feldhay, Renn, Schemmel, and Valleriani (2018). *Emergence and Expansion of Pre-classical Mechanics*.

Hobbes, Thomas (1651). *Leviathan, or The Matter, Forme, & Power of a Common-Wealth Ecclesiasticall and Civill*. London: Andrew Crooke.

Hobsbawm, Eric J. (1975). *The Age of Capital, 1848–1875*. London: Weidenfeld & Nicolson.

Hoffmann, Dieter, Birgit Kolboske, and Jürgen Renn, eds. (2015). *"Dem Anwenden muss das Erkennen vorausgehen": Auf dem Weg zu einer Geschichte der Kaiser-Wilhelm/Max-Planck-Gesellschaft*. 2nd ed. Proceedings 6. Berlin: Edition Open Access. http://edition-open-access.de/proceedings/6/index.html.

Hoffmann, Hans Falk (2012). "The Role of Open and Global Communication in Particle Physics." In *The Globalization of Knowledge in History*, ed. J. Renn, 713–36. Studies 1. Berlin: Edition Open Access. http://edition-open-access.de/studies/1/32/index.html.

Höhler, Sabine (2015). *Spaceship Earth in the Environmental Age, 1960–1990*. London: Routledge.

———(2018). "Ecospheres: Model and Laboratory for Earth's Environment." *Technosphere Magazine*. Accessed July 17, 2018. https://technosphere-magazine.hkw.de/p/Ecospheres-Model-and-Laboratory-for-Earths-Environment-qfrCXdpGUyenDt224wXyjV.

Holbrook, Jarita (2012). "Celestial Navigation and Technological Change on Moce Island." In *The Globalization of Knowledge in History*, ed. J. Renn, 439–57. Studies 1. Berlin: Edition Open Access. http://edition-open-access.de/studies/1/23/index.html.

Hölderlin, Friedrich (1990). *Selected Poems—Including Hölderlin's Sophocles*. Translated by David Constantine. Hexham: Bloodaxe Books.

Holdermann, Karl (1953). *Im Banne der Chemie: Carl Bosch—Leben und Werk*. Düsseldorf: Econ-Verlag.

Holmes, Arthur (1926). "Contributions to the Theory of Magmatic Cycles." *Geological Magazine* 63 (7): 306–29.

Holzkamp, Klaus (1968). *Wissenschaft als Handlung: Versuch einer neuen Grundlegung der Wissenschaftslehre*. Berlin: De Gruyter.

Holzkamp, Klaus, and Volker Schurig ([1973] 2015). "Zur Einführung in A. N. Leontjew 'Probleme des Psychischen.' " In *Kritische Psychologie als Subjektwissenschaft: Marxistische Begründung der kritischen Psychologie*, ed. F. Haug, W. Maiers, and U. Osterkamp, 33–74. Schriften 6. Hamburg: Argument Verlag.

Hooker, James T., Christopher B. F. Walker, W. V. Davies, John Chadwick, John F. Healey, B. F. Cook, and Larissa Bonfante (1990). *Reading the Past: Ancient Writing from Cuneiform to the Alphabet*. Berkeley: University of California Press.

Hooykaas, Reijer (1963). *Natural Law and Divine Miracle: The Principle of Uniformity in Geology, Biology and Theology*. 2nd ed. Leiden: Brill.

Houston, Stephen D., ed. (2004). *The First Writing: Script Invention as History and Process*. Cambridge: Cambridge University Press.

Howell, David L. (1992). "Proto-industrial Origins of Japanese Capitalism." *Journal of Asian Studies* 51 (2): 269–86.

Howlett, Peter, and Mary S. Morgan, eds. (2011). *How Well Do Facts Travel? The Dissemination of Reliable Knowledge*. Cambridge: Cambridge University Press.

Hoyningen-Huene, Paul (1993). *Reconstructing Scientific Revolutions: Thomas S. Kuhn's Philosophy of Science*. Chicago: University of Chicago Press.

Høyrup, Jens (2007). "The Roles of Mesopotamian Bronze Age Mathematics Tool for State Formation and Administration: Carrier of Teachers' Professional Intellectual Autonomy." *Educational Studies in Mathematics* 66 (2): 257–71.

———(2009). "State, 'Justice,' Scribal Culture and Mathematics in Ancient Mesopotamia." Sarton Chair Lecture. *Sartoniana* 22:13–45.

———(2012). "Was Babylonian Mathematics Created by 'Babylonian Mathematicians'?" In *Wissenskultur im Alten Orient: Weltanschauung, Wissenschaften, Techniken, Technologien*, ed. H. Neumann, 105–19. Wiesbaden: Harrassowitz Verlag.

———(2015). "Written Mathematical Traditions in Ancient Mesopotamia: Knowledge, Ignorance, and Reasonable Guesses." In *Traditions of Written Knowledge in Ancient Egypt and Mesopotamia: Proceedings of Two Workshops Held at Goethe-University, Frankfurt/Main in December 2011 and May 2012*, ed. D. Bawanypeck and A. Imhausen, 189–213. Alter Orient und Altes Testament 403. Münster: Ugarit-Verlag.

———(2017a). *Algebra in Cuneiform: Introduction to an Old Babylonian Geometrical Technique*. Textbooks 2. Berlin: Edition Open Access. http://edition-open-access.de/textbooks/2/index.html.

———(2017b). *From Hesiod to Saussure, from Hippocrates to Jevons: An Introduction to the History of Scientific Thought between Iran and the Atlantic*. Roskilde: Roskilde Universitet.

Høyrup, Jens, and Peter Damerow, eds. (2001). *Changing Views of Ancient Near Eastern Mathematics*. Berlin: Dietrich Reimer Verlag.

Hu, Danian (2005). *China and Albert Einstein: The Reception of the Physicist and His Theory in China 1917–1979*. Cambridge, MA: Harvard University Press.

Hu, Shih (1922). *The Development of the Logical Method in Ancient China*. Shanghai: Oriental Book Co.

Huber, Veronika, Hans Joachim Schellnhuber, Nigel W. Arnell, Katja Frieler, Andrew D. Friend, Dieter Gerten, Ingjerd Haddeland, Pavel Kabat, Hermann Lotze-Campen, Wolfgang Lucht, Martin Parry, Franziska Piontek, Cynthia Rosenzweig, Jacob Schewe, and Lila Warszawski (2014). "Climate Impact Research: Beyond Patchwork." *Earth System Dynamics* 5:399–408.

Hublin, Jean-Jacques, Abdelouahed Ben-Ncer, Shara E. Bailey, Sarah E. Freidline, Simon Neubauer, Matthew M. Skinner, Inga Bergmann, Adeline Le Cabec, Stefano Benazzi,

Katerina Harvati, and Philipp Gunz (2017). "New Fossils from Jebel Irhoud, Morocco and the Pan-African Origin of Homo Sapiens." *Nature* 546 (7657): 289–92.

Huerta, Santiago, ed. (2003). *Proceedings of the First International Congress on Construction History: Madrid, 20th–24th January 2003*. 3 vols. Madrid: Institutio Juan de Herrera; Escuela Técnica Superior de Arquitectura.

Huff, Toby E. (2017). *The Rise of Early Modern Science: Islam, China and the West*. 3rd ed. Cambridge: Cambridge University Press.

Hufnagel, Henning, and Anne Eusterschulte, eds. (2013). *Turning Traditions Upside Down: Rethinking Giordano Bruno's Enlightenment*. Budapest: Central European University Press.

Hughes, Thomas P. (1983). *Networks of Power: Electrification in Western Society, 1880–1930*. Baltimore, MD: Johns Hopkins University Press. http://hdl.handle.net/2027/fulcrum.w6634365g.

———(1994). "Technological Momentum." In *Does Technology Drive History? The Dilemma of Technological Determinism*, ed. L. Marx and M. Roe Smith, 101–13. Cambridge, MA: MIT Press.

Hull, David L. (1988). *Science as a Process: An Evolutionary Account of the Social and Conceptual Development of Science*. Chicago: University of Chicago Press.

———(2001). *Science and Selection: Essays on Biological Evolution and the Philosophy of Science*. Cambridge: Cambridge University Press.

Humboldt, Alexander von (1849–58). *Cosmos: A Sketch of a Physical Description of the Universe*. Translated by E. C. Otté, B. H. Paul, and W. S. Dallas. 5 vols. London: Henry G. Bohn.

Humboldt, Wilhelm von ([1809/10] 1993). "Über die innere und äußere Organisation der höheren wissenschaftlichen Anstalten in Berlin." In *Werke in fünf Bänden*. Vol. 4: *Schriften zur Politik und zum Bildungswesen*, ed. A. Flitner and K. Giel, 255–66. Darmstadt: Wissenschaftliche Buchgesellschaft.

———([1820] 1994). "Über das vergleichende Sprachstudium in Beziehung auf die verschiedenen Epochen der Sprachentwickelung." In *Werke in fünf Bänden*. Vol. 3: *Schriften zur Sprachphilosophie*, ed. A. Flitner and K. Giel, 1–25. Darmstadt: Wissenschaftliche Buchgesellschaft.

Hume, David ([1748] 2000). "Sceptical Solution of These Doubts." In *An Enquiry concerning Human Understanding: A Critical Edition*, ed. T. L. Beauchamp, 35–45. Oxford: Clarendon Press.

Hunt, Terry L., and Carl Philipp Lipo (2012). "Ecological Catastrophe and Collapse: The Myth of 'Ecocide' on Rapa Nui (Easter Island)." *PERC Research Paper* 12 (3).

Husserl, Edmund (1939). *Erfahrung und Urteil: Untersuchungen zur Genealogie der Logik*. Prague: Academia Verlag.

———(2001). *Analyses concerning Passive and Active Synthesis: Lectures on Transcendental Logic*. Translated by Anthony J. Steinbock. Husserliana: Edmund Husserl—Collected Works 9. Dordrecht: Springer.

Hutchins, Edwin (1996). *Cognition in the Wild*. 2nd ed. Cambridge, MA: MIT Press.

Huurdeman, Anton A. (2005). *The Worldwide History of Telecommunications*. Hoboken, NJ: John Wiley & Sons.

Huxley, Julian (1948). *Evolution: The Modern Synthesis*. 1942. 5th ed. London: George Allen & Unwin.

Hyman, Arthur, ed. (1986). *Averroes' de Substantia Orbis*. Cambridge, MA: Medieval Academy of America and Israel Academy of Sciences and Humanities.

Hyman, Isabelle, ed. (1974). *Brunelleschi in Perspective*. Upper Saddle River, NJ: Prentice Hall.

Hyman, Ludmilla (2012). "The Soviet Psychologists and the Path to International Psychology." In *The Globalization of Knowledge in History*, ed. J. Renn, 631–68. Studies 1. Berlin: Edition Open Access. http://edition-open-access.de/studies/1/30/index.html.

Hyman, Malcom D. (2006). "Of Glyphs and Glottography." *Language & Communication* 26 (3–4): 231–49.

———(2007). "Semantic Networks: A Tool for Investigating Conceptual Change and Knowledge Transfer in the History of Science." In *Übersetzung und Transformation*, ed. H. Böhme, C. Rapp, and W. Rösler, 355–67. Transformationen der Antike 1. Berlin: De Gruyter.

Hyman, Malcolm D., and Jürgen Renn (2012a). "Survey 1: From Technology Transfer to the Origins of Science." In *The Globalization of Knowledge in History*, ed. J. Renn, 75–104. Studies 1. Berlin: Edition Open Access. http://edition-open-access.de/studies/1/7/index.html.

———(2012b). "Toward an Epistemic Web." In *The Globalization of Knowledge in History*, ed. J. Renn, 821–38. Studies 1. Berlin: Edition Open Access. http://edition-open-access.de /studies/1/36/index.html.

Iannaccone, Isaia (1998). *Johann Schreck Terrentius: Le scienze rinascimentali e lo spirito dell'Accademia dei Lincei nella Cina dei Ming.* Series Minor 54. Naples: Istituto Universitario Orientale, Dipartimento di Studi Asiatici.

Ibañez, Juan José, Patricia C. Anderson, Jesús Gonzalez-Urquijo, and Juan Gibaja (2016). "Cereal Cultivation and Domestication as Shown by Microtexture Analysis of Sickle Gloss through Confocal Microscopy." *Journal of Archaeological Science* 73:62–81.

Infeld, Leopold, ed. (1964). *Relativistic Theories of Gravitation: Proceedings of a Conference Held in Warsaw and Jablonna, July, 1962.* Oxford: Pergamon Press.

Inkster, Ian (2001). *Japanese Industrialisation: Historical and Cultural Perspectives.* London: Routledge.

IPCC (2018). "History of the IPCC." Intergovernmental Panel on Climate Change (IPCC). Accessed January 4, 2019. https://www.ipcc.ch/about/history.

IPCC and Robert T. Watson (2001). *Climate Change 2001: Synthesis Report; A Contribution of Working Groups I, II, and III to the Third Assessment Report of the Intergovernmental Panel on Climate Change.* Cambridge: Cambridge University Press.

Ippolito, Lamberto, and Chiara Peroni (1997). *La cupola di Santa Maria del Fiore.* Rome: Nuova Italia Scientifica.

Isett, Christopher, and Stephen Miller (2016). *The Social History of Agriculture: From the Origins to the Current Crisis.* Lanham, MD: Rowman & Littlefield.

Ishikawa, Eisuke (2000a). "Japan in the Edo Period: An Ecologically-Conscious Society." Japan for Sustainability. https://www.japanfs.org/en/edo/index.html.

———(2000b). *Ōedo ekorojī jijō.* Tokyo: Kodansha.

Jack, Mary Ann (1976). "The Accademia del Disegno in Late Renaissance Florence." *Sixteenth Century Journal* 7 (2): 3–20.

Jackson, Peter, and David Morgan (1990). *The Mission of Friar William of Rubruck: His Journey to the Court of the Great Khan Möngke, 1253–1255.* Translated by Peter Jackson. London: Routledge / Hakluyt Society.

Jacob, Margaret C. (1997). *Scientific Culture and the Making of the Industrial West.* New York: Oxford University Press.

———(2014). *The First Knowledge Economy: Human Capital and the European Economy, 1750–1850.* Cambridge: Cambridge University Press.

Jamme, Christoph, and Helmut Schneider, eds. (1984). *Mythologie der Vernunft: Hegels ältestes Systemprogramm des deutschen Idealismus.* Frankfurt am Main: Suhrkamp.

Janssen, Michel (1995). "A Comparison between Lorentz's Ether Theory and Special Relativity in the Light of the Experiments of Trouton and Noble." Ph.D. diss., University of Pittsburgh.

———(2002). "Reconsidering a Scientific Revolution: The Case of Einstein versus Lorentz." *Physics in Perspective* 4 (4): 421–46.

————(2014). "'No Success like Failure . . .': Einstein's Quest for General Relativity, 1907–1920." In *The Cambridge Companion to Einstein*, ed. C. Lehner and M. Janssen, 167–227. Cambridge: Cambridge University Press.

————(2019). "Arches and Scaffolds: Bridging Continuity and Discontinuity in Theory Change." In *Beyond the Meme: Articulating Dynamic Structures in Cultural Evolution*, ed. A. C. Love and W. C. Wimsatt, 95–199. Minneapolis: University of Minnesota Press.

Janssen, Michel, and Christoph Lehner, eds. (2014). *The Cambridge Companion to Einstein*. New York: Cambridge University Press.

Janssen, Michel, and Jürgen Renn (2015). "Arch and Scaffold: How Einstein Found His Field Equations." *Physics Today* 68 (11): 30–36.

Janssen, Michel, and John Stachel (2004). *The Optics and Electrodynamics of Moving Bodies*. Preprint 265. Berlin: Max Planck Institute for the History of Science. https://www.mpiwg-berlin.mpg.de/sites/default/files/Preprints/P265.pdf.

Jasanoff, Sheila, ed. (2004). *States of Knowledge: The Co-production of Science and the Social Order*. London: Routledge.

Jaspers, Karl ([1949] 1953). *The Origin and Goal of History*. Translated by Michael Bullock. London: Routledge & Kegan Paul.

————([1949] 1983). *Vom Ursprung und Ziel der Geschichte*. 8th ed. Munich: Piper Verlag.

Jeschke, Jonathan M., Sophie Lokatis, Isabelle Bartram, and Klement Tockner (2018). *Knowledge in the Dark: Scientific Challenges and Ways Forward*. EarthArXiv Preprints. https://eartharxiv.org/qrt6p.

Jevons, William Stanley (1865). *The Coal Question: An Inquiry Concerning the Progress of the Nation, and the Probable Exhaustion of Our Coal Mines*. London: Macmillan & Co.

Joas, Hans, and Jens Beckert (2001). "Action Theory." In *Handbook of Sociological Theory*, ed. J. H. Turner, 269–85. Dordrecht: Springer.

Johnson, Christopher N., John Alroy, Nicholas J. Beeton, Michael I. Bird, Barry W. Brook, Alan Cooper, Richard Gillespie, Salvador Herrando-Pérez, Zenobia Jacobs, Gifford H. Miller, Gavin J. Prideaux, Richard G. Roberts, Marta Rodríguez-Rey, Frédérik Saltré, Chris S. M. Turney, and Corey J. A. Bradshaw (2016). "What Caused Extinction of the Pleistocene Megafauna of Sahul?" *Proceedings of the Royal Society B: Biological Sciences* 283 (1824): 20152399.

Johnson, Noor, Carolina Behe, Finn Danielsen, Eva-Maria Krummel, Scot Nickels, and Peter L. Pulsifer (2016). *Community-Based Monitoring and Indigenous Knowledge in a Changing Arctic: A Review for the Sustaining Arctic Observing Networks*. Ottawa: Inuit Circumpolar Council Canada. http://www.inuitcircumpolar.com/community-based-monitoring.html.

Johnson-Laird, Philip N. (1983). *Mental Models: Towards a Cognitive Science of Language, Inference, and Consciousness*. Cambridge, MA: Harvard University Press.

Johnson-Laird, Philip N., and Peter Cathcart Wason, eds. (1977). *Thinking: Readings in Cognitive Science*. Cambridge: Cambridge University Press.

Johnston, Harold (1971). "Reduction of Stratospheric Ozone by Nitrogen Oxide Catalysts from Supersonic Transport Exhaust." *Science* 173 (3996): 517–22.

Jordanus de Nemore (1960). "Elementa Jordani super demonstrationem ponderum." In *The Medieval Science of Weights (Scienta de Ponderibus): Treatises Ascribed to Euclid, Archimedes, Thabit Ibn Qurra, Jordanus de Nemore and Blasius of Parma*, ed. E. A. Moody and M. Clagett, 119–42. Madison: University of Wisconsin Press.

Judson, Olivia P. (2017). "The Energy Expansions of Evolution." *Nature Ecology & Evolution* 1:0138.

Jungnickel, Christa, and Russell McCormmach (1986a). *Intellectual Mastery of Nature: Theoretical Physics from Ohm to Einstein*. 2 vols. (Chicago: University of Chicago Press.

————(1986b). *The Second Physicist: On the History of Theoretical Physics in Germany*. Cham: Springer.

Kahn, Julius (1891). *Münchens Großindustrie und Großhandel*. Munich: Verlag von A. Ackermanns.

Kaiser, David I. (2005). *Drawing Theories Apart: The Dispersion of Feynman Diagrams in Postwar Physics*. Chicago: University of Chicago Press.

————(2012). "Booms, Busts, and the World of Ideas: Enrollment Pressures and the Challenge of Specialization." *Osiris* 27 (1): 276–302.

Kaiser, Walter, and Wolfgang König, eds. (2006). *Geschichte des Ingenieurs: Ein Beruf in sechs Jahrtausenden*. Munich: Hanser.

Kaldeway, David, and Désirée Schauz, eds. (2018). *Basic and Applied Research: The Language of Science Policy in the Twentieth Century*. New York: Berghahn Books.

Kamen, Henry (2003). *Spain's Road to Empire: The Making of a World Power, 1492–1763*. London: Penguin Books.

Kant, Immanuel ([1746–49] 2012). "Thoughts on the True Estimation of Living Forces and Assessment of the Demonstrations That Leibniz and Other Scholars of Mechanics Have Made Use of in This Controversial Subject, Together with Some Prefatory Considerations Pertaining to the Force of Bodies in General." In *Kant: Natural Science*, ed. E. Watkins, 1–155. Cambridge Edition of the Works of Immanuel Kant in Translation. Cambridge: Cambridge University Press.

————([1781] 1998). "On the Deduction of the Pure Concepts of the Understanding." In *Critique of Pure Reason*, ed. P. Guyer and A. W. Wood, 219–66. Cambridge: Cambridge University Press.

————([1786] 2002). "Metaphysical Foundations of Natural Science." In *Theoretical Philosophy after 1781*, ed. H. Allison, 171–270. Cambridge Edition of the Works of Immanuel Kant in Translation. Cambridge: Cambridge University Press.

Kaplan, Jed O., Kristen M. Krumhardt, Erle C. Ellis, William F. Ruddiman, Carsten Lemmen, and Kees Klein Goldewijk (2011). "Holocene Carbon Emissions as a Result of Anthropogenic Land Cover Change." *Holocene* 21 (5): 775–91.

Kaufmann, Doris (2017). "'Gas, Gas, Gaas!' The Poison Gas War in the Literature and Visual Arts of Interwar Europe." In *One Hundred Years of Chemical Warfare: Research, Deployment, Consequences*, ed. B. Friedrich, D. Hoffmann, J. Renn, F. Schmaltz, and M. Wolf. Cham: Springer. https://www.springer.com/de/book/9783319516639.

Kaufmann, Stefan H. E. (2009). *The New Plagues: Pandemics and Poverty in a Globalized World*. London: Haus.

Kaufmann, Stefan H. E., and Shreemanta K. Parida (2007). "Changing Funding Patterns in Tuberculosis." *Nature Medicine* 13 (3): 299–303.

Kavanagh, Patrick H., Bruno Vilela, Hannah J. Haynie, Ty Tuff, Matheus Lima-Ribeiro, Russell D. Gray, Carlos A. Botero, and Michael C. Gavin (2018). "Hindcasting Global Population Densities Reveals Forces Enabling the Origin of Agriculture." *Nature Human Behaviour* 2 (7): 478–84.

Keen for Green (2018). "Pakistani Children Work in Our Discarded Illegal E-waste." Accessed February 5, 2019. http://keenforgreen.pk/pakistani-children-work-in-our-discarded -illegal-e-waste.

Keller, Monika (2007). "Moralentwicklung und moralische Sozialisation." In *Moralentwicklung von Kindern und Jugendlichen*, ed. D. Horster, 17–49. Wiesbaden: VS Verlag für Sozialwissenschaften.

Keller, Monika, and Wolfgang Edelstein (1991). "The Development of Socio-Moral Meaning Making: Domains, Categories, and Perspective-Taking." In *Handbook of Moral Behavior and Development.* Vol. 2: *Research,* ed. W. M. Kurtines and J. L. Gewirtz, 89–114. Hillsdale, NJ: Erlbaum.

Keller, Vera (2015). *Knowledge and the Public Interest, 1575–1725.* Cambridge: Cambridge University Press.

Kennett, Douglas J., and Bruce Winterhalder, eds. (2006). *Behavioral Ecology and the Transition to Agriculture.* Berkeley: University of California Press.

Kenney, Martin, and John Zysman (2016). "The Rise of the Platform Economy." *Issues in Science and Technology* 32 (3): 61–69.

Kepler, Johannes (1609). *Astronomia nova: Aitiologetos, seu physica coelestis, tradita commentariis de motibus stellae martis, ex observationibus G. V. Tychonis Brahe.* Heidelberg: Voegelin.

———([1619] 1997). *The Harmony of the World.* Translated by E. J. Aiton, Alistair M. Duncan, and Judith V. Field. Philadelphia: American Philosophical Society.

Kerr, Roy P. (1963). "Gravitational Field of a Spinning Mass as an Example of Algebraically Special Metrics." *Physical Review Letters* 11 (5): 237–38.

Kim, Samuel S. (2000). "East Asia and Globalization: Challenges and Responses." In *East Asia and Globalization,* ed. S. S. Kim, 1–30. Lanham, MD: Rowman & Littlefield.

Kintsch, Walter (1998). *Comprehension: A Paradigm for Cognition.* Cambridge: Cambridge University Press.

Kirk, Geoffrey S., and John E. Raven (1957). *The Presocratic Philosophers: A Critical History with a Selection of Texts.* 1st ed. Cambridge: Cambridge University Press.

Kirk, Geoffrey S., John Earle Raven, and M. Schofield (1983). *The Presocratic Philosophers: A Critical History with a Selection of Texts.* 2nd ed. Cambridge: Cambridge University Press.

Kirkhusmo Pharo, Lars (2013). *The Ritual Practice of Time: Philosophy and Sociopolitics of Mesoamerican Calendars.* The Early Americas: History and Culture 4. Leiden: Brill.

Kleidon, Axel (2016). *Thermodynamic Foundations of the Earth System.* Cambridge: Cambridge University Press.

Klein, Ursula (1994). *Verbindung und Affinität: Die Grundlegung der neuzeitlichen Chemie an der Wende vom 17. zum 18. Jahrhundert.* Basel: Birkhäuser.

———(2001). "The Creative Power of Paper Tools in Early Nineteenth-Century Chemistry." in *Tools and Modes of Representation in the Laboratory Sciences,* ed. U. Klein, 13–34. Boston Studies in the Philosophy and History of Science 222. Dordrecht: Kluwer Academic.

———(2003). *Experiments, Models, Paper Tools: Cultures of Organic Chemistry in the Nineteenth Century.* Stanford, CA: Stanford University Press.

———(2012a). "Artisanal-Scientific Experts in Eighteenth-Century France and Germany." *Annals of Science* 69 (3): 303–6.

———(2012b). "The Prussian Mining Official Alexander von Humboldt." *Annals of Science* 69 (1): 27–68.

———(2015a). "A Revolution That Never Happened." *Studies in History and Philosophy of Science Part A* 49:80–90.

———(2015b). *Humboldts Preußen: Wissenschaft und Technik im Aufbruch.* Darmstadt: Wissenschaftliche Buchgesellschaft.

———(2016a). "Abgesang on Kuhn's 'Revolutions.'" In *Shifting Paradigms: Thomas S. Kuhn and the History of Science,* ed. A. Blum, K. Gavroglu, C. Joas, and J. Renn, 223–31. Proceedings 8. Berlin: Edition Open Access. http://edition-open-access.de/proceedings/8/18/index.html.

———(2016b). "Alexander von Humboldt: Vater der Umweltbewegung?" In *Achtsamer Umgang mit Ressourcen und miteinander—gestern und heute,* ed. Kunst und Bildung e.V. Humboldt-Gesellschaft für Wissenschaft, 115–29. Abhandlungen der Humboldt-Gesellschaft für Wissenschaft, Kunst und Bildung e.V. 37. Roßdorf: TZ-Verlag.

————(2016c). *Nützliches Wissen: Die Erfindung der Technikwissenschaften.* Göttingen: Wallstein Verlag.

————(2017). "Hybrid Experts." In *The Structures of Practical Knowledge*, ed. M. Valleriani, 287–306. Cham: Springer.

Klein, Ursula, and Wolfgang Lefèvre (2007). *Materials in Eighteenth-Century Science: A Historical Ontology.* London: Routledge.

Klingan, Katrin, Ashkan Sepahvand, Christoph Rosol, and Bernd M. Scherer, eds. (2014). *Textures of the Anthropocene: Grain Vapor Ray.* 4 vols. Cambridge, MA: MIT Press.

Klingenfeld, Daniel, and Hans Joachim Schellnhuber (2012). "Climate Change as a Global Challenge—and Its Implications for Knowledge Generation and Dissemination." In *The Globalization of Knowledge in History*, ed. J. Renn, 795–820. Studies 1. Berlin: Edition Open Access. http://edition-open-access.de/studies/1/35/index.html.

Kocka, Jürgen (2016). *Capitalism: A Short History.* Translated by Jeremiah Riemer. Princeton, NJ: Princeton University Press.

————(2017a). "Globalisierung als Motor des Fortschritts in der Geschichtswissenschaft?" *Nova Acta Leopoldina NF* 414:215–26.

————(2017b). "Schöpferische Zerstörung: Joseph Schumpeter über Kapitalismus." *Mittelweg 36* 26 (6): 45–54.

Kohler, Robert E. (1994). *Lords of the Fly: Drosophila Genetics and the Experimental Life.* Chicago: University of Chicago Press.

Kohlrausch, Martin, and Helmuth Trischler (2014). *Building Europe on Expertise: Innovators, Organizers, Networks.* Making Europe: Technology and Transformations, 1850–2000. New York: Palgrave Macmillan.

Kolbert, Elizabeth (2014). *The Sixth Extinction: An Unnatural History.* New York: Henry Holt.

Kollmorgen, Raj, Hans-Jürgen Wagener, and Wolfgang Merkel, eds. (2015). *Handbuch Transformationsforschung.* Wiesbaden: Springer.

Komlos, John (2016). "Has Creative Destruction Become More Destructive?" *B.E. Journal of Economic Analysis & Policy* 16 (4): 20160179.

Koyré, Alexandre (1939). *Études galiléennes.* Paris: Éditions Hermann.

————([1939] 1978). *Galileo Studies.* Translated by John Mepham. Atlantic Highlands, NJ: Humanities Press.

Krampf, Arie (2012). "Translation of Central Banking to Developing Countries in the Post–World War II Period: The Case of the Bank of Israel." In *The Globalization of Knowledge in History*, ed. J. Renn, 459–81. Studies 1. Berlin: Edition Open Access. http://edition-open-access.de/studies/1/24/index.html.

Krause, Johannes, and Wolfgang Haak (2017). "Neue Erkenntnisse zur genetischen Geschichte Europas." In *Tagungen des Landesmuseum für Vorgeschichte Halle (Saale)*. Band 17: *Migration und Integration von der Urgeschichte bis zum Mittelalter*, ed. H. Meller, F. Daim, J. Krause, and R. Risch, 21–38. Halle (Saale): Landesamt für Denkmalpflege und Archäologie Sachsen-Anhalt/Landesmuseum für Vorgeschichte.

Krause, Johannes, Carles Lalueza-Fox, Ludovic Orlando, Wolfgang Enard, Richard E. Green, Hernán A. Burbano, Jean-Jacques Hublin, Catherine Hänni, Javier Fortea, Marco de la Rasilla, Jaume Bertranpetit, Antonio Rosas, and Svante Pääbo (2007). "The Derived FOXP2 Variant of Modern Humans Was Shared with Neandertals." *Current Biology* 17 (21) :1908–12.

Kräutli, Florian, and Matteo Valleriani (2018). "CorpusTracer: A CIDOC Database for Tracing Knowledge Networks." *Digital Scholarship in the Humanities* 33 (2): 336–46.

Krebernik, Manfred (1998). "Die Texte aus Fara und Tell Abu Salabikh." In *Mesopotamien: Späturuk-Zeit und Frühdynastische Zeit*, ed. P. Attinger and M. Wäfler, 237–27. Annäherungen 1. Orbis Biblicus et Orientalis 160. Göttingen: Vandenhoeck & Ruprecht.

———(2007a). "Buchstabennamen, Lautwerte und Alphabetgeschichte." In *Getrennte Wege? Kommunikation, Raum und Wahrnehmung in der alten Welt*, ed. R. Rollinger, A. Luther, and J. Wiesehöfer, 108–75. Frankfurt am Main: Verlag Antike.

———(2007b). "Zur Entwicklung des Sprachbewusstseins im Alten Orient." In *Das geistige Erfassen der Welt im Alten Orient*, ed. C. Wilcke, 39–62. Wiesbaden: Harrassowitz Verlag.

Krementsov, Nikolai (1997). "Russian Science in the Twentieth Century." In *Science in the Twentieth Century*, ed. J. Krige and D. Pestre, 777–94. Amsterdam: Harwood Academic.

Krige, John (2006). "Atoms for Peace, Scientific Internationalism, and Scientific Intelligence." *Osiris* 21 (1): 161–81.

Krige, John, and Dominique Pestre, eds. (1997). *Science in the Twentieth Century*. Amsterdam: Harwood Academic.

Kroeber, Alfred L. (1940). "Stimulus Diffusion." *American Anthropologist* 42 (1): 1–20.

Kühn, Karl Gottlob, ed. (1828). *Claudii Galeni opera omnia*. Vol. 17a: *Medicorum graecorum opera quae extant*. Leipzig: Car. Cnoblochii. http://www.biusante.parisdescartes.fr /histoire/medica/resultats/?cote=45674x17a&do=chapitre.

Kuhn, Thomas S. (1959). *The Copernican Revolution: Planetary Astronomy in the Development of Western Thought*. New York: Random House.

———(1970). *The Structure of Scientific Revolutions*. 2nd enlg. ed. International Encyclopedia of Unified Science 2.2. Chicago: University of Chicago Press.

———(1979). Foreword to *Genesis and Development of a Scientific Fact*, by Ludwik Fleck (1935), ed. T. J. Trenn and R. K. Merton, vii–xi. Chicago: University of Chicago Press.

———(1996). *The Structure of Scientific Revolutions*. 1970. 3rd ed. Chicago: University of Chicago Press.

———(2000). *The Road since Structure: Philosophical Essays, 1970–1993, with an Autobiographical Interview*. Chicago: University of Chicago Press.

Kurapkat, Dietmar (2014). "Bauwissen im Neolithikum Vorderasiens." In *Wissensgeschichte der Architektur*. Vol. 1: *Vom Neolithikum bis zum Alten Orient*, ed. J. Renn, W. Osthues, and H. Schlimme, 57–127. Studies 3. Berlin: Edition Open Access. http://edition-open-access.de /studies/3/4/index.html.

Kusukawa, Sachiko, and Ian Maclean, eds. (2006). *Transmitting Knowledge: Words, Images, and Instruments in Early Modern Europe*. Oxford: Oxford University Press.

Lagrange, Joseph Louis de ([1811] 1997). *Analytical Mechanics*. Translated by Auguste Boissonnade and Victor N. Vagliente. Boston Studies in the Philosophy of Science 191. Dordrecht: Kluwer Academic.

Laland, Kevin N. (2017). *Darwin's Unfinished Symphony: How Culture Made the Human Mind*. Princeton, NJ: Princeton University Press.

Lalli, Roberto (2014). "A New Scientific Journal Takes the Scene: The Birth of Reviews of Modern Physics." *Annalen der Physik* 526 (9–10): A83–A87.

———(2016). " 'Dirty Work,' but Someone Has to Do It: Howard P. Robertson and the Refereeing Practices of *Physical Review* in the 1930s." *Notes and Records* 70 (2): 151–74.

———(2017). *Building the General Relativity and Gravitation Community during the Cold War*. SpringerBriefs in History of Science and Technology. Cham: Springer.

Lamarck, Jean-Baptiste Pierre Antoine de Monet de ([1809] 1963). *Zoological Philosophy: An Exposition with Regard to the Natural History of Animals—The Diversity of Their Organisation and the Faculties Which They Derive from It—The Physical Causes Which Maintain Life within Them and Give Rise to Their Various Movements—Lastly, Those Which Produce*

Feeling and Intelligence in Some among Them. Translated by Hugh Elliot. New York: Hafner.

Lambert, Wilfred G. (1957). "Ancestors, Authors and Canonicity." *Journal of Cuneiform Studies* 11 (1): 1–14.

Lampedusa, Giuseppe Tomasi di ([1958] 1963). *The Leopard.* Translated by Archibald Colquhoun. Rev. ed. London: Collins/Fontana Books.

Lamprecht, Heinz-Otto (1996). *Opus Caementitium: Bautechnik der Römer.* Düsseldorf: Beton-Verlag.

Landecker, Hannah (2016). "Antibiotic Resistance and the Biology of History." *Body & Society* 22 (4): 19–52.

Landes, David S. (1969). *The Unbound Prometheus: Technological Change and Industrial Development in Western Europe from 1750 to the Present.* London: Cambridge University Press.

Lane, Jan-Erik (2000). *New Public Management: An Introduction.* London: Routledge.

Larner, John (1999). *Marco Polo and the Discovery of the World.* New Haven, CT: Yale University Press.

Latour, Bruno (2014). "Agency at the Time of the Anthropocene." *New Literary History* 45 (1): 1–18.

Lattery, Mark J. (2016). *Deep Learning in Introductory Physics: Exploratory Studies of Model-Based Reasoning.* Charlotte, NC: Information Age.

Laubichler, Manfred D. (2009). "Evolutionary Developmental Biology Offers a Significant Challenge to the Neo-Darwinian Paradigm." In *Contemporary Debates in Philosophy of Biology,* ed. F. J. Ayala and R. Arp, 199–212. Hoboken, NJ: Wiley-Blackwell.

Laubichler, Manfred D., and Jane Maienschein (2013). "Developmental Evolution." In *The Cambridge Encyclopedia of Darwin and Evolutionary Thought,* ed. M. Ruse, 375–82. Cambridge: Cambridge University Press.

Laubichler, Manfred D., Jane Maienschein, and Jürgen Renn (2013). "Computational Perspectives in the History of Science: To the Memory of Peter Damerow." *Isis* 104 (1): 119–30.

Laubichler, Manfred D., Sonja J. Prohaska, and Peter F. Stadler (2018). "Toward a Mechanistic Explanation of Phenotypic Evolution: The Need for a Theory of Theory Integration." *Journal of Experimental Zoology Part B: Molecular and Developmental Evolution* 330 (1): 5–14.

Laubichler, Manfred D., and Jürgen Renn (2015). "Extended Evolution: A Conceptual Framework for Integrating Regulatory Networks and Niche Construction." *Journal of Experimental Zoology Part B: Molecular and Developmental Evolution* 324 (7): 565–77.

———(2019). "Daring to Ask the Big Questions." In *Looking Back as We Move Forward: The Past, Present, and Future of the History of Science; Liber amicorum for Jed Z. Buchwald on His 70th Birthday,* 202–12. Pasadena, CA: N.p. [Caltech], ISBN 978-0-578-45417-7.

Laue, Max von (1934). "Fritz Haber." *Die Naturwissenschaften* 22 (7): 97.

Lavoisier, Antoine-Laurent de (1778). "Considérations générales sur la nature des acides, et sur les principes dont ils sont composés." In *Histoire de l'Académie Royale des Sciences: Avec les mémoires de mathématique & de physique, pour la même anneé—Tirés des registres de cette Académie,* ed. J. Boudot, 535–47. Paris: Imprimerie Royale.

———(1862–93). *Oeuvres de Lavoisier: Publiées par les soins de son excellence le Ministre de l'instruction publique et des cultes.* Vol. 3 of 6. Paris: Imprimerie Impériale.

Lavoisier, Antoine-Laurent de, Louis Bernard Guyton de Morveau, Claude-Louis Berthollet, and Antoine François de Fourcroy (1787). *Méthode de nomenclature chimique.* Paris: Cuchet.

Lax, Gregor (2018a). *From Atmospheric Chemistry to Earth System Science: Contributions to the Recent History of the Max Planck Institute for Chemistry (Otto- Hahn- Institute), 1959–2000.* Translated by Sabine Wacker. Diepholz: GNT-Verlag.

———(2018b). *Von der Atmosphärenchemie zur Erforschung des Erdsystems: Beiträge zur jüngeren Geschichte des Max-Planck-Instituts für Chemie (Otto-Hahn-Institut), 1959–2000.*

Preprint 5. Berlin: Forschungsprogramm Geschichte der Max-Planck-Gesellschaft. http://gmpg.mpiwg-berlin.mpg.de/media/cms_page_media/2/GMPG-Preprint_05_Lax_2018_gejUMBp.pdf.

Lazega, Emmanuel, and Tom A. B. Snijders, eds. (2016). *Multilevel Network Analysis for the Social Sciences: Theory, Methods and Applications*. Methodos Series 12. Cham: Springer.

Leeuw, Sander van der (2012). "For Every Solution There Are Many Problems: The Role and Study of Technical Systems in Socio-environmental Coevolution." *Geografisk Tidsskrift* 112 (2): 105–16.

Leeuwen, Joyce van (2016). *The Aristotelian Mechanics: Text and Diagrams*. Dordrecht: Springer.

Lefèvre, Wolfgang (1978). *Naturtheorie und Produktionsweise: Probleme einer materialistischen Wissenschaftsgeschichtsschreibung; Eine Studie zur Genese der neuzeitlichen Naturwissenschaft*. Darmstadt: Luchterhand Literaturverlag.

———(1981). "Rechensteine und Sprache: Zur Begründung der wissenschaftlichen Mathematik durch die Pythagoreer." In *Rechenstein, Experiment, Sprache: Historische Fallstudien zur Entstehung der exakten Wissenschaften*, ed. P. Damerow and W. Lefèvre, 115–221. Stuttgart: Klett-Cotta.

———(2000). "Material and Social Conditions in a Historical Epistemology of Scientific Thinking." In *Science and Power: The Historical Foundations of Research Policies in Europe*, ed. L. Guzzetti, 239–46. Brussels: European Communities.

———(2003). "Darwin, Marx, and Warranted Progress." In *Revisiting the Foundations of Relativistic Physics: Festschrift in Honor of John Stachel*, ed. J. Renn, A. Ashtekar, R. S. Cohen, D. Howard, S. Sarkar, and A. Shimony, 593–613. Dordrecht: Springer.

———(2004). *Picturing Machines: 1400–1700*. Cambridge, MA: MIT Press.

———(2007). "Der Darwinismus-Streit der Evolutionsbiologen." In *Der Darwinismus-Streit*, ed. K. Bayertz, M. Gerhard, and W. Jaeschke, 19–46. Weltanschauung, Philosophie und Naturwissenschaft im 19. Jahrhundert 2. Hamburg: Felix Meiner Verlag.

———(2009a). *Die Entstehung der biologischen Evolutionstheorie*. 2nd ed. Frankfurt am Main: Suhrkamp.

———(2009b). "Epilog nach 25 Jahren: Facetten einer Revolution." In *Die Entstehung der biologischen Evolutionstheorie*, ed. W. Lefèvre, 294–317. Frankfurt am Main: Suhrkamp.

Lefèvre, Wolfgang, and Marcus Popplow (2006–9). "Database Machine Drawings." Accessed August 22, 2018. http://dmd.mpiwg-berlin.mpg.de/home.

Lefèvre, Wolfgang, Jürgen Renn, and Urs Schoepflin, eds. (2003). *The Power of Images in Early Modern Science*. Basel: Birkhäuser.

Lehner, Christoph, and Helge Wendt (2017). "Mechanics in the Querelle des Anciens et des Modernes." *Isis* 108 (1): 26–39.

Leibniz, Gottfried Wilhelm ([1686] 1989). "A Brief Demonstration of a Notable Error of Descartes and Others concerning a Natural Law." In *Philosophical Papers and Letters*, ed. L. E. Loemker, 296–302. Dordrecht: Kluwer Academic.

———([1695] 1989). "Specimen dynamicum." In *Philosophical Papers and Letters*, ed. L. E. Loemker, 435–52. Dordrecht: Kluwer Academic.

———([1714] 1889). "The Monadology." In *The Monadology and Other Philosophical Writings*, 215–77. Oxford: Clarendon Press.

Lemay, Richard (1962). *Abu Ma'shar and Latin Aristotelianism in the Twelfth Century: The Recovery of Aristotle's Natural Philosophy through Arabic Astrology*. Oriental Series 38. Beirut: American University of Beirut.

———(1963). "Dans l'Espagne du XIIe siècle: Les traductions de l'arabe au latin." *Annales: Histoire, Sciences Sociales* 18 (4): 639–65.

———(1977). "The Hispanic Origin of Our Present Numeral Forms." *Viator* (8): 435–77.

Lemercier, Claire (2015). "Formal Network Methods in History: Why and How?" In *Social Networks, Political Institutions, and Rural Societies*, ed. G. Fertig, 281–304. Rural History in Europe 11. Turnhout: Brepols.

Lenzen, Manuela (2018). *Künstliche Intelligenz: Was sie kann & was uns erwartet*. Munich: C. H. Beck.

Leontiev, Aleksej N. (1978). *Activity, Consciousness, and Personality*. Translated by Marie J. Hall. Englewood Cliffs, NJ: Prentice Hall.

———(1981). *Problems of the Development of the Mind* [*Problemy razvitija psichiki*, originally published 1959]. Translated by M. Kopylova. Moscow: Progress.

Lepik, Andres (1994). *Das Architekturmodell in Italien 1335–1550*. Worms: Wernersche Verlagsgesellschaft.

———(1995). "Das Architekturmodell der frühen Renaissance: Die Erfindung eines Mediums." In *Architekturmodelle der Renaissance: Die Harmonie des Bauens von Alberti bis Michelangelo*, ed. B. Evers and S. Benedetti, 10–20. Munich: Prestel-Verlag.

Levine, Baruch A. (2014). "Global Monotheism: The Contribution of the Israelite Prophets." In *Melammu: The Ancient World in an Age of Globalization*, ed. M. J. Geller, 29–47. Proceedings 7. Berlin: Edition Open Access. http://edition-open-access.de/proceedings/7/4/index.html.

Levinson, Stephen C. (2003). *Space in Language and Cognition: Explorations in Cognitive Diversity*. Language, Culture and Cognition 5. Cambridge: Cambridge University Press.

———(2014). "Language and Wallace's Problem." *Science* 344 (6191): 1458–59.

———(in press). "Interactional Foundations of Language: The Interaction Engine Hypothesis." In *Human Language: From Genes and Brains to Behavior*, ed. P. Hagoort. Cambridge, MA: Harvard University Press.

Levinson, Stephen C., and Dan Dediu (2018). "Neanderthal Language Revisited: Not Only Us," *Current Opinion in Behavioral Sciences* 21:49–55.

Levinson, Stephen C., and Judith Holler (2014). "The Origin of Human Multi-Modal Communication" *Philosophical Transactions of the Royal Society B: Biological Sciences* 369 (1651): 20130302.

Levit, Georgi S. (2001). *Biogeochemistry—Biosphere—Noosphere: The Growth of the Theoretical System of Vladimir Ivanovich Vernadsky*. Berlin: Verlag für Wissenschaft und Bildung.

Lewis, Cherry (2000). *The Dating Game: One Man's Search for the Age of the Earth*. Cambridge: Cambridge University Press.

Lewis, David (1972). *We, the Navigators: The Ancient Art of Landfinding in the Pacific*. Canberra: Australian National University Press.

Lewis, Simon L., and Mark A. Maslin (2015). "Defining the Anthropocene." *Nature* 519 (7542): 171–80.

———(2018). *The Human Planet: How We Created the Anthropocene*. New Haven, CT: Yale University Press.

Liebig, Justus von (1840). *Organic Chemistry in Its Applications to Agriculture and Physiology*. London: Taylor and Walton.

Lipphardt, Veronika, and David Ludwig (2011). "Wissens- und Wissenschaftstransfer." Institut für Europäische Geschichte (IEG). Accessed June 27, 2018. http://www.ieg-ego.eu/lipphardtv-ludwigd-2011-de.

Lisheng, Feng (2017). "On the Structure and Functions of the Multiplication Table in the Tsinghua Collection of Bamboo Slips." *Chinese Annals of History of Science and Technology* 1 (1): 1–23.

Lloyd, Geoffrey E. R. (1964). *Aristotle: The Growth and Structure of His Thought*. Cambridge: Cambridge University Press.

————(1970). *Early Greek Science: Thales to Aristotle*. New York: W. W. Norton.

————(1973). *Greek Science after Aristotle*. New York: W. W. Norton.

————(1996). *Adversaries and Authorities: Investigations into Ancient Greek and Chinese Science*. Cambridge: Cambridge University Press.

————(2002). *The Ambitions of Curiosity: Understanding the World in Ancient Greece and China*. Cambridge: Cambridge University Press.

Lloyd, Geoffrey E. R., and Nathan Sivin (2003). *The Way and the Word: Science and Medicine in Early China and Greece*. New Haven, CT: Yale University Press.

Lock, Andrew J. (2000). "Phylogenetic Time and Symbol Creation: Where Do Zopeds Come From?" *Culture & Psychology* 6 (2): 105–29.

Locke, John (1689). *Two Treatises of Government: In the Former, the False Principles, and Foundation of Sir Robert Filmer, and His Followers, Are Detected and Overthrown, The Latter Is an Essay concerning the True Original, Extent, and End of Civil Government*. London: Awnsham Churchill. https://en.wikisource.org/wiki/Two_Treatises_of_Government.

Logan, Corina J., Alexis J. Breen, Alex H. Taylor, Russell D. Gray, and William J. E. Hoppitt (2016). "How New Caledonian Crows Solve Novel Foraging Problems and What It Means for Cumulative Culture." *Learning & Behavior* 44 (1): 18–28.

Lorentz, Hendrik Antoon (1892). *La théorie électromagnétique de Maxwell et son application aux corps mouvants*. Leiden: Brill.

————([1892] 1937). "The Relative Motion of the Earth and the Ether." In *Collected Papers*. Vol. 4, ed. P. Zeeman and A. D. Fokker, 219–223. The Hague: Martinus Nijhoff Publishers.

————(1895). *Versuch einer Theorie der electrischen und optischen Erscheinungen in bewegten Körpern*. Leiden: Brill.

————(1899). "Simplified Theory of Electrical and Optical Phenomena in Moving Systems." *Proceedings of the Royal Netherlands Academy of Arts and Sciences* 1:427–42.

————(1904). "Electromagnetic Phenomena in a System Moving with Any Velocity Smaller Than That of Light." *Proceedings of the Royal Netherlands Academy of Arts and Sciences* 6:809–31.

Losee, John (2004). *Theories of Scientific Progress: An Introduction*. London: Routledge.

Lovelock, James E. ([1979] 2000). *Gaia: A New Look at Life on Earth*. Reissued, with a new preface and corrections. Oxford: Oxford University Press.

Lovelock, James E., and Lynn Margulis (1974). "Atmospheric Homeostasis by and for the Biosphere: The Gaia Hypothesis." *Tellus* 26 (1–2): 2–10.

Lucas, Adam Robert (2005). "Industrial Milling in the Ancient and Medieval Worlds: A Survey of the Evidence for an Industrial Revolution in Medieval Europe." *Technology and Culture* 46 (1): 1–30.

Lucht, Wolfgang (2010). "Commentary: Earth System Analysis and Taking a Crude Look at the Whole." In *Global Sustainability: A Nobel Cause*, ed. H. J. Schellnhuber, N. Molina, M. Stern, V. Huber, and S. Kadner, 19–32. Cambridge: Cambridge University Press.

Lucht, Wolfgang, and Rajendra Kumar Pachauri (2004). "The Mental Component of the Earth System." In *Earth System Analysis for Sustainability*, ed. H. J. Schellnhuber, P. J. Crutzen, W. C. Clark, M. Claussen, and H. Held, 341–65. Cambridge, MA: MIT Press.

Lurija, Aleksandr R. (1979). *The Making of Mind: A Personal Account of Soviet Psychology*. Cambridge, MA: Harvard University Press.

Lyn, Heidi, Patricia M. Greenfield, E. Sue Savage-Rumbaugh, Kristen Gillespie-Lynch, and William D. Hopkins (2011). "Nonhuman Primates Do Declare! A Comparison of Declarative Symbol and Gesture Use in Two Children, Two Bonobos, and a Chimpanzee." *Language & Communication* 31 (1): 63–74.

Lyth, Peter, and Helmuth Trischler, eds. (2003). *Wiring Prometheus: History, Globalisation and Technology*. Aarhus: Aarhus University Press.

Mach, Ernst ([1905] 1976). *Knowledge and Error*. Translated by Thomas J. McCormack and Paul Foulkes. Dordrecht: D. Reidel.

———(1910). "Die Leitgedanken meiner naturwissenschaftlichen Erkenntnislehre und ihre Aufnahme durch die Zeitgenossen." *Physikalische Zeitschrift* 11 (1): 599–606.

———([1910] 1992). "The Leading Thoughts of My Scientific Epistemology and Its Acceptance by Contemporaries." In *Ernst Mach: A Deeper Look—Documents and New Perspectives*, ed. J. Blackmore, 133–39. Dordrecht: Kluwer Academic.

MacLean, Ian (2009). *Learning and the Market Place: Essays in the History of the Early Modern Book*. Leiden: Brill.

———(2012). *Scholarship, Commerce, Religion: The Learned Book in the Age of Confessions, 1560–1630*. Cambridge, MA: Harvard University Press.

Maier, Anneliese (1949–58). *Studien zur Naturphilosophie der Spätscholastik*. 5 vols. Rome: Edizioni di Storia e Letteratura.

———(1964). *Ausgehendes Mittelalter: Gesammelte Aufsätze zur Geistesgeschichte des 14. Jahrhunderts*. 3 vols. Rome: Edizioni di Storia e Letteratura.

———(1982). *On the Threshold of Exact Science: Selected Writings of Anneliese Maier on Late Medieval Natural Philosophy*. Translated by Steven D. Sargent. Philadelphia: University of Pennsylvania Press.

Maier, John (1998). *Gilgamesh: A Reader*. Mundelein, IL: Bolchazy-Carducci.

Malinowski, Bronislaw (1947). *Freedom and Civilization*. London: George Allen & Unwin.

———(2002). *Collected Works*. 10 vols. Reprint, London: Routledge.

Malkin, Irad (2011). *A Small Greek World: Networks in the Ancient Mediterranean*. Oxford: Oxford University Press.

Malm, Andreas (2016). *Fossil Capital: The Rise of Steam Power and the Roots of Global Warming*. London: Verso Books.

———(2018). *The Progress of This Storm: Nature and Society in a Warming World*. New York: Verso Books.

Malpangotto, Michaela (2016). "The Original Motivation for Copernicus's Research." *Archive for History of Exact Sciences* 70 (3): 361–411.

Manetti, Antonio (1970). *The Life of Brunelleschi*. University Park: Penn State University Press.

———(1976). *Vita di Filippo Brunelleschi, preceduta da la novella del Grasso*. Milan: Edizioni Il Polifilo.

Marcon, Federico (2015). *The Knowledge of Nature and the Nature of Knowledge in Early Modern Japan*. Chicago: University of Chicago Press.

Marin, Louis (1988). *Portrait of the King*. Translated by Martha M. Houle. Theory and History of Literature 57. Minneapolis: University of Minnesota Press.

Markovits, Claude, Jacques Pouchepadass, and Sanjay Subrahmanyam, eds. (2006). *Society and Circulation: Mobile People and Itinerant Cultures in South Asia, 1750–1950*. London: Anthem Press.

Marx, Karl ([1843] 1975). "Contribution to the Critique of Hegel's Philosophy of Law." In *Karl Marx / Frederick Engels: Collected Works*. Vol. 3: *Marx and Engels: March 1843–August 1844*, 3–129. London: Lawrence & Wishart.

———([1844] 1970). *Economic and Philosophic Manuscripts of 1844*. Translated by Martin Milligan. London: Lawrence & Wishart.

———([1852] 1979). "The Eighteenth Brumaire of Louis Bonaparte." In *Karl Marx / Frederick Engels: Collected Works*. Vol. 11: *Marx and Engels: 1851–53*, 99–197. London: Lawrence & Wishart.

———([1859] 1987). "A Contribution to the Critique of Political Economy: Part One." In *Karl Marx / Frederick Engels: Collected Works*. Vol. 29: *Marx: 1857–61*, 257–417. London: Lawrence & Wishart.

————([1867] 1990). *Capital, a Critical Analysis of Capitalist Production*. Translated by Samuel Moore and Edward Aveling (London, 1887). Originally published in German, 1867. *Marx-Engels-Gesamtausgabe (MEGA)*. Abt. 2: "Das Kapital" und Vorarbeiten 9.1. Berlin: Dietz Verlag.

Marx, Karl, and Frederick Engels ([1845–46] 1976). "The German Ideology: Critique of Modern German Philosophy According to Its Representatives Feuerbach, B. Bauer and Stirner, and of German Socialism According to Its Various Prophets." In *Karl Marx / Frederick Engels: Collected Works*. Vol. 5: *Marx and Engels: 1845–47*, 19–539. London: Lawrence & Wishart.

Matthews, Thomas A., and Allan R. Sandage (1963). "Optical Identification of 3c 48, 3c 196, and 3c 286 with Stellar Objects." *Astrophysical Journal* 138 (1): 30–56.

Mauelshagen, Franz (2003). "Netzwerke des Vertrauens: Gelehrtenkorrespondenzen und wissenschaftlicher Austausch in der Frühen Neuzeit." In *Vertrauen: Historische Annäherungen*, ed. U. Frevert, 119–51. Göttingen: Vandenhoeck & Ruprecht.

————(2016). "Der Verlust der (bio-)kulturellen Diversität im Anthropozän." In *Die Welt im Anthropozän: Erkundungen im Spannungsfeld zwischen Ökologie und Humanität*, ed. W. Haber, M. Held, and M. Vogt, 39–55. Munich: Oekom Verlag).

————(2018a). "Climate as a Scientific Paradigm: Early History of Climatology to 1800." In *The Palgrave Handbook of Climate History*, ed. S. White, C. Pfister, and F. Mauelshagen, 565–88. London: Palgrave Macmillan.

————(2018b). "Migration and Climate in World History." In *The Palgrave Handbook of Climate History*, ed. S. White, C. Pfister, and F. Mauelshagen, 413–44. London: Palgrave Macmillan.

Maynard Smith, John, and Eörs Szathmáry (1995). *The Major Transitions in Evolution*. Oxford: Oxford University Press.

Mayr, Otto (1971). "Maxwell and the Origins of Cybernetics." *Isis* 62:424–44.

Max Planck Institute for the History of Science (1994–). "Research Reports." https://www.mpiwg-berlin.mpg.de/research-reports.

Maxwell, James Clerk (1865). "VIII. A Dynamical Theory of the Electromagnetic Field." *Philosophical Transactions of the Royal Society of London* 155:459–512.

McCloskey, Michael (1983). "Naive Theories of Motion." In *Mental Models*, ed. D. Gentner and A. L. Stevens, 299–324. Hillsdale, NJ: Erlbaum.

McCloskey, Michael, Alfonso Caramazza, and Bert Green (1980). "Curvilinear Motion in the Absence of External Forces: Naive Beliefs about the Motion of Objects." *Science* 210 (4474): 1139–41.

McCorduck, Pamela (2004). *Machines Who Think: A Personal Inquiry into the History and Prospects of Artificial Intelligence*. 2nd ed. Natick, MA: A K Peters.

McCraw, Thomas K. (2007). *Prophet of Innovation: Joseph Schupmeter and Creative Destruction*. Cambridge, MA: Harvard University Press.

McDermott, Drew, and Jon Doyle (1980). "Non-monotonic Logic I." *Artificial Intelligence* 13 (1–2): 41–72.

McLaughlin, Peter, and Jürgen Renn (2018). "The Balance, the Lever and the Aristotelian Origins of Mechanics." In *Emergence and Expansion of Pre-classical Mechanics*, ed. R. Feldhay, J. Renn, M. Schemmel, and M. Valleriani. Boston Studies in the Philosophy and History of Science 333. Dordrecht: Springer.

McNeill, John R. (1994). "Of Rats and Men: A Synoptic Environmental History of the Island Pacific." *Journal of World History* 5 (2): 299–349.

————(2000). *Something New under the Sun: An Environmental History of the Twentieth-Century World*. New York: W. W. Norton.

McNeill, John R., and Peter Engelke (2014). *The Great Acceleration: An Environmental History of the Anthropocene since 1945*. Cambridge, MA: Belknap Press of Harvard University Press.

McNeill, John R., Mahesh Rangarajan, and Jose Augusto Padua, eds. 2009. *Environmental History: As if Nature Existed*. New Delhi: Oxford University Press.

McNeill, William H. (1993). "The Age of Gunpowder Empires, 1450–1800." In *Islamic and European Expansion: The Forging of a Global Order*, ed. M. Adas, 103–40. Philadelphia: Temple University Press.

Meadows, Donella H., Dennis L. Meadows, Jørgen Randers, and William W. III Behrens (1972). *The Limits to Growth: A Report for the Club of Rome's Project on the Predicament of Mankind*. New York: Universe Books.

Medhurst, Walter Henry (1838). *China: Its State and Prospects, with Especial Reference to the Spread of the Gospel: Containing Allusions to the Antiquity, Extent, Population, Civilization, Literature, and Religion of the Chinese*. London: John Snow.

Medick, Hans (2016). "Turning Global? Microhistory in Extension." *Historische Anthropologie* 24 (2): 241–52.

Meier, Christian, and Reinhard Koselleck (1975). "Fortschritt." In *Geschichtliche Grundbegriffe*. Vol. 2, ed. C. Meier and R. Koselleck, 371–423. Stuttgart: Klett-Cotta.

Meli, Domenico Bertoloni (1993). *Equivalence and Priority: Newton versus Leibniz—Including Leibniz's Unpublished Manuscripts on the "Principia."* Oxford: Clarendon Press.

Mercier, André, and Michel Kervaire, eds. (1956). *Fünfzig Jahre Relativitätstheorie: Verhandlungen—Cinquantenaire de la théorie de la relativité: Actes—Jubilee of Relativity Theory: Proceedings*. Helvetica Physica Acta 4. Basel: Birkhäuser.

Merton, Robert K. (1968). "The Matthew Effect in Science." *Science* 159 (3810): 56–63.

Mesoudi, Alex, Andrew Whiten, and Kevin N. Laland (2006). "Towards a Unified Science of Cultural Evolution." *Behavioral and Brain Sciences* 29 (4): 329–47.

Meyer, John W., and Brian Rowan (1977). "Institutionalized Organizations: Formal Structure as Myth and Ceremony." *American Journal of Sociology* 83 (2): 340–63.

Michelson, Albert A. (1881). "The Relative Motion of the Earth and the Luminiferous Ether." *American Journal of Science* 3/22 (122): 120–29.

Michelson, Albert A., and Edward W. Morley (1887). "On the Relative Motion of the Earth and the Luminiferous Ether." *American Journal of Science* 3/34 (134): 333–45.

Middleton, Guy D. (2017). *Understanding Collapse: Ancient History and Modern Myths*. Cambridge: Cambridge University Press.

Mignolo, Walter D. (2000). *Local Histories / Global Designs: Coloniality, Subaltern Knowledges, and Border Thinking*. Princeton, NJ: Princeton University Press.

Milgram, Stanley (1967). "The Small-World Problem." *Psychology Today* 1 (1): 61–67.

Miller, Seumas (2014). "Social Institutions: The Stanford Encyclopedia of Philosophy." Metaphysics Research Lab, Center for the Study of Language and Information, Stanford University. Accessed August 23, 2018. https://plato.stanford.edu/archives/win2014/entries/social-institutions.

Milne, Alice E., Jutta L. Mueller, Claudia Männel, Adam Attaheri, Angela D. Friederici, and Christopher I. Petkov (2016). "Evolutionary Origins of Non-adjacent Sequence Processing in Primate Brain Potentials." *Scientific Reports* 6 (36259).

Minsky, Marvin (1975). "A Framework for Representing Knowledge." In *The Psychology of Computer Vision*, ed. P. H. Winston, 211–76. New York: McGraw-Hill.

———(1986). *The Society of Mind*. New York: Simon & Schuster.

Mirowski, Philip, and Esther-Mirjam Sent (2007). "The Commercialization of Science and the Response of STS." In *The Handbook of Science and Technology Studies*, ed. E. J. Hackett, O. Amsterdamska, M. E. Lynch, and J. Wajcman, 719–40. Cambridge, MA: MIT Press.

Misak, Cheryl (2013). *The American Pragmatists*. Oxford: Oxford University Press.

Mitchell, Timothy (2011). *Carbon Democracy: Political Power in the Age of Oil*. London: Verso Books.

Mittasch, Alwin (1951). *Geschichte der Ammoniaksynthese*. Weinheim: Verlag Chemie.

Mody, Cyrus C.M. (2016). *The Long Arm of Moore's Law: Microelectronics and American Science*. Cambridge: MIT Press.

Moiseev, Nikita N. (1993). "A New Look at Evolution: Marx, Teilhard de Chardin, Vernadsky." *World Futures* 36 (1): 1–19.

Mokyr, Joel (1999). "The Second Industrial Revolution, 1870–1914." In *L'età della rivoluzione industriale*, ed. V. Castronovo, 219–45. Storia dell'economia mondiale. Rome: Editori Laterza.

———(2002). *The Gifts of Athena: Historical Origins of the Knowledge Economy*. Princeton, NJ: Princeton University Press.

Molina, Mario J., and F. Sherman Rowland (1974). "Stratospheric Sink for Chlorofluoromethanes: Chlorine Atom-Catalysed Destruction of Ozone." *Nature* 249:810–12.

Möllers, Nina, Christian Schwägerl, and Helmuth Trischler, eds. (2015). *Willkommen im Anthropozän: Unsere Verantwortung für die Zukunft der Erde*. Munich: Deutsches Museum.

Monatheuil, Henri de, ed. (1599). *Aristotelis mechanica: Graeca, emendata, latina facta, & commentariis illustrata*. Paris: Perier. http://echo.mpiwg-berlin.mpg.de/MPIWG:GYZ68VPF.

Montesinos Sirera, José , and Jürgen Renn (2003). "Expediciones científicas a las Islas Canarias en el periodo romántico (1770–1830)" In *Ciencia y romanticismo 2002*, ed. J. Montesinos, J. Ordónez, and S. Toledo, 329–53. Maspalomas: Fundación Canaria Orotava de Historia de la Ciencia.

Moody, Ernest A., and Marshall Clagett, eds. (1960). *The Medieval Science of Weights (Scienta de ponderibus): Treatises Ascribed to Euclid, Archimedes, Thabit Ibn Qurra, Jordanus de Nemore and Blasius of Parma*. 2nd ed. University of Wisconsin Publications in Medieval Science 1. Madison: University of Wisconsin Press.

Moore, Gordon E. (1965). "Cramming More Components onto Integrated Circuits." *Electronics* 38 (8): 114–17.

Moore, Jason W., ed. (2016). *Anthropocene or Capitalocene? Nature, History, and the Crisis of Capitalism*. Oakland, CA: PM Press.

Morford, Jill P. (2002). "Why Does Exposure to Language Matter?" In *The Evolution of Language out of Pre-language*, ed. T. Givón and B. F. Malle, 329–41. Typological Studies in Language 53. Amsterdam: John Benjamins.

Morgan, Mary S., and Margaret Morrison, eds. (1999). *Models as Mediators: Perspectives on Natural and Social Science*. Cambridge: Cambridge University Press.

Morozov, Evgeny (2013). *To Save Everything, Click Here: Technology, Solutionism, and the Urge to Fix Problems That Don't Exist* . London: Allen Lane.

———(2014a). "Don't Believe the Hype, the 'Sharing Economy' Masks a Failing Economy." *Guardian*, September 28. https://www.theguardian.com/commentisfree/2014/sep/28/sharing-economy-internet-hype-benefits-overstated-evgeny-morozov.

———(2014b). "The Rise of Data and the Death of Politics." *Guardian*. July 19. https://www.theguardian.com/technology/2014/jul/20/rise-of-data-death-of-politics-evgeny-morozov-algorithmic-regulation.

———(2018). "Silicon Valley oder die Zukunft des digitalen Kapitalismus." *Blätter für deutsche und internationale Politik* 1:93–104.

Mosley, Layna (2003). *Global Capital and National Governments*. New York: Cambridge University Press.

Most, Glenn W., ed. (1999). *Commentaries—Kommentare: Aporemata—Kritische Studien zur Philologiegeschichte*. Göttingen: Vandenhoeck & Ruprecht.

Moulier-Boutang, Yann (2012). *Cognitive Capitalism*. Cambridge: Polity Press.

Mowery, David C., and Nathan Rosenberg (1999). *Paths of Innovation: Technological Change in 20th-Century America*. Cambridge: Cambridge University Press.

Müller-Hansen, Finn, Maja Schlüter, Michael Mäs, Jonathan F. Donges, Jakob J. Kolb, Kirsten Thonicke, and Jobst Heitzig (2017). "Towards Representing Human Behavior and Decision Making in Earth System Models: An Overview of Techniques and Approaches." *Earth System Dynamics* 8:977–1007.

Müller-Wille, Staffan (1999). *Botanik und weltweiter Handel: Zur Begründung eines Natürlichen Systems der Pflanzen durch Carl von Linné (1707–1778)*. Studien zur Theorie der Biologie 3. Berlin: Verlag für Wissenschaft und Bildung.

Müller-Wille, Staffan, and Hans-Jörg Rheinberger (2012). *A Cultural History of Heredity*. Chicago: University of Chicago Press.

Mulvany, Patrick (2005). "Corporate Control over Seeds: Limiting Access and Farmers' Rights." *IDS Bulletin* 36 (2): 68–74.

Münch, Richard (2016). "Academic Capitalism." In *Oxford Research Encyclopedia of Politics*. Oxford: Oxford University Press. http://politics.oxfordre.com/view/10.1093/acrefore/9780190228637.001.0001/acrefore-9780190228637-e-15.

National Human Genome Research Institute (n.d.). "All about the Human Genome Project (HGP)." Accessed November 29, 2018. https://www.genome.gov/hgp.

Naveh, Zev (1982). "Landscape Ecology as an Emerging Branch of Human Ecosystem Science." In *Advances in Ecological Research*. Vol. 12, ed. A. Macfadyen Ford, 189–237. London: Academic Press.

Needham, Joseph, ed. (1954–2015). *Science and Civilisation in China*. 26 vols. Cambridge: Cambridge University Press.

———(1964). "Chinese Priorities in Cast Iron Metallurgy." *Technology and Culture* 5 (3): 398–404.

———(1986). *Chemistry and Chemical Technology 7: Military Technology—The Gunpowder Epic*. Reprint ed. Science and Civilization in China 5. Cambridge: Cambridge University Press.

Nellen, Henk (1990). "Editing Seventeenth-Century Scholarly Correspondence: Grotius, Huygens and Mersenne." *LIAS: Sources and Documents Relating to the Early Modern History of Ideas* 17 (1): 9–20.

Nelson, Anitra (2001). "The Poverty of Money: Marxian Insights for Ecological Economists." *Ecological Economics* 36 (3): 499–511.

Nelson, Richard R. (1959). "The Simple Economics of Basic Scientific Research." *Journal of Political Economy* 67 (3): 297–306.

———(1986). "Institutions Supporting Technical Advance in Industry." *American Economic Review* 76 (2): 186–89.

Nelson, Sara, Christoph Rosol, and Jürgen Renn, eds. (2017). "Perspectives on the Technosphere." Special issue of *Anthropocene Review* 4 (1–2).

Nerdinger, Winfried, ed. (2012). *Der Architekt: Geschichte und Gegenwart eines Berufstandes*. Vol. 1. Munich: Prestel-Verlag.

Nernst, Walther (1907). "Über das Ammoniakgleichgewicht." *Zeitschrift für Elektrochemie* 13 (32): 521–24.

Netz, Reviel (1999). *The Shaping of Deduction in Greek Mathematics: A Study in Cognitive History*. Cambridge: Cambridge University Press.

Neumann, John von (1927a). "Mathematische Begründung der Quantenmechanik." *Nachrichten von der Gesellschaft der Wissenschaften zu Göttingen, Mathematisch-Physikalische Klasse*: 1–57.

———(1927b). "Thermodynamik quantenmechanischer Gesamtheiten." *Nachrichten von der Gesellschaft der Wissenschaften zu Göttingen, Mathematisch-Physikalische Klasse*: 273–91.

————(1927c). "Wahrscheinlichkeitstheoretischer Aufbau der Quantenmechanik." *Nachrichten von der Gesellschaft der Wissenschaften zu Göttingen, Mathematisch-Physikalische Klasse*: 245–72.

Newcombe, Nora S., and Janellen Huttenlocher (2003). *Making Space: The Development of Spatial Representation and Reasoning*. Cambridge, MA: MIT Press.

Newman, Mark E. J., and Duncan J. Watts (1999). "Scaling and Percolation in the Small-World Network Model." *Physical Review E* 60 (6): 7332–42.

Newton, Isaac ([1687] 2016). *The Principia: Mathematical Principles of Natural Philosophy—The Authoritative Translation and Guide*. Translated by I. Bernard Cohen and Anne Whitman. Berkeley: University of California Press.

————([1730] 1952). *Opticks: Or, A Treatise of the Reflections, Refractions, Inflections & Colours of Light*. Based on the London 4th ed. New York: Dover. https://archive.org/details /Optics_285.

Niles, Daniel (2018). "Agricultural Heritage and Conservation beyond the Anthropocene." In *The Oxford Handbook of Public Heritage Theory and Practice*, ed. A. M. Labrador and N. A. Silberman, n.p. Oxford: Oxford Handbooks Online.

Niles, Daniel, and Sander van der Leeuw (2018). "The Material Order." *Technosphere Magazine*. Accessed July 17, 2018. https://technosphere-magazine.hkw.de/p/The-Material-Order -4gK5EMpZ3SzB79aTePfJo7.

Nilsson, Nils J. (2010). *The Quest for Artificial Intelligence: A History of Ideas and Achievements*. Cambridge: Cambridge University Press.

Nissen, Hans Jörg (1988). *The Early History of the Ancient Near East, 9000–2000 B.C.* Chicago: University of Chicago Press.

Nissen, Hans Jörg, Peter Damerow, and Robert K. Englund (1993). *Archaic Bookkeeping: Early Writing and Techniques of Economic Administration in the Ancient Near East*. Translated by Paul Larsen. Chicago: University of Chicago Press.

North, Douglass C. (1981). *Structure and Change in Economic History*. New York: W. W. Norton.

————(1992). *Transaction Costs, Institutions, and Economic Performance*. San Francisco: ICS Press.

Norton, John (2014). "Einstein's Special Theory of Relativity and the Problems in the Electrodynamics of Moving Bodies That Led Him to It." In *The Cambridge Companion to Einstein*, ed. C. Lehner and M. Janssen, 72–102. Cambridge: Cambridge University Press.

Nöth, Winfried (1995). *Handbook of Semiotics*. Advances in Semiotics. Bloomington: Indiana University Press.

Nunberg, Geoffrey (1996). "Farewell to the Information Age." In *The Future of the Book*, ed. G. Nunberg, 103–38. Berkeley: University of California Press.

Nuovo, Angela (2013). *The Book Trade in the Italian Renaissance*. Leiden: Brill.

Nye, Mary Jo, ed. (2003). *The Modern Physical and Mathematical Sciences*. The Cambridge History of Science 5. Cambridge: Cambridge University Press.

Nyerere, Julius K. (1968). *Ujamaa: Essays on Socialism*. London: Oxford University Press.

O'Brien, Patricia J. (1972). "The Sweet Potato: Its Origin and Dispersal." *American Anthropologist* 74 (3): 342–65.

O'Brien, Patrick K. (2009). "The Needham Question Updated: A Historiographical Survey and Elaboration." In *History of Technology*. Vol. 29, ed. I. Inkster, 7–28. London: Bloomsbury.

Ochiai, Eiichiro (2007). "Japan in the Edo Period: Global Implications of a Model of Sustainability." *Asia-Pacific Journal: Japan Focus* 5 (2): 1–10.

Odling-Smee, F. John, Douglas H. Erwin, Eric P. Palkovacs, Marcus W. Feldman, and Kevin N. Laland (2013). "Niche Construction Theory: A Practical Guide for Ecologists." *Quarterly Review of Biology* 88 (1): 4–28.

Oldroyd, David (1996). *Thinking about the Earth: A History of Ideas in Geology*. London: Athlone Press.

Olschki, Leonardo (1919). *Die Literatur der Technik und der angewandten Wissenschaften: Vom Mittelalter bis zur Renaissance.* Geschichte der neusprachlichen wissenschaftlichen Literatur 1. Heidelberg: Carl Winter's Universitätsbuchhandlung.

———(1922). *Bildung und Wissenschaft im Zeitalter der Renaissance in Italien.* Geschichte der neusprachlichen wissenschaftlichen Literatur 2. Leipzig: Leo S. Olschki.

———(1927). *Galilei und seine Zeit.* Geschichte der neusprachlichen wissenschaftlichen Literatur 3. Halle (Saale): Max Niemeyer.

Omodeo, Pietro D. (2014). *Copernicus in the Cultural Debates of the Renaissance: Reception, Legacy, Transformation.* Leiden: Brill.

———(2016). "Kuhn's Paradigm of Paradigms: Historical and Epistemological Coordinates of 'The Copernican Revolution.'" In *Shifting Paradigms: Thomas S. Kuhn and the History of Science,* ed. A. Blum, K. Gavroglu, C. Joas, and J. Renn, 71–104. Proceedings 8. Berlin: Edition Open Access. http://edition-open-access.de/proceedings/8/7/index.html.

———(2017). "The Politics of Apocalypse: The Immanent Transcendence of Anthropocene." *Stvar: Casopis Za Teoriskje Prakse* 9:1–11.

———(2018). "Soggettività, strutture, egemonie: Questioni politico-culturali in epistemologia storica." *Studi Culturali* 15(2): 211–234.

———, ed. (2019). *Bernardino Telesio and the Natural Sciences in the Renaissance.* Leiden: Brill.

———(in press a). *Political Epistemology: The Problem of Ideology in Science Studies.* Cham: Springer.

———(in press b). *Presence-Absence of Alexander of Aphrodisias in Renaissance Cosmo-Psychology.*

Omodeo, Pietro D., and Massimiliano Badino, eds. (forthcoming in 2019). Cultural Hegemony in a Scientific World. Gramscian Concepts for the History of Science. Leiden: Brill.

Omodeo, Pietro D., and Jürgen Renn (2019). *Science in Court Society: Giovanni Battista Benedetti's "Diversarum speculationum mathematicarum et physicarum liber" (Turin, 1585).* Berlin: Edition Open Access.

Ordine, Nuccio (2017). *The Usefulness of the Useless.* Translated by Alastair McEwen. Philadelphia: Paul Dry Books.

O'Reilly, Tim (2013). "Open Data and Algorithmic Regulation." In *Beyond Transparency: Open Data and the Future of Civic Innovation,* ed. B. Goldstein and L. Dyson, 289–300. San Francisco: Code for America Press. http://beyondtransparency.org/chapters/part-5 /open-data-and-algorithmic-regulation.

Oreskes, Naomi (1999). *The Rejection of Continental Drift: Theory and Method in American Earth Science.* Oxford: Oxford University Press.

———(2004). "The Scientific Consensus on Climate Change." *Science* 306 (5702): 1686.

Oreskes, Naomi, and Erik M. Conway (2010). *Merchants of Doubt: How a Handful of Scientists Obscured the Truth on Issues from Tobacco Smoke to Global Warming.* New York: Bloomsbury.

Oreskes, Naomi, Erik M. Conway, David J. Karoly, Joelle Gergis, Urs Neu, and Christian Pfister (2018). "The Denial of Global Warming." In *The Palgrave Handbook of Climate History,* ed. S. White, C. Pfister, and F. Mauelshagen, 149–71. London: Palgrave Macmillan.

Oresme, Nicolas ([1350s] 1968). "Tractatus de configurationibus qualitatum et motuum." In *Nicole Oresme and the Medieval Geometry of Qualities and Motions: A Treatise on the Uniformity and Difformity of Intensities Known as "Tractatus de configurationibus qualitatum et motuum,"* ed. M. Clagett, 157–517. Madison: University of Wisconsin Press.

Osborn, Fairfield (1948). *Our Plundered Planet.* New York: Little, Brown and Co.

Ossendrijver, Mathieu (2011). "Science in Action: Networks in Babylonian Astronomy." In *Babylon: Wissenskultur in Orient und Okzident/Science Culture Between Orient and Occident,* ed. E. Cancik-Kirschbaum, M. van Ess, and J. Marzahn, 213–21. Berlin: De Gruyter.

Osterhammel, Jürgen (2009). *Die Verwandlung der Welt: Eine Geschichte des 19. Jahrhunderts.* Munich: C. H. Beck.

Osthues, Wilhelm (2014a). "Bauwissen im Antiken Griechenland." In *Wissensgeschichte der Architektur.* Vol. 2: *Vom Alten Ägypten bis zum Antiken Rom,* ed. J. Renn, W. Osthues, and H. Schlimme, 127–264. Studies 4. Berlin: Edition Open Access. http://edition-open-access.de/studies/4/4/index.html.

———(2014b). "Bauwissen im Antiken Rom." In *Wissensgeschichte der Architektur.* Vol. 2: *Vom Alten Ägypten bis zum Antiken Rom,* ed. J. Renn, W. Osthues, and H. Schlimme, 265–422. Studies 4. Berlin: Edition Open Access. http://edition-open-access.de/studies/4/5/index.html.

Östling, Johan, Erling Sandmo, David Larsson Heidenblad, Anna Nilsson Hammar, and Kari Nordberg, eds. (2018). *Circulation of Knowledge: Explorations in the History of Knowledge.* Lund: Nordic Academic Press.

Ostrom, Elinor (2012). *The Future of the Commons: Beyond Market Failure and Government Regulation.* London: Institute of Economic Affairs.

———(2015). *Governing the Commons: The Evolution of Institutions for Collective Action.* Canto Classics. Cambridge: Cambridge University Press.

Ostrom, Elinor, Joanna Burger, Christopher B. Field, Richard B. Norgaard, and David Policansky (1999). "Revisiting the Commons: Local Lessons, Global Challenges." *Science* 284 (5412): 278–82.

Ostwald, Martin (1992). "Athens as a Cultural Centre." In *The Fifth Century,* ed. D. M. Lewis, J. Boardman, J. K. Davies, and M. Ostwald, 306–69. The Cambridge Ancient History 5. Cambridge: Cambridge University Press.

Ostwald, Martin, and John P. Lynch (1994). "The Growth of Schools and the Advance of Knowledge." In *The Fourth Century B.C.,* ed. D. M. Lewis, J. Boardman, S. Hornblower, and M. Ostwald, 592–633. The Cambridge Ancient History 6. Cambridge: Cambridge University Press.

Ostwald, Wilhelm (1926–27). *Lebenslinien: Eine Selbstbiographie.* Berlin: Klasing.

Ottoni, Federica, and Carlo Blasi (2014). "Results of a 60-Year Monitoring System for Santa Maria del Fiore Dome in Florence." *International Journal of Architectural Heritage* 9 (1): 7–24.

Ouyang, Xiaoli, and Christine Proust (forthcoming). "Place Value Notations in the Ur III Period: Marginal Numbers in Administrative Texts." In *Cultures of Computation and Quantification,* ed. K. Chemla, A. Keller, and C. Proust. Dordrecht: Springer.

Overmann, Karenleigh A. (2017). "Thinking Materially: Cognition as Extended and Enacted." *Journal of Cognition and Culture* 17 (3–4): 354–73.

Özdoğan, Mehmet (2014). "A New Look at the Introduction of the Neolithic Way of Life in Southeastern Europe: Changing Paradigms of the Expansion of the Neolithic Way of Life." *Documenta Praehistorica* 41:33–49.

———(2016). "The Earliest Farmers of Europe: Where Did They Come From?" In *Southeast Europe and Anatolia in Prehistory: Essays in Honor of Vassil Nikolov on His 65th Anniversary,* ed. K. Bacvarov and R. Gleser. Bonn: Rudolf Habelt Verlag.

Ozone Secretariat (1987). "The Montreal Protocol on Substances That Deplete the Ozone Layer." United Nations Environment Programme (UNEP). Accessed May 24, 2019. https://ozone.unep.org/montreal-protocol-substances-deplete-ozone-layer/79705/5.

Pääbo, Svante (2014). *Neanderthal Man: In Search of Lost Genomes.* New York: Basic Books.

Pacey, Arnold (1990). *Technology in World Civilization: A Thousand-Year History.* Cambridge, MA: MIT Press.

Padgett, John F., and Christopher K. Ansell (1993). "Robust Action and the Rise of the Medici, 1400–1434." *American Journal of Sociology* 98 (6): 1259–319.

Pagano, Ugo, and Maria A. Rossi (2009). "The Crash of the Knowledge Economy." *Cambridge Journal of Economics* 33 (4): 665–83.

Paillard, Didier (2008). "From Atmosphere, to Climate, to Earth System Science." *Interdisciplinary Science Reviews* 33 (1): 25–35.

Pantin, Isabelle (1998). "Les problèmes de l'édition des livres scientifiques: L'exemple de Guillaume Cavellat." In *Le livre dans l'Europe de la Renaissance: Actes du XXVIIIe Colloque international d'études humanistes de Tours*, ed. Bibliothèque nationale, 240–52. Paris: Promodis, Éditions du Cercle de la Librairie.

———(2006). "Teaching Mathematics and Astronomy in France: The Collège Royal (1550–1650)." *Science & Education* 15:189–207.

Pantin, Isabelle, and Philippe Renouard, eds. (1986). *Imprimeurs et libraires parisiens du XVIe siècle: Cavellat—Marnef et Cavellat*. Paris: Bibliothèque nationale.

Parker, Geoffrey (2013). *Global Crisis: War, Climate Change and Catastrophe in the Seventeenth Century*. New Haven, CT: Yale University Press.

Parsons, Talcott (1949). *The Structure of Social Action: A Study in Social Theory with Special Reference to a Group of Recent European Writers*. 1937. 2nd ed. Glencoe, IL: Free Press.

Parzinger, Hermann (2014). *Die Kinder des Prometheus: Eine Geschichte der Menschheit vor der Erfindung der Schrift*. Munich: C. H. Beck.

Patiniotis, Manolis (2013). "Between the Local and the Global: History of Science in the European Periphery Meets Post-colonial Studies." *Centaurus* 55 (4): 361–84.

Pedersen, Olaf (1983). "The Ecclesiastical Calendar and the Life of the Church." In *Gregorian Reform of the Calendar: Proceedings of the Vatican Conference to Commemorate its 400th Anniversary, 1582–1982*, ed. G. V. Coyne, M. S. Hoskin, and O. Pedersen, 17–74. Vatican City: Specolo Vaticano.

———(1993). *Early Physics and Astronomy: A Historical Introduction*. 2nd ed. Cambridge: Cambridge University Press.

Peirce, Charles Sanders (1967). *The Simplest Mathematics*. 3rd ed. Collected Papers of Charles Sanders Peirce 4. Cambridge, MA: Belknap Press of Harvard University Press.

Penrose, Roger (1965). "Gravitational Collapse and Space-Time Singularities." *Physical Review Letters* 14 (3): 57–59.

Penzias, Arno A., and Robert Woodrow Wilson (1965). "A Measurement of Excess Antenna Temperature at 4080 Mc/s." *Astrophysical Journal* 142 (1): 419–21.

Peroni, Adriano (2006). "Le ricostruzioni grafiche della Santa Maria del Fiore di Arnolfo: Un bilancio." In *Arnolfo di Cambio e la sua epoca: Costruire, scolpire, dipingere, decorare*, ed. V. F. Pardo, 381–94. Atti del Convegno Internazionale di Studi, Firenze-Colle di Val d'Elsa. Rome: Viella Libreria Editrice.

Pestre, Dominique (1997). "Science, Political Power and the State." In *Science in the Twentieth Century*, ed. J. Krige and D. Pestre, 61–75. Amsterdam: Harwood Academic.

Peter, Isabelle S., and Eric H. Davidson (2015). *Genomic Control Process: Development and Evolution*. London: Academic Press.

Petit, Jean R., Jean Jouzel, Dominique Raynaud, Nartsiss I. Barkov, Jean-Marc Barnola, Isabelle Basile, Michael Bender, Jérôme Chappellaz, M. Davis, Gilles Delaygue, Marc F. Delmotte, Vladimir M. Kotlyakov, Michel Legrand, V. Y. Lipenkov, Claude Lorius, Laurence Pépin, Catherine Ritz, Eric Saltzman, and Michel Stievenard (1999). "Climate and Atmospheric History of the Past 420,000 Years from the Vostok Ice Core, Antarctica." *Nature* 399 (6735): 429–36.

Peuerbach, Georg von (1473). *Theoricae novae planetarum*. Nuremberg: [Regiomontanus].

Piaget, Jean (1951). *Play, Dreams and Imitation in Childhood*. [*La formation du symbole chez l'enfant: Imitation, jeu et reve, image et représentation* (1945)]. Translated by C. Gattegno and F. M. Hodgson. London: Routledge & Kegan Paul.

520 REFERENCES

————(1965). *The Moral Judgment of the Child*. [*Le jugement moral chez l'enfant* (1932)]. Translated by M. Gabain. Glencoe, IL: Free Press.

————(1969). *The Child's Conception of Time*. [*Le développement de la notion de temps chez l'enfant* (1927)]. Translated by A. J. Pomerans. London: Routledge & Kegan Paul.

————(1970). *Genetic Epistemology*. New York: Columbia University Press.

————(1981). *The Psychology of Intelligence*. [*La psychologie de l'intelligence* (1947)]. Translated by M. Piercy and D. E. Berlyne. Totowa, NJ: Littlefield Adams.

————(1982). *The Essential Piaget: An Interpretative Reference and Guide*. London: Routledge & Kegan Paul.

————(1983). *Biologie und Erkenntnis: Über die Beziehung zwischen organischen Regulationen und kognitiven Prozessen*. Translated by A. Greyer. Frankfurt am Main: Fischer.

————(1999). *Child's Conception of Physical Causality*. [*La causalite physique chez l'enfant* (1927)]. Translated by Marjorie Gabain. London: Routledge.

Piaget, Jean, and Rolando Garcia (1989). *Psychogenesis and the History of Science*. Translated by Helga Feider. New York: Columbia University Press.

Piaget, Jean, and Bärbel Inhelder (1956). *The Child's Conception of Space*. [*La représantation de l'espace chez l'enfant* (1948).] Translated by F. J. Langdon and J. L. Lunzer. International Library of Psychology, Philosophy and Scientific Method. London: Routledge & Kegan Paul.

Pinhasi, Ron, and Jay Stock (2011). *Human Bioarchaeology of the Transition to Agriculture*. Chichester: Wiley-Blackwell.

Pinto, Giuliano (1984). "L'organizzazione del lavoro nei cantieri edili (Italia centro-settentrionale)." In *Artigiani e salariati: Il mondo del lavoro in Italia nei secoli XII–XV*, 69–101. Pistoia: Presso la sede del Centro.

————(1991). "I lavoratori salariati nell'Italia bassomedievale: Mercato del lavoro e livelli di vita." In *Travail et travailleurs en Europe au Moyen Âge et au debut des temps modernes*, ed. C. Dolan, 47–62. Toronto: Pontifical Institute of Mediaeval Studies.

Planck, Max (1900). "Zur Theorie des Gesetzes der Energieverteilung im Normalspectrum." *Verhandlungen der Deutschen Physikalischen Gesellschaft* 2 (1): 237–45.

————([1900] 1967). "On the Theory of the Energy Distribution Law of the Normal Spectrum." In *The Old Quantum Theory*, ed. D. Ter Haar, 82–90. Oxford: Pergamon Press.

————([1920] 1967). "The Genesis and Present State of Development of the Quantum Theory." In *Nobel Lectures: Physics, 1901–1921*, 407–20. Amsterdam: Elsevier.

Plato (1997a). *Phaedo*. In *Plato: Complete Works*. Translated by G. M. A. Grube, ed. J. M. Cooper, 49–100. Indianapolis, IN: Hackett.

————(1997b). *Republic*. In *Plato: Complete Works*. Translated by G. M. A. Grube, ed. J. M. Cooper, 971–1223. Indianapolis, IN: Hackett.

————(1997c). *Theaetetus*. In *Plato: Complete Works*. Translated by G. M. A. Grube, ed. J. M. Cooper, 157–234. Indianapolis, IN: Hackett.

Plofker, Kim (2009). *Mathematics in India*. Princeton, NJ: Princeton University Press.

Po-chia Hsia, Ronnie (2015). "The End of the Jesuit Mission in China." In *The Jesuit Suppression in Global Context: Causes, Events, and Consequences*, ed. J. D. Burson and J. Wright, 100–116. Cambridge: Cambridge University Press.

Poe, Marshall T. (2011). *History of Communications: Media and Society from the Evolution of Speech to the Internet*. Cambridge: Cambridge University Press.

Poincaré, Henri (1905). "Sur la dynamique de l'électron." *Comptes Rendus de l'Académie des Sciences* 140:1504–8.

————(1907). *The Value of Science*. Translated by George Bruce Halsted. New York: Science Press.

Polanyi, Michael (1983). *The Tacit Dimension*. Gloucester, MA: Peter Smith.

Polich, Laura (2000). "The Search for Proto-NSL: Looking for the Roots of the Nicaraguan Deaf Community." In *Bilingualism and Identity in Deaf Communities*, ed. M. Metzger, 255–305. Washington, DC: Gallaudet University Press.

———(2005). *The Emergence of the Deaf Community in Nicaragua: "With Sign Language You Can Learn So Much."* Washington, DC: Gallaudet University Press.

Pomeranz, Kenneth (2000). *The Great Divergence: China, Europe, and the Making of the Modern World Economy.* Princeton, NJ: Princeton University Press.

Poole, Reginald Lane, and Austin Lane Poole (1934). *Studies in Chronology and History.* Oxford: Clarendon Press.

Poole, Robert (2010). *Earthrise: How Man First Saw the Earth.* New Haven, CT: Yale University Press.

Popper, Karl R. ([1959] 2002). *The Logic of Scientific Discovery.* London: Routledge.

Popplow, Marcus (2015). "Formalization and Interaction: Toward a Comprehensive History of Technology-Related Knowledge in Early Modern Europe." *Isis* 106 (4): 848–56.

Popplow, Marcus, and Jürgen Renn (2002). "Ingegneria e macchine." In *La rivoluzione scientifica*, ed. M. Bray, 258–74. Storia della scienza 5. Rome: Istituto della Enciclopedia Italiana.

Porter, Philip W. (2006). *Challenging Nature: Local Knowledge, Agroscience, and Food Security in Tanga Region, Tanzania.* Chicago: University of Chicago Press.

Porter, Theodore M. (1997). "The Management of Society by Numbers." In *Science in the Twentieth Century*, ed. J. Krige and D. Pestre, 97–110. Amsterdam: Harwood Academic.

Portz, Helga (1994). *Galilei und der heutige Mathematik-Unterricht: Ursprüngliche Festigkeitslehre und Ähnlichkeitsmechanik und ihre Bedeutung für die mathematische Bildung.* Lehrbücher und Monographien zur Didaktik der Mathematik 22. Mannheim: BI-Wissenschaftsverlag.

Posner, Gerald L., and John Ware (1986). *Mengele: The Complete Story.* New York: McGraw-Hill.

Potts, Daniel T. (2007). "Differing Modes of Contact between India and the West: Some Achaemenid and Seleucid Examples." In *Memory as History: The Legacy of Alexander in Asia*, ed. H. P. Ray and D. T. Potts, 122–30. New Delhi: Aryan Books International.

———(2011). "Equus Asinus in Highland Iran: Evidence Old and New." In *Between Sand and Sea: The Archaeology and Human Ecology of Southwestern Asia—Festschrift in Honor of Hans-Peter Uerpmann*, ed. N. J. Conard, P. Drechsler, and A. Morales, 167–75. Tübingen: Kerns Verlag.

———(2012). "Technological Transfer and Innovation in Ancient Eurasia." In *The Globalization of Knowledge in History*, ed. J. Renn, 105–23. Berlin: Edition Open Access. http:// edition-open-access.de/studies/1/8/index.html.

Powell, Walter W., and Kaisa Snellman (2004). "The Knowledge Economy." *Annual Review of Sociology* 30:199–220.

Pradhan, Gauri R., Claudio Tennie, and Carel P. van Schaik (2012). "Social Organization and the Evolution of Cumulative Technology in Apes and Hominins." *Journal of Human Evolution* 63 (1): 180–90.

Prager, Frank D., and Gustina Scaglia (1970). *Brunelleschi: Studies of His Technology and Inventions.* Cambridge, MA: MIT Press.

Prantl, Carl von, ed. (1881). *De coelo et De generatione et corruptione: Aristotelis quae feruntur de coloribus, de audibilibus, physiognomonica.* Leipzig: Teubner.

Pratt, John Henry, and James Challis (1855). "I. On the Attraction of the Himalaya Mountains, and of the Elevated Regions beyond Them, upon the Plumb-Line in India." *Philosophical Transactions of the Royal Society of London* 145:53–100.

Preiser-Kapeller, Johannes (2015). "Harbours and Maritime Networks as Complex Adaptive Systems." In *Harbours and Maritime Networks as Complex Adaptive Systems: A Thematic*

Introduction, ed. J. Preiser-Kapeller and F. Daim, 1–24. Regensburg: Verlag Schnell & Steiner.

Preston, Richard (1996). *First Light: The Search for the Edge of the Universe*. New York: Random House.

Price, Derek J. de Solla (1965). "Networks of Scientific Papers." *Science* 149 (3683): 510–15.

———(1976). "A General Theory of Bibliometric and Other Cumulative Advantage Processes." *Journal of the American Society for Information Science* 27 (5): 292–306.

Prinz, Wolfgang (2012). *Open Minds: The Social Making of Agency and Intentionality*. Cambridge, MA: MIT Press.

Proctor, Robert N. (2011). *Golden Holocaust: Origins of the Cigarette Catastrophe and the Case for Abolition*. Berkeley: University of California Press.

Proust, Joseph-Louis (1794). "Extrait d'un mémoire intitulé: Recherches sur le Bleu de Prusse." *Journal de Physique, de Chimie, d'Histoire Naturelle et des Arts* 2:334–41.

———(1799). "Recherches sur le cuivre." *Annales de Chimie ou Recueil de Mémoires concernant la Chimie et les Arts Qui en Dépendent* 32:26–54.

Pulte, Helmut (2005). *Axiomatik und Empirie: Eine wissenschaftstheoriegeschichtliche Untersuchung zur Mathematischen Naturphilosophie von Newton bis Neumann*. Darmstadt: Wissenschaftliche Buchgesellschaft.

Pyarelal (1958). *The Last Phase*. Part 2. Mahatma Gandhi 10. Ahmedabad: Navajiva.

Pyers, Jennie E., Anna Shustermann, Ann Senghas, Elizabeth S. Spelke, and Karen Emmorey (2010). "Evidence from an Emerging Sign Language Reveals That Language Supports Spatial Cognition." *Proceedings of the National Academy of Sciences of the United States of America* 107 (27): 12116–20.

Raffnsøe, Sverre (2016). *Philosophy of the Anthropocene: The Human Turn*. Basingstoke: Palgrave Macmillan.

Rahmsdorf, Lorenz (2011). "Re-integrating 'Diffusion': The Spread of Innovations among the Neolithic and Bronze Age Civilizations of Europe and the Near East." In *Interweaving Worlds: Systemic Interactions in Eurasia, 7th to 1st Millenium BC*, ed. T. C. Wilkinson, S. Sheratt, and J. Bennet, 100–120. Oxford: Oxbow Books.

Rahwan, Iyad, and Manuel Cebrian (2018). *Machine Behavior Needs to Be an Academic Discipline*. New York: Nautilus.

Raina, Dhruv (1999). "From West to Non-West? Basalla's Three-Stage Model Revisited." *Science as Culture* 8 (4): 497–516.

Raj, Kapil (2013). "Beyond Postcolonialism . . . and Postpositivism: Circulation and the Global History of Science." *Isis* 104 (2): 337–47.

Ramelli, Agostino (1588). *Le diverse et artificiose machine*. Paris: In casa del'autore.

Rapoport, Amos (1969). *House Form and Culture*. Foundations of Cultural Geography Series. Englewood Cliffs, NJ: Prentice Hall.

Rapp, Friedrich (1981). *Analytical Philosophy of Technology*. Translated by Stanley R. Carpenter and Theodor Langenbruch. Boston Studies in the Philosophy and History of Science 63. Dordrecht: D. Reidel.

Reckwitz, Andreas (2003). "Grundelemente einer Theorie sozialer Praktiken: Eine sozialtheoretische Perspektive." *Zeitschrift für Soziologie* 32 (4): 282–301.

Regier, Jonathan, and Pietro D. Omodeo (in press). "Celestial Physics." In *The Cambridge History of Philosophy of the Scientific Revolution*, ed. D. Jalobeanu and D. M. Miller. Cambridge: Cambridge University Press.

Reich, David (2018). *Who We Are and How We Got Here: Ancient DNA and the New Science of the Human Past*. New York: Pantheon Books.

Reinhard, Wolfgang (1979). *Freunde und Kreaturen: "Verflechtung" als Konzept zur Erforschung historischer Führungsgruppen—Römische Oligarchie um 1600.* Schriften der Philosophischen Fachbereiche der Universität Augsburg 14. Munich: Verlag Ernst Vögel.

Reinhardt, Carsten (2010). "Historische Wissenschaftsforschung, heute: Überlegungen zu einer Geschichte der Wissensgesellschaft." *Berichte zur Wissenschaftsgeschichte* 33 (1): 81–99.

Renfrew, Colin (2008). "Neuroscience, Evolution and the Sapient Paradox: The Factuality of Value and of the Sacred." *Philosophical Transactions of the Royal Society B: Biological Sciences* 363 (1499): 2041–47.

Renn, Jürgen (1993). "Einstein as a Disciple of Galileo: A Comparative Study of Concept Development in Physics." *Science in Context* 6 (1): 311–41.

———(1994). *Historical Epistemology and Interdisciplinarity.* Preprint 2. Berlin: Max Planck Institute for the History of Science.

———(1995). "Historical Epistemology and Interdisciplinarity." In *Physics, Philosophy, and the Scientific Community: Essays in the Philosophy and History of the Natural Sciences and Mathematics in Honor of Robert S. Cohen,* ed. K. Gavroglu, J. Stachel, and M. W. Wartofsky, 241–51. Boston Studies in the Philosophy and History of Science 163. Dordrecht: Kluwer Academic.

———(1996). *Historical Epistemology and the Advancement of Science.* Preprint 36. Berlin: Max Planck Institute for the History of Science.

———(2001a). "Editor's Introduction: Galileo in Context—An Engineer-Scientist, Artist, and Courtier at the Origins of Classical Science." In *Galileo in Context,* ed. J. Renn, 1–8. Cambridge: Cambridge University Press.

———, ed. (2001b). *Galileo in Context.* Cambridge: Cambridge University Press.

———(2005). "Wissenschaft als Lebensorientierung: Eine Erfolgseschichte?" In *Leben: Verständnis—Wissenschaft—Technik,* ed. E. Herms, 15–31. Veröffentlichungen der Wissenschaftlichen Gesellschaft für Theologie 24. Tübingen: Gütersloher Verlagshaus.

———(2007a). "Classical Physics in Disarray: The Emergence of the Riddle of Gravitation." In *Einstein's Zurich Notebook: Introduction and Source,* ed. M. Janssen, J. Norton, J. Renn, T. Sauer, and J. Stachel, 21–80. The Genesis of General Relativity 1. Dordrecht: Springer.

———, ed. (2007b). *The Genesis of General Relativity.* 4 vols. Boston Studies in the Philosophy and History of Science 250. Dordrecht: Springer.

———(2012a). "Survey 2: Knowledge as a Fellow Traveler." In *The Globalization of Knowledge in History,* ed. J. Renn, 205–43. Studies 1. Berlin: Edition Open Access. http://edition-open-access.de/studies/1/13/index.html.

———(2012b). "Survey 3: The Place of Local Knowledge in the Global Community." In *The Globalization of Knowledge in History,* ed. J. Renn, 369–97. Studies 1. Berlin: Edition Open Access. http://edition-open-access.de/studies/1/20/index.html.

———, ed. (2012c). *The Globalization of Knowledge in History.* Studies 1. Berlin: Edition Open Access. http://edition-open-access.de/studies/1/index.html.

———(2013a). "Florenz: Matrix der Wissenschaft." In *Florenz!,* ed. Kunst- und Ausstellungshalle der Bundesrepublik Deutschland (Bonn), 100–111. Munich: Hirmer Verlag.

———(2013b). "Schrödinger and the Genesis of Wave Mechanics." In *Erwin Schrödinger: 50 Years After,* ed. W. L. Reiter and J. Yngvason, 9–36. Zurich: European Mathematical Society.

———(2014a). "Beyond Editions: Historical Sources in the Digital Age." In *Internationalität und Interdisziplinarität der Editionswissenschaft,* ed. M. Stolz and Y.-C. Chen, 9–28. Berlin: De Gruyter.

———(2014b). "Preface: The Globalization of Knowledge in the Ancient Near East." In *Melammu: The Ancient World in an Age of Globalization,* ed. M. J. Geller, 1–3. Proceedings

7. Berlin: Edition Open Access. http://edition-open-access.de/proceedings/7/1/index .html.

———(2014c). "The Globalization of Knowledge in History and Its Normative Challenges." *Rechtsgeschichte—Legal History* 22:52–60.

———(2015a). "Die Globalisierung des Wissens in der Geschichte." In *Welt-Anschauungen: Interdisziplinäre Perspektiven auf die Ordungen des Globalen*, ed. O. Breidbach, A. Christoph, and R. Godel, 137–48. Acta Historica Leopoldina 67. Stuttgart: Wissenschaftliche Verlagsgesellschaft.

———(2015b). "From the History of Science to the History of Knowledge—and Back." *Centaurus* 57 (1): 37–53.

———(2015c). "Learning from Kushim about the Origin of Writing and Farming." In *Textures of the Anthropocene: Grain, Vapor, Ray*, ed. K. Klingan, A. Sepahvand, C. Rosol, and B. M. Scherer, 241–59. Cambridge, MA: MIT Press.

———(2015d). "Was wir von Kuschim über die Evolution des Wissens und die Ursprünge des Anthropozäns lernen können." In *Das Anthropozän: Zum Stand der Dinge*, ed. J. Renn and B. Scherer, 184–209. Berlin: Matthes & Seitz.

———(2016). "Q Quest." In *Wissen Macht Geschlecht: Ein ABC der transnationalen Zeitgeschichte*, ed. B. Kolboske, A. C. Hüntelmann, I. Heumann, S. Heim, R. Fritz, and R. Birke, 95–104. Proceedings 9. Berlin: Edition Open Access. http://edition-open-access.de /proceedings/9/18/index.html.

———(2017). "On the Construction Sites of the Anthropocene." In *Out There: Landscape Architecture on the Global Terrain*, ed. A. Lepik, 16–19. Berlin: Hatje Cantz.

Renn, Jürgen, Giuseppe Castagnetti, and Simone Rieger (2001). "Adolf von Harnack und Max Planck." In *Adolf von Harnack: Theologe, Historiker, Wissenschaftspolitiker*, ed. K. Nowak and G. Oexle, 127–55. Veröffentlichungen des Max-Planck-Instituts für Geschichte 161. Göttingen: Vandenhoeck & Ruprecht.

Renn, Jürgen, and Peter Damerow (2007). "Mentale Modelle als kognitive Instrumente der Transformation von technischem Wissen." In *Übersetzung und Transformation*, ed. H. Böhme, C. Rapp, and W. Rösler, 311–31. Transformationen der Antike 1. Berlin: De Gruyter.

———(2012). *The Equilibrium Controversy: Guidobaldo del Monte's Critical Notes on the Mechanics of Jordanus and Benedetti and Their Historical and Conceptual Background*. Sources 2. Berlin: Edition Open Access. http://www.edition-open-sources.org/sources/2 /index.html.

Renn, Jürgen, Peter Damerow, and Peter McLaughlin (2003). "Aristotle, Archimedes, Euclid, and the Origin of Mechanics: The Perspective of Historical Epistemology." In *Symposium Arquímedes Fundación Canaria Orotava de Historia de la Ciencia*, ed. J. Montesinos, 43–59. Preprint 239. Berlin: Max Planck Institute for the History of Science.

Renn, Jürgen, Peter Damerow, and Simone Rieger (2001). "Hunting the White Elephant: When and How Did Galileo Discover the Law of Fall? (with an Appendix by Domenico Giulini)." In *Galileo in Context*, ed. J. Renn, 29–149. Cambridge: Cambridge University Press.

Renn, Jürgen, Peter Damerow, Matthias Schemmel, Christoph Lehner, and Matteo Valleriani (2018). "Mental Models as Cognitive Instruments in the Transformation of Knowledge." In *Emergence and Expansion of Pre-classical Mechanics*, ed. R. Feldhay, J. Renn, M. Schemmel, and M. Valleriani. Boston Studies in the Philosophy and History of Science 333. Cham: Springer.

Renn, Jürgen, and Malcolm D. Hyman (2012a). "Survey 4: The Globalization of Modern Science." In *The Globalization of Knowledge in History*, 561–604. Studies 1. Berlin: Edition Open Access. http://edition-open-access.de/studies/1/28/index.html.

————(2012b). "The Globalization of Knowledge in History: An Introduction." In *The Global-ization of Knowledge in History*, ed. J. Renn, 15–44. Studies 1. Berlin: Edition Open Access. http://edition-open-access.de/studies/1/5/index.html.

Renn, Jürgen, Benjamin Johnson, and Benjamin Steininger (2017). "Ammoniak: Wie eine epochale Erfindung das Leben der Menschen und die Arbeit der Chemiker verändert." *Naturwissenschaftliche Rundschau* 70 (10): 507–14.

Renn, Jürgen, and Manfred D. Laubichler (2017). "Extended Evolution and the History of Knowledge: Problems, Perspectives, and Case Studies." In *Integrated History and Philoso-phy of Science*, ed. F. Stadler, 109–25. Vienna Circle Institute Yearbook 20. Cham: Springer.

Renn, Jürgen, Manfred D. Laubichler, and Helge Wendt (2014). "Energietransformationen zwischen Kaffee und Koevolution." In *Willkommen im Anthropozän! Unsere Verantwor-tung für die Zukunft der Erde, Katalog zur Sonderausstellung am Deutschen Museum*, ed. N. Möllers, C. Schwägerl, and H. Trischler, 81–84. Munich: Deutsches Museum.

Renn, Jürgen, Wilhelm Osthues, and Hermann Schlimme, eds. (2014a). *Wissensgeschichte der Architektur*. 3 vols. Studies 3–5. Berlin: Edition Open Access. http://edition-open-access .de/studies/index.html.

————, eds. (2014b). *Wissensgeschichte der Architektur*. Vol. 1: *Vom Neolithikum bis zum Alten Orient*. Studies 3. Berlin: Edition Open Access. http://edition-open-access.de/studies/3 /index.html.

————, eds. (2014c). *Wissensgeschichte der Architektur*. Vol. 2: *Vom Alten Ägypten bis zum Anti-ken Rom*. Studies 4. Berlin: Edition Open Access. http://edition-open-access.de/studies /4/index.html.

————, eds. (2014d). *Wissensgeschichte der Architektur*. Vol. 3: *Vom Mittelalter bis zur Frühen Neuzeit*. Studies 5. Berlin: Edition Open Access. http://edition-open-access.de/studies/5 /index.html.

Renn, Jürgen, and Robert Rynasiewicz (2014). "Einstein's Copernican Revolution." In *The Cambridge Companion to Einstein*, ed. C. Lehner and M. Janssen, 38–71. Cambridge Com-panions to Philosophy. Cambridge: Cambridge University Press.

Renn, Jürgen, and Tilman Sauer (2007). "Pathways out of Classical Physics: Einstein's Double Strategy in Searching for the Gravitational Field Equation." In *Einstein's Zurich Notebook: Introduction and Source*, ed. M. Janssen, J. Norton, J. Renn, T. Sauer, and J. Stachel. The Genesis of General Relativity 1. Dordrecht: Springer.

Renn, Jürgen, and Matthias Schemmel (2000). *Waagen und Wissen in China*. Preprint 136. Ber-lin: Max Planck Institute for the History of Science.

————(2006). "Mechanics in the Mohist Canon and Its European Counterparts." In *Studies on Ancient Chinese Scientific and Technical Texts: Proceedings of the 3rd ISACBRST, March 31– April 3, Tübingen, Germany*, ed. H. U. Vogel, C. Moll-Murata, and G. Xuan, 24–31. Zheng-zhou: Elephant Press.

————(2012). "The Encounter of Two Systems of Knowledge in the Life and Works of the Jesuit Scholar Johannes Schreck." In *The Art of Enlightenment*, ed. L. Zhangshen, 74–82. Beijing: National Museum of China.

————(2017). "Wie oft sind die Naturwissenschaften entstanden?" *Nova Acta Leopoldina NF* 414:1–13.

————, eds. (2019). *Culture and Cognition: Essays in Honor of Peter Damerow*. Berlin: Edition Open Access.

Renn, Jürgen, and Bernd Scherer, eds. (2015a). *Das Anthropozän: Zum Stand der Dinge*. Berlin: Matthes & Seitz.

————(2015b). "Einführung." In *Das Anthropozän: Zum Stand der Dinge*, ed. J. Renn and B. Scherer, 7–23. Berlin: Matthes & Seitz.

Renn, Jürgen, Robert Schlögl, Christoph Rosol, and Benjamin Steininger (2017). "A Rapid Transition of the World's Energy Systems." *Nature Outlook*. Available for download at: https://www.mpg.de/11825144/socio-technical-energy-systems.

Renn, Jürgen, Robert Schlögl, and Hans-Peter Zenner, eds. (2011). *Herausforderung Energie: Ausgewählte Vorträge der 126. Versammlung der Gesellschaft Deutscher Naturforscher und Ärzte e.V.* Proceedings 1. Berlin: Edition Open Access. http://edition-open-access.de /proceedings/1/index.html.

Renn, Jürgen, Urs Schoepflin, and Milena Wazeck (2002). "The Classical Image of Science and the Future of Science Policy, Ringberg Symposium October 2000." In *Innovative Structures in Basic Research*, ed. U. Opolka and H. Schoop, 11–21. Munich: Max Planck Society.

Renn, Jürgen, and Matteo Valleriani (2001). "Galileo and the Challenge of the Arsenal." *Nuncius* 2:481–503.

———(2014). "Elemente einer Wissensgeschichte der Architektur." In *Wissensgeschichte der Architektur*. Vol. 1: *Vom Neolithikum bis zum Alten Orient*, ed. J. Renn, W. Osthues, and H. Schlimme, 7–53. Studies 3. Berlin: Edition Open Access. http://edition-open-access.de /studies/3/3/index.html.

Renn, Jürgen, Dirk Wintergrün, Roberto Lalli, Manfred Laubichler, and Matteo Valleriani (2016). "Netzwerke als Wissensspeicher." In *Die Zukunft der Wissensspeicher: Forschen, Sammeln und Vermitteln im 21. Jahrhundert*, ed. J. Mittelstraß, 35–79. Konstanzer Wissenschaftsforum 7. Konstanz: Universitätsverlag Konstanz.

Rheinberger, Hans-Jörg (1997). *Toward a History of Epistemic Things: Synthesizing Proteins in the Test Tube*. Stanford, CA: Stanford University Press.

———(2012). "Internationalism and the History of Molecular Biology." In *The Globalization of Knowledge in History*, ed. J. Renn, 737–44. Studies 1. Berlin: Edition Open Access. http:// edition-open-access.de/studies/1/33/index.html.

Rich, Nathaniel (2019). *Losing Earth: A Recent History*. New York: MCD.

Richards, Edward Graham (1998). *Mapping Time: The Calendar and Its History*. Oxford: Oxford University Press.

Richerson, Peter J., and Robert Boyd (2005). *Not by Genes Alone: How Culture Transformed Human Evolution*. Chicago: University of Chicago Press.

———(2010). "Why Possibly Language Evolved." *Biolinguistics* 4 (2–3): 289–306.

Richerson, Peter J., and Morten H. Christiansen (2013). *Cultural Evolution: Society, Technology, Language, and Religion*. Cambridge, MA: MIT Press.

Ringer, Fritz K. (1990). *The Decline of the German Mandarins: The German Academic Community, 1890–1933*. Middletown, CT: Wesleyan University Press.

Ritt-Benmimoun, Veronika, ed. (2014). *Wiener Zeitschrift für die Kunde des Morgenlandes*. Vol. 104. Vienna: Selbstverlag des Instituts für Orientalistik.

Ritter, James (2000). "Egyptian Mathematics." In *Mathematics across Cultures: The History of Non-Western Mathematics*, ed. H. Selin, 115–36. Dordrecht: Kluwer Academic.

Robson, Eleanor (2008). *Mathematics in Ancient Iraq: A Social History*. Princeton, NJ: Princeton University Press.

Rocchi Coopmans de Yoldi, Giuseppe (2006). *Santa Maria del Fiore: Teorie e storie dell'archeologia e del restauro nella città delle fabbriche arnolfiane*. Florence: Alinea.

Rochberg, Francesca (2004). *The Heavenly Writing: Divination, Horoscopy, and Astronomy in Mesopotamian Culture*. Cambridge: Cambridge University Press.

———(2017). *Before Nature: Cuneiform Knowledge and the History of Science*. Chicago: University of Chicago Press.

Rochot, Bernard (1967). "Le P. Mersenne et les relations intellectuelles dans l'Europe du XVIIe siècle." *Cahiers d'Histoire Mondiale* 10:55–73.

Rockström, Johan, Guy Brasseur, Brian Hoskins, Wolfgang Lucht, Hans Joachim Schellnhuber, Pavel Kabat, Nebojsa Nakicenovic, Peng Gong, Peter Schlosser, Maria Máñez Costa, April Humble, Nick Eyre, Peter Gleick, Rachel James, Andre Lucena, Omar Masera, Marcus Moench, Roberto Schaeffer, Sybil Seitzinger, Sander van der Leeuw, Bob Ward, Nicholas Stern, James Hurrell, Leena Srivastava, Jennifer Morgan, Carlos Nobre, Youba Sokona, Roger Cremades, Ellinor Roth, Diana Liverman, and James Arnott (2014). "Climate Change: The Necessary, the Possible and the Desirable: Earth League Climate Statement on the Implications for Climate Policy from the 5th IPCC Assessment." *Earth's Future* 2 (12): 606–11.

Rockström, Johan, Will Steffen, Kevin Noone, Åsa Persson, F. Stuart Chapin III, Eric Lambin, Timothy M. Lenton, Marten Scheffer, Carl Folke, Hans Joachim Schellnhuber, Björn Nykvist, Cynthia A. de Wit, Terry Hughes, Sander van der Leeuw, Henning Rodhe, Sverker Sörlin, Peter K. Snyder, Robert Costanza, Uno Svedin, Malin Falkenmark, Louise Karlberg, Robert W. Corell, Victoria J. Fabry, James Hansen, Brian Walker, Diana Liverman, Katherine Richardson, Paul Crutzen, and Jonathan Foley (2009). "Planetary Boundaries: Exploring the Safe Operating Space for Humanity." *Ecology and Society* 14 (2): 32.

Rogers Ackermann, Rebecca, Alex Mackay, and Michael L. Arnold (2016). "The Hybrid Origin of 'Modern' Humans." *Evolutionary Biology* 43 (1): 1–11.

Rohbeck, Johannes (1987). *Die Fortschrittstheorie der Aufklärung.* Frankfurt am Main: Campus Verlag.

Romer, Paul M. (1990). "Endogenous Technological Change." *Journal of Political Economy* 98 (5, pt. 2): S71–S102.

Rosińska-Balik, Karolina, Agnieszka Ochał-Czarnowicz, Marcin Czarnowicz, and Joanna Dębowska-Ludwin, eds. (2015). *Copper and Trade in the South-Eastern Mediterranean: Trade Routes of the Near East in Antiquity.* British Archaeological Reports International Series 2753. Oxford: Archaeopress.

Rosol, Christoph (2007). *RFID: Vom Ursprung einer (All)gegenwärtigen Kulturtechnologie.* Berlin: Kadmos.

———(2017). "Data, Models and Earth History in Deep Convolution: Paleoclimate Simulations and their Epistemological Unrest." *Berichte zur Wissenschaftsgeschichte* 40 (2): 120–139. https://onlinelibrary.wiley.com/doi/full/10.1002/bewi.201701822.

Rosol, Christoph, Sara Nelson, and Jürgen Renn (2017). "Introduction: In the Machine Room of the Anthropocene." *Anthropocene Review* 4 (1): 2–8.

Rosol, Christoph, Benjamin Steininger, Jürgen Renn, and Robert Schlögl (2018). "On the Age of Computation in the Epoch of Humankind." Sponsor feature. *Nature Research.* Accessed November 29, 2018. https://www.nature.com/articles/d42473-018-00286-8.

Rossabi, Morris (2010). *Voyager from Xanadu: Rabban Sauma and the First Journey from China to the West.* Berkeley: University of California Press.

Rossi, Paolo Alberto (1982). *Le cupole del Brunelleschi: Capire per conservare.* Bologna: Calderini.

Rottenburg, Richard (2012). "On Juridico-Political Foundations of Meta-Codes." In *The Globalization of Knowledge in History,* ed. J. Renn, 483–500. Studies 1. Berlin: Edition Open Access. http://edition-open-access.de/studies/1/25/index.html.

Rousseau, Jean-Jacques ([1762] 1964). "Du contrat social: Ou, principes du droit politique." In *Du contrat social—Écrits politiques.* Vol. 3 of *Oeuvres complètes,* ed. B. Gagnebin and M. Raymond, 347–470. Paris: Éditions Gallimard.

Ruddiman, William F. (2003). "The Anthropogenic Greenhouse Era Began Thousands of Years Ago." *Climatic Change* 61 (3): 261–93.

———(2005). *Plows, Plagues, and Petroleum: How Humans Took Control of Climate.* Princeton, NJ: Princeton University Press.

————(2013). "The Anthropocene." *Annual Review of Earth and Planetary Sciences* 41:45–68.

Rudwick, Martin J. S. (2014). *Earth's Deep History*. Chicago: University of Chicago Press.

Rüegg, Walter, ed. (1996). *Geschichte der Universität in Europa*. 2 vols. Munich: C. H. Beck.

Ruffini, Remo, and John A. Wheeler (1971). "Introducing the Black Hole." *Physics Today* 24 (1): 30–41.

Rushkoff, Douglas (2018). "Survival of the Richest: The Wealthy Are Plotting to Leave Us Behind." Medium: Future Human. Accessed July 18, 2018. https://medium.com/s /futurehuman/survival-of-the-richest-9ef6cdddocc1.

Russell-Wood, Anthony J. R. (1998). *The Portuguese Empire, 1415–1808: A World on the Move*. Baltimore, MD: Johns Hopkins University Press.

Russo, Lucio (2004). *The Forgotten Revolution: How Science Was Born in 300 BC and Why It Had to Be Reborn*. Heidelberg: Springer.

Saalman, Howard (1980). *Filippo Brunelleschi: The Cupola of Santa Maria del Fiore*. Studies in Architecture 20. London: A. Zwemmer.

Sachse, Carola, ed. (2003). *Die Verbindung nach Auschwitz: Biowissenschaften und Menschenversuche an Kaiser-Wilhelm-Instituten*. Geschichte der Kaiser-Wilhelm-Gesellschaft im Nationalsozialismus 6. Göttingen: Wallstein Verlag.

Sacrobosco, Johannes de, and Pedro Nunes (1537). *Tratado da sphera com a theorica do sol & da lua e ho primeiro livro da geographia de Claudio Ptolomeo Alexandrino: Tirados novamente de Latim em lingoagem pello Doutor Pero Nunez cosmographo del Rey Don Joano. Et acrescentatos de muitas annotações i figuras per que mays facilmente se podem entender. Tratado que ho doutor per nunez fez em defensam de carta de marear. Item dous tratados quo mesmo Doutor fez sobra a carta de marear. Em os quaes se decrerano todas as principaes du vidas da navegação. Con as pavoa do movimento do sol: I sua declinação. Eo regimento de altura*. Lisbon: Galharde.

Sacrobosco, Johannes de, Èlie Vinet, Petrus Valerianus, Petrus Nuñes, and Joannes Regiomontanus (1556). *Sphaera Ioannis de Sacro Bosco emendata: Eliae Vineti Santonis scholia in eandem sphaeram, ab ipso auctore restituta. Adiunximus huic libro compendium in sphaeram, per Pierium Valerianum Bellunensem, et, Petri Nonii Salaciensis demonstrationem eorum, quae in extremo capite de Climatibus Sacroboscius scribit, de inaequali climatum latitudine, eodem Vineto interprete*. Paris: Gulielmum Cavellat.

Sadler, John Edward (2014). *J. A. Comenius and the Concept of Universal Education*. New York: Routledge.

Sagan, Carl (1985). *Contact: A Novel*. New York: Simon & Schuster.

Sagan, Carl, Linda Salzman Sagan, and Frank Drake (1972). "A Message from Earth." *Science* 175 (4024): 881–84.

Sahagún, Fray Bernardino de (1577). *Historia general de las cosas de nueva España* [General history of the things of New Spain]. *The Florentine Codex*. Vol. 3, book 12: *The Conquest of Mexico*. Florence: Biblioteca Medicea Laurenziana. World Digital Library. https://www .wdl.org/en/item/10096/#institution=laurentian-library.

Sahlins, Marshall (1994). "Cosmologies of Capitalism: The Trans-Pacific Sector of the 'World-System.'" In *Culture / Power / History: A Reader in Contemporary Social Theory*, ed. N. B. Dirks, G. Eley, and S. B. Ortner, 412–56. Princeton, NJ: Princeton University Press.

Saito, Osamu (2002). "The Frequency of Famines as Demographic Correctives in the Japanese Past." In *Famine Demography: Perspectives from the Past and Present*, ed. T. Dyson and C. Ó Gráda, 218–39. Oxford: Oxford University Press.

Sallaberger, Walther (1999). "Ur III-Zeit." In *Mesopotamien: Akkade-Zeit und Ur-III Zeit*, ed. P. Attinger and M. Wäfler, 121–390. Annäherungen 3. Orbis Biblicus et Orientalis 160. Göttingen: Vandenhoeck & Ruprecht.

Sandom, Christopher, Søren Faurby, Brody Sandel, and Jens-Christian Svenning (2014). "Global Late Quaternary Megafauna Extinctions Linked to Humans, Not Climate Change." *Proceedings of the Royal Society B: Biological Sciences* 281 (1787): 20133254.

Sasson, Jack M., ed. (1995). *Civilizations of the Ancient Near East*. Vol. 4 of 4 vols. New York: Charles Scribner's Sons.

Savage-Rumbaugh, E. Sue, William M. Fields, Pär Segerdahl, and Duane M. Rumbaugh (2005). "Culture Prefigures Cognition in 'Pan/Homo' Bonobos." *Theoria: An International Journal for Theory, History and Foundations of Science, SEGUNDA EPOCA* 20 (3[54]): 311–28.

Savage-Rumbaugh, E. Sue, and Roger Lewin (1994). *Kanzi: The Ape at the Brink of the Human Mind*. New York: John Wiley & Sons.

Savage-Rumbaugh, E. Sue, Jeannine Murphy, Rose A. Sevic, Karen E. Brakke, Shelly L. Williams, Duane M. Rumbaugh, and Elizabeth Bates (1993). "Language Comprehension in Ape and Child." *Monographs of the Society for Research in Child Development* 58 (3–4): 1–252.

Savage-Rumbaugh, E. Sue, Rose A. Sevic, Duane M. Rumbaugh, and Elizabeth Rubert (1985). "The Capacity of Animals to Acquire Language: Do Species Differences Have Anything to Say to Us?" *Philosophical Transactions of the Royal Society B: Biological Sciences* 308 (1135): 177–85.

Schaper, Joachim (2019). " 'Real Abstraction' and the Origins of Intellectual Abstraction in Ancient Mesopotamia: Ancient Economic History as a Key to the Understanding and Evaluation of Marx's Labour Theory of Value." In *Culture and Cognition: Essays in Honor of Peter Damerow*, ed. J. Renn and M. Schemmel. Berlin: Edition Open Access.

Scharf, Peter M., and Malcolm D. Hyman (2012). *Linguistic Issues in Encoding Sanskrit*. Delhi: Motilal Banarsidass.

Schellnhuber, Hans Joachim, Olivia Maria Serdeczny, Sophie Adams, Claudia Köhler, Ilona Magdalena Otto, and Carl-Friedrich Schleussner (2016). "The Challenge of a 4°C World by 2100." In *Handbook on Sustainability Transition and Sustainable Peace*, ed. H. G. Brauch, Ú. Oswald Spring, J. Grin, and J. Scheffran, 267–83. Hexagon Series on Human and Environmental Security and Peace 10. Cham: Springer.

Schelsky, Helmut, ed. (1970). *Zur Theorie der Institution*. Interdisziplinäre Studien 1. Düsseldorf: Bertelsmann-Universitätsverlag.

Schemmel, Matthias (2001a). "England's Forgotten Galileo: A View on Thomas Harriot's Ballistic Parabolas." In *Largo campo di filosofare: Eurosymposium Galileo 2001*, ed. J. Montesinos and C. Solís, 269–80. La Orotava: Fundación Canaria Orotava de Historia de la Ciencia.

———(2001b). *The Sections on Mechanics in the Mohist Canon*. Preprint 182. Berlin: Max Planck Institute for the History of Science.

———(2006). "The English Galileo: Thomas Harriot and the Force of Shared Knowledge in Early Modern Mechanics." *Physics in Perspective* 8 (4): 360–80.

———(2008). *The English Galileo: Thomas Harriot's Work on Motion as an Example of Preclassical Mechanics*. Boston Studies in the Philosophy and History of Science 268. Dordrecht: Springer.

———(2012). "The Transmission of Scientific Knowledge from Europe to China in the Early Modern Period." In *The Globalization of Knowledge in History*, ed. J. Renn, 269–93. Studies 1. Berlin: Edition Open Access. http://edition-open-access.de/studies/1/15/index.html.

———(2013). "Stevin in Chinese: Aspects of the Transformation of Early Modern European Science in Its Transfer to China." In *Translating Knowledge in the Early Modern Low Countries*, ed. H. J. Cook and S. Dupré, 369–85. Münster: LIT Verlag.

———(2014). "Medieval Representations of Change and Their Early Modern Application." *Foundations of Science* 19 (1): 11–34.

———(2016a). *Historical Epistemology of Space: From Primate Cognition to Spacetime Physics*. Springer Briefs in History of Science and Technology. Cham: Springer.

————, ed. (2016b). *Spatial Thinking and External Representation: Towards a Historical Epistemology of Space*. Studies 8. Berlin: Edition Open Access. http://edition-open-access.de/studies/8/index.html.

————(2016c). "Towards a Historical Epistemology of Space: An Introduction." In *Spatial Thinking and External Representation: Towards a Historical Epistemology of Space*, ed. M. Schemmel, 1–33. Studies 8. Berlin: Edition Open Access. http://edition-open-access.de/studies/8/2/index.html.

————(2019). *Everyday Language and Technical Terminology: Reflective Abstractions in the Long-Term History of Spatial Terms*. Preprint 491. Berlin: Max Planck Institute for the History of Science.

Schemmel, Matthias, and William G. Boltz (2016). "Theoretical Reflections on Elementary Actions and Instrumental Practices: The Example of the Mohist Canon." In *Spatial Thinking and External Representation: Towards a Historical Epistemology of Space*, ed. M. Schemmel, 121–44. Studies 8. Berlin: Edition Open Access. http://edition-open-access.de/studies/8/5/index.html.

Schiefenhövel, Wulf (1991). "Eipo." In *Oceania*, ed. T. E. Hays, 55–59. Vol. 2 of *Encyclopedia of World Cultures*. Boston: G. K. Hall.

————(2013). "Biodiversity through Domestication: Examples from New Guinea." *Revue d'ethnoécologie* 3:1–16.

Schiefenhövel, Wulf, Volker Heeschen, and Irenäus Eibl-Eibesfeldt (1980). "Requesting, Giving and Taking: The Relationship between Verbal and Nonverbal Behavior in the Speech Community of the Eipo, Irian Jaya (West New Guinea)." In *The Relationship of Verbal and Nonverbal Communication*, ed. M. R. Key, 139–66. The Hague: Mouton.

Schiefsky, Mark (2007). "Theory and Practice in Heron's Mechanics." In *Mechanics and Natural Philosophy before the Scientific Revolution*, ed. W. R. Laird and S. Roux, 15–49. Boston Studies in the Philosophy of Science 254. New York: Springer.

————(2012). "The Creation of Second-Order Knowledge in Ancient Greek Science as a Process in the Globalization of Knowledge." In *The Globalization of Knowledge in History*, ed. J. Renn, 177–86. Studies 1. Berlin: Edition Open Access. http://edition-open-access.de/studies/1/12/index.html.

Schlimme, Hermann, ed. (2006). *Practice and Science in Early Modern Italian Building: Towards an Epistemic History of Architecture*. Milan: Electa.

————(2009). "Die frühe 'Accademia et Compagnia dell'Arte del Disegno' in Florenz und die Architekturausbildung." In *Entwerfen: Architektenausbildung in Europa von Vitruv bis Mitte des 20. Jahrhunderts—Geschichte, Theorie, Praxis*, ed. R. Johannes, 326–43. Hamburg: Junius Verlag.

Schlimme, Hermann, Dagmar Holste, and Jens Niebaum (2014). "Bauwissen im Italien der Frühen Neuzeit." In *Wissensgeschichte der Architektur*. Vol. 3: *Vom Mittelalter bis zur Frühen Neuzeit*, ed. J. Renn, W. Osthues, and H. Schlimme, 97–368. Studies 5. Berlin: Edition Open Access. http://edition-open-access.de/studies/5/4/index.html.

Schlögl, Robert (2012). "The Role of Chemistry in the Global Energy Challenge." In *The Globalization of Knowledge in History*, ed. J. Renn, 745–94. Studies 1. Berlin: Edition Open Access. http://edition-open-access.de/studies/1/34/index.html.

Schmandt-Besserat, Denise (1992a). *Before Writing*. Vol. 1: *From Counting to Cuneiform*. Austin: University of Texas Press.

————(1992b). *How Writing Came About*. Abridged ed. of *Before Writing*. Vol. 1: *From Counting to Cuneiform*. Austin: University of Texas Press.

Schmelz, Martin, Sebastian Grueneisen, Alihan Kabalak, Jürgen Jost, and Michael Tomasello (2017). "Chimpanzees Return Favors at a Personal Cost." *Proceedings of the National Academy of Sciences of the United States of America* 114 (28): 7462–67.

Schmieg, Greogor, Esther Meyer, Isabell Schrickel, Jeremias Herberg, Guido Caniglia, Ulli Vilsmaier, Manfred Laubichler, Erich Hörl, and Daniel Lang (2017). "Modeling Normativity
in Sustainability: A Comparison of the Sustainable Development Goals, the Paris Agreement, and the Papal Encyclical." *Sustainability Science* 13 (3): 785–96.

Schmuhl, Hans-Walter (2005). *Grenzüberschreitungen: Das Kaiser-Wilhlem-Institut für Anthropologie, menschliche Erblehre und Eugenik 1927–1945.* Geschichte der Kaiser-Wilhelm-
Gesellschaft im Nationalsozialismus 9. Göttingen: Wallstein Verlag.

Schneider, Birgit (2017). "The Future Face of the Earth: The Visual Semantics of the Future in
the Climate Change Imagery of the IPCC." In *Cultures of Prediction in Atmospheric and
Climate Science: Epistemic and Cultural Shifts in Computer-Based Modelling and Simulation*, ed. M. Heymann, G. Gramelsberger, and M. Mahony, 231–52. London: Routledge.

Schrödinger, Erwin (1946). *What Is Life? The Physical Aspect of the Living Cell: Based on Lectures Delivered under the Auspices of the Institute for Advanced Studies at Trinity College,
Dublin, in February 1943.* Reprint, Cambridge: Cambridge University Press.

Schülein, Johann August (1987). *Theorie der Institution: Eine Dogmengeschichtliche und Konzep-
tionelle Analyse.* Opladen: Westdeutscher Verlag.

Schumpeter, Joseph A. ([1934] 2008). *The Theory of Economic Development: An Inquiry into
Profits, Capital, Credit, Interest, and the Business Cycle.* Translated by Redvers Opie.
14th ed. Social Science Classics Series. New Brunswick, NJ: Transaction.

———(1949). "Science and Ideology." *American Economic Review* 39 (2): 346–59.

Schwägerl, Christian (2014). *The Anthropocene: The Human Era and How It Shapes Our Planet.*
Santa Fe, NM: Synergetic Press.

Schweber, Silvan S. (2015). "Hacking the Quantum Revolution: 1925–1975." *European Physical
Journal H* 40 (1): 53–149.

Scott, James C. (2017). *Against the Grain: A Deep History of the Earliest States.* New Haven, CT:
Yale University Press.

Scott, John T., ed. (2006). *Jean-Jacques Rousseau: Critical Assessments of Leading Political Philosophers.* London: Routledge.

Searle, John R. (1995). *The Construction of Social Reality.* London: Penguin Books.

Secord, James A. (2004). "Knowledge in Transit." *Isis* 95 (4): 654–72.

Seipel, Wilfried, ed. (2003). *Der Turmbau zu Babel: Ursprung der Vielfalt von Sprache und
Schrift.* Vol. 3: *Schrift.* Vienna: Kunsthistorisches Museum.

Selin, Helaine, ed. (2008). *Encyclopaedia of the History of Science, Technology, and Medicine in
Non-Western Cultures.* 2nd ed. Dordrecht: Springer.

Senft, Gunter (2016). " 'Masawa—bogeokwa si tuta!': Cultural and Cognitive Implications of
the Trobriand Islanders' Gradual Loss of Their Knowledge of How to Make a Masawa
Canoe." In *Ethnic and Cultural Dimensions of Knowledge*, ed. P. Meusburger, T. Freytag,
and L. Suarsana, 229–56. Knowledge and Space 8. Dordrecht: Springer.

Senghas, Ann (1995). "Children's Contribution to the Birth of Nicaraguan Sign Language."
Ph.D. diss., Massachusetts Institute of Technology.

———(2000). "The Development of Early Spatial Morphology in Nicaraguan Sign Language."
In *Proceedings of the 24th Annual Boston University Conference on Language Development*, ed.
S. C. Howell, S. A. Fish, and T. Keith-Lucas, 696–707. Somerville, MA: Cascadilla Press.

Senghas, Ann, and Marie Coppola (2001). "Children Creating Language: How Nicaraguan Sign
Language Acquired a Spatial Grammar." *Psychological Science* 12 (4): 323–28.

Senghas, Richard J. (1997). "An 'Unspeakable, Unwriteable' Language: Deaf Identity, Language
and Personhood among the First Cohort of Nicaraguan Signers." Ph.D. diss., University
of Rochester.

Senghas, Richard J., Ann Senghas, and Jennie E. Pyers (2005). "The Emergence of Nicaraguan
Sign Language: Questions of Development, Acquisition, and Evolution." In *Biology and*

Knowledge Revisited: From Neurogenesis to Psychogenesis, ed. S. T. Parker, J. Langer, and C. Milbrath, 287–306. Mahwah, NJ: Erlbaum.

Sevic, Rose A., and E. Sue Savage-Rumbaugh (1994). "Language Comprehension and Use by Great Apes." *Language & Communication* 14 (1): 37–58.

Seyfarth, Robert M., Dorothy L. Cheney, and Peter Marler (1980). "Vervet Monkey Alarm Calls: Semantic Communication in a Free-Ranging Primate." *Animal Behavior* 28 (4): 1070–94.

Shannon, Claude E., and Warren Weaver (1949). *The Mathematical Theory of Communication*. Urbana: University of Illinois Press.

Shapin, Steven, and Simon Schaffer (1985). *Leviathan and the Air-Pump: Hobbes, Boyle, and the Experimental Life*. Princeton, NJ: Princeton University Press.

Sieferle, Rolf Peter (2001). *The Subterranean Forest: Energy Systems and the Industrial Revolution*. Translated by Michael P. Osman. Knapwell: White Horse Press.

Sievertsen, Uwe (2014). "Bauwissen im Alten Orient." In *Wissensgeschichte der Architektur*. Vol. 1: *Vom Neolithikum bis zum Alten Orient*, ed. J. Renn, W. Osthues, and H. Schlimme, 131–280,. Studies 3. Berlin: Edition Open Access. http://edition-open-access.de/studies /3/5/index.html.

Silva da Silva, Circe Mary, and Ligia Arantes Sad (2012). "The Transformations of Knowledge through Cultural Interactions in Brazil: The Case of the Tupinikim and the Guarani." In *The Globalization of Knowledge in History*, ed. J. Renn, 525–58. Studies 1. Berlin: Edition Open Access. http://edition-open-access.de/studies/1/27/index.html.

Simmel, Georg (2004). *The Philosophy of Money*. Translated by Tom Bottomore. London: Taylor & Francis.

Simon, Dagmar, Andreas Knie, Stefan Hornbostel, and Karin Zimmermann, eds. (2016). *Handbuch Wissenschaftspolitik*. Springer Reference Sozialwissenschaften. Wiesbaden: Springer.

Singh, Shree N., and Amitosh Verma (2007). "Environmental Review: The Potential of Nitrification Inhibitors to Manage the Pollution Effect of Nitrogen Fertilizers in Agricultural and Other Soils: A Review." *Environmental Practice* 9 (4): 266–79.

Sivasundaram, Sujit (2010a). "Focus: Global Histories of Science: Introduction." *Isis* 101 (1): 95–97.

——— (2010b). "Sciences and the Global: On Methods, Questions, and Theory." *Isis* 101 (1): 146–58.

Slaughter, Sheila, and Larry L. Leslie (1997). *Academic Capitalism: Politics, Policies and the Entrepreneurial University*. Baltimore, MD: Johns Hopkins University Press.

Slobin, Dan I. (2004). "From Ontogenesis to Phylogenesis: What Can Child Language Tell Us about Language Evolution?" In *Biology and Knowledge Revisited: From Neurogenesis to Psychogenesis*, ed. J. Langer, S. T. Parker, and C. Milbrath, 255–85. Mahwah, NJ: Erlbaum.

Smith, Bruce D., and Melinda A. Zeder (2013). "The Onset of the Anthropocene." *Anthropocene* 4:8–13.

Smith, Cyril Stanley (1977). Reviews of *Metallurgical Remains of Ancient China*, by Noel Bernard and Tamotsu Sato; and *The Cradle of the East: An Enquiry into the Indigenous Origins of Techniques and Ideas of Neolithic and Early Historic China, 5000–1000 B.C.*, by Ping-Ti Ho. *Technology and Culture* 18 (1): 80–86.

Smith, George E. (2006). "The *Vis Viva* Dispute: A Controversy at the Dawn of Dynamics." *Physics Today* 59 (10): 31–36.

Smith, Pamela H. (2004). *The Body of the Artisan: Art and Experience in the Scientific Revolution*. Chicago: University of Chicago Press.

Smitka, Michael, ed. (1998). *The Japanese Economy in the Tokugawa Era 1600–1868*. Japanese Economic History 1600–1960 6. London: Routledge.

Solnit, Rebecca (2014). *Men Explain Things to Me*. New York: Haymarket Books.

Speer, Andreas, and Lydia Wegener (2008). *Wissen über Grenzen: Arabisches Wissen und Lateinisches Mittelalter*. Miscellanea Mediaevalia 33. Berlin: De Gruyter.

Sperber, Dan, ed. (2000). *Metarepresentations: A Multidisciplinary Perspective*. Vancouver Studies in Cognitive Science 10. New York: Oxford University Press.

Spicer, Dag, Gwen Bell, Jan Zimmerman, Jacqueline Boas, Bill Boas, and Dan Lythcott-Haims (1997). "Internet History 1962 to 1992." Computer History Museum. Accessed February 7, 2018. http://www.computerhistory.org/internethistory.

Spindler, Martin (2014). "The Center and the Edge." Accessed July 26, 2015. http://mjays.net /the-center-and-the-edge.

Sprenger, Florian, and Christoph Engemann (2015a). "Im Netz der Dinge: Zur Einleitung." In *Internet der Dinge: Über Smarte Objekte, Intelligente Umgebungen und die Technische Durchdringung der Welt*, ed. F. Sprenger and C. Engemann, 7–58. Bielefeld: Transcript Verlag.

———, eds. (2015b). *Internet der Dinge: Über Smarte Objekte, Intelligente Umgebungen und die Technische Durchdringung der Welt*. Bielefeld: Transcript Verlag.

Srnicek, Nick (2017). *Platform Capitalism*. Cambridge: Polity Press.

Stadler, Friedrich (2001). *The Vienna Circle: Studies in the Origins, Development, and Influence of Logical Empiricism*. Translated by Camilla Nielsen, Joel Golb, Sabine Schmidt, and Thomas Ernst. Vienna: Springer.

Staley, Richard (2009). *Einstein's Generation: The Origins of the Relativity Revolution*. Chicago: University of Chicago Press.

Star, Susan Leigh (2010). "This Is Not a Boundary Object: Reflections on the Origin of a Concept." *Science, Technology, & Human Values* 35 (5): 601–17.

Stearns, Peter N. (1993). "Interpreting the Industrial Revolution." In *Islamic and European Expansion: The Forging of a Global Order*, ed. M. Adas. Philadelphia: Temple University Press.

Steels, Luc (2011). "Modeling the Cultural Evolution of Language." *Physics of Life Reviews* 8 (4): 339–56.

Steffen, Will, Wendy Broadgate, Lisa Deutsch, Owen Gaffney, and Cornelia Ludwig (2015). "The Trajectory of the Anthropocene: The Great Acceleration." *Anthropocene Review* 2 (1): 81–98.

Steffen, Will, Paul J. Crutzen, and John R. McNeill (2007). "The Anthropocene: Are Humans Now Overwhelming the Great Forces of Nature?" *AMBIO: A Journal of the Human Environment* 36 (8): 614–21.

Steffen, Will, Jacques Grinevald, Paul Crutzen, and John McNeill (2011). "The Anthropocene: Conceptual and Historical Perspectives." *Philosophical Transactions of the Royal Society A: Mathematical, Physical and Engineering Sciences* 369 (1938): 842–67.

Steffen, Will, Reinhold Leinfelder, Jan Zalasiewicz, Colin N. Waters, Mark Williams, Colin Summerhayes, Anthony D. Barnosky, Alejandro Cearreta, Paul Crutzen, Matt Edgeworth, Erle C. Ellis, Ian J. Fairchild, Agnieszka Galuszka, Jacques Grinevald, Alan Haywood, Juliana Ivardo Sul, Catherine Jeandel, J. R. McNeill, Eric Odada, Naomi Oreskes, Andrew Revkin, Daniel deB. Richter, James Syvitski, Davor Vidas, Michael Wagreich, Scott L. Wing, Alexander P. Wolfe, and H. J. Schellnhuber (2016). "Stratigraphic and Earth System Approaches to Defining the Anthropocene." *Earth's Future* 4 (8): 324–45.

Steffen, Will, Katherine Richardson, Johan Rockström, Sarah E. Cornell, Ingo Fetzer, Elena M. Bennett, Reinette Biggs, Stephen R. Carpenter, Wim de Vries, Cynthia A. de Wit, Carl Folke, Dieter Gerten, Jens Heinke, Georgina M. Mace, Linn M. Persson, Veerabhadran Ramanathan, Belinda Reyers, and Sverker Sörlin (2015). "Planetary Boundaries: Guiding Human Development on a Changing Planet." *Science* 347 (6223): 1259855.

Steffen, Will, Johan Rockström, Katherine Richardson, Timothy M. Lenton, Carl Folke, Diana Liverman, Colin P. Summerhayes, Anthony D. Barnosky, Sarah E. Cornell, Michel

Crucifix, Jonathan F. Donges, Ingo Fetzer, Steven J. Lade, Marten Scheffer, Ricarda Win-kelmann, and Hans Joachim Schellnhuber (2018). "Trajectories of the Earth System in the Anthropocene." *Proceedings of the National Academy of Sciences of the United States of America* 115 (33): 8252–59.

Steffen, Will, Angelina Sanderson, Peter Tyson, Jill Jäger, Pamela Matson, Berrien Moore III, Frank Oldfield, Kathrine Richardson, John Schellnhuber, B. L. Turner, and Robert Wasson (2004). *Global Change and the Earth System: A Planet under Pressure.* Heidelberg: Springer.

Steinberg, Sigfrid Henry ([1955] 2017). *Five Hundred Years of Printing.* Reprint, Mineola, NY: Dover.

Steininger, Benjamin (2014). "Refinery and Catalysis." In *Textures of the Anthropocene: Grain, Vapor, Ray.* Vol. 2, ed. K. Klingan, A. Sepahvand, C. Rosol, and B. M. Scherer, 105–18. Cambridge, MA: MIT Press.

———(2018). "Petromoderne-Petromonströs." *Azimuth* 12 (VI): 15–29.

Sterelny, Kim, and Trevor Watkins (2015). "Neolithization in Southwest Asia in a Context of Niche Construction Theory." *Cambridge Archaeological Journal* 25 (3): 673–91.

Sterling, Bruce (2015). *The Epic Struggle of the Internet of Things.* Moscow: Strelka Press.

Stern, Steve J. (1988). "Feudalism, Capitalism, and the World-System in the Perspective of Latin America and the Caribbean." *American Historical Review* 93 (2): 829–72.

Sternberger, Dolf (1977). *Panorama of the Nineteenth Century.* Translated by Joachim Neugros-chel. New York: Urizen Books.

Stevenson, Christopher M., Cedric O. Puleston, Peter M. Vitousek, Oliver A. Chadwick, Sonia Haoa, and Thegn N. Ladefoged (2015). "Variation in Rapa Nui (Easter Island) Land Use Indicates Production and Population Peaks prior to European Contact." *Proceedings of the National Academy of Sciences of the United States of America* 112 (4): 1025–30.

Stichweh, Rudolf (1984). *Zur Entstehung des modernen Systems wissenschaftlicher Disziplinen: Physik in Deutschland 1740–1890.* Frankfurt am Main: Suhrkamp.

Streeck, Wolfgang (2014). *Buying Time: The Delayed Crisis of Democratic Capitalism.* Translated by Patrick Camiller and David Fernbach. London: Verso Books.

Streeck, Wolfgang, and Kathleen Thelen (2005). "Introduction: Institutional Change in Advanced Political Economies." In *Beyond Continuity: Institutional Change in Advanced Political Economies*, ed. W. Streeck and K. Thelen, 1–39. Oxford: Oxford University Press.

Su, Ching (1996). "The Printing Presses of the London Missionary Society among the Chinese." Ph.D. diss., University College London.

Suchak, Malini, Timothy M. Eppley, Matthew W. Campbell, Rebecca A. Feldman, Luke F. Quarles, and Frans B. M. de Waal (2016). "How Chimpanzees Cooperate in a Competitive World." *Proceedings of the National Academy of Sciences of the United States of America* 113 (36): 10215–20.

Suess, Eduard (1904–24). *The Face of the Earth [Das Anlitz der Erde].* Translated by Her-tha B. C. Sollas. 5 vols. Oxford: Clarendon Press.

Sukhdev, Pavan, Peter May, and Alexander Müller (2016). "Fix Food Metrics." *Nature* 540 (7631): 33–34.

Swerdlow, Noel (1973). "The Derivation and First Draft of Copernicus's Planetary Theory: A Translation of the *Commentariolus* with Commentary." *Proceedings of the American Philo-sophical Society* 117 (6): 423–512.

———(2004). "An Essay on Thomas Kuhn's First Scientific Revolution: The Copernican Revo-lution." *Proceedings of the American Philosophical Society* 148 (1): 64–120.

Swerdlow, Noel M., and Otto Neugebauer (1984). *Mathematical Astronomy in Copernicus's "De revolutionibus."* 2 vols. Studies in the History of Mathematics and Physical Sciences 10. New York: Springer.

Szabó, Árpád (1978). *The Beginnings of Greek Mathematics*. Dordrecht: D. Reidel.

Szabó, Árpád, and Erkka Maula (1982). *Enklima: Untersuchungen zur Frühgeschichte der Griechischen Astronomie, Geographie und der Sehnentafeln*. Athens: Akademie Athen, Forschungsinstitut für Griechische Philosophie.

Szerszynski, Bronislaw (2017). "The Anthropocene Monument: On Relating Geological and Human Time." *European Journal of Social Theory* 20 (1): 111–31.

Szöllösi-Janze, Margit (1998). *Fritz Haber 1868–1934: Eine Biographie*. Munich: C. H. Beck.

———(2000). "Losing the War, but Gaining Ground: The German Chemical Industry during World War I." In *The German Chemical Industry in the Twentieth Century*, ed. J. E. Lesch, 91–121. Chemists and Chemistry 18. Dordrecht: Springer.

Taisbak, Christian Marinus (2003). *Euclid's Data: The Importance of Being Given*. Acta Historica Scientiarum Naturalium et Medicinalium 45. Copenhagen: Museum Tusculanum Press.

Tajima, Kayo (2007). *The Marketing of Urban Human Waste in the Early Modern Edo/Tokyo Metropolitan Era*. Environnement Urbain: Cartographie d'un Concept 1. Quebec: Institut National de Recherche Scientifique Centre Urbanisation Culture et Société.

Tallerman, Maggie (2005). *Language Origins: Perspectives on Evolution*. Oxford: Oxford University Press.

Terenzi, Pierluigi (2015). "Maestranze e organizzazione del lavoro negli Anni della Cupola." "The Years of the Cupola—Studies." http://duomo.mpiwg-berlin.mpg.de/STUDIES /study004/study004.html.

Thagard, Paul (1993). *Conceptual Revolutions*. Princeton, NJ: Princeton University Press.

Thelen, Kathleen (2004). *How Institutions Evolve: The Political Economy of Skills in Germany, Britain, the United States, and Japan*. Cambridge: Cambridge University Press.

Thelen, Kathleen, and Sven Steinmo (1992). "Historical Institutionalism in Comparative Politics." In *Structuring Politics: Historical Institutionalism in Comparative Analysis*, ed. K. Thelen, S. Steinmo, and F. Longstreth, 1–32. Cambridge: Cambridge University Press.

Thiering, Martin, and Wulf Schiefenhövel (2016). "Spatial Concepts in Non-literate Societies: Language and Practice in Eipo and Dene Chipewyan." In *Spatial Thinking and External Representation: Towards a Historical Epistemology of Space*, ed. M. Schemmel, 35–92. Studies 9. Berlin: Edition Open Access. http://edition-open-access.de/studies/8/3/index.html.

Thomas, Brinley (1985). "Escaping from Constraints: The Industrial Revolution in a Malthusian Context." *Journal of Interdisciplinary History* 15 (4): 729–53.

Thomas, Julia Adeney (2017). "The Historians' Task in the Age of the Anthropocene: Finding Hope in Japan? Presentation 2, October 12, 2017." Presentation, Max Planck Institute for the History of Science. Accessed March 28, 2018. https://www.mpiwg-berlin.mpg.de /video/historians-task-age-anthropocene-finding-hope-japan-presentation-2.

———(forthcoming). *The Historian's Task in the Anthropocene: Theory, Practice, and the Case of Japan*. Princeton: Princeton University Press.

———(forthcoming). "Practicing Hope in the Anthropocene." *American Historical Review*.

Thompson, William R. (2006). "Climate, Water, and Political-Economic Crises in Ancient Mesopotamia and Egypt." In *The World System and the Earth System*, ed. A. Hornborg and C. Crumley, 163–79. Walnut Creek, CA: Left Coast Press.

Thorne, Kip (1994). *Black Holes and Time Warps: Einstein's Outrageous Legacy*. New York: W. W. Norton.

Tilley, Helen (2010). "Global Histories, Vernacular Science, and African Genealogies: Or, Is the History of Science Ready for the World?" *Isis* 101 (1): 110–19.

Toepfer, Georg (2011). *Analogie—Ganzheit*. Historisches Wörterbuch der Biologie: Geschichte und Theorie der biologischen Grundbegriffe 1. Stuttgart: J. B. Metzler.

Tomasello, Michael (1999). *The Cultural Origins of Human Cognition.* Cambridge, MA: Harvard University Press.

———(2003). *Constructing a Language: A Usage-Based Approach to Language.* Cambridge, MA: Harvard University Press.

———(2014). *A Natural History of Human Thinking.* Cambridge, MA: Harvard University Press.

Tomasello, Michael, and Joseph Call (1997). *Primate Cognition.* Oxford: Oxford University Press.

Tomasello, Michael, Ann C. Kruger, and Hilary H. Ratner (1993). "Cultural Learning." *Behavioral and Brain Sciences* 16 (3): 495–511.

Tomlinson, Gary (2015). *A Million Years of Music: The Emergence of Human Modernity.* Cambridge, MA: Zone Books.

———(2018). *Culture and the Course of Human Evolution.* Chicago: University of Chicago Press.

Torricelli, Evangelista (1919). "De motu gravium naturaliter descendentium et proiectorum (1644)." In *Lezioni accademiche—Meccanica—Scritti vari,* ed. G. Vassura, 101–232. Opera di Evangelista Torricelli 2. Faenza: Danilo Montanari Editore.

Toulmin, Stephen (1972). *Human Understanding.* Vol. 1 of 3. Princeton, NJ: Princeton University Press.

Toynbee, Arnold J. (1954). *A Study of History.* Vol. 7. London: Oxford University Press.

Travis, Anthony S. (1993). *The Rainbow Makers: The Origins of the Synthetic Dyestuffs Industry in Western Europe.* Bethlehem, PA: Lehigh University Press.

Trischler, Helmuth (2016). "The Anthropocene: A Challenge for the History of Science, Technology, and the Environment." *NTM Zeitschrift für Geschichte der Wissenschaften, Technik und Medizin* 24 (3): 309–35.

Tsuen-Hsuin, Tsien (1987). *Chemistry and Chemical Technology Part 1: Paper and Printing.* 3rd ed. Science and Civilisation in China 5.1. Cambridge: Cambridge University Press.

Turner, Billie Lee, II, Pamela A. Matson, James J. McCarthy, Robert W. Corell, Lindsey Christensen, Noelle Eckley, Grete K. Hovelsrud-Broda, Jeanne X. Kasperson, Roger E. Kasperson, Amy Luers, Marybeth L. Martello, Svein Mathiesen, Rosamond Naylor, Colin Polsky, Alexander Pulsipher, Andrew Schiller, Henrik Selin, and Nicholas Tyler (2003). "Illustrating the Coupled Human–Environment System for Vulnerability Analysis: Three Case Studies." *Proceedings of the National Academy of Sciences of the United States of America* 100 (14): 8080–85.

Ufano, Diego (1628). *Artillerie, ou vraye instruction de l'artillerie et de ses appartenances: Contenant une declaration de tout ce qui est de l'office du General d' icelle, tant en un siege qu'en un lieu assiegé; Item des batteries, contre-batteries, ponts, mines & galleries, & de toutes fortes de machines requises au train.* Rouen: Jean Berthelin.

Uhrqvist, Ola (2014). "Seeing and Knowing the Earth as a System: An Effective History of Global Environmental Change Research as Scientific and Political Practice." Ph.D. diss., Linköping University.

Uhrqvist, Ola, and Björn-Ola Linnér (2015). "Narratives of the Past for Future Earth: The Historiography of Global Environmental Change Research." *Anthropocene Review* 2 (2): 159–73.

Valente, Thomas W. (1995). *Network Models of the Diffusion of Innovations.* Cresskill, NJ: Hampton Press.

Valleriani, Matteo (2010). *Galileo Engineer.* Boston Studies in the Philosophy and History of Science 269. Dordrecht: Springer.

———(2013). *Metallurgy, Ballistics and Epistemic Instruments: The "Nova scientia" of Nicolò Tartaglia—a New Edition.* Translated by Matteo Valleriani, Lindy Divarci, and Anna Siebold. Sources 6. Berlin: Edition Open Access. http://www.edition-open-sources.org/sources/6/index.html.

———(2017a). "The Epistemology of Practical Knowledge." In *The Structures of Practical Knowledge,* ed. Matteo Valleriani, 1–19. Cham: Springer.

————, ed. (2017b). *The Structures of Practical Knowledge*. Cham: Springer.

————(2017c). "The Tracts on the Sphere: Knowledge Restructured over a Network." In *The Structures of Practical Knowledge*, ed. M. Valleriani, 421–73. Cham: Springer.

Vasari, Giorgio (1550). *Le vite de' più eccellenti architetti, pittori, et scultori italiani, da Cimabue infino a' tempi nostri: Descritte in lingua toscana da Giorgio Vasari, pittore arentino—Con una sua utile et necessaria introduzione a le arti loro*. 2 vols. Florence: Lorenzo Torrentino.

————(1878–85). *Le vite de' più eccellenti pittori, scultori ed architettori, scritte da Giorgio Vasari pittore aretino, con nuove annotazioni e commenti di Gaetano Milanesi*. 9 vols. Florence: G. C. Sansoni.

————(1987). *Lives of the Artists*. Translated by George Bull. 2 vols. Harmondsworth: Penguin Books.

Vecce, Carlo (2017). *La biblioteca perduta: I libri di Leonardo*. Rome: Salerno Editrice.

Veenhof, Klaas R., and Jesper Eidem (2008). *Mesopotamia: The Old Assyrian Period*. Annäherungen 5; Orbis Biblicus et Orientalis 160. Göttingen: Vandenhoeck & Ruprecht.

Vernadsky, Vladimir I. ([1938] 1997). *Scientific Thought as a Planetary Phenomenon*. Translated by B. A. Starostin. Moscow: Nongovernmental Ecological V.I. Vernadsky Foundation.

Vickers, Brian (1992). "Francis Bacon and the Progress of Knowledge." *Journal of the History of Ideas* 53 (3): 495–518.

Vierck, Henning (2001). "Der Comenius-Garten in Berlin als philosophische Praxis." *Zeitschrift für Didaktik der Philosophie und Ethik* 23 (2): 160–64.

Vilsmaier, Ulli, and Daniel Lang (2014). "Transdisziplinäre Forschung." In *Nachhaltigkeitswissenschaften*, ed. H. Heinrichs and G. Michelsen, 87–113. Berlin: Springer.

Vleuten, Erik van der, and Arne Kaijser (2005). "Networking Europe." *History and Technology* 21 (1): 21–48.

Vogel, Ezra F. (1991). *The Four Little Dragons: The Spread of Industrialization in East Asia*. Cambridge, MA: Harvard University Press.

Vogel, Jakob (2004). "Von der Wissenschafts- zur Wissensgeschichte: Für eine Historisierung der 'Wissensgesellschaft.'" *Geschichte und Gesellschaft* 30 (4): 639–60.

Vogel, Klaus Anselm (1995). "Sphaera terrae: Das mittelalterliche Bild der Erde und die kosmographische Revolution." Ph.D. diss., Fachbereich Historisch-Philologische Wissenschaften, Georg-August-Universität zu Göttingen. https://ediss.uni-goettingen.de/bitstream/handle/11858/00-1735-0000-0022-5D5F-5/vogel_re.pdf.

Voosen, Paul (2016). "Anthropocene Pinned to Postwar Period." *Science* 353 (6302): 852–53.

Vygotskij, Lev S. (1978). *Mind in Society: The Development of Higher Psychological Processes*. Cambridge, MA: Harvard University Press.

————(1987). *Problems of General Psychology*. Vol. 1 of *The Collected Works of L. S. Vygotsky*. Translated by Norris Minick. New York: Plenum Press.

————(1987–99). *The Collected Works of L. S. Vygotsky*. 6 vols. Cognition and Language: A Series in Psycholinguistics. New York: Plenum Press.

Vygotskij, Lev S., and Aleksandr R. Lurija (1994). "Tool and Symbol in Child Development." 1930. In *The Vygotsky Reader*, ed. J. Valsiner and R. van der Veer, 99–174. Oxford: Blackwell Publishing.

Wackernagel, Mathis, and William Rees (1996). *Our Ecological Footprint: Reducing Human Impact on the Earth*. Gabriola Island, BC: New Society Press.

Wallerstein, Immanuel (2011). *The Modern World-System*. 4 vols. Berkeley: University of California Press.

Wallis, Faith (1999). *Bede: The Reckoning of Time*. Translated Texts for Historians 29. Liverpool: Liverpool University Press.

Walsby, Malcolm, and Natasha Constantinidou, eds. (2013). *Documenting the Early Modern Book World: Inventories and Catalogues in Manuscript and Print*. Leiden: Brill.

Want, Roy (2010). "An Introduction to Ubiquitous Computing." In *Ubiquitous Computing Fundamentals*, ed. J. Krumm, 1–35. Boca Raton, FL: Chapman & Hall.

Ward, Peter, and Joe Kirschvink (2015). *A New History of Life: The Radical New Discoveries about the Origins and Evolution of Life on Earth*. New York: Bloomsbury.

Warde, Paul, and Sverker Sörlin (2015). "Expertise for the Future: The Emergence of Environmental Prediction c. 1920–1970." In *The Struggle for the Long Term in Transnational Science and Politics: Forging the Future*, ed. J. Anderson and E. Rindzevičiūtė, 38–62. Abingdon: Routledge.

Wasserman, Stanley, and Katherine Faust (1997). *Social Network Analysis: Methods and Applications*. Structural Analysis in the Social Sciences 8. Reprint with corr., Cambridge: Cambridge University Press.

Waters, Colin N., Jan A. Zalasiewicz, Colin Summerhayes, Ian J. Fairchild, Neil L. Rose, Neil J. Loader, William Shotyk, Alejandro Cearreta, Martin J. Head, James P. M. Syvitski, Mark Williams, Michael Wagreich, Anthony D. Barnosky, An Zhisheng, Reinhold Leinfelder, Catherine Jeandel, Agnieszka Gałuszka, Juliana A. Ivar do Sul, Felix Gradstein, Will Steffen, John R. McNeill, Scott Wing, Clément Poirier, and Matt Edgeworth (2018). "Global Boundary Stratotype Section and Point (GSSP) for the Anthropocene Series: Where and How to Look for Potential Candidates." *Earth-Science Reviews* 178:379–429.

Waters, Colin N., Jan A. Zalasiewicz, Mark Williams, Michael A. Ellis, and Andrea Snelling (2014). "A Stratigraphical Basis for the Anthropocene?" In *A Stratigraphical Basis for the Anthropocene*, ed. C. N. Waters, J. A. Zalasiewicz, and M. Williams, 1–21. Special Publications 395. London: Geological Society.

Watkins, Trevor (2010). "New Light on Neolithic Revolution in South-West Asia." *Antiquity* 84 (325): 621–34.

———(2013). "Neolithisation Needs Evolution, as Evolution Needs Neolithisation." *Neo-Lithics* 2:5–10.

Watts, Duncan J. (2003). *Six Degrees: The Science of a Connected Age*. New York: W. W. Norton.

Watts, Duncan J., and Steven H. Strogatz (1998). "Collective Dynamics of 'Small-World' Networks." *Nature* 393 (6684): 440–42.

Weart, Spencer (2003). *The Discovery of Global Warming*. New Histories of Science, Technology, and Medicine. Cambridge, MA: Harvard University Press.

———(2004). "Reflections on the Scientific Process, as Seen in Climate Studies." AIP Publishing. Accessed April 6, 2018. https://history.aip.org/climate/pdf/Reflect.pdf.

Weber, Max ([1917/19] 2004). "Science as a Vocation." In *The Vocation Lectures*, ed. D. Owen and T. B. Strong, 1–31. Indianapolis, IN: Hackett.

Wegener, Alfred L. (1912). "Die Entstehung der Kontinente." *Geologische Rundschau* 3 (4): 276–92.

———(1915). *Die Entstehung der Kontinente und Ozeane*. Braunschweig: Friedrich Vieweg.

———(1966). *The Origin of Continents and Oceans: Translated from the Fourth Revised German Edition*. 1929. Translated by John Biram. New York: Dover.

Weiser, Mark (1991). "The Computer for the 21st Century." *Scientific American* 265 (3): 94–104.

Wendt, Helge (2016a). "Kohle in Arcadien: Transformationen von Energiesystemen und Kolonialregimen (ca. 1630–1730)." In *Francia: Forschungen zur Westeuropäischen Geschichte (Sonderdruck)*. Vol. 43, ed. Deutsches Historisches Institut Paris, 119–36. Ostfildern: Jan Thorbecke Verlag.

———, ed. (2016b). *The Globalization of Knowledge in the Iberian Colonial World*. Proceedings 10. Berlin: Edition Open Access. http://edition-open-access.de/proceedings/10/index.html.

Wendt, Helge, and Jürgen Renn (2012). "Knowledge and Science in Current Discussions of Globalization." In *The Globalization of Knowledge in History*, ed. J. Renn, 45–72. Studies 1. Berlin: Edition Open Access. http://edition-open-access.de/studies/1/6/index.html.

Werner, Michael, and Benedicte Zimmermann (2006). "Beyond Comparison: Histoire Croisee and the Challenge of Reflexivity." *History and Theory* 45 (1): 30–50.

Wertheimer, Max ([1912] 2012). "Experimental Studies on Seeing Motion." In *On Perceived Motion and Figural Organization*, ed. L. Spillmann, 1–9. Cambridge, MA: MIT Press.

———([1959] 1978). *Productive Thinking*. Enlarged ed. Westport, CT: Greenwood Press.

Wesolowski, Amy, Taimur Qureshi, Maciej F. Boni, Pål Roe Sundsøy, Michael A. Johansson, Syed Basit Rasheed, Kenth Engø-Monsen, and Caroline O. Buckee (2015). "Impact of Human Mobility on the Emergence of Dengue Epidemics in Pakistan." *Proceedings of the National Academy of Sciences of the United States of America* 112 (38): 11887–92.

West, Candace, and Don H. Zimmerman (1987). "Doing Gender." *Gender & Society* 1 (2): 125–51.

Westenholz, Aage (1999). "The Old Akkadian Period: History and Culture." In *Mesopotamien: Akkade-Zeit und Ur-III Zeit*, ed. P. Attinger and M. Wäfler, 17–117. Annäherungen 3. Orbis Biblicus et Orientalis 160. Göttingen: Vandenhoeck & Ruprecht.

Weyer, Johannes, ed. (2012). *Soziale Netzwerke: Konzepte und Methoden der sozialwissenschaftlichen Netzwerkforschung*. 2nd ed. Munich: Oldenbourg Verlag.

Whitmee, Sarah, Andy Haines, Chris Beyrer, Frederick Boltz, Anthony G. Capon, Braulio Ferreira de Souza Dias, Alex Ezeh, Howard Frumkin, Peng Gong, Peter Head, Richard Horton, Georgina M. Mace, Robert Marten, Samuel S. Myers, Sania Nishtar, Steven A. Osofsky, Subhrendu K. Pattanayak, Montira J. Pongsiri, Cristina Romanelli, Agnes Soucat, Jeanette Vega, and Derek Yach, July 16, (2015). "Safeguarding Human Health in the Anthropocene Epoch: Report of the Rockefeller Foundation–Lancet Commission on Planetary Health." *Lancet* 386 (10007): 1973–2028. Accessed July 28, 2019. https://www.thelancet.com/commissions/planetary-health.

Wien, Wilhelm (1894). "Temperatur und Entropie der Strahlung." *Annalen der Physik* 288 (5): 132–65.

Will, Clifford M. (1986). *Was Einstein Right? Putting General Relativity to the Test*. New York: Basic Books.

———(1989). "The Renaissance of General Relativity." In *The New Physics*, ed. P. Davies, 7–33. Cambridge: Cambridge University Press.

Willcox, George, Sandra Fornite, and Linda Herveux (2008). "Early Holocene Cultivation before Domestication in Northern Syria." *Vegetation History and Archaeobotany* 17 (3): 313–25.

Winchell, Frank, Chris J. Stevens, Charlene Murphy, Louis Champion, and Dorian Q. Fuller (2017). "Evidence for Sorghum Domestication in Fourth Millennium BC Eastern Sudan Spikelet Morphology from Ceramic Impressions of the Butana Group." *Current Anthropology* 58 (5): 673–83.

Wintergrün, Dirk (2019). "Netzwerkanalysen und semantische Datenmodellierung als heuristische Instrumente für die historische Forschung." Ph.D. diss., Technischen Fakultät der Friedrich-Alexander-Universität Erlangen-Nürnberg (FAU): https://opus4.kobv.de/opus4-fau/frontdoor/index/index/docId/11189.

Wissemeier, Alexander H. (2015). "Können neue, innovative Düngemitteltypen das moderne Stickstoffproblem lösen?" In *N: Stickstoff—Ein Element schreibt Weltgeschichte*, ed. G. Ertl and J. Soentgen, 205–16. Munich: Oekom Verlag.

Wittfogel, Karl August (1957). *Oriental Despotism: A Comparative Study of Total Power*. New York: Random House.

Wöhrle, Georg, ed. (2014). *The Milesians*. Vol. 1: *Thales*. Traditio Praesocratica 1. Berlin: De Gruyter.

Wolff, Michael (1978). *Geschichte der Impetustheorie: Untersuchungen zum Ursprung der klassischen Mechanik*. Frankfurt am Main: Suhrkamp.

Woodard, Roger D. (1997). *Greek Writing from Knossos to Homer: A Linguistic Interpretation of the Origin of the Greek Alphabet and the Continuity of Ancient Greek Literacy*. Oxford: Oxford University Press.

Woods, Christopher, ed. (2015). *Visible Language: Inventions of Writing in the Ancient Middle East and Beyond*. 2nd ed. Oriental Institute Museum Publications 32. Chicago: Oriental Institute of the University of Chicago.

———(2017). "The Abacus in Mesopotamia: Considerations from a Comparative Perspective." In *The First Ninety Years: A Sumerian Celebration in Honor of Miguel Civil*, ed. L. Feliu, F. Karahashi, and G. Rubio, 416–78. Studies in Ancient Near Eastern Records 12. Berlin: De Gruyter.

World Health Organization (2014). "Media Centre: WHO's First Global Report on Antibiotic Resistance Reveals Serious, Worldwide Threat to Public Health." Accessed April 6, 2018. https://www.who.int/mediacentre/news/releases/2014/amr-report/en.

Wright, David K. (2017). "Humans as Agents in the Termination of the African Humid Period." *Frontiers in Earth Science* 5 (4): 1–14.

Wrigley, Edward A. (2010). *Energy and the English Industrial Revolution*. Cambridge: Cambridge University Press.

Xenophanes (2016). "Testimonia, Part 2: Doctrine (D)." In *Early Greek Philosophy*. Vol. 3: *Early Ionian Thinkers*, pt. 2, ed. A. Laks and G. W. Most, 23–73. Loeb Classical Library 526. Cambridge, MA: Harvard University Press.

Yasnitsky, Anton (2011). "Vygotsky Circle as a Personal Network of Scholars: Restoring Connections between People and Ideas." *Integrative Psychological and Behavioral Science* 45:422–57.

———, ed. (2018a). *Questioning Vygotsky's Legacy: Scientific Psychology or Heroic Cult*. Oxford: Routledge.

———(2018b). *Vygotsky: An Intellectual Biography*. Oxford: Routledge.

Yule, Henry (1903). *The Book of Ser Marco Polo, the Venetian, concerning the Kingdoms and Marvels of the East*. Translated by Henry Yule. 2 vols. Rev. 3rd ed. London: John Murray. https://archive.org/details/bookofsermarcopo001polo/page/n7; https://archive.org/details/bookofsermarcopo002polo/page/n9.

Yusoff, Kathryn (2016). "Anthropogenesis: Origins and Endings in the Anthropocene." *Theory, Culture & Society* 33 (2): 3–28.

Zabaglia, Niccola, and Domenico Fontana (1743). *Castelli, e ponti: Con alcune ingegnose pratiche e con la descrizione del trasporto dell'obelisco vaticano e di altri del cavaliere Domenico Fontana*. Rome: Niccolò e Marco Pagliarini.

Zalasiewicz, Jan. 2008. *The Earth After Us: What Legacy Will Humans Leave In the Rocks?* New York: Oxford University Press.

Zalasiewicz, Jan, Colin N. Waters, Colin Summerhayes, Alexander P. Wolfe, Anthony D. Barnosky, Alejandro Cearreta, Paul Crutzen, Erle C. Ellis, Ian J. Fairchild, Agnieszka Galuszka, Peter Haff, Irka Hajdas, Martin J. Head, Juliana Ivardo Sul, Catherine Jeandel, Reinhold Leinfelder, John R. McNeill, Cath Neal, Eric Odada, Naomi Oreskes, Will Steffen, James Syvitski, Davor Vidas, Michael Wagreich, and Mark Williams (2017). "The Working Group on the Anthropocene: Summary of Evidence and Interim Recommendations." *Anthropocene* 19:55–60.

Zalasiewicz, Jan, Colin N. Waters, Mark Williams, Anthony D. Barnosky, Alejandro Cearreta, Paul Crutzen, Erle C. Ellis, Michael A. Ellis, Ian J. Fairchild, Jacques Grinevald, Peter K. Haff, Irka Hajdas, Reinhold Leinfelder, John McNeill, Eric O. Odada, Clément Poirier, Daniel Richter, Will Steffen, Colin Summerhayes, James P. M. Syvitski, Davor Vidas, Michael Wagreich, Scott L. Wing, Alexander P. Wolfe, An Zhisheng, and Naomi Oreskes

(2015). "When Did the Anthropocene Begin? A Mid-twentieth Century Boundary Level Is Stratigraphically Optimal." *Quaternary International* 383:196–203.

Zalasiewicz, Jan, Colin N. Waters, Mark Williams, and Colin Summerhayes, eds. (2018). *The Anthropocene as a Geological Time Unit: A Guide to the Scientific Evidence and Current Debate*. Cambridge: Cambridge University Press.

Zalasiewicz, Jan, and Mark Williams (2009). "A Geological History of Climate Change." In *Climate Change: Observed Impacts on Planet Earth*, ed. T. M. Letcher, 127–42. Amsterdam: Elsevier.

———(2012). *The Goldilocks Planet: The 4 Billion Year Story of Earth's Climate*. New York: Oxford University Press.

Zalasiewicz, Jan, Mark Williams, Will Steffen, and Paul Crutzen (2010). "The New World of the Anthropocene." *Environmental Science & Technology* 44 (7): 2228–31.

Zalasiewicz, Jan A., Mark Williams, Colin N. Waters, Anthony D. Barnosky, and Peter Haff (2014). "The Technofossil Record of Humans." *Anthropocene Review* 1 (1): 34–43.

Zeder, Melinda A. (2009). "The Neolithic Macro-(R)evolution: Macroevolutionary Theory and the Study of Culture Change." *Journal of Archaeological Research* 17 (1): 1–63.

———(2011). "The Origins of Agriculture in the Near East." *Current Anthropology* 52 (S4): 221–35.

Zeder, Melinda A., Daniel Bradley, Eve Emshwiller, and Bruce D. Smith, eds. (2006). *Documenting Domestication: New Genetic and Archaeological Paradigms*. Berkeley: University of California Press.

Zeder, Melinda A., and Bruce D. Smith (2009). "A Conversation on Agricultural Origins: Talking Past Each Other in a Crowded Room." *Current Anthropology* 50 (5): 681–90.

Zeller, Kirsten (2016). "The Privatisation of Seeds." Reset: Digital for Good. Accessed March 20, 2018. https://en.reset.org/knowledge/privatisation-seeds.

Zemon Davis, Natalie (2011). "Decentering History: Local Stories and Cultural Crossings in a Global World." *History and Theory* 50 (2): 188–202.

Zhang, Baichun, and Jürgen Renn (2006). *Transformation and Transmission: Chinese Mechanical Knowledge and the Jesuit Intervention*. Preprint 313. Berlin: Max Planck Institute for the History of Science.

Zhang, Baichun, Miao Tian, Matthias Schemmel, Jürgen Renn, and Peter Damerow, eds. (2008). *Chuanbo yu huitong: Qiqi tushuo yanjiu yu jiaoyi* [Transmission and integration: "Qiqi tushuo"]. Nanjing: Jiangsu kexue jishu chubanshe.

Zhang, Baichun, Jiuchun Zhang, and Yao Fang. (2006). "Technology Transfer from the Soviet Union to the People's Republic of China 1949–1966." *Comparative Technology Transfer and Society* 4 (2): 105–167.

Zilsel, Edgar ([1976] 2000). *The Social Origins of Modern Science*. Boston Studies in the Philosophy and History of Science 200. Dordrecht: Kluwer Academic.

Ziman, John, ed. (2000). *Technological Innovation as an Evolutionary Process*. Cambridge: Cambridge University Press.

Zuboff, Shoshana (2019). *The Age of Surveillance Capitalism: The Fight for a Human Future at the New Frontier of Power*. New York: PublicAffairs.

Index

Page numbers in italics refer to figures; numbers in bold to tables.